NEUROMETHODS

AF768567

Series Editor
Wolfgang Walz
University of Saskatchewan
Saskatoon, SK, Canada

For further volumes:
http://www.springer.com/series/7657

Animal Models of Brain Tumors

Edited by

Ricardo Martínez-Murillo

CSIC, Instituto Cajal, Neurovascular Research Group, Madrid, Spain

Alfredo Martínez

Centro de Investigación Biomédica de la Rioja (CIBIR), Logroño, Spain

Humana Press

Editors
Ricardo Martínez Murillo
CSIC Instituto Cajal
Neurovascular Research Group
Madrid, Spain

Alfredo Martínez
Centro de Investigación Biomédica
de la Rioja (CIBIR)
Logroño, Spain

Unidad Asociada CIBIR-CSIC

ISSN 0893-2336
ISSN 1940-6045 (electronic)
ISBN 978-1-4939-5894-8
ISBN 978-1-62703-209-4 (eBook)
DOI 10.1007/978-1-62703-209-4
Springer New York Heidelberg Dordrecht London

Softcover re-print of the Hardcover 1st edition 2013

Printed on acid-free paper

Humana Press is a brand of Springer
Springer is part of Springer Science+Business Media (www.springer.com)

Preface to the Series

Under the guidance of its founders Alan Boulton and Glen Baker, the Neuromethods series by Humana Press has been very successful since the first volume appeared in 1985. In about 17 years, 37 volumes have been published. In 2006, Springer Science+Business Media made a renewed commitment to this series. The new program will focus on methods that are either unique to the nervous system and excitable cells or which need special consideration to be applied to the neurosciences. The program will strike a balance between recent and exciting developments like those concerning new animal models of disease, imaging, *in vivo* methods, and more established techniques. These include immunocytochemistry and electrophysiological technologies. New trainees in neurosciences still need a sound footing in these older methods in order to apply a critical approach to their results. The careful application of methods is probably the most important step in the process of scientific inquiry. In the past, new methodologies led the way in developing new disciplines in the biological and medical sciences. For example, Physiology emerged out of Anatomy in the 19th century by harnessing new methods based on the newly discovered phenomenon of electricity. Nowadays, the relationships between disciplines and methods are more complex. Methods are now widely shared between disciplines and research areas. New developments in electronic publishing also make it possible for scientists to download chapters or protocols selectively within a very short time of encountering them. This new approach has been taken into account in the design of individual volumes and chapters in this series.

Neuherberg, Germany *Wolfgang Walz*

Preface

Experimental neuro-oncology demands the creation of appropriate animal models to assess the efficacy of innovate approaches for the treatment of human brain tumors (1; 2). These models may be used for therapeutic drug screening, in preclinical trials, or to study the basic biology of tumors, making possible the transition from scientific concepts to understanding the reality of a cancer in its natural context.

Understanding the cellular and molecular features of specific human cancers is needed to accelerate both our current knowledge of the molecular pathogenesis of cancer and the discovery of therapeutic targets. Recent advances in molecular biology have led to the identification of some molecular and genetic causes of brain tumors (3–7). Although the available models, whether at the cellular, tissue, or animal level, do not exactly represent the biology of human brain tumors, animal models might offer significant insights into these tumors, thus providing a better understanding of biological mechanisms underlying tumor generation, growth, angiogenesis, invasion, and metastasis.

As knowledge of the molecular biology of neoplasia advances, transgenic technology has been utilized in elucidating the complex molecular mechanisms underlying oncogenesis and progression of brain tumors. Experimental animal models for brain tumors using genetically modified mice, with overexpressed and/or deleted genes have been characterized. However, since there are no adequate mechanisms to correlate genetic alterations in mouse tumors with their proposed human counterparts, to what extent the existing models mimic the biology of primary brain cancer in humans remains uncertain and should be clarified. Gene expression profiles, controlling elements, pathways, and cell of origin for brain tumors remain largely unknown and the molecular elements, such as tissue-specific promoters and enhancers to create genetically accurate models of brain tumors, are still poorly understood. Standardized orthotopic models of mouse infiltrating brain tumors that mimic closely the human clinical situation in terms of histopathological characteristics and angiogenic features were also developed by means of transplant-based models of mouse and rat glioma derived from cultured tumor cells (8;9). It should be noted that established human glioma cell lines experience over time numerous morphologic and genetic changes and do not reproduce the biology of the original human tumor. Thus, implantation animal models with human cancer cells do not accurately reflect the interaction between tumor and host that occurs in patients. Finally, models of chemically induced "in vivo" brain tumors might provide clues on the effect of carcinogens in the development of certain types of cerebral neoplasms.

The goal of Animal Models of Brain Tumors is to bring together developments and discoveries in "in vivo" experimental tumor research of the last two decades that have provided advances in our understanding of the cellular and molecular mechanisms involved in the generation, progression, and clinical outcome of brain neoplasms. The utility of the different animal tumor models in pre-clinical studies requires new methods to validate, compare, and contrast the animal model with its proposed human counterpart. Readers will be referred to sections, including: i.) Animal Models Offer Insights into Human Brain Neoplasms, ii) Cellular, Molecular, and Genetic Basis of Brain Tumors: data from animal models, iii). Therapies in the Treatment of Malignant Glioma: Data from animal models, and iv). Imaging Technologies in Animal Tumor Models: Measuring Brain Tumor Growth and Metabolism. A full description of the laboratory protocols will be given, thus allowing their validation by independent investigators.

Ricardo Martínez-Murillo
Alfredo Martínez

References

1. Jacobs VL, Valdes PA, Hickey WF, De Leo JA (2011) Current review of in vivo GBM rodent models: emphasis on the CNS-1 tumour model. ASN Neuro 3:e00063
2. Westphal M, Lamszus K (2011) The neurobiology of gliomas: from cell biology to the development of therapeutic approaches. Nat Rev Neurosci. 12:495–508
3. Momota H, Holland EC (2009) Mouse models of CNS embryonal tumors. Brain Tumor Pathol 26:43–50
4. Hoenerhoff MJ, Hong HH, Ton TV, Lahousse SA, Sills RC (2009) A review of the molecular mechanisms of chemically induced neoplasia in rat and mouse models in national toxicology program bioassays and their relevance to human cancer. Toxicol Pathol 37:835–848
5. Bleeker FE, Molenaar RJ, Leenstra S (2012) Recent advances in the molecular understanding of glioblastoma. J Neurooncol 108:11–27
6. Mladkova N, Chakravarti A (2009) Molecular profiling in glioblastoma: prelude to personalized treatment. Curr Oncol Rep 11:53–61
7. Wee B, Charles N, Holland EC (2011) Animal models to study cancer-initiating cells from glioblastoma. Front Biosci 16:2243–2258
8. Lai CP, Bechberger JF, Naus CC (2009) Pannexin2 as a novel growth regulator in C6 glioma cells. Oncogene 28:4402–4408
9. Jiang Y, Boije M, Westermark B, Uhrbom L (2011) PDGF-B Can sustain self-renewal and tumorigenicity of experimental glioma-derived cancer-initiating cells by preventing oligodendrocyte differentiation. Neoplasia 13:492–503

Contents

List of Contributors

CARLES ARÚS • *Departament de Bioquímica i Biologia Molecular, Unitat de Bioquímica de Biociències, Edifici Cs, Universitat Autònoma de Barcelona (UAB), Centro de Investigación Biomédica en Red en Bioingeniería, Biomateriales y Nanomedicina (CIBER-BBN), Institut de Biotecnologia i de Biomedicina (IBB), Universitat Autònoma de Barcelona (UAB), Cerdanyola del Vallès, Spain*

HIKMAT ASSI • *Departments of Neurosurgery and Cell & Developmental Biology, University of Michigan Medical School, Ann Arbor, MI, USADepartment of Molecular and Medical Pharmacology, David Geffen School of Medicine, UCLA, Los Angeles, CA, USA*

KRISTEN A. BATICH • *Department of Pathology, Duke University Medical Center, Durham, NC, USA*

ELENA BOGDANOVIC • *Arthur & Sonia Labatt Brain Tumor Center, Hospital for Sick Children Research Institute, University of Toronto, Toronto, ON, Canada*

KELLY BURRELL • *Arthur & Sonia Labatt Brain Tumor Center, Hospital for Sick Children Research Institute, University of Toronto, Toronto, ON, Canada*

ANA PAULA CANDIOTA • *Centro de Investigación Biomédica en Red en Bioingeniería, Biomateriales y Nanomedicina (CIBER-BBN), Departament de Bioquímica i Biologia Molecular, Unitat de Bioquímica de Biociències, Edifici Cs, Universitat Autònoma de Barcelona (UAB), Institut de Biotecnologia i de Biomedicina (IBB), Universitat Autònoma de Barcelona (UAB), Cerdanyola del Vallès, Spain*

MARIANELA CANDOLFI • *Facultad de Medicina, Instituto de Investigaciones Biomedicas, Universidad de Buenos Aires, Buenos Aires, Argentina*

MARIA G. CASTRO • *Departments of Neurosurgery and Cell & Developmental Biology, University of Michigan Medical School, Ann Arbor, MI, USA*

AL CHAREST • *Department of Neurosurgery, Tufts Medical Center, Molecular Oncology Research Institute, Tufts University School of Medicine, Boston, MA, USA*

BRYAN D. CHOI • *Duke Brain Tumor Immunotherapy Program, Division of Neurosurgery, Department of Surgery and Department of Pathology, Duke University Medical Center, Durham, NC, USA*

PHILIPPE CHOQUET • *Service de Biophysique et Médecine Nucléaire, Hôpitaux Universitaires & IMFS CNRS FRE 3240, Strasbourg, France*

DAVID COBRINIK • *Department of Pediatrics, Memorial Sloan-Kettering Cancer Center, New York, NY, USA*

ANDRÉ CONSTANTINESCO • *Service de Biophysique et Médecine Nucléaire, Hôpitaux Universitaires & IMFS CNRS FRE 3240, Strasbourg, France*

WILLIAM T. COULDWELL • *Department of Neurosurgery, Clinical Neurosciences Center, University of Utah, Salt Lake City, UT, USA*

ANDREAS FAUST • *European Institute for Molecular Imaging (EIMI), Westfälische Wilhelms-Universität Münster (WWU), Münster, Germany*

CHRISTIAN GOETZ • *Service de Biophysique et Médecine Nucléaire, Hôpitaux Universitaires & IMFS CNRS FRE 3240, Strasbourg, France*

ABHIJIT GUHA • *Division of Neurosurgery, Department of Surgery, Arthur & Sonia Labatt Brain Tumor Center, Hospital for Sick Children Research Institute, Western Hospital, University of Toronto, Toronto, ON, Canada*

DAVID H. GUTMANN • *Department of Neurology, Washington University School of Medicine, St. Louis, MO, USA*

IBRAHIM HUSSAIN • *Department of Neurology, Washington University School of Medicine, St. Louis, MO, USA*

LEANNE C. HUYSENTRUYT • *Department of Medicine, Hematology, and Oncology, University of California, San Francisco, San Francisco, CA, USA*

ANDREAS H. JACOBS • *Department of Nuclear Medicine, Interdisciplinary Center for Clinical Research (IZKF), European Institute for Molecular Imaging (EIMI), Westfälische Wilhelms-Universität Münster (WWU), Münster, Germany*

SHAHRZAD JALALI • *Arthur & Sonia Labatt Brain Tumor Center, Hospital for Sick Children Research Institute, University of Toronto, Toronto, ON, Canada*

RANDY L. JENSEN • *Department of Neurosurgery, University of Utah, Salt Lake City, UT, USA*

MARGARIDA JULIÀ-SAPÉ • *Centro de Investigación Biomédica en Red en Bioingeniería, Biomateriales y Nanomedicina (CIBER-BBN), Departament de Bioquímica i Biologia Molecular, Unitat de Bioquímica de Biociències, Edifici Cs, Universitat Autònoma de Barcelona (UAB), Institut de Biotecnologia i de Biomedicina (IBB), Universitat Autònoma de Barcelona (UAB), Cerdanyola del Vallès, Spain*

APARNA KAUL • *Department of Neurology, Washington University School of Medicine, St. Louis, MO, USA*

KLAUS KOPKA • *Department of Nuclear Medicine, Interdisciplinary Center for Clinical Research (IZKF), Westfälische Wilhelms-Universität Münster (WWU), Münster, Germany*

LOIS A. LAMPSON • *Department of Neurosurgery, Brigham and Women's Hospital & Harvard Medical School, Boston, MA, USA*

JASMINE LAU • *The W.A. Weiss Lab, University of California, San Francisco, San Francisco, CA, USA*

JULIE LESSARD • *Department of Neurosurgery, Tufts Medical Center, Molecular Oncology Research Institute, Tufts University School of Medicine, Boston, MA, USA*

NANNA LINDBERG • *Department of Immunology, Genetics, and Pathology, Uppsala University, Uppsala, Sweden*

PEDRO R. LOWENSTEIN • *Departments of Neurosurgery and Cell & Developmental Biology, University of Michigan Medical School, Ann Arbor, MI, USA*

DUANE A. MITCHELL • *Duke Brain Tumor Immunotherapy Program, Division of Neurosurgery, Department of Surgery, The Preston Robert Tisch Brain Tumor Center, Duke University Medical Center, Durham, NC, USA*

JOSÉ M. ORELLANA-MURIANA • *University of Alcalá de Henares & Cajal Institute of Neurobiology, Madrid, Spain*

JESSICA RAINEY • *Department of Neurosurgery, University of Utah, Salt Lake City, UT, USA*

BURKHARD RIEMANN • *Department of Nuclear Medicine, Westfälische Wilhelms-Universität Münster (WWU), Münster, Germany*

Stephen M. Robbins • *Hughes Childhood Cancer Research Program, Department of Oncology, Southern Alberta Cancer Research Institute, University of Calgary, Calgary, AB, Canada*
Klaus Schäfers • *European Institute for Molecular Imaging (EIMI), Westfälische Wilhelms-Universität Münster (WWU), Münster, Germany*
Sonja Schäfers • *European Institute for Molecular Imaging (EIMI), Westfälische Wilhelms-Universität Münster (WWU), Münster, Germany*
Otmar Schober • *Department of Nuclear Medicine, European Institute for Molecular Imaging (EIMI), Westfälische Wilhelms-Universität Münster (WWU), Münster, Germany*
Donna L. Senger • *Hughes Childhood Cancer Research ProgramDepartment of Oncology, Southern Alberta Cancer Research Institute, University of Calgary, Calgary, AB, Canada*
Thomas N. Seyfried • *Biology Department, Boston College, Chestnut Hill, MA, USA*
Laura M. Shelton • *Minerva Biotechnologies, Waltham, MA, USA*
Rui V. Simões • *Dept. Medical Physics, Memorial Sloan Kettering Cancer Center, New York, NY, USA*
Walavan Sivakumar • *Department of Neurosurgery, Clinical Neurosciences Center, University of Utah, Salt Lake City, UT, USA*
Lene Uhrbom • *Department of Immunology, Genetics, and Pathology, Uppsala University, Uppsala, Sweden*
Jesús Vaquero • *Neurosurgical Department, Hospital Universitario Puerta de Hierro-Majadahonda, Autonomous University, Madrid, Spain*
Thomas Viel • *European Institute for Molecular Imaging (EIMI), Westfälische Wilhelms-Universität Münster (WWU), Münster, Germany*
Yannic Waerzeggers • *European Institute for Molecular Imaging (EIMI), Westfälische Wilhelms-Universität Münster (WWU), Münster, Germany*
Stefan Wagner • *Department of Nuclear Medicine, Westfälische Wilhelms-Universität Münster (WWU), Münster, Germany*
William A. Weiss • *Departments of Neurology, Pediatrics, and Neurological Surgery, University of California, San Francisco, San Francisco, CA, USA*
Gelareh Zadeh • *Division of Neurosurgery, Department of Surgery, Arthur & Sonia Labatt Brain Tumor Center, Hospital for Sick Children Research Institute, Western Hospital, University of Toronto, Toronto, ON, Canada*
Mercedes Zurita • *Neurosurgical Department, Hospital Universitario Puerta de Hierro-Majadahonda, Autonomous University, Madrid, Spain*

Part I

Introduction

Neuromethods (2013) 77: 3–20
DOI 10.1007/7657_2011_24

Published online: 13 March 2012

Brain Tumor Models to Predict Clinical Outcome: Like the Phoenix?

Lois A. Lampson

Abstract

Small animal models have given great insight into tumor biology, but have been less successful at predicting clinical outcome. It is expected that newer models will be more predictive because they are better tumor mimics. Unfortunately, other impediments to clinical prediction remain, as reviewed in this chapter. Given these limits, plus advances in other approaches, how can small animal models best be exploited for developing new tumor therapies? Suggestions include focus on targets that are shared by many different tumors, and use of small animal models to reveal broad principles or answer specific questions, rather than aiming for rodent cures.

Key words: Clinical trial, Predict, Disappoint, Glioma, Brain metastases

Come l'arabia fenice...
Non è questa, non è quella/Non fu mai, non vi sarà...

Like the phoenix of Arabia...
Is not this one, is not that one/Never was, never will be...

Cosi' Fan Tutte, Act 1, scene 1

1. Introduction

Small animal models have been invaluable for probing all aspects of tumor biology, from tumor initiation to the mechanisms underlying the response—or resistance—to therapy. This chapter explores one area where the place and best use of small animal models is still evolving: predicting therapeutic outcome in human patients. For brain tumors, just as for many other tumors and disorders, clinical trials have often been disappointing, despite promising results in small animal models (1–5). In seeking the reasons, much of the discussion centers on the models themselves, especially their failings as biological mimics of the relevant tumor or other pathology.

Recent work has made real advances in the extent to which small animal models mimic human disease, for brain tumors as well as for other types of cancer (2, 6) and disorders. Complementary advances have occurred in parallel. Enhanced ability to analyze human tissue and new strategies for clinical trials aid translation in their own right (7) and contribute to the iterative process of refining the animal models.

Appreciation of the challenges to successful tumor therapy has also advanced. Deeper understanding of tumor biology has, as was hoped, led to identification of rational targets and development of new drugs and strategies against them. Unfortunately, not all seemingly eligible patients respond and, among those who do, the response is often transient (8). There is good reason to expect that new combination therapies will give further improvements, but the very wealth of potential targets, agents, combinations, doses, and schedules increases the challenge of efficiently identifying successful strategies.

In developing the next round of therapies, it is useful to review the different kinds of reasons that have made it hard to predict clinical outcome from small animal models. As will be brought out, the need for good tumor mimics is but one among many concerns.

2. Model Challenges: A Long Chain of Events

A long-held ideal has been that one would select a small animal model for a particular tumor (or other disorder), use the model to demonstrate efficacy of a novel therapy, and, ultimately, find that the therapy was indeed beneficial to human patients. Disappointments in many fields show how far we are from this ideal (1, 3–5).

One general problem is that, in the ideal scenario, the chain of events that must take place in the model is so long (Fig. 1). Any discrepancy between the rodent tumor and human tumor, or the rodent response and human response, can mean that the outcome will not be predictive for human patients. As tumor mimics, as well as in other ways, traditional brain tumor models are weak in many links of this long chain.

3. Model Challenges: Weaknesses as Tumor Mimics

Two kinds of examples illustrate the challenges to developing good brain tumor mimics. Among primary brain tumors, glioma, especially glioblastoma multiforme (GBM), the most common and aggressive primary brain tumor of adults, has been a major

PRIMARY BRAIN TUMOR initiation → **PRIMARY BRAIN TUMOR** growth

Primary tumor initiation → Primary tumor growth → **BRAIN METASTASES**

↘ ↙

Spontaneous host response to tumor
↓
TEST THERAPY
↓
Initial response to therapy
or initial resistance
↓
Initial host response to therapy and to tumor damage
(side-effects, toxicity)
↓
Longer-term response to therapy
or acquired resistance
↓
Longer-term host response to therapy and to tumor damage
(side-effects, toxicity)

Fig. 1. A long chain of events. This depicts the chain of events from tumor initiation through the final effects of tumor therapy, for both primary brain tumors and brain metastases. Mismatches at any point between a small animal model and human patient can mean that results in the model will not be predictive for the patient.

focus for work in small animal models. Brain metastases, from tumors that originate in other organs, present an equally difficult clinical problem, and somewhat different challenges for tumor models.

3.1. Primary Brain Tumors

In traditional models for primary brain tumor, typically a tumor-derived cell line is implanted in the brain or even under the skin, and the test therapy is administered within a few days or even before the tumor. Among many mismatches, taking glioma as an example: A homogeneous, long-term cell line does not model the complex composition of an endogenous tumor; a rapidly expanding cell mass does not model glioma patterns of growth and spread; a subcutaneous site does not model the brain microenvironment; the short time frame curtails reciprocal interactions between tumor and host; the use of immunodeficient hosts, as is necessary for study of human cell lines, prevents an appropriate contribution by the host immune response; the criteria for efficacy, typically slowed tumor growth, do not match clinical endpoints (2, 9, 10).

Some of these problems have been addressed by improvements to the traditional models (10–13), and some are avoided altogether in genetically engineered mouse (GEM) models. It seems only logical that, as better mimics of tumor biology, the improved models will be more predictive of clinical outcome, as many authors suggest. In practice, it is not yet known how much more predictive

Table 1
Why results from small animal models have not predicted clinical outcome

Topic	Challenge
Long chain	Full sequence of steps, from tumor growth to therapy, allows many chances for mismatch between model and human
Poor mimic	Basic features of tumor biology may differ from human Criteria for success in the model may differ from clinical endpoints
Heterogeneity	*Tumor related* Variations among tumors of the same name (No one model mimics all) Variations among sites of the same tumor Variations among cells at the same tumor site *Host related* Variations in host response, and in effect of therapy, among species, etc.[a] Biochemical differences, among species, etc.[a] Species differences in scale
Promiscuity	A given target molecule or pathway may affect many functions and many cell types. Net effect may vary among species, etc.[a]
Unknowns	Key effector mechanisms. May vary among species, etc.[a] Drug concentration delivered to tumor cell. Impediments may vary among species, etc.[a]

[a]May vary among species, among inbred rodent strains, among human individuals, among tumor sites, and among individual tumor cells

the newest GEM models will be, or how they will compare to alternatives. Currently, some positive examples are cases where the GEM model mirrors the known efficacy—or lack of efficacy—of a treatment that has already been tested in the clinic, while often GEM models are simply not used (2, 9, 14, 15). Some of the factors that may still limit successful prediction of clinical outcome, even in the newest GEM models, are listed in Table 1.

3.2. Brain Metastases

The major effort for GEM brain tumor models has focused on primary brain tumors. Even though brain metastases are many-fold more common, corresponding GEM models are not yet well-developed. The natural starting point for mimics of brain metastases would be models of the solid tumor of origin. For GEM models of solid tumors in general, however, metastasis to appropriate sites has been more difficult to achieve than other aspects of tumor growth (6, 15, 16).

The typical model for parenchymal brain metastases is still to inject tumor-derived cell lines into the blood (17–20) or even directly into the brain (21). Similarly for meningeal tumor, whether primary or metastatic, the most common method is to inject cells into the cerebrospinal fluid (CSF) (21–23). It is possible to address many specific questions using these methods, and individual cell lines may show appropriate distributions after delivery to the blood (24). Many other aspects of the multistep, interactive process that leads to the establishment, growth, and possible response to therapy of brain metastases are lost. For example, injecting cells into the blood obscures not only the timing of the genetic changes that predispose to metastasis (25, 26) but also the normal timing of tumor cells' entry to the brain.

The timing of tumor dissemination is of particular interest in the brain, for primary tumor as well as metastases. Although brain metastases are often detected late in the clinical course, it is not known when the cells first entered the brain. Similarly, although distantly infiltrating tumor is characteristic of glioma, it is not known when the cells first became so widely distributed. Thus, it is not known how to weigh the importance of preventing new metastases or new infiltration, as opposed to attacking disseminated tumor that is already in place (27).

Although the need for better tumor mimics has been a major challenge, it is not the only one. Even as GEM models for primary or metastatic tumor improve, one major source of complexity lies in the many facets of tumor heterogeneity, some being harder than others to incorporate into small animal models.

4. Model Challenges: A Multifaceted Heterogeneity

Brain tumors, like other tumors, display a multifaceted heterogeneity: among tumors of the same type, among—and within—sites of the same tumor, and among different small animal hosts. This heterogeneity is reflected in tumor growth, in the spontaneous response to the tumor, and in the response to therapy. Although some aspects of this heterogeneity are well handled, or even used to advantage, by current models, others are more challenging.

4.1. Heterogeneity Within a Tumor Type

Increasingly, tumors and other disorders are being subdivided, in response to new analysis (26, 28, 29). Brain tumors are no exception (30–32). This affects the path from bench to bedside in more than one way. It means that, even for tumors of the same name, no single small animal model will be appropriate for all patients. Further along the path, it means that, when a clinical trial is ultimately conducted, a positive response in a small subset of patients may go undetected, while other patients may be unaffected or even harmed.

At one level, GEM models are well-adapted to take this kind of heterogeneity into account. As human tumor variants are identified, corresponding changes can be introduced into the models. As information accumulates, about the tumor itself and about the response to therapy among different patients, the model can be further refined.

One factor that complicates this iterative process is that, even within the same rodent species, the same genetic manipulation can have very different effects in different inbred strains. GEM models of glioma and of metastatic tumor both provide examples of such strain-specific differences (15, 26).

4.2. Heterogeneity Among Inbred Strains

Although initially a source of complexity, differences among inbred strains can be an advantage, to the extent that they help reveal or explain variation among individual humans. Comparisons among inbred rodent strains have led to important discoveries in various fields, and there are ongoing efforts to improve their usefulness (15, 33, 34).

Although it can give key insights, comparison among inbred strains is not a sharply focused approach, nor is it all-inclusive. There is no way to know a priori that any particular point that is revealed by the inbred strains will be important in humans, or that the most important points will emerge. Rather, just as for other findings, the early and frequent analysis of human tissue is an essential part of an iterative process for confirming human relevance (35–37) as well as refining the animal models (15–26).

4.3. Heterogeneity Among Species

As fruitful as the iterative process of refining animal models can be, it is ultimately limited by heterogeneity among species. Biochemical differences and differences in scale pose different kinds of challenges.

Biochemical differences. For any given property, the potential for species differences between humans and mice or rats is well appreciated. Species differences can also mislead in a more subtle way. Brain tumors do grow in mice and rats; novel therapies do affect them. The rodent biology will be internally consistent, informative, and fascinating in its own right—and yet may not mimic the biology of human patients. Examples are seen at all levels of tumor study.

At the level of the tumor itself, not all tumor behaviors seen in GEM models have their counterparts in human tumors. Conversely, not all aspects of human tumor biology have been easily modeled. The difficulty of obtaining metastases in GEM models of appropriate primary tumors is a pertinent example (6, 15, 16). At the level of the host contribution to therapy, many examples of species differences are seen in the immune response (38–41).

These examples draw attention to the many possibilities for species differences in a complex, unfolding pathway (Fig. 1). They point to the potential value of focusing on discrete points, rather than the final effect on tumor growth, in the small animal host. Equally, the examples point up the importance of frequent cross-checking of preclinical findings against human tissue.

Responding to biochemical differences: humanized small animal hosts. Turning to nonhuman primates is not realistic as a general response to the problem of species differences (42). Biochemical differences are still present, problems of ethics and cost are exacerbated, and the advantages of smaller hosts are lost.

A more common response to biochemical species differences is to cause the rodent to express relevant human genes, cells, or tissues (2, 43, 44). Tumor xenografts are an example, but the applications are much broader. For example, mice can be made to express human tumor antigens or human immunocompetent cells.

Use of such *humanized* mice does greatly expand the range of human properties that can be studied in vivo. The limitation is that biological processes are interwoven, not compartmentalized (45). This means that host/human boundaries can be pushed back, but not avoided entirely (43, 46).

For example, human lymphocytes will not interact appropriately with murine homing receptors. If the mouse's homing receptors are humanized, then human lymphocytes can show more appropriate traffic patterns. The complication is that other kinds of functions, unrelated to lymphocyte traffic, are also served by the homing receptors, and these may also be affected.

The integrin family of cell surface proteins illustrates these points. The integrins, like so many protein families (45), serve multiple functions, in a variety of cell types. Integrins are important not only for lymphocyte homing but also for tumor growth and metastasis. They also serve functions that are unrelated to tumors or host defense, including normal brain functions. These different functions are served by multiple possible interactions between the different integrins and their counter-receptors (47–52). Humanizing mice with respect to one particular integrin receptor or function may thus affect other functions, not directly related to tumor therapy. Side effects that are not relevant to humans may be created. Conversely, effects that would be clinically relevant may not be revealed.

Differences in scale. One kind of species difference that cannot be humanized is a difference in scale. The life span of a mouse or rat is simply not long enough for the full chain of events that includes tumor initiation, interplay between the tumor and the environment, initial response and ultimate resistance to therapy, and development of long-term side effects in human patients (Fig. 1). Compressing the first steps may still allow development of tumors, but may also increase the divergence from human tumor biology (2, 15). The different time frames thus give a second reason, in addition to biochemical differences, to focus on specific points, rather than attempting to follow a longer chain of events, in a single rodent model or single experiment.

A complementary difference in scale between species is the difference in size. As compared to humans, the absolute limit of tumor size is much smaller in the rodent brain, while an intracerebral injection of test agents may affect a relatively larger proportion of the brain. Either factor may contribute to results in the rodent model that are not predictive for human patients. While interpreting results from small animal models, simply keeping in mind possible consequences of the differences in size may be helpful.

4.4. Heterogeneity Among—and Within—Tumor Sites

Brain tumors provide many examples of heterogeneity among tumor sites within a single patient. This is a defining feature of GBM, and there can also be great variation among individual brain metastases (53). Both the tumor itself and the local environment contribute to this heterogeneity. A tumor is plastic, changing at each site as it grows. The local regulatory environment is site-specific, even at baseline (54–56), and is also plastic. It changes as part of the spontaneous response to the tumor and in response to therapy, as well as to unrelated events (27, 43, 46, 57).

The interaction of a tumor with its environment, and the relevance of specific features of the local microenvironment, is increasingly appreciated. The vasculature and angiogenesis, components of the innate or adaptive immune response, and physical and metabolic impediments to drug delivery (as discussed in Sect. 6.2) are features of the environment that have been stressed by different authors. For the brain as a whole, characteristic features known to affect tumor growth or response to therapy include the blood–brain barrier (BBB) (as reviewed in (27)) and differences among major anatomical regions, such as parenchyma vs. meninges or gray matter vs. white matter.

A still finer level of organization is relevant for the smallest tumor foci. Microenvironments within the brain are well-mapped in terms of their neurobiological functions, such as the distributions of particular neuropeptides and neurotransmitters. Despite their names, these neurochemicals, like most other biologically active molecules, affect many different functions, in many cell types (45, 54, 55). They can affect the growth of a tumor, the spontaneous response to the tumor, and the response to therapy. For example, the widespread neuropeptide, substance P (SP), can affect both tumor growth (58, 59) and cytokine-mediated therapy (54, 55).

The different rodent species and inbred strains each have their own variations at the level of the local environment. For example, the maps of neuropeptides and neurotransmitters vary among species and among strains (60). Local regulatory differences thus contribute to heterogeneity among tumor sites and also among small animal hosts. The smaller the tumor focus, the more relevant these local differences would be.

Even within a single tumor focus, individual cells vary in their properties, including their response to any given therapy.

One consequence is that acquired resistance to a given therapy can reflect outgrowth of tumor variants that were already present; it need not reflect new mutations (61).

4.5. Tumor Size and Other Variables

The larger a tumor mass, the greater the extent to which it will have created its own regulatory environment. The smaller the focus, the more relevant is the regulatory environment of the surrounding tissue. Important examples of microscopic tumor in the brain include residual, disseminated, or infiltrative glioma and the earliest blood-borne micrometastases from other organs. Despite the importance of microscopic tumor, few studies of brain tumor therapy have directly evaluated microtumor targets (8, 17–19, 27, 62, 63).

Age and gender are two other variables that are known to be important for human brain tumors, but are not usually studied directly, when therapy is tested in small animal models. Less often mentioned is that the needle wound and anesthesia that are necessary for implanting tumor or injecting test drugs can also affect tumor growth or the antitumor response; while interpreting preclinical tests of therapy, these factors are seldom taken into account.

Physiological changes that may be unrelated to the tumor can affect the outcome of a given therapy (43, 46, 57). For example, there may be inflammatory changes or, for a variety of reasons, the levels of drug efflux proteins or drug-metabolizing enzymes may change (64–68). The effect of co-morbidities is of special concern for older patients (43). If new therapies are tested in young adult hosts that display the tumor of interest but are otherwise healthy, these factors are obscured (43–46).

5. Model Challenges: Promiscuity of Function

Yet another aspect of heterogeneity is seen in the functional diversity that is characteristic of so many biological molecules, with an individual molecule able to affect many functions and many cell types (45). Specific examples discussed in this chapter include the many functions of the integrins (47–52), the neurotransmitter, substance P (54, 55, 58–60); the complement cascade (69–73); and the P450 family of drug-metabolizing enzymes (64, 74). In each case, tumor growth, the response to tumor, and normal neurobiological functions are all affected. In each case, species and strain differences further complicate the task of predicting the net effect, for human patients, of manipulations that are first tested in animal models.

6. Model Challenges: Two Unknowns

6.1. Unknown Effector Mechanisms

Even for the most successful tumor therapies, it can be difficult to identify the key effector mechanisms (27). For example, although the monoclonal antibody rituximab has now been used successfully for several years, it is not yet known which of the many possible effector mechanisms are most important in different patients (75).

Evolving experience with farnesyltransferase inhibitors provides a complementary example; in this case, the target itself has been hard to define (76–78). As another kind of example, it is often not known whether apparent efficacy of new treatments against metastatic tumor reflects attack of metastases per se or simply attack of the primary tumor mass (27, 79).

The many aspects of tumor heterogeneity complicate the challenge of identifying—or modifying—underlying effector mechanisms. The most important mechanism can vary among species, strains, or individuals, or among different tumor sites. To take an example from immunotherapy, an important set of immune effector functions is mediated by the complement cascade, which differs among organs (69) and also displays polymorphisms among individuals. Moreover, complement has nonimmunological functions, including those in the brain (70–73), which may be affected if complement is manipulated.

Thus, the most relevant mechanisms—or toxicities—may not be constant from patient to patient or site to site. More subtly, the small animal model that is the best mimic for a given tumor will not necessarily be the best model for the effector mechanism that is most relevant to that particular patient or to human patients in general. As information from human tissue accumulates, relevant mechanisms for individual patients and sites should become clarified. Small animal models can then be evaluated and modified accordingly.

6.2. Unknown Doses

Tumor resistance to therapy is often discussed in biochemical terms (such as use of an alternative metabolic pathway), but failure to deliver an appropriate dose to the tumor can underlie resistance as well (3, 10, 11, 80–85). The ability of the BBB to block drug access is well-known (27), but more general impediments are just as relevant. Drug efflux proteins, drug-metabolizing enzymes, the extracellular matrix, and interstitial pressure are examples of factors that can impede delivery of therapeutics within solid tumors at any site (3, 10, 11, 80, 81, 84, 85).

Information about drug doses from rodent models can mislead in different ways. The dose used to control tumor in the model may be too high for an equivalent dose to be safely used in humans or, conversely, too low to reveal toxicity (11, 12). The different types of heterogeneity are relevant as well. Both the optimal dose and the efficiency of delivery may vary with the tumor size or site (54–56), as well as among individuals (86), rodent species, or inbred strains. For example, effective doses and also toxicities may vary between sick and healthy individuals (9, 43, 46), and drug-metabolizing enzymes have different distributions in different hosts (2). Thus, results in a rodent model will not necessarily predict whether, in humans, the intended drug dose reaches the tumor site (87) or tumor cell (82, 84).

7. Model Challenges: A Wealth of Alternatives

Targeted therapies have only begun to fulfill their promise: They can control tumor, but not for all patients, and often not indefinitely (8). Goals for confronting tumor resistance, whether initial or acquired, include identification of responsive patients, definition of underlying mechanisms of resistance and response, and strategies for preventing or responding to the resistance. The first two goals benefit from ongoing analysis of human tissue, complemented by growing insight into the kinds of factors that are relevant, as more cases are solved (37, 75, 88–94). As important factors are defined, small animal models can be adapted to display them, as part of the iterative process described in earlier sections.

Concerning the third goal, preventing or responding to the resistance, it is logical that combination therapies could prevent outgrowth of resistant tumor, and in vivo testing should give important insights, especially with good tumor mimics. One challenge is to limit the number of possibilities to be tested in vivo. Among many variables, established properties of single drugs may be altered in combination (43, 95); the sequence with which the components are given can alter the outcome (95); and conclusions may change as the field evolves, for example, as new trials are conducted or other components of the therapy improve (96, 97). Overlaid on all of this are the many aspects of heterogeneity reviewed above.

Balanced against the abundance of variables are developments that aid choice and analysis. As mechanisms of tumorigenesis, and of resistance to therapy, become better understood, and as more information is obtained directly from human tissue, rational choices can be made more easily. Focus on targets that are shared by many tumors can further control the number of possibilities to be studied. Fortunately, many of the mechanisms that support tumor growth are indeed widely shared, often cutting across conventional tumor types. Even if a particular molecular change is not shared, the same regulatory pathway may be altered (98).

As a well-known example, of epidermal growth factor receptor (EGFR) is an important target for many different cancer types, and many different targeted therapeutics, with different patterns of reactivity, have been developed against EGFR family members (12, 95, 97, 99–104). Of course, any particular tumor may be resistant, or become resistant, to any particular therapy. Thus, a given therapeutic may be effective against non-small cell lung cancer (NSCLC), breast cancer, and glioma, but not every NSCLC, breast cancer, or glioma, and not indefinitely. As the bases for resistance become better understood, the explanations and responses are also likely to apply to many tumors and to cut across tumor types.

Another factor that may help to limit the variables to be tested is that resistance can be cyclical, with a given agent showing renewed benefit, beyond progression. One possible mechanism, applicable to many tumors, is that different preexisting subpopulations may wax or wane, as the therapy is continued or halted (61). For brain tumors in particular, variations in the BBB, as the tumor grows or responds to therapy, can be an additional factor (27).

Emphasis on common tumor targets can help control the number of possibilities to be tested, but there are still many variables. Even for shared targets, the context is important. The net effect of the abnormalities in a given tumor, the normal regulatory pathways in that cell type, and the microenvironment can all influence tumor growth and response to therapy (105, 106). This complexity does not necessarily mandate a unique targeted therapeutic—or a unique animal model—for each tumor. As understanding and experience evolve, it should be increasingly possible to develop a basic set of therapeutics that are each appropriate for many different tumors, including different tumor types. Individual patients would receive personalized combinations, rather than unique agents.

7.1. A Specific Example

The cytochrome P450 family of drug-metabolizing enzyme provides specific examples of current challenges to tumor therapy and the evolving role of small animal models. Efflux pumps and metabolizing enzymes can each contribute to tumor resistance, by preventing therapeutics from achieving their intended dose at tumor sites (82, 84). The P450 enzymes themselves are heterogeneous at many levels. Different family members interact with different drugs and have characteristic distributions among cells, tissues, individuals, strains, and species (43, 64, 66, 107).

A wealth of background information is available about the structure, function, and distribution of the P450 enzymes and enzyme/drug interactions (43, 64–66, 107). For a promising new drug or new combination, this body of information can delay the need for small animal models, and allow more efficient use when they are required. From the drug structure, one may be able to predict which P450 enzyme will be most important for human patients (43, 65). The enzyme/drug interaction can be confirmed in cell lines that express the correct P450 variant, and the distribution of the appropriate enzyme in humans can be confirmed in archived tissue. Strategies to modulate enzyme expression or function can first be tested in vitro.

To take into account the full environment, animal models are needed. For example, enzyme levels may be altered in the context of growing tumor, tumor therapy, or unrelated processes (64, 68). Modulating drug metabolism can have complex effects within solid tumors, and small animal models can reveal effects on drug distribution that apply beyond the specific drug/enzyme interaction being studied (82, 84).

When in vivo testing is needed, one value of the earlier steps is that they can prevent a misleading mismatch between the P450 variant and distribution in the model as compared to human patients. Rather, a rodent model can be selected, humanized, or otherwise modified to express the correct P450 variant in the appropriate cell types (107). Insight can be gained from focus on specific questions, such as the distribution of a particular drug within the tumor (82, 84); it is not necessary to follow a full sequence of steps leading to rodent tumor therapy (Fig. 1).

8. Summary: Challenges and Responses

A long chain of steps, imperfect mimicry, multifaceted heterogeneity, and an abundance of choices all complicate the use of small animal models to develop new therapies. Fortunately, these challenges are balanced by improvements that can help to simplify the task.

The chain of steps is shortened because, as mechanisms of tumor growth and resistance are better understood, analysis can be focused on specific steps, rather than the endpoint of a longer chain (Fig. 1). Often, these more pointed questions can be answered without resort to rodent models: from human tissue; in less complex animals or in vitro; in silico; or even in biblio, given the greater ease of accessing already-published work.

Improved ease and power of analysis allow one to get more information directly from human tissue. In parallel, improvements in medical imaging aid study of ongoing responses in brain tumor patients, while new clinical trial designs allow more efficient comparisons among alternatives (7). As the search for biomarkers bears fruit, direct information will be even more accessible. The same improvements allow for more frequent cross-checking of preclinical findings against human material.

When animal models are needed, a variety of technical improvements provide better *tumor mimics.* More sophisticated genetic manipulations are complemented by developments in other areas. For example, an original reason to study subcutaneous tumor was that tumor growth could easily be followed within a single animal, by measuring the expanding mass with calipers. Improvements in imaging techniques allow tumor growth to be followed noninvasively even in the brain (10), although microtumor still presents challenges (17–19, 62, 63).

Heterogeneity is better accounted for as its underlying sources are identified and incorporated into experimental work. *The abundance of choices* should be reduced as growing understanding allows focus on targets and mechanisms that are shared among many tumors and patients, often cutting across conventional tumor types.

9. Looking Ahead

Even as small animal models become more sophisticated, their role in predicting clinical outcome can be reduced and refined, as other approaches also evolve. When small animals are needed, their increased validity as tumor mimics will be complemented by growing ability to focus on specific questions and to cross-check against human tissue.

The promise of new tumor therapies is just starting to be realized. Just because so many of the challenges are so widespread, shared by many tumor types, it is likely that the solutions, as they evolve, will also be shared. The multifaceted tumor heterogeneity does not necessarily require a unique therapy—or a unique animal model—for each case. Rather, a given insight or therapeutic is likely to be relevant for many tumors, including tumors of different types.

In terms of the opening quotation, there may be no perfect animal model, no phoenix, for predicting clinical outcome for brain tumor therapy. No single model can fully correspond to a given human tumor, nor can it show the full course of events. Instead, both specific points and general principles, learned from a variety of models, will complement other sources of insight, and may apply to many different tumors of different types. One will be able to say for appropriate sets of tumors: these tumors may be diverse, but with respect to this abnormality or this therapeutic, *Cosi' Fan Tutti* (they all act the same).

Acknowledgment

I thank Cara Tripp McClallen for helping to prepare the manuscript.

References

1. Codarri L, Fontana A, Becher B (2010) Cytokine networks in multiple sclerosis: lost in translation. Curr Opin Neurol 23:205–211
2. Frese KK, Tuveson DA (2007) Maximizing mouse cancer models. Nat Rev Cancer 7:645–658
3. Hermann DM (2007) Biodistribution processes as underestimated confounders in translational stroke research. Curr Med Chem 14:3179–3184
4. Lohle M, Reichmann H (2010) Clinical neuroprotection in Parkinson's disease—still waiting for the breakthrough. J Neurol Sci 289:104–114
5. Ludman AJ, Yellon DM, Hausenloy DJ (2010) Cardiac preconditioning for ischaemia: lost in translation. Dis Model Mech 3:35–38
6. Kasper S (2005) Survey of genetically engineered mouse models for prostate cancer: analyzing the molecular basis of prostate cancer development, progression, and metastasis. J Cell Biochem 94:279–297
7. Berry DA (2006) Bayesian clinical trials. Nat Rev Drug Discov 5:27–36
8. Lampson LA (2009) Targeted therapy for neuro-oncology: reviewing the menu. Drug Discov Today 14:185–191
9. Talmadge JE, Singh RK, Fidler IJ et al (2007) Murine models to evaluate novel and conventional therapeutic strategies for cancer. Am J Pathol 170:793–804
10. Damia G, D'Incalci M (2009) Contemporary pre-clinical development of anticancer agents—what are the optimal preclinical models? Eur J Cancer 45:2768–2781
11. Kerbel RS (2003) Human tumor xenografts as predictive preclinical models for anticancer drug activity in humans: better than commonly perceived-but they can be improved. Cancer Biol Ther 2:S134–S139
12. Troiani T, Schettino C, Martinelli E et al (2008) The use of xenograft models for the selection of cancer treatments with the EGFR as an example. Crit Rev Oncol Hematol 65:200–211
13. Varticovski L, Hollingshead MG, Robles AI et al (2007) Accelerated preclinical testing using transplanted tumors from genetically engineered mouse breast cancer models. Clin Cancer Res 13:2168–2177
14. Gopinathan A, Tuveson DA (2008) The use of GEM models for experimental cancer therapeutics. Dis Model Mech 1:83–86
15. Walrath JC, Hawes JJ, Van Dyke T et al (2010) Genetically engineered mouse models in cancer research. Adv Cancer Res 106:113–164
16. Ursini-Siegel J, Schade B, Cardiff RD et al (2007) Insights from transgenic mouse models of ERBB2-induced breast cancer. Nat Rev Cancer 7:389–397
17. Lampson LA, Tripp CA (2007) Where do blood-borne cells enter the brain? J Immunol 178:S195
18. Tripp CA, Lampson LA (2007) A rat model to study early micro-metastases in the brain: LB308. J Neuropathol Exp Neurol 66:457–458
19. Tripp CA, Hurley ML, Lampson LA (2006) Reproducible blood-borne micro-metastases in the brain proper: a syngeneic rat model. Neuro Oncol 8:391–500
20. Yin JJ, Zhang L, Munasinghe J et al (2010) Cediranib/AZD2171 inhibits bone and brain metastasis in a preclinical model of advanced prostate cancer. Cancer Res 70:8662–8673
21. Yang WQ, Senger DL, Lun XQ et al (2004) Reovirus as an experimental therapeutic for brain and leptomeningeal metastases from breast cancer. Gene Ther 11:1579–1589
22. Brandsma D, Taphoorn MJ, Reijneveld JC et al (2004) MR imaging of mouse leptomeningeal metastases. J Neurooncol 68: 123–130
23. Gu C, Li S, Tokuyama T et al (2010) Therapeutic effect of genetically engineered mesenchymal stem cells in rat experimental leptomeningeal glioma model. Cancer Lett 291:256–262
24. Turner WJ, Chatten J, Lampson LA (1990) Human neuroblastoma cell growth in xenogeneic hosts: comparison of T cell-deficient and NK-deficient hosts, and subcutaneous or intravenous injection routes. J Neurooncol 8:121–132
25. Husemann Y, Klein CA (2009) The analysis of metastasis in transgenic mouse models. Transgenic Res 18:1–5
26. Shoushtari AN, Michalowska AM, Green JE (2007) Comparing genetically engineered mouse mammary cancer models with human breast cancer by expression profiling. Breast Dis 28:39–51
27. Lampson L (2011) Monoclonal antibodies in neuro-oncology: getting past the blood–brain barrier. MAbs 3(2):153–160

28. Krug N, Rabe KF (2008) Animal models for human asthma: the perspective of a clinician. Curr Drug Targets 9:438–442
29. Spanagel R, Kiefer F (2008) Drugs for relapse prevention of alcoholism: ten years of progress. Trends Pharmacol Sci 29:109–115
30. Johnson RA, Wright KD, Poppleton H et al (2010) Cross-species genomics matches driver mutations and cell compartments to model ependymoma. Nature 466:632–636
31. Gibson P, Tong Y, Robinson G et al (2010) Subtypes of medulloblastoma have distinct developmental origins. Nature 468: 1095–1099
32. Parsons DW, Li M, Zhang X et al (2011) The genetic landscape of the childhood cancer medulloblastoma. Science 331:435–439
33. Hunter KW, Crawford NP (2008) The future of mouse QTL mapping to diagnose disease in mice in the age of whole-genome association studies. Annu Rev Genet 42:131–141
34. Peters LL, Robledo RF, Bult CJ et al (2007) The mouse as a model for human biology: a resource guide for complex trait analysis. Nat Rev Genet 8:58–69
35. Andrechek ER, Cardiff RD, Chang JT et al (2009) Genetic heterogeneity of Myc-induced mammary tumors reflecting diverse phenotypes including metastatic potential. Proc Natl Acad Sci U S A 106:16387–16392
36. Wertheim GB, Yang TW, Pan TC et al (2009) The Snf1-related kinase, Hunk, is essential for mammary tumor metastasis. Proc Natl Acad Sci U S A 106:15855–15860
37. Nazarian R, Shi H, Wang Q et al (2010) Melanomas acquire resistance to B-RAF (V600E) inhibition by RTK or N-RAS upregulation. Nature 468:973–977
38. Hayakawa Y, Huntington ND, Nutt SL et al (2006) Functional subsets of mouse natural killer cells. Immunol Rev 214:47–55
39. Rehli M (2002) Of mice and men: species variations of Toll-like receptor expression. Trends Immunol 23:375–378
40. Song ES, Young K, Sears DW (1990) Rat and human natural killers exhibit contrasting immunoglobulin G subclass specificities in antibody-dependent cellular cytotoxicity reflecting differences in their Fc receptors (Fc gamma R). J Leukoc Biol 48:524–530
41. Ward ES, Ober RJ (2009) Chapter 4: multitasking by exploitation of intracellular transport functions the many faces of FcRn. Adv Immunol 103:77–115
42. Chapman K, Pullen N, Coney L et al (2009) Preclinical development of monoclonal antibodies: considerations for the use of nonhuman primates. MAbs 1:505–516
43. Caccia S, Garattini S, Pasina L et al (2009) Predicting the clinical relevance of drug interactions from pre-approval studies. Drug Saf 32:1017–1039
44. Capt A, Luzy AP, Esdaile D et al (2007) Comparison of the human skin grafted onto nude mouse model with in vivo and in vitro models in the prediction of percutaneous penetration of three lipophilic pesticides. Regul Toxicol Pharmacol 47:274–287
45. Lampson L (1984) Molecular bases of neuronal individuality: lessons from anatomical and biochemical studies with monoclonal antibodies. In: Kennett R et al (eds) Monoclonal antibodies and functional cell lines: progress and applications. Plenum, New York
46. Dixit R, Boelsterli UA (2007) Healthy animals and animal models of human disease(s) in safety assessment of human pharmaceuticals, including therapeutic antibodies. Drug Discov Today 12:336–342
47. Cox D, Brennan M, Moran N (2010) Integrins as therapeutic targets: lessons and opportunities. Nat Rev Drug Discov 9:804–820
48. Constantin G (2008) Chemokine signaling and integrin activation in lymphocyte migration into the inflamed brain. J Neuroimmunol 198:20–26
49. Gril B, Evans L, Palmieri D et al (2010) Translational research in brain metastasis is identifying molecular pathways that may lead to the development of new therapeutic strategies. Eur J Cancer 46:1204–1210
50. Juhasz G, Vass G, Bozso Z et al (2008) Integrin activation modulates NMDA and AMPA receptor function of CA1 cells in a dose-related fashion in vivo. Brain Res 1233:20–26
51. Milner R, Campbell IL (2002) The integrin family of cell adhesion molecules has multiple functions within the CNS. J Neurosci Res 69:286–291
52. Pylayeva Y, Guo W, Giancotti FG (2007) Analysis of integrin signaling in genetically engineered mouse models of mammary tumor progression. Methods Enzymol 426:439–461
53. Takahashi JA, Llena JF, Hirano A (1996) Pathology of cerebral metastases. Neurosurg Clin N Am 7:345–367
54. McCluskey LP, Lampson LA (2000) Local neurochemicals and site-specific immune regulation in the CNS. J Neuropathol Exp Neurol 59:177–187
55. McCluskey LP, Lampson LA (2001) Local immune regulation in the central nervous system by substance P vs. glutamate. J Neuroimmunol 116:136–146

56. Phillips LM, Simon PJ, Lampson LA (1999) Site-specific immune regulation in the brain: differential modulation of major histocompatibility complex (MHC) proteins in brainstem vs. hippocampus. J Comp Neurol 405:322–333
57. Brezski RJ, Vafa O, Petrone D et al (2009) Tumor-associated and microbial proteases compromise host IgG effector functions by a single cleavage proximal to the hinge. Proc Natl Acad Sci U S A 106:17864–17869
58. Esteban F, Munoz M, Gonzalez-Moles MA et al (2006) A role for substance P in cancer promotion and progression: a mechanism to counteract intracellular death signals following oncogene activation or DNA damage. Cancer Metastasis Rev 25:137–145
59. Nowicki M, Ostalska-Nowicka D, Kondraciuk B et al (2007) The significance of substance P in physiological and malignant haematopoiesis. J Clin Pathol 60:749–755
60. Baker H (1986) Species differences in the distribution of substance P and tyrosine hydroxylase immunoreactivity in the olfactory bulb. J Comp Neurol 252:206–226
61. Sharma SV, Lee DY, Li B et al (2010) A chromatin-mediated reversible drug-tolerant state in cancer cell subpopulations. Cell 141:69–80
62. Lampson LA, Wen P, Roman VA et al (1992) Disseminating tumor cells and their interactions with leukocytes visualized in the brain. Cancer Res 52:1018–1025
63. Lampson LA, Lampson MA, Dunne AD (1993) Exploiting the lacZ reporter gene for quantitative analysis of disseminated tumor growth within the brain: use of the lacZ gene product as a tumor antigen, for evaluation of antigenic modulation, and to facilitate image analysis of tumor growth in situ. Cancer Res 53:176–182
64. Rodriguez-Antona C, Gomez A, Karlgren M et al (2010) Molecular genetics and epigenetics of the cytochrome P450 gene family and its relevance for cancer risk and treatment. Hum Genet 127:1–17
65. Hisaka A, Ohno Y, Yamamoto T et al (2010) Prediction of pharmacokinetic drug–drug interaction caused by changes in cytochrome P450 activity using in vivo information. Pharmacol Ther 125:230–248
66. Hrycay EG, Bandiera SM (2009) Expression, function and regulation of mouse cytochrome P450 enzymes: comparison with human P450 enzymes. Curr Drug Metab 10:1151–1183
67. Gomez A, Ingelman-Sundberg M (2009) Epigenetic and microRNA-dependent control of cytochrome P450 expression: a gap between DNA and protein. Pharmacogenomics 10:1067–1076
68. Morgan ET (2009) Impact of infectious and inflammatory disease on cytochrome P450-mediated drug metabolism and pharmacokinetics. Clin Pharmacol Ther 85:434–438
69. Diepenhorst GM, van Gulik TM, Hack CE (2009) Complement-mediated ischemia-reperfusion injury: lessons learned from animal and clinical studies. Ann Surg 249:889–899
70. Gendrel M, Rapti G, Richmond JE et al (2009) A secreted complement-control-related protein ensures acetylcholine receptor clustering. Nature 461:992–996
71. Perry VH, O'Connor V (2008) C1q: the perfect complement for a synaptic feast? Nat Rev Neurosci 9:807–811
72. Rutkowski MJ, Sughrue ME, Kane AJ et al (2010) Complement and the central nervous system: emerging roles in development, protection and regeneration. Immunol Cell Biol 88:781–786
73. van Beek J, Elward K, Gasque P (2003) Activation of complement in the central nervous system: roles in neurodegeneration and neuroprotection. Ann N Y Acad Sci 992:56–71
74. Dutheil F, Beaune P, Loriot MA (2008) Xenobiotic metabolizing enzymes in the central nervous system: contribution of cytochrome P450 enzymes in normal and pathological human brain. Biochimie 90:426–436
75. Stolz C, Schuler M (2009) Molecular mechanisms of resistance to rituximab and pharmacologic strategies for its circumvention. Leuk Lymphoma 50:873–885
76. Martinelli G, Iacobucci I, Paolini S et al (2008) Farnesyltransferase inhibition in hematologic malignancies: the clinical experience with tipifarnib. Clin Adv Hematol Oncol 6:303–310
77. Sousa SF, Fernandes PA, Ramos MJ (2008) Farnesyltransferase inhibitors: a detailed chemical view on an elusive biological problem. Curr Med Chem 15:1478–1492
78. Zhou J, Vos CC, Gjyrezi A et al (2009) The protein farnesyltransferase regulates HDAC6 activity in a microtubule-dependent manner. J Biol Chem 284:9648–9655
79. Gandhi L, McNamara KL, Li D et al (2009) Sunitinib prolongs survival in genetically engineered mouse models of multistep lung carcinogenesis. Cancer Prev Res (Phila) 2:330–337
80. Baker JH, Lindquist KE, Huxham LA et al (2008) Direct visualization of heterogeneous extravascular distribution of trastuzumab in

human epidermal growth factor receptor type 2 overexpressing xenografts. Clin Cancer Res 14:2171–2179

81. Kyle AH, Huxham LA, Yeoman DM et al (2007) Limited tissue penetration of taxanes: a mechanism for resistance in solid tumors. Clin Cancer Res 13:2804–2810
82. Minchinton AI, Tannock IF (2006) Drug penetration in solid tumours. Nat Rev Cancer 6:583–592
83. Olive KP, Jacobetz MA, Davidson CJ et al (2009) Inhibition of Hedgehog signaling enhances delivery of chemotherapy in a mouse model of pancreatic cancer. Science 324:1457–1461
84. Tredan O, Galmarini CM, Patel K et al (2007) Drug resistance and the solid tumor microenvironment. J Natl Cancer Inst 99:1441–1454
85. Vlashi E, Sturgis JE, Thomas M et al (2009) Real time, noninvasive imaging and quantitation of the accumulation of ligand-targeted drugs into receptor-expressing solid tumors. Mol Pharm 6:1868–1875
86. Cartron G, Blasco H, Paintaud G et al (2007) Pharmacokinetics of rituximab and its clinical use: thought for the best use? Crit Rev Oncol Hematol 62:43–52
87. Zhou Q, Guo P, Kruh GD et al (2007) Predicting human tumor drug concentrations from a preclinical pharmacokinetic model of temozolomide brain disposition. Clin Cancer Res 13:4271–4279
88. Bivona TG, Hieronymus H, Parker J et al (2011) FAS and NF-kappaB signalling modulate dependence of lung cancers on mutant EGFR. Nature 471:523–526
89. Edwards SL, Brough R, Lord CJ et al (2008) Resistance to therapy caused by intragenic deletion in BRCA2. Nature 451:1111–1115
90. Hurtado A, Holmes KA, Geistlinger TR et al (2008) Regulation of ERBB2 by oestrogen receptor-PAX2 determines response to tamoxifen. Nature 456:663–666
91. Naka K, Hoshii T, Muraguchi T et al (2010) TGF-beta-FOXO signalling maintains leukaemia-initiating cells in chronic myeloid leukaemia. Nature 463:676–680
92. Sakai W, Swisher EM, Karlan BY et al (2008) Secondary mutations as a mechanism of cisplatin resistance in BRCA2-mutated cancers. Nature 451:1116–1120
93. Sotillo R, Schvartzman JM, Socci ND et al (2010) Mad2-induced chromosome instability leads to lung tumour relapse after oncogene withdrawal. Nature 464:436–440
94. Tran C, Ouk S, Clegg NJ et al (2009) Development of a second-generation antiandrogen for treatment of advanced prostate cancer. Science 324:787–790
95. Mierzwa ML, Nyati MK, Morgan MA et al (2010) Recent advances in combined modality therapy. Oncologist 15:372–381
96. Pirker R (2010) Bevacizumab in lung cancer: hurdles in clinical development. Lancet Oncol 11:703–704
97. Okines A, Cunningham D (2009) Current perspective: bevacizumab in colorectal cancer—a time for reappraisal? Eur J Cancer 45:2452–2461
98. Wood LD, Parsons DW, Jones S et al (2007) The genomic landscapes of human breast and colorectal cancers. Science 318:1108–1113
99. Debucquoy A, Machiels JP, McBride WH et al (2010) Integration of epidermal growth factor receptor inhibitors with preoperative chemoradiation. Clin Cancer Res 16:2709–2714
100. Gerber DE, Choy H (2010) Cetuximab in combination therapy: from bench to clinic. Cancer Metastasis Rev 29:171–180
101. Wheatley-Price P, Shepherd FA (2008) Epidermal growth factor receptor inhibitors in the treatment of lung cancer: reality and hopes. Curr Opin Oncol 20:162–175
102. Pegram M, Ngo D (2006) Application and potential limitations of animal models utilized in the development of trastuzumab (Herceptin): a case study. Adv Drug Deliv Rev 58:723–734
103. Zaidi SH, Huddart RA, Harrington KJ (2009) Novel targeted radiosensitisers in cancer treatment. Curr Drug Discov Technol 6:103–134
104. Magne N, Chargari C, Castadot P et al (2008) The efficacy and toxicity of EGFR in the settings of radiotherapy: focus on published clinical trials. Eur J Cancer 44:2133–2143
105. Hatzivassiliou G, Song K, Yen I et al (2010) RAF inhibitors prime wild-type RAF to activate the MAPK pathway and enhance growth. Nature 464:431–435
106. Poulikakos PI, Zhang C, Bollag G et al (2010) RAF inhibitors transactivate RAF dimers and ERK signalling in cells with wild-type BRAF. Nature 464:427–430
107. Cheung C, Gonzalez FJ (2008) Humanized mouse lines and their application for prediction of human drug metabolism and toxicological risk assessment. J Pharmacol Exp Ther 327:288–299

Neuromethods (2013) 77: 21–36
DOI 10.1007/7657_2012_59

Published online: 19 July 2012

Animal Models in Cancer Research: Assessment of Severity and the Application of Humane Endpoints

José M. Orellana-Muriana

Abstract

Refinement is one of the three foundations of the 3Rs concept, a concept that underpins good animal-based research. For ethical and legal reasons, the severity of a procedure or an experiment on the welfare and health of an experimental animal needs to be estimated before starting the experiment. This estimation needs to be done for two main reasons. First, any inherent negative effects that a procedure or an experiment has on an experimental animal can be minimized by establishing humane endpoints. Second, the severity of all procedures in a scientific experiment contributes to the cumulative suffering that an experimental animal will experience during its lifespan. The aim of this chapter is to explain refinement and humane endpoints and their use in reducing the severity and thereby assure the welfare of an experimental animal while it participates in a scientific investigation.

Key words: Animal welfare, Severity, Suffering, Humane endpoint, Animal research, Animal models, Refinement, Cancer, Neurosciences

1. Introduction

The conduct of animal research entails the establishment of a relationship between the scientist and the experimental animal. Since the experimental animal is the weaker party in this relationship, today's scientists have a legal and moral duty to protect the animal and ensure that its welfare is compromised as little as possible.

Since a scientific experiment can affect an animal's welfare, the following points should always be borne in mind when doing animal-based research:

1. The severity of any procedure, which is to be done or was done on an animal, should be assessed during the procedure and after completion of the procedure.
2. The severity of any procedure can be reduced and minimized by establishing a humane scientific endpoint.
3. All procedures and experiments can contribute to the cumulative suffering that the animal will experience during its lifespan.

4. Researchers should familiarize themselves with the concept of refinement in order to minimize the actual or potential pain, distress, and other adverse effects of a procedure and/or an experiment on an experimental animal.

These four points are particularly relevant for experiments that have the potential for an unusually high impact on the animal's welfare, such as an investigation in which an animal model of cancer (brain tumours) is wanted, or in an animal-based study in the neurosciences.

2. Procedure Versus Experiment

Since the words "procedure" and "experiment" will be frequently used in this chapter, it is important to distinguish the differences between them. Since the use of animals in research, teaching, and testing is an important political issue, many definitions of the words "procedure" and "experiment" can be found in government documents that pertain to animal-based research. For example, the word "procedure" is defined in Directive 2010/63/EU of the European Parliament and the Council (1) as any use, invasive or noninvasive, of an animal for experimental or other scientific purposes, with known or unknown outcome, or educational purposes, which may cause the animal a level of pain, suffering, distress, or lasting harm equivalent to, or higher than, that caused by the introduction of a needle in accordance with good veterinary practice. For the purposes of this chapter, (a) a procedure will be defined as a course of action on an animal that is intended to achieve a result during a scientific experiment and (b) an experiment will be defined as an act doing a controlled test or investigation. Although an experiment could comprise several procedures, its purpose is to acquire new knowledge or confirm existing knowledge. For the animal and researcher, a procedure and the experiment itself should ideally cause no harm to the experimental animal and not compromise its welfare and/or its physical and psychological well-being.

3. Severity

In animal-based research, the word "severity" is used to describe the amount of suffering that an animal experiences during or following a procedure in a scientific experiment. Severity can be expressed in terms of the level to which its welfare has been compromised during and following a procedure or an experiment and during its lifespan. Many articles on how to assess severity and animal welfare and on the ways to reduce the severity, such as that written by Hawkins and others (2), have been published in the scientific literature.

Before starting any animal-based experiment, the potential severity of all the procedures that will be done in the entire experiment should be assessed in order to avoid unnecessary animal suffering. In many cases, the severity of the procedures and experiment will also need to be assessed at the end of the experiment in order to determine the animal's actual suffering, discomfort, distress, and cumulative suffering. This second assessment, which is called a retrospective assessment of severity, enables both experienced and inexperienced researchers to make adjustments to those procedures that will be done in their future investigation in order to preserve the animal's welfare in these future investigations. This retrospective assessment can also allow investigators to consider the possibility of reusing the animals in another scientific experiment or rehoming the animals when the level of severity for the experiment in which the animals were used is low. When making these assessments, the level of severity should be assigned to different categories depending on the expected or actual level of suffering that the animal experiences during or following a procedure in an experiment. Although the number and name of these categories vary from country to country, severity is usually graded in terms of "mild," "moderate," and "severe." These assessments can be made using score sheets, which score clinical signs and measures of animal welfare. An example of a score sheet and the scoring of clinical signs and aberrations in animal welfare are given in Sect. 59.13. In addition, provision for both types of assessment should be included in an application form to perform an animal-based experiment that will be submitted to an institution's Animal Ethics Committee (AEC) or Animal Care and Use Committee (IACUC).

4. Maximum Limit of Severity

The maximum limit of severity is the highest level of severity above which the carrying out of a procedure is not permitted. Determination of this maximum limit should be done using indicators of animal health and welfare and under the conditions in which the procedure is to be performed. For example, the maximum limit of severity could be determined by the appearance and/or disappearance of several clinical signs or when the measures of animal welfare are lessened during a procedure or an experiment.

5. How to Assign a Category of Severity

The level of severity of a procedure or an experiment can be determined by appraising the level of discomfort that an individual experimental animal experiences during a procedure or a scientific project. This appraisal should be done on an individual animal and not on a group of animals and after all the refinement techniques have been applied. The appraisal is usually made by scoring and recording the clinical signs that appear or disappear on a score sheet (see later). The severity of each procedure that is to be done in an experiment should be separately scored, and procedure with the highest severity score will define the level of severity of the entire experiment. In addition, the maximum level of severity for a particular procedure or experiment can then be determined by reviewing the scores that were obtained in each animal that was subject to that procedure or was used in the experiment.

6. Factors That Can Influence the Severity of a Procedure

Numerous factors can influence the level of pain, suffering, and distress that an animal can experience during or following a procedure or cause lasting harm to an animal as a result of a procedure and experiment. These factors can be categorized as being procedure-related, husbandry-related, animal-related, and personnel-related factors.

6.1. Procedure-Related Factors

- The type and number of treatments and the method of its application or administration.
- The physical and chemical characteristics of any solution that is to be injected or administered to the animal, such as its pH, consistency, density, or concentration.
- The intensity, duration, and frequency of the procedure, and the interval between successive procedures.

6.2. Husbandry-Related Factors

- The method of housing the animals during the experiment, such as single housing for metabolic studies and isolation in an infectious microorganism study.
- The method of restraining the animal in order to perform the procedure, such as the use of restraining devices or anaesthesia.

6.3. Animal-Related Factors

- The animal species, its size, age, and sex. For example, the size and age of the animal can determine the actual volume and frequency of a blood sample that can be collected from the animal provided the volume of the sample does not exceed 10% of the animal's body weight. Repeated blood sampling can be more easily done in cattle than in mice, and the physiological impact of that blood sample will be less in cattle than in mice.

- The animal's genotype. For example, a genetic modification could have its own inherent level of severity and can affect the welfare and/or health of a genetically modified animal at any time during its lifespan.

6.4. Personnel-Related Factors

Since the animal-handling skills (stockmanship) of researchers, animal carers, and technicians can influence the severity of a procedure, the person who is assigned to carry out the procedure should be aware that all vertebrate animal species are sentient animals. In addition, this person should treat the experimental animals on an individual basis, and not as a part of a group. This person should also have appropriate knowledge of:

- Animal welfare.
- Normality. The researcher should be familiar with what is normal for the animal species that is used in a particular investigation and should know which animal species is the most suitable for use in a particular animal-based experiment. For example, the males of some mice strains are inherently more aggressive than others, and their aggressive behaviour towards their cage mates when housed together is considered normal even though they fight and injure each other. Although this aggressive behaviour is "normal," it is not considered beneficial for the animal's welfare.
- The expected effects of a specific procedure on an animal's welfare.

The severity of a procedure or experiment can be lowered by:

- Using experienced staff to train the individual(s) who will be doing the procedure so that this person will not cause unnecessary discomfort, pain, and suffering to the animal.
- Training the animal before carrying out the procedure in order to make the animal cooperative while the procedure is being done.
- Applying refinement methods in order to minimize the animal's discomfort, pain, and suffering.
- Applying early humane endpoints in order to avoid causing unnecessary discomfort, pain, and suffering to the animal.

7. Refinement

Refinement is any approach which avoids or minimizes the actual or potential pain, distress, and other adverse effects that an animal experiences during its life or enhances its physical and psychological well-being. The purpose of refinement in animal-based research is to maximize the benefits of the research (3) because happy animals make good science (4). Today, it is ethically mandatory that an investigator applies all available methods of refinement in order to maximally minimize the level of severity of a procedure or experiment. Anaesthesia and analgesia are well-known and frequently used methods of refinement. Other refinement methods include:

1. Environmental enrichment of the animal's housing.
2. Modifying the design of animal experiments using statistical software in order to reduce the number of animals that will be used in an experiment.
3. Improving the performance of a specific procedure or intervention, such as the surgical method for implanting a tumour or tumour cells.
4. The training of staff for performing a particular procedure.
5. The implementation of a humane endpoint.

8. Humane Endpoints

Direct or necessary inhumanities, such as pain, distress, and suffering, may sometimes be unavoidable in animal-based research, and contingent or avoidable inhumanities must always be avoided in animal experiments. A humane endpoint has been defined in different ways by different organisations and individuals (5–8). For purposes of this chapter, a humane endpoint will be defined as the earliest indicator in an animal-based experiment where an experimental animal is experiencing severe pain, distress, and suffering or close to impending death. A humane endpoint should reflect the actual or imminent deterioration of an animal's physical and psychological condition. A humane endpoint is also the point when the procedure or experiment must be stopped because the animal's physical and/or psychological states make it difficult and sometimes not possible to continue or complete the procedure or experiment. A humane endpoint is also the point at which the animal's pain, distress, and suffering exceed the maximum limit of severity and when humane killing of the animal should be considered. Hence, the aims of establishing a humane endpoint for a procedure in an experiment are to (a) minimize the welfare costs of justifiable animal-based research by reducing or eliminating avoidable animal suffering and (b) predict the time of onset of severe pain, distress, and suffering and impending death before the animal will experience these events.

Collecting data from an animal in pain or in distress can compromise the quality of the data and distort the results. Since a humane endpoint is one that occurs before the animal becomes morbid, the death of an animal should not be used as a humane endpoint. For example, the humane endpoint in an animal with a brain tumour could be the onset of hind limb paresis or paralysis or when a tumour reaches a determinate size after which it begins to vacuolated, become necrotic, or ulcerate. In both cases, it is not necessary to keep the animal alive because allowing them to live will (a) severely compromise the animal's welfare by causing unnecessary suffering and (b) increase the risk of developing unwanted complications, such as infections.

9. Severity and Humane Endpoints in Animal-Based Cancer Research

In cancer research, animal experiments are essential for understanding the fundamental mechanisms that underpin malignancy and for discovering improved methods to prevent, diagnose, and treat cancer. In such studies, it is also clearly important that the welfare of animals in cancer research is protected. In animal-based cancer research, two different situations can be identified. In the first situation, the animal is being used to maintain a colony of tumour cells or a transplanted tumour. In the second situation, a novel chemotherapeutic or antitumour molecule is being evaluated in a tumour-bearing animal. In both cases, the animal's clinical signs and level of suffering may be different.

Irrespective of the situation, the following points should be borne in mind when doing animal-based cancer studies:

1. For studies that assess the efficacy of a novel chemotherapeutic molecule in mice that are used for maintaining a tumour, the maximum permitted size of a tumour before starting the treatment should be larger in the treated mouse (15-mm diameter) than in the non-treated mouse (12-mm diameter) because it is anticipated that the treatment will reduce tumour size. These sizes are recommended because tumours whose sizes exceed these dimensions can spontaneously vacuolize or become necrotic and place a high burden on the animal.
2. A novel chemotherapeutic molecule can exert unwanted effects on the animal that are unrelated to its antitumour effects, such as causing diarrhoea or the faeces to become bloody.
3. When the treatment is effective, changes in the skin that overlies the tumour area or even inside the tumour should be expected. These changes include dermatitis, skin dryness and/or darkening, scab formation, necrosis, and ulceration.

More information on humane endpoints in animal-based cancer research can be found in an issue of the ILAR journal (9).

10. General Clinical Signs in Animal-Based Cancer Research That Could Be Used as a Humane Endpoint

Several guidelines for the welfare and use of animals in cancer research have been published and some are now available as downloadable documents from the internet (5, 9–12).

Although numerous clinical signs, when considered alone or together, are listed as possible humane endpoints, I have included humane endpoints in this list that I established in my professional experience as a laboratory animal veterinarian and think that readers could find useful.

10.1. Behaviour and Appearance

1. Emaciation and/or dehydration due to failure to eat or drink over a 24- to 48-h period.
2. Consistent or rapid body weight loss that can be as much as 20% at any time or a 15% loss of body weight over 72 h.
3. Persistent hypothermia.
4. A bloody or mucopurulent discharge from any orifice.
5. Significant abdominal distension so that the animal's appearance becomes similar to that of a pregnant animal and the cancerous burden exceeds 10% of the animal's bodyweight.
6. Long-standing diarrhoea (>48 h).
7. Enlarged lymph nodes.
8. Infection of the tumour.
9. Ulceration of the tumour.

10.2. Physiological Indicators

1. Laboured (dyspnoea) or very rapid respiration (tachypnoea), particularly if accompanied by a nasal discharge and/or cyanosis.
2. Anaemia, which can be determined by the presence of pale feet, a pale skin in nude rodents, or haematological measures.
3. Tumours that interfere with locomotion or cause abnormal vocalization.
4. Hypoxia and cyanosis which can be diagnosed by the presence of a bluish-coloured skin.

10.3. Other General and Clinical Signs

1. Trembling.
2. Progressive dermatitis.
3. Rough hair coat, hunched posture, lethargy, or persistent recumbency.
4. Jaundice.
5. Self-induced trauma.
6. Fight wounds and injuries caused by aggressive cage mates.
7. Isolation from cage mates.

When any one of these signs is present in an individual animal, then the animal should be euthanized immediately and any remaining animals observed closely for changes in their condition.

11. Neurological Clinical Signs That Could Be Used to Establish a Humane Endpoint in Animal-Based Cancer Research That Involves the Central Nervous System

The clinical signs in animals with brain tumours are usually quite similar to those found in animals that are used in other animal research on the central nervous system, such as stroke. Accordingly, neurological testing of the animal can be done using previously described protocols (13, 14).

In these tests, four neurological indices are scored: (a) spontaneous activity (moving/exploring = 0, moving without exploration = 1, no moving or only moves when the tail is pulled = 2); (b) circling to the left or right (no circling = 0, circling when the tail is pulled or used to lift the animal or when the tail is used to push the animal = 1, spontaneous circling = 2, circling without displacement (the animal resembles a spinning top) = 3); (c) resistant to allowing the forelimb to be caught and held (animal does not allow the limb to be held = 0, animal allows the limb to be held = 1, the animal does not resist the operator from holding the limb = 2); and (d) the presence of the parachute reflex (symmetrical = 0, asymmetrical = 1, the contralateral forelimb is retracted = 2). In this assessment system, the total severity score ranges from 0, which means the procedure or experiment does not exert any negative effects on the animal, to 9, which is the maximum limit of severity and the score for implementing a humane endpoint.

Other specific neurological signs that could be considered:

1. Hindlimb paralysis or weakness.
2. Frequent seizure activity.
3. Inability to maintain an upright position or to move.
4. Morbundity, lethargy, or failure to respond to gentle stimuli.
5. Unconsciousness or comatoseness.
6. Consistent loss of body weight (14, 15).

12. Suffering, Distress, and Cumulative Suffering

Suffering is defined as the conscious endurance of pain and stress, and is the negative emotional state that is produced by persistent pain and stress, and usually cannot be alleviated. In contrast to this simple definition of suffering, there are many definitions of distress (16).

It has been defined as (a) an aversive, negative state in which coping and adaptation processes in response to stressors fail to return an organism to physiological and/or psychological homeostasis, (b) an aversive state in which an animal is unable to adapt completely to stressors, and can result in maladaptive behaviours, (c) an aversive state resulting from maladaption or inability to adapt to stressors, and (d) a state in which an animal cannot escape from or adapt to the internal or external stressors or conditions that it is experiencing and results in negative effects on its well-being. Distress is also usually associated with a change in motility or locomotion and can result in stereotypic behaviours.

Since the absence of evidence is not evidence of absence and if something is known to cause suffering in humans, it should be assumed that this something will also cause suffering in animals. Hence, there are no reasons to think that an experimental animal with a well-developed nervous system does not suffer and/or experience distress during or following a procedure or experiment.

Cumulative suffering is the maximum amount of suffering that any animal experiences during its lifespan. For experimental animals, cumulative suffering is the maximum suffering that is permissible in an animal due to the cumulative effect of several procedures that it may undergo in a scientific investigation, even after applying refinement techniques, during its lifespan. Cumulative suffering is particularly relevant in research when animals are alive at the end of an experiment, a situation in which experiment is common for control animals and large animal species. In these instances, the animal can be re-homed or considered for reuse in another scientific project on the assumption that the level of severity was graded as "mild."

13. Score Sheets

Since (a) an assessment of the severity of a procedure and an experiment is subjective, and (b) the tools for determining the actual feelings of animals are still lacking, score sheets, which score clinical signs and measures of animal welfare, have been developed in order to surmount these obstacles.

A score sheet is a table in which an individual can score a specific or particular clinical sign in an experimental animal during the course of a procedure or an experiment according to the change from normal. Score sheets can also be a very useful tool when assessing the severity of a procedure or an experiment, and can be used during and after each procedure, for the follow-up of an individual animal, and the retrospective assessment of severity.

Table 1
Instructions for the actions that should be done on the animal for each level of severity

Total severity score	Actions to take for the different level of severity	Severity level
0	No action to be taken because animal is normal	Mild
1–5	Monitor more closely	Mild
6–8	Increase frequency of monitoring; consider corrective measures for improving welfare and well-being (supplementary fluids/care, consider relief measures), notify researcher	Moderate
9–11	Significant suffering likely, provide appropriate relief, observe frequently, consider euthanasia as a humane endpoint	Severe
>12	Highest severity score. Severe suffering, euthanize. Repeating the experiment is not recommended. If repeating the experiment is essential, ask for exemption and provide full justification on the need to repeat	Maximum limit of severity

The use of score sheets for scoring clinical signs was first suggested by Morton and Griffiths in 1985 (17) and then further developed by others, such as Morton and Hau (18). Score sheets are also useful for assessing the severity of procedures and the determination of cumulative suffering. When using such a sheet, it is again crucial that the researcher be familiar with what is "normal" for that animal species when scoring the animal's responses to a particular procedure. Assistance from the principal investigator, the attending veterinarian, the animal welfare officer, an expert technician, or animal caretaker can help a new and inexperienced observer score the animal's clinical signs and behaviours. Lastly, the sheet should also include instructions for the actions that should be done on the animal for each level of severity (Table 1).

A good example of a score sheet is the one that was included in the report of the BVAAWF/FRAME/RSPCA/UFAW Joint Working Group on Refinement (19). This score sheet was developed to score clinical signs in rats when they are used in inflammatory bowel disease studies. A table, such as this, could be modified to make it suitable for use in animal-based cancer studies. It is recommended that score sheets should be in an electronic format because this format allows the sheet to be modified at any time in order to permit the recording of unexpected events and any new events when they occur during the procedure or experiment. It is also recommended that a score sheet be structured in such a way that the results of different ways of observing the experimental animals can be recorded, such as on handling, close up, or from a distance.

Items that should be listed on the score sheet include:

1. All relevant data about the animal, researcher, and the project's title.
2. Details of all procedures and/or techniques that will be performed on the animal.
3. Details of the day and/or time of day when each procedure was done.
4. The number of times that each procedure will be done on a single animal.
5. Details of the clinical signs, which are to be expected according to previous experience or have been published in the scientific literature.

Using my personal experience on animals that were used in cancer research and the reports from other authors, such as Workman et al. (10), and organisations, such as the Canadian Council on Animal Care (6) and the Netherlands Inspectorate for Health Protection, Commodities and Veterinary Public Health (20), an example of a score sheet that lists the clinical signs and aberrations in animal welfare and their scores in an animal-based cancer study is shown in Table 2. These clinical signs and aberrations in animal welfare are scored according to the change from normal, and the scores are not specific for a particular type of tumour. The user can modify, if needed, the list of clinical signs and aberrations in animal welfare and the scoring system in order to make the score sheet more relevant for the investigator's particular research.

14. Severity Scores and the Acceptable Limits of Severity

For severity and its limits, the grade and the limit reflect the time that a humane endpoint should be implemented, such as stopping the project or killing the animal. For example, when a clinical sign or behaviour is given the maximum severity score, on a scale of 0–4, where "0" is normal and "4" is the maximum, a humane endpoint should be applied when a feature of the animal's behaviour or appearance or a clinical sign is given a score of 4. This endpoint should be applied even if the severity scores for the other clinical signs and behaviours are lower than their maximum scores, such as the appearance of ulcers on a tumour or on the skin that overlies a tumour.

Table 2
General example of a score of clinical signs and measures of animal welfare in a cancer research study using nude mice

	Signs of compromised animal welfare	Score
Distance observations		
Appearance		
Coat	Normal	0
	General lack of grooming	1
	Dry coat and/or ocular or nasal discharge	3
	Cyanosis as indicated by bluish-coloured skin/limbs	3
	Anaemia as indicated by a pale skin/feet	3
Dehydration	"Pinched" skin, ridge lines in skin	3
	Dehydration >48 h	4
Abnormal body posture	Hunched abdomen so that its appearance is similar to that of a pregnant mouse	1
	Stretching	2
Locomotion	Tumours that interfere with locomotion	4
	Hind limb paralysis or weakness	4
Clinical signs		
After injection of tumour cells or agent	Transient discomfort after injection	1
Feeding and drinking	Transient post-injection anorexia	1
	Recurrent anorexia	2
	Not drinking	3
Breathing	Normal breathing	0
	Rapid breathing (tachypnoea)	1
	Slightly laboured breathing (dyspnoea)	2
	Very laboured breathing (severe dyspnoea)	3
	Laboured respiration accompanied by nasal discharge and/or cyanosis (bluish-coloured skin)	4
Faecal characteristics	Normal or slightly soft faeces	0
	Soft faeces	1
	Diarrhoea	2
	Bloody diarrhoea	3
	Persistent diarrhoea >48 h	4
Behaviour		
Unprovoked behaviour	Normal behaviour	0
	Transient lethargy after injection	1
	Moderate change in behaviour and/or isolated from cage mates	2
	Persistent lethargy	3
	Reacts violently to handle/vocalization	3
	Repetitive movements (stereotypic behaviours)	1
	Fight wounds or injuries	1
Provoked behaviour	Reluctance to move	1
	No response to sound(s)	2
On handling observations		
Appearance		
Body condition	Emaciation (severe muscular atrophy)	4
Ascites	No ascites	0
	Abdominal distension	2
	20 % increase in body girth, similar to the appearance of a pregnant mouse	4

(continued)

Table 2 (continued)

	Signs of compromised animal welfare	Score
Clinical signs		
Body weight (% weight loss)	Normal or up to 5 % loss	0
	5–10 % loss	1
	11–15 % loss	2
	16–20 % loss	3
	>20 % loss in more than 72 h	4
Body temperature	Normal	0
	Fever	2
	Hypothermia	1
	Persistent hypothermia	4
Specific signs of tumoural changes		
Tumour size (diameter)	0.5–8 mm if the tumour is not treated	1
	9–11 mm if the tumour is not treated	2
	>12 mm if the tumour is not treated	4
	0.5–9 mm if it is treated	1
	10–14 mm if the tumour is treated	2
	>15 mm if the tumour is treated	4
Skin changes	No changes	0
	Necrosis of superficial tissues, dry	2
	Necrosis of superficial tissues, suppurating	3
	Necrosis of superficial tissues, exudative	3
	Necrosis of superficial tissues, suppurating >48 h	4
	Ulceration >20 % surface area	4
Other effects	Enlarged lymph nodes	4
Total severity score		

The highest permissible severity score for a procedure or experiment (maximum limit of severity) should be set before the start of the procedure or experiment. Since the maximum level of severity of an experiment is the sum of all the scores of all the clinical signs at a determined point in time, a high total score means that the level of severity that the animal experienced during the course of an experiment was great in both its degree and its intensity. For example, if the highest permissible severity score was set at 12, this score means that the procedure or experiment should be stopped (the humane endpoint) when the severity of the procedure or experiment is scored as 12. If the procedure or experiment is to be repeated in another investigation, its use in this future investigation will have to be fully justified because the procedure or experiment has the potential to compromise the experimental animal's welfare.

15. Conclusion

In many cases of animal-based research, the research has a negative impact on the welfare of experimental animals. Refinement in animal-based research requires knowledge and an assessment of those elements in a research programme that can adversely affect the research animals and should be applied in order to reduce and minimize this negative impact. In this chapter, I have tried to explain refinement and the establishment of humane endpoints, using examples from my own experience and the published scientific literature. By doing so, it is hoped that this information will be useful for assessing the severity of procedures and determining the time point at which humane endpoints should be instituted in animal-based research.

Acknowledgement

The author thanks Dr. Arieh Bomzon, ConsulWrite (www.consulwrite.com) for his scientific input and editorial assistance in preparing this chapter.

References

1. European Parliament and the Council (2010) Directive 2010/63/EU of 22 September 2010 on the protection of animals used for scientific purposes. OJEU 20.10.2010 L 276: 33–79
2. Hawkins P, Dennison N, Goodman G, Hetherington S, Llywelyn-Jones S, Ryder K, Smith AJ (2011) Guidance on the severity classification of scientific procedures involving fish: report of a Working Group appointed by the Norwegian Consensus-Platform for the Replacement, Reduction and Refinement of animal experiments (Norecopa). Lab Anim 45:219–224
3. Buchanan-Smith HM, Rennie AE, Vitale A, Pollo S, Prescott MJ, Morton DB (2005) Harmonising the definition of refinement. Anim Welfare 14:379–384
4. Poole T (1997) Happy animals make good science. Lab Anim 31:116–124
5. Organisation for Economic Co-operation and Development (OECD) (2000) Guidance document on the recognition, assessment and use of clinical signs as humane endpoints for experimental animals used in safety evaluation. http://www.oecd-ilibrary.org/docserver/download/fulltext/9750191e.pdf?expires=1340025329&id=id&accname=guest&checksum=75EB34E03062C68DF9F7773FF5913A17
6. Canadian Council on Animal Care (CCAC) (1998) Guidelines on choosing an appropriate endpoint in experiments using animals for research teaching and testing. http://www.ccac.ca/Documents/Standards/Guidelines/Appropriate_endpoint.pdf
7. Hendriksen CFM (2009) Replacement, reduction and refinement alternatives to animal use in vaccine potency measurement. Expert Rev. Vaccines 8:313–322. http://ntp.niehs.nih.gov/NTP/About_NTP/SACATM/2010/June/Materials/KeyReferences.pdf
8. National Research Council of National Academies (2011). Guide for the care and use of laboratory animals, 8th edn. http://grants.nih.gov/grants/olaw/Guide-for-the-Care-and-Use-of-Laboratory-Animals.pdf
9. ILAR Journal V41(2) (2000) Humane endpoints for animals used in biomedical research and testing. http://dels-old.nas.edu/ilar_n/ilarjournal/41_2/CancerResearch.shtml
10. Workman P, Aboagye EO, Balkwill F, Balmain A, Bruder G, Chaplin DJ, Double JA, Everitt J,

Farningham DAH, Glennie MJ, Kelland LR, Robinson V, Stratford IJ, Tozer GM, Watson S, Wedge SR, Eccles S et al (2010) An ad hoc committee of the National Cancer Research Institute. Guidelines for the welfare and use of animals in cancer research. Br J Cancer 102:1555–1577

11. Humane endpoints in laboratory animal experimentation (webpage) http://www.humane-endpoints.info/eng/index.php?option=com_content&view=article&id=48&Itemid=27
12. United Kingdom Coordinating Committee on Cancer Research (UKCCCR) (1998) Guidelines for the welfare of animals in experimental neoplasia (2nd ed.). Br J Cancer 77(1):1–10.
13. Pérez-Asensio F, De la Rosa X, Jiménez-Altayó F, Gorina R, Martínez E, Messeguer A, Vila E, Chamorro A, Planas AM (2010) Antioxidant CR-6 protects against reperfusion injury after a transient episode of focal brain ischemia in rats. J Cereb Blood Flow Metab 30:638–652
14. Justicia C, Martín A, Rojas S, Gironella M, Cervera A, Panés J, Chamorro A, Planas AM (2006) Anti-VCAM-1 antibodies did not protect against ischemic damage neither in rats nor in mice. J Cereb Blood Flow Metab 26:321–332
15. Redgate ES, Deutsch M, Boggs SS (1991) Time of death of CNS tumor-bearing rats can be reliably predicted by body weight-loss patterns. Lab Anim Sci 41(3):269–273
16. Committee on Regulatory Issues in Animal Care and Use, Institute for Laboratory Animal Research, National Research Council (2000). Definition of pain and distress and reporting requirements for laboratory animals: proceedings of the workshop held June 22. National Academy Press. ISBN-10: 0-309-07698-6
17. Morton DB, Griffiths PH (1985) Guidelines on the recognition of pain, distress and discomfort in experimental animals and a hypothesis for assessment. Vet Rec 116(16):431–436
18. Morton DB, Hau J (2010) Welfare assessment and human endpoints. In: Hau J, Schapiro S (eds) Handbook of laboratory animal science, vol I, 3rd edn. CRC Press, pp 535–572. ISBN: 978-1-4200-8455-9. e-Book ISBN: 978-1-4200-8456-6.
19. BVAAWF/FRAME/RSPCA/UFAW Joint Working Group on Refinement (2011) A guide to defining and implementing protocols for the welfare assessment of laboratory animals. Lab Anim 45(1):1–13
20. Netherlands Inspectorate for Health Protection, Commodities and Veterinary Public Health (1999) Code of Practice on animal experiments in cancer research. http://www.proefdierkunde.nl/db/upload/documents/CofP_-_Cancer_Research.pdf

Part II

Animal Models Offer Insights into Human Brain Neoplasms

Neuromethods (2013) 77: 39–55
DOI 10.1007/7657_2012_34

Published online: 3 April 2012

The VM Mouse Model of Glioblastoma Multiforme

Thomas N. Seyfried, Laura M. Shelton, and Leanne C. Huysentruyt

Abstract

Information is presented on the new VM-M3 mouse model for glioblastoma multiforme (GBM). The VM-M3 tumor arose in the brain of inbred VM strain, which is known to have a high incidence of spontaneous brain tumors. The failure to develop effective treatments for GBM has been due in part to the failure of animal models to manifest the key invasive properties of the disease. Scherer originally described the properties of malignant brain tumors in terms of their invasive behavior independent of cell classification. These properties are referred to as Secondary Structures and involve growth and invasion along blood vessels, through ventricles, white matter tracts, through the corpus callosum and across pial membranes. While extracranial metastasis is not often reported, numerous reports show that GBM can be highly metastatic if the cells gain access to extracranial sites. The VM-M3 GBM model is unique in displaying the Secondary Structures of Scherer and showing systemic metastasis when grown outside the CNS. The VM-M3 cells express the luciferase gene and can be used to assess quantitatively distal CNS invasion. Evidence is presented showing the calorie restriction reduces VM-M3 CNS invasion. The VM-M3 model will be useful for evaluating new therapies for GBM.

Key words: Metastasis, Sialic acid, Distal invasion, Xenograft, Tumor-associated macrophages, Microglia, Glioblastoma, Bioluminescent imaging

1. Glioblastoma Multiforme

Glioblastoma Multiforme (GBM) represents ~65% of all new malignant gliomas diagnosis and is the most common form of primary brain cancer in adults (1, 2). GBM portends an extremely poor outcome with only about 3–5% of patients surviving 5 years after diagnosis, and a median survival of ~12 months (3, 4). The poor prognosis is due largely to the highly invasive nature of GBM. Some neoplastic GBM invade throughout the brain and often produce multicentric secondary legions at sites distant from the primary tumor (5–7). Complete surgical resection of GBM is extremely rare. Radiation therapy, which enhances the necrotic microenvironment, often results in further tissue damage and more aggressive tumors (8–11). There have been no major advances in GBM management for over 50 years, though the use of Temozolomide has produced marginal improvement in survival (12, 13).

The large emphasis on histological and genetic classification of GBM has had little if any impact on patient survival (14–16). No reliable gene defects are common to all GBM (16, 17). Hence, effective therapeutic options are desperately needed for GBM patients.

GBM generally contain multiple morphologically diverse cell types that express neural, glial, and myeloid markers (5, 18–22). In fact, mesenchymal cells with characteristics of tumor-associated macrophages (TAMs) and/or microglia can comprise up to 70% of some GBM (21, 22). It has been difficult to determine with certainty, however, the origin of all the mesenchymal and macrophage-like cells that appear within GBM (23, 24). Based on numerous similarities between macrophages and invasive neoplastic cells in non-CNS tumors, we proposed that some neoplastic cells within GBM arise from transformed macrophage/microglia (2, 25). As cells of the myeloid/macrophage lineage naturally embody the capacity to invade (26, 27), we suggested that some of the highly invasive mesenchymal-type cells within GBM may arise from resident or infiltrating myeloid cells of the tumor stroma that then become neoplastic during disease progression (2). The basis for our suggestion emerged from our studies on highly invasive tumors that arose spontaneously in the inbred VM mouse strain. These cells express multiple properties of macrophages (2, 18, 25).

The invasive qualities of GBM include subpial and ventricular spread, perivascular, perineuronal, peri- and intrafascicular growth, and interhemispheric invasion. These routes of invasion have been referred to as the Secondary Structures of Scherer (5, 7, 28). These secondary structures were identified in highly invasive human brain tumors to include GBM, astrocytoma, and oligodendroglioma (29, 30). The secondary structures describe the growth of invasive tumor cells along the preexisting structures of the nervous system and are independent of histological grade or cellular classification. The growth characteristics of GBM will be more important for eventual management than will histological classification, as histological classification of GBM cells can be ambiguous (7, 23, 24).

Besides the secondary structures of Scherer, extracranial metastasis is also seen occasionally in human GBM. Although extracranial GBM metastasis is considered rare, there are numerous reports documenting this phenomenon (2, 5, 18, 31–34). If GBM cells can gain access to extracranial sites, they can be metastatic. One report showed that recipients of organs from a donor with GBM developed metastatic cancer (35). This indicates that neoplastic cells from this GBM metastasized from the brain and infiltrated extraneural tissues without detection. As extraneural tissues are not often examined in patients dying from GBM, it is not clear if this was a rare event or was part of a more general phenomenon. This tragedy might have been prevented with greater awareness that GBM cancer cells can metastasize outside the CNS.

Dr. Brent Reynolds, a leader in the stem cell field, mentioned to us that circulating metastatic cells are not uncommon in GBM patients (personal communication). Moreover, extracranial metastasis of brain tumors portends an extremely poor survival, with the vast majority of patients surviving less than 6 months from the diagnosis of metastatic GBM disease (36). The widely held view that metastasis does not occur for GBM should be reevaluated. Many GBM patients often die prior to detection of systemic metastasis. The development of effective GBM therapies will therefore depend on rodent models that can replicate the multiple growth characteristics seen in the human disease including extracranial metastasis.

2. GBM Mouse Models

Good rodent models of GBM can provide insight on disease mechanisms and identify effective new therapies. Many available GBM rodent models, however, fall short of fully recapitulating all major characteristics seen in human GBM (37, 38). Few rodent GBM models replicate the invasive or metastatic properties seen in human GBM. Thus, a good rodent model of GBM will show extensive intracranial invasion and metastatic spread to multiple organs when grown in extracranial sites.

3. Problems with Xenograft GBM Models

Xenograft models involve growth of human glioma cells in nude mice or some other mice with a compromised innate and/or adaptive immune system. It is not possible to grow human tumors in mice that have normal T-cell and B-cell immunity due to antibody production and host tumor rejection. Additionally, functional innate immunity derived from natural killer cells (NK), complement, etc. may contribute to tumor–host interaction. Most knowledgeable investigators in the cancer field know that xenograft models are unrepresentative of the real world situation (39, 40). Many xenograft models used as glioma models are locally invasive, but rarely show the distal brain invasion of the secondary structures seen in most human GBM (41, 42). Moreover, no xenograft GBM model is known to our knowledge that shows systemic metastasis when grown outside the CNS.

The characteristics of xenograft glioma models can deviate significantly from typical GBM behavior, as human glioma cells implanted into a mouse host gradually take on biochemical characteristics of mouse cells. We showed that human U87MG brain cancer cells express mouse carbohydrates on their surface when grown as a xenograft in immune-deficient mice (43). Over 65% of the sialic acid composition on the U87 tumor cells consisted of the nine-carbon sugar *N*-glycolylneuraminic acid. Humans, however, are unable to synthesize *N*-glycolylneuraminic acid due to a mutation in the gene that encodes a common mammalian hydroxylase enzyme (43, 44). The hydroxylase mutation occurred in the human genome sometime after our evolutionary split with the great apes (44). The acquisition of mouse carbohydrates and lipids will likely occur in any human tumor grown in the body of a mouse or rat. *N*-glycolylneuraminic acid also alters the characteristics of human embryonic stem cells when grown on nonhuman feeder cells. This has been a confounding variable in stem cell research (45).

Expression of mouse carbohydrates and lipids on human glioma cells when grown as xenografts can alter gene expression and growth behavior of the tumor cells, thus altering their response to changes in the microenvironment. Also troubling is the fact the basal metabolic rate of the mouse is also sevenfold greater than in humans (46). It is not clear how the striking differences in basal metabolic rate between the mouse and human will influence tumor biology. Fifty million years of evolution separate human and mouse physiology. Many cancer researchers are unaware of these complications. If researchers are aware of these problems, little attention is given to them in the scientific literature. It might be reasonable to view the human xenograft tumor models as a type of human–mouse centaur! (72).

We also found that food consumption is substantially greater in immunocompromised SCID mice, a common xenograft host, than in the C57BL/6J mouse strain that has a normal immune system (47). Differences in food consumption are indicative of differences in energy metabolism. The NOD-SCID mice are also commonly used as a host for growing xenograft human tumors (48). The acronym stands for mice that are *N*on-*O*bese *D*iabetic and *S*everely *C*ompromised *I*mmuno-*D*eficient. These mice not only have an abnormal immune system, but they also express characteristics of both type-1 diabetes and type-2 diabetes (49). This is not a usual situation for most cancer patients. This experimental model might be useful for those individuals who have malignant glioma, are genetically immunodeficient, and also suffer from both type-1 and type-2 diabetes. It is naive to assume that the growth behavior and response to therapies of human brain tumors grown as xenografts would be similar to the situation in the natural host.

If most glioma xenograft models are flawed in representing the real world situation, then why does the brain cancer field persist with requirements for showing therapeutic efficacy in these animal models? The short answer is because xenograft models are often required by reviewers in order to get papers published in top scientific journals or to get research grants funded. Many investigators believe that xenograft models are more representative of the human disease than are natural animal models of cancer simply because the tumor cells are derived from humans. Consequently, many cancer researchers use xenografts to demonstrated therapeutic effects. Many clinical drug trials have been initiated in patients based on information generated from xenograft models. Many of these drugs are later discontinued due to lack of efficacy, unacceptable toxicity, or some combination of these. Should this be surprising considering the unnatural nature of the model system? (72).

4. The VM Inbred Mouse

In contrast to the relatively high incidence of brain tumors in humans, the incidence of spontaneous brain tumors in mice is low (50). The VM (VM/dk) inbred mouse strain is unique in expressing a relatively high incidence (1.5%) of spontaneous brain tumors (51, 52). The incidence of spontaneous brain tumor formation in VM mice is about 210-fold greater than the incidence in C57BL/6J mice (53). The brain tumors in VM mice develop over a broad age range between 4 and 18 months (51). Most of the VM brain tumors were characterized as astrocytomas, but microgliomas and neural stem cell tumors are also found (18, 51, 52). We found that the composition of fatty acid molecular species in the mitochondrial-enriched phospholipid, cardiolipin, was different in brains of VM and C57BL/6J mice (53). In addition, the activities of Complex I, I/III, and II/III enzymes were lower, whereas the activity of complex IV was higher in the mitochondria of VM mice than in B6 mice. We suggested that the high glioma incidence and alterations in electron transport chain activities in VM mice compared to C57BL/6J mice could be related to the unusual composition of mitochondrial lipids in the VM mouse brain (53). Further studies will be needed to confirm this possibility.

5. The Growth and Invasive Properties of VM Brain Tumors

We initially characterized three independent, spontaneous brain tumors in the VM mouse strain (18, 41). The tumors are therefore syngeneic with the VM host background. Two of the tumors, VM-M2 and VM-M3 were highly invasive, whereas another one, VM-NM1, was mostly noninvasive (Fig. 1). The invasive VM-M2

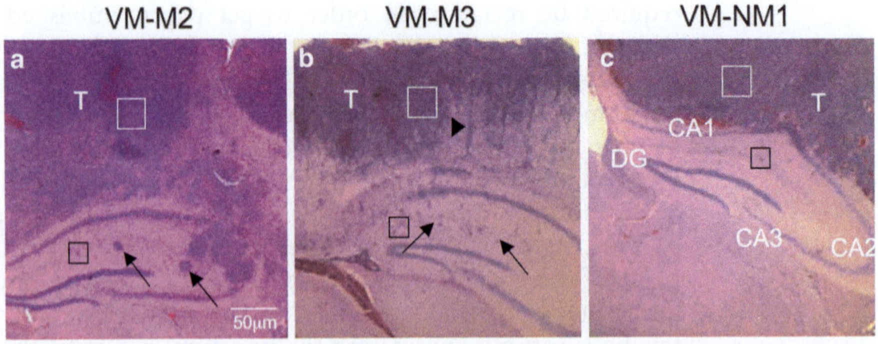

Fig. 1. VM tumor cells express extensive brain invasion. The VM-M2 and VM-M3 tumors are diffusely invasive into the neural parenchyma showing multiple tumor foci in the hippocampal region of the brain well in advance of the main tumor mass. The VM-M2 tumor cells invade through the hippocampal CA2 region and overlying cerebral cortex and the VM-M3 tumor cells invade as streams through the cerebral cortex (*arrowhead*). Despite having independent origins, the invasive properties of the VM-M2 and VM-M3 cells are similar (18). In contrast to the invasive VM-M2 and the VM-M3 tumors, the VM-NM1 tumor is noninvasive and forms a sharp boundary between the main tumor mass and the neural parenchyma. No tumor foci are seen in advance of the main tumor mass. Also, no perivascular pseudorosettes occur in brains containing the NM1 tumor (18). *T* tumor, CA1-3 (hippocampal regions). Reprinted with permission from (18).

and VM-M3 tumors expressed multiple markers of microglia/macrophages (18). These markers were not seen in the noninvasive/metastatic VM-NM1 cells, which expressed markers of neural stem cells. The invasive VM-M3 cells, like invasive human GBM cells, express high levels of the chemokine receptor, CXCR4 (41, 54, 55). CXCR4 is generally considered a marker for GBM cells with the highest invasive potential (54, 56, 57).

We transduced VM-M3 cells with a lentivirus vector containing the firefly luciferase gene under control of the cytomegalovirus promoter (VM-M3/Fluc) (18). The lentivirus construct was obtained as a gift from Miguel Sena-Esteves, U. Massachusetts Medical School. We used the luciferase labeled VM-M3 tumor and monitored tumor growth noninvasively over time (Fig. 2). The bioluminescent signal could be evaluated both qualitatively and quantitatively during tumor progression (Fig. 2a, b).

To quantitate the level of tumor invasion, we removed the brains at the end of the study and imaged ex vivo (41). We sectioned the brains through the midline and imaged each brain half separately. As shown in Fig. 3a, bioluminescence was detected in both the ipsilateral and the contralateral hemispheres. The level of invasion into the contralateral hemisphere was also measured. In addition, histology was used to confirm the presence of invading tumor cells in both the ipsilateral and contralateral hemispheres (Fig. 3b). We also were able to quantitatively and qualitatively analyze distal spread of VM-M3/Fluc cells to the brain stem and cerebellum (Fig. 4). Viewed collectively, these findings indicate that the VM-M3/Fluc is an excellent in vivo brain invasion assay.

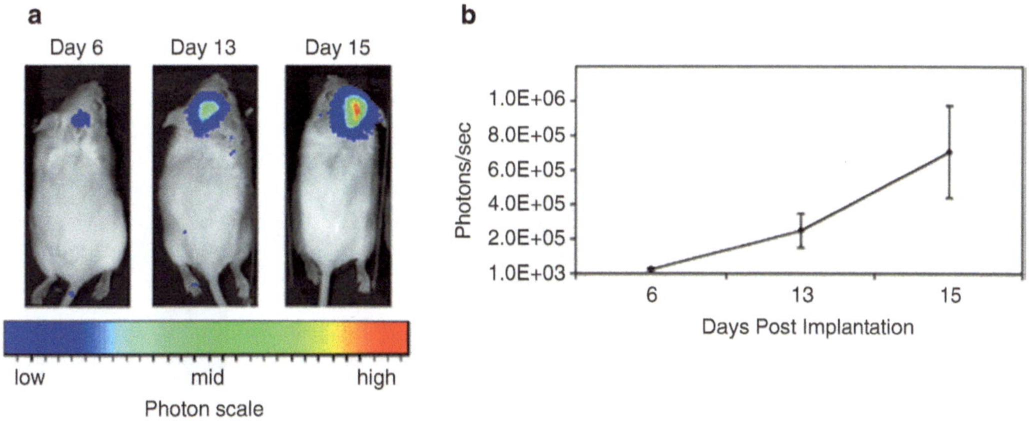

Fig. 2. Growth of the VM-M3/Fluc tumor with bioluminescence imaging. (**a**) VM-M3/Fluc tumor fragments were implanted as we previously described (41). Cranial images were taken over a period of 15 days (representative mouse shown). (**b**) Bioluminescence from the whole mouse was quantified and plotted on a linear scale. All values are expressed as the mean of 6 independent samples $\pm$ SEM. Bioluminescence, above the lower limit of the imaging system (approximately 1×10^3 photons/s), could be detected as early as day 6. Reprinted with permission from (41).

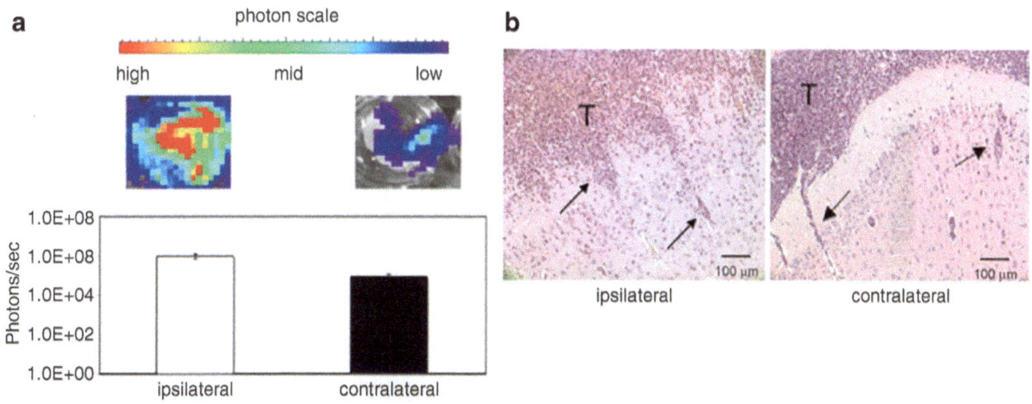

Fig. 3. VM-M3 in vivo brain invasion assay. Small tissue fragments from the VM-M3/Fluc cells, containing an established microenvironment, were implanted into the right cerebral hemisphere of adult VM mice (3 months) according to our standard procedures (41, 58). Whole brains were removed 17 days post implantation and were dissected into ipsilateral and contralateral cerebral hemispheres. (**a**) Each hemisphere was imaged for bioluminescence (photons/s) ex vivo. Brain tissue was imaged with the Xenogen IVIS system. Data acquisition and analysis was performed with Living Image®software (Xenogen). Bioluminescence from each brain half was quantified and plotted on a *log scale*. Bioluminescence in the contralateral cerebral hemisphere is indicative of tumor spread. (**b**) Histological analysis was used to confirm the presence of invasive tumor cells in the contralateral hemisphere. The tumor cells spread to the contralateral hemisphere using the corpus callosum and pial membranes. Invasive tumor cells (*arrows*) are shown in the neural parenchyma of the ipsilateral and the contralateral hemispheres. Similar results are obtained if cultured VM-M3 cells are implanted instead of tissue fragments, but time to invasion is more for cells than for fragments (41).

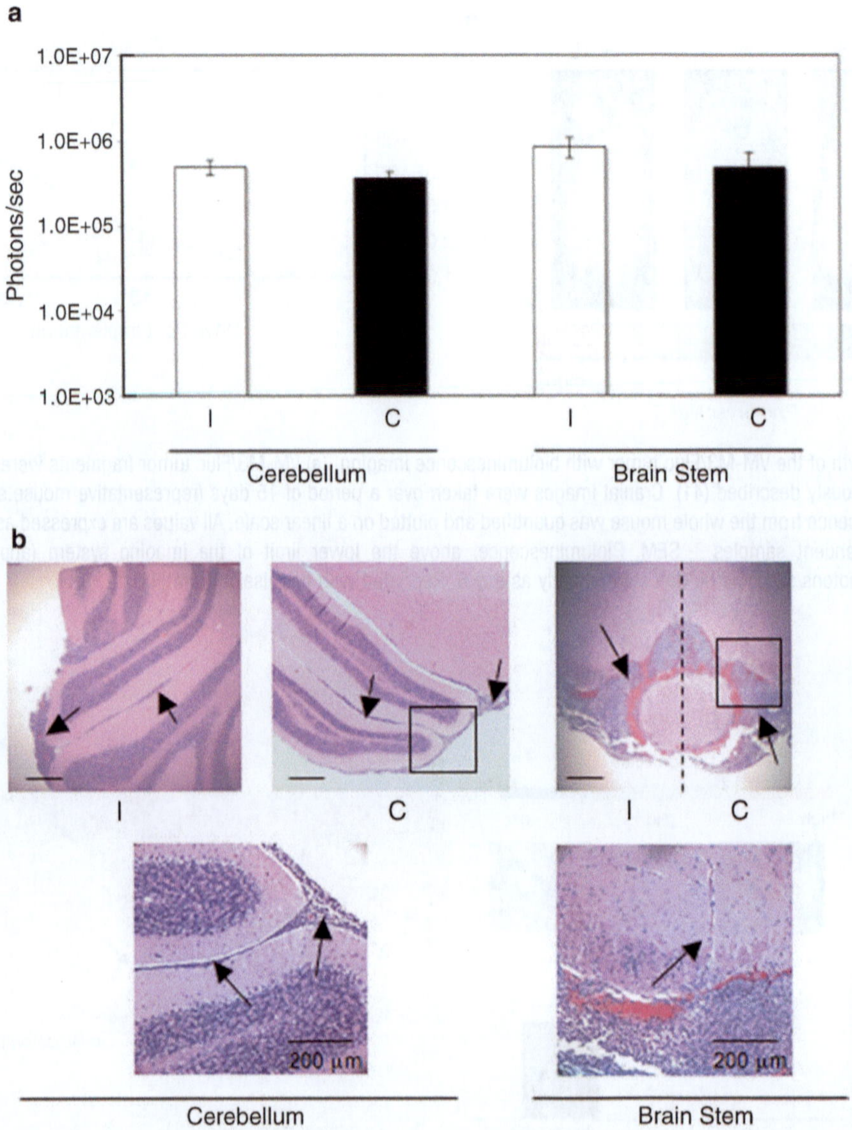

Fig. 4. Quantitative and qualitative analysis of distal spread of VM-M3/Fluc cells to the brain stem and cerebellum. Removed brains were sectioned through the midline and were further dissected into cerebellum and brain stem. (**a**) Bioluminescence was quantified and plotted on a log scale. All values are expressed as the mean ± SEM of 10 independent samples. (**b**) Histological analysis (H&E) was used to validate the presence of tumor cells as described (41). *Top panel images* are shown at ×50. The scale bar represents 250 μm. The *black boxes* from the *top panel* images are shown in higher power in the *bottom panel. Bottom panel* images are shown at ×200. *Arrows* identify subpial tumor cell spread in the cerebellum and tumor cell invasion in the brain stem. The *dashed line* demarcates the ipsilateral from the contralateral side of brain stem (41).

6. The VM-M3 GBM Cells Invade the Brain Using Secondary Structures of Scherer

We found that the VM-M3 brain tumor cells migrated through the CNS using the Secondary Structures of Scherer (Fig. 5). Besides H&E staining, we showed that the VM-M3 cells invading through the Scherer structures also express the Ki-67 marker of proliferation (41). To our knowledge, the VM-M3 cells are the only mouse GBM cells that invade the CNS using the Secondary Structures of Scherer.

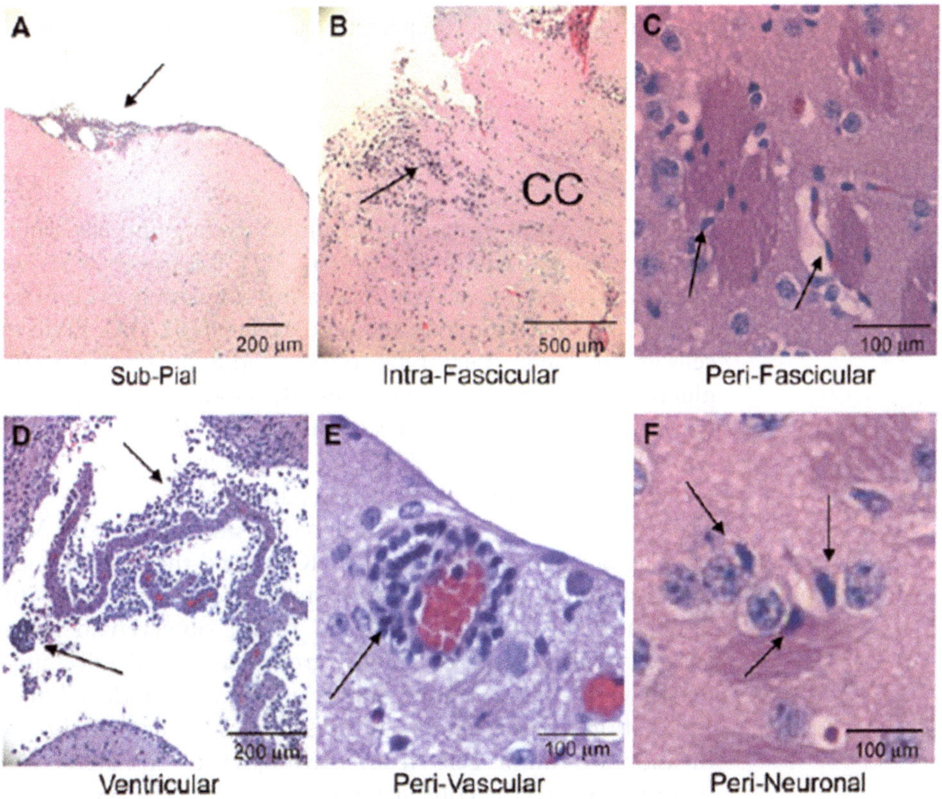

Fig. 5. Migratory routes of the VM-M3/Fluc brain tumor cells. VM-M3/Fluc tumor fragments or cells were implanted as described. Histological analysis (H&E) was used to validate the presence of tumor cells (41). The VM-M3/Fluc tumor cells are shown invading using the Secondary Structures of Scherer including the pial surface (*arrow*, **a**), within the corpus callosum (CC, *arrow*, **b**), along myelinated axons crossing through the striatum (*arrow*, **c**), through the ventricular system (*arrows*, **d**), around the blood vessels (*arrow*, **e**), and around neurons (*arrow*, **f**). Images are shown at ×100 (**a**), ×50 (**b**), ×400 (**c**), ×200 (**d**), and ×400 (**e**, **f**). *Arrows* identify regions containing tumor cells. Reprinted with permission from (41).

7. The VM GBM Cells Show Systemic and Brain Metastasis When Implanted Subcutaneously

Human GBM cells can show systemic metastasis if the cells gain access to extracranial sites (discussed above). We found that the VM-M2 and VM-M3 cells were highly metastatic if the cells were inoculated subcutaneously in the flank of the syngeneic VM mouse host (18). The primary metastatic sites were lung, liver, kidney, spleen, and brain (Fig. 6). We also found that the VM-M2 and VM-M3 cells metastasized to bone marrow. Diffuse bone marrow metastasis has been reported in some GBM patients (59, 60). The VM-M2 and VM-M3 brain tumors are the only rodent models that manifest systemic metastatic behavior seen in human GBM. We suggest that diffuse brain invasion involving Scherer's secondary structures and systemic metastasis can be used identify those brain tumor models that best represent the true nature of human GBM.

8. Calorie Restriction Reduces Brain Invasion of VM-M3 GBM Cells

Calorie restriction (CR) involves under feeding and has long been known to inhibit tumor growth (61, 62). CR lowers body weight and blood glucose levels while elevating circulating levels of ketone bodies (β-hydroxybutyrate and acetoacetate) (62, 63). Brain tumor cells are unable to metabolize ketone bodies for energy under low glucose and glutamine conditions (8, 64–66). CR targets angiogenesis, inflammation, and signaling through the IGF-1, PI3K, AKT, Hif-1a pathways (47, 62, 67, 68). Hence, CR targets tumor growth through multiple signaling pathways.

We also showed that CR inhibits invasion of the VM-M3 glioblastoma. As seen in Fig. 7, CR reduced the growth and local invasion of the VM-M3 primary tumor. Compared to the diffuse, ill-defined border of the VM-M3 tumor observed in the unrestricted control mice, the tumor grown in the CR mice appeared denser with a more defined border, indicative of reduced invasion. In addition to inhibiting local invasion, CR also reduced the distal invasion of tumor cells from the implanted ipsilateral cerebral hemisphere into the contralateral hemisphere (Fig. 8).

While invading tumor cells were identified in all regions of the contralateral hemisphere of the control *ad libitum*-fed (AL) mice, only subpial invasion was found in the contralateral hemisphere of the CR group (Fig. 9). These findings indicate that CR blocks VM-M3 invasion through Scherer's secondary structures (69).

We also found that the total percentage of Ki-67-stained VM-M3 cells within the primary tumor and the total number of blood vessels was also significantly lower in the CR-treated mice

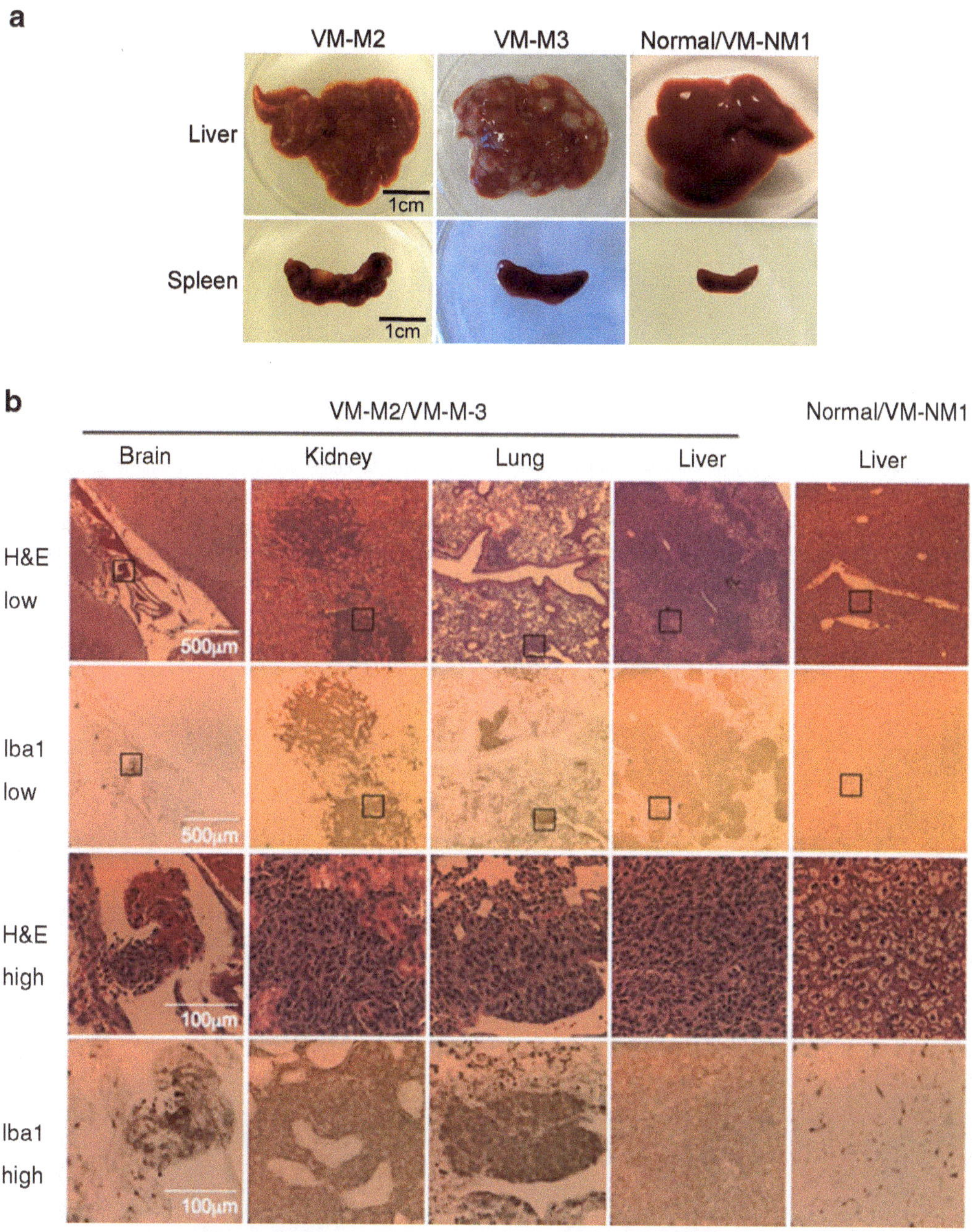

Fig. 6. VM-M2 and VM-M3 tumors show systemic metastasis. Appearance of gross lesions (**a**), and micrometastatic lesions (**b**) from mice bearing VM tumors. H&E and Iba-1 staining (microglial marker) revealed numerous micrometastatic lesions in the kidney, lung, brain meninges, and liver in mice bearing the VM-M3 tumor (**b**). The *black boxes* in the lower power images (×100) in the *top 2 rows* are shown at higher power (×400) in the *bottom 2 rows*. The distribution, morphology, and staining of the micrometastatic lesions in mice bearing the VM-M3 appeared the same as that of the mice bearing the VM-M2 tumor (not shown). At least three independent mice were analyzed for each tumor. Reprinted with permission from (18).

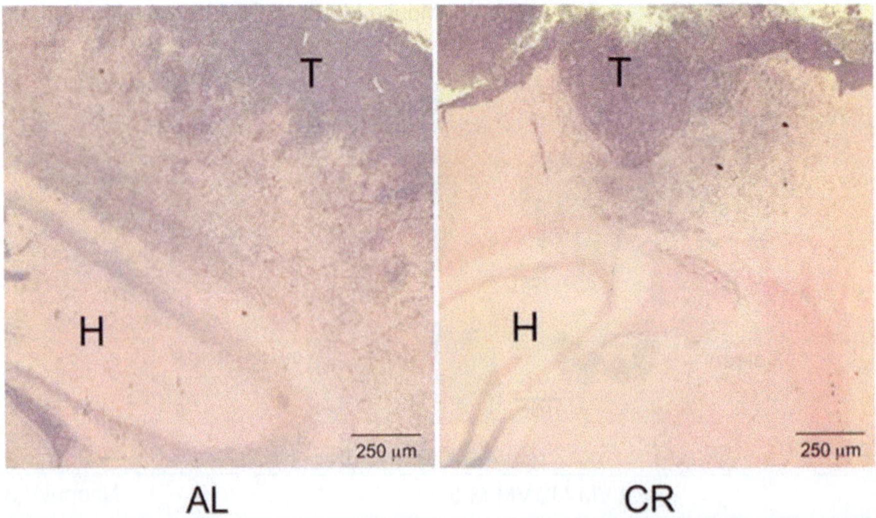

Fig. 7. Calorie restriction reduces diffuse brain invasion of VM-M3 GBM cells. VM-M3/Fluc tumor fragments were implanted into the cerebral cortex, were fixed, and then stained with hematoxylin and eosin (H & E) as described (69). Images are shown at 50× (*T* tumor, *H* Hippocampus). Tumor cell invasion through the neural parenchyma (*dark blue cells*) is less in the CR-fed mice than in the ad libitum AL-fed mice. The border between tumor tissue and normal brain tissue is more sharply defined in the CR-fed mice than in the AL-fed mice. The results show that CR reduces VM-M3/Fluc tumor cell invasion. Reprinted with permission from ASN Neuro (69).

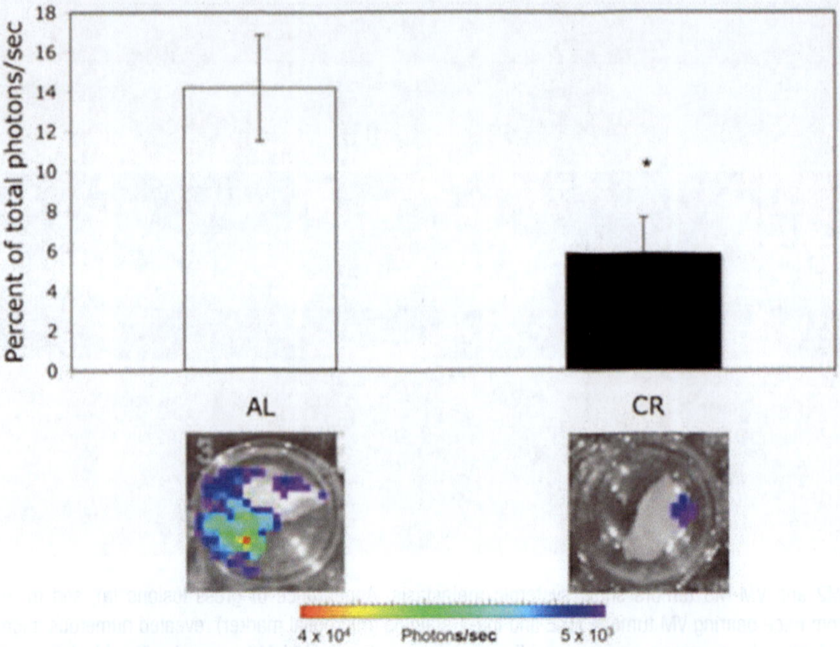

Fig. 8. Influence of CR on bioluminescence in the contralateral hemisphere. VM-M3/Fluc tumor fragments were implanted as described (69). Each hemisphere was imaged for bioluminescence ex vivo as described (69). The bioluminescence from each hemisphere was added together to obtain a total bioluminescence value (photons/s) for each brain. Data for the contralateral hemisphere was then expressed as the percentage of the total brain photons/s. Values represent the means ± SEM for 9–10 mice per group. Representative bioluminescence images are shown. The *asterisk* indicates that the CR values differ significantly from the AL control group at $P < 0.05$ using the two-tailed Student's *t* test. The results show that CR reduces interhemispheric invasion. Reprinted with permission from (69).

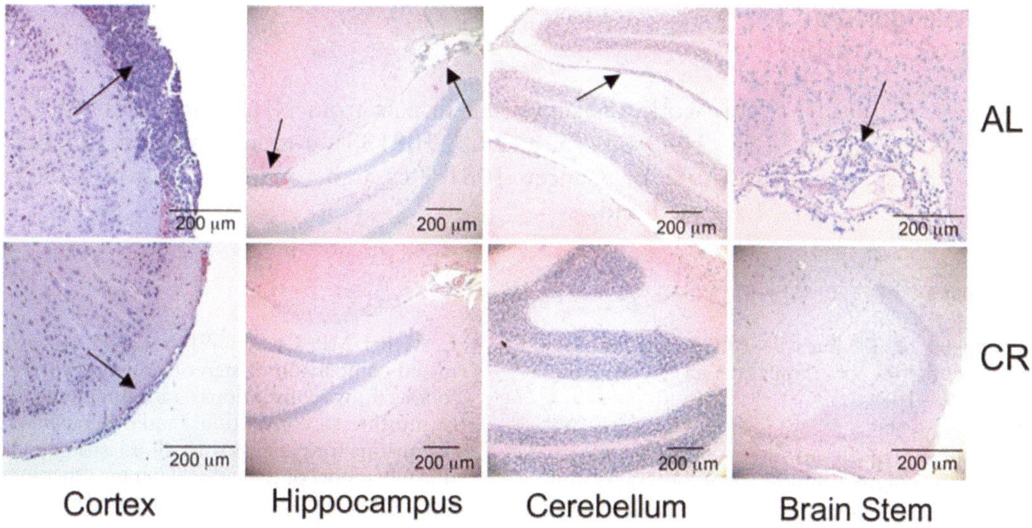

Fig. 9. CR inhibits VM-M3 tumor cell invasion to the contralateral hemisphere. VM-M3/Fluc tumor fragments were implanted into the right cerebral hemisphere as described (69). Histological analysis (H&E) was used to validate the presence of tumor cells under AL (*top*) and CR (*bottom*) in cerebral cortex (200×), hippocampus (100×), cerebellum (100×), and brain stem (200×) of the contralateral hemisphere. *Arrows* indicate the presence of tumor cells. At least three samples were examined per group. Reprinted with permission from ASN Neuro (69).

than in the mice fed AL, indicating that CR is also anti-proliferative and anti-angiogenic in this invasive tumor model (8, 69). These findings clearly indicate that calorie restriction alone can inhibit proliferation, as well as both local and distal invasion of the VM-M3 mouse GBM. We would also expect similar findings using restricted ketogenic diets. Indeed, previous studies showed that ketogenic diets, which lower glucose and elevate ketones, inhibit growth and invasion in children and an adult with malignant brain cancer including GBM (70, 71). Hence, the VM-M3 mouse GBM model can be a powerful preclinical tool for assessing translational benefit of new GBM therapies.

9. Conclusions

A new model of GBM is described in the VM mouse strain. The VM-M3 GBM model shows CNS growth patterns similar to those described in invasive malignant brain cancers. Like human GBM, the VM-M3 cells showed systemic metastasis when grown outside the brain. The model will be useful for preclinical evaluation of therapies that can target distal tumor cell invasion in the CNS.

Acknowledgments

This work was supported in part from NIH grants (HD-39722, NS- 1080 55195, and CA-102135), a grant from the American Institute of Cancer 1081 Research, and the Boston College Expense Fund.

References

1. Ohgaki H, Kleihues P (2005) Epidemiology and etiology of gliomas. Acta Neuropathol 109:93–108
2. Huysentruyt LC, Akgoc Z, Seyfried TN (2011) Hypothesis: are neoplastic macrophages/microglia present in glioblastoma multiforme? ASN Neuro. doi:10.1042/AN20110011
3. Krex D, Klink B, Hartmann C, von Deimling A, Pietsch T, Simon M, Sabel M, Steinbach JP, Heese O, Reifenberger G, Weller M, Schackert G (2007) Long-term survival with glioblastoma multiforme. Brain 130:2596–2606
4. Preusser M, de Ribaupierre S, Wohrer A, Erridge SC, Hegi M, Weller M, Stupp R (2011) Current concepts and management of glioblastoma. Ann Neurol 70:9–21
5. Rubinstein LJ (1972) Tumors of the central nervous system. Armed Forces Institute of Pathology, Washington, D.C
6. Laws ER Jr, Goldberg WJ, Bernstein JJ (1993) Migration of human malignant astrocytoma cells in the mammalian brain: Scherer revisited. Int J Dev Neurosci 11:691–697
7. Scherer HJ (1940) A critical review: the pathology of cerebral gliomas. J Neurol Neuropsychiatry 3:147–177
8. Seyfried TN, Kiebish MA, Marsh J, Shelton LM, Huysentruyt LC, Mukherjee P (2010) Metabolic management of brain cancer. Biochim Biophys Acta 1807:577–594
9. Seyfried TN, Shelton LM, Mukherjee P (2010) Does the existing standard of care increase glioblastoma energy metabolism? Lancet Oncol 11:811–813
10. Kargiotis O, Geka A, Rao JS, Kyritsis AP (2010) Effects of irradiation on tumor cell survival, invasion and angiogenesis. J Neurooncol 100:323–338
11. Duffner PK (2006) The long term effects of chemotherapy on the central nervous system. J Biol 5:21
12. Souhami L, Seiferheld W, Brachman D, Podgorsak EB, Werner-Wasik M, Lustig R, Schultz CJ, Sause W, Okunieff P, Buckner J, Zamorano L, Mehta MP, Curran WJ Jr (2004) Randomized comparison of stereotactic radiosurgery followed by conventional radiotherapy with carmustine to conventional radiotherapy with carmustine for patients with glioblastoma multiforme: report of Radiation Therapy Oncology Group 93-05 protocol. Int J Radiat Oncol Biol Phys 60:853–860
13. Stupp R, Hegi ME, Mason WP, van den Bent MJ, Taphoorn MJ, Janzer RC, Ludwin SK, Allgeier A, Fisher B, Belanger K, Hau P, Brandes AA, Gijtenbeek J, Marosi C, Vecht CJ, Mokhtari K, Wesseling P, Villa S, Eisenhauer E, Gorlia T, Weller M, Lacombe D, Cairncross JG, Mirimanoff RO (2009) Effects of radiotherapy with concomitant and adjuvant temozolomide versus radiotherapy alone on survival in glioblastoma in a randomised phase III study: 5-year analysis of the EORTC-NCIC trial. Lancet Oncol 10:459–466
14. Vitucci M, Hayes DN, Miller CR (2010) Gene expression profiling of gliomas: merging genomic and histopathological classification for personalised therapy. Br J Cancer 104(4):545–53
15. von Deimling A, Louis DN, Wiestler OD (1995) Molecular pathways in the formation of gliomas. Glia 15:328–338
16. Parsons DW, Jones S, Zhang X, Lin JC, Leary RJ, Angenendt P, Mankoo P, Carter H, Siu IM, Gallia GL, Olivi A, McLendon R, Rasheed BA, Keir S, Nikolskaya T, Nikolsky Y, Busam DA, Tekleab H, Diaz LA Jr, Hartigan J, Smith DR, Strausberg RL, Marie SK, Shinjo SM, Yan H, Riggins GJ, Bigner DD, Karchin R, Papadopoulos N, Parmigiani G, Vogelstein B, Velculescu VE, Kinzler KW (2008) An integrated genomic analysis of human glioblastoma multiforme. Science 321:1807–1812
17. Colman H, Zhang L, Sulman EP, McDonald JM, Shooshtari NL, Rivera A, Popoff S, Nutt CL, Louis DN, Cairncross JG, Gilbert MR, Phillips HS, Mehta MP, Chakravarti A, Pelloski CE, Bhat K, Feuerstein BG, Jenkins RB, Aldape K (2010) A multigene predictor of outcome in glioblastoma. Neuro-oncology 12:49–57

18. Huysentruyt LC, Mukherjee P, Banerjee D, Shelton LM, Seyfried TN (2008) Metastatic cancer cells with macrophage properties: evidence from a new murine tumor model. Int J Cancer 123:73–84
19. Seyfried TN (2001) Perspectives on brain tumor formation involving macrophages, glia, and neural stem cells. Perspect Biol Med 44:263–282
20. Yuan X, Curtin J, Xiong Y, Liu G, Waschsmann-Hogiu S, Farkas DL, Black KL, Yu JS (2004) Isolation of cancer stem cells from adult glioblastoma multiforme. Oncogene 23:9392–9400
21. Wood GW, Morantz RA (1979) Immunohistologic evaluation of the lymphoreticular infiltrate of human central nervous system tumors. J Natl Cancer Inst 62:485–491
22. Wood GW, Morantz RA, Tilzer SA, Gollahon KA (1980) Immunoglobulin bound in vivo to Fc receptor-positive cells in human central nervous system tumors. J Natl Cancer Inst 64:411–418
23. Ricci-Vitiani L, Pallini R, Larocca LM, Lombardi DG, Signore M, Pierconti F, Petrucci G, Montano N, Maira G, De Maria R (2008) Mesenchymal differentiation of glioblastoma stem cells. Cell Death Differ 15:1491–1498
24. Tso CL, Shintaku P, Chen J, Liu Q, Liu J, Chen Z, Yoshimoto K, Mischel PS, Cloughesy TF, Liau LM, Nelson SF (2006) Primary glioblastomas express mesenchymal stem-like properties. Mol Cancer Res 4:607–619
25. Huysentruyt LC, Seyfried TN (2010) Perspectives on the mesenchymal origin of metastatic cancer. Cancer Metastasis Rev 29:695–707
26. Banaei-Bouchareb L, Peuchmaur M, Czernichow P, Polak M (2006) A transient microenvironment loaded mainly with macrophages in the early developing human pancreas. J Endocrinol 188:467–480
27. Mallat M, Marin-Teva JL, Cheret C (2005) Phagocytosis in the developing CNS: more than clearing the corpses. Curr Opin Neurobiol 15:101–107
28. Zagzag D, Esencay M, Mendez O, Yee H, Smirnova I, Huang Y, Chiriboga L, Lukyanov E, Liu M, Newcomb EW (2008) Hypoxia- and vascular endothelial growth factor-induced stromal cell-derived factor-1alpha/CXCR4 expression in glioblastomas: one plausible explanation of Scherer's structures. Am J Pathol 173:545–560
29. Scherer HJ (1938) Structural development in gliomas. Am J Cancer 34(3):333–351
30. Scherer HJ (1940) The forms of growth in gliomas and their practical significance. Brain 63:1–34
31. Hoffman HJ, Duffner PK (1985) Extraneural metastases of central nervous system tumors. Cancer 56:1778–1782
32. Taha M, Ahmad A, Wharton S, Jellinek D (2005) Extra-cranial metastasis of glioblastoma multiforme presenting as acute parotitis. Br J Neurosurg 19:348–351
33. Laerum OD, Bjerkvig R, Steinsvag SK, de Ridder L (1984) Invasiveness of primary brain tumors. Cancer Metastasis Rev 3:223–236
34. Gotway MB, Conomos PJ, Bremner RM (2010) Pleural metastatic disease from glioblastoma multiforme. J Thorac Imaging 26(2): W54–W58
35. Armanios MY, Grossman SA, Yang SC, White B, Perry A, Burger PC, Orens JB (2004) Transmission of glioblastoma multiforme following bilateral lung transplantation from an affected donor: case study and review of the literature. Neurooncology 6:259–263
36. Ng WH, Yeo TT, Kaye AH (2005) Spinal and extracranial metastatic dissemination of malignant glioma. J Clin Neurosci 12:379–382
37. Candolfi M, Curtin JF, Nichols WS, Muhammad AG, King GD, Pluhar GE, McNiel EA, Ohlfest JR, Freese AB, Moore PF, Lerner J, Lowenstein PR, Castro MG (2007) Intracranial glioblastoma models in preclinical neurooncology: neuropathological characterization and tumor progression. J Neurooncol 85:133–148
38. Rubinstein LJ (1977) Correlation of animal brain tumor models with human neuro-oncology. Natl Cancer Inst Monogr 46:43–49
39. Peterson JK, Houghton PJ (2004) Integrating pharmacology and in vivo cancer models in preclinical and clinical drug development. Eur J Cancer 40:837–844
40. Kim IS, Baek SH (2010) Mouse models for breast cancer metastasis. Biochem Biophys Res Commun 394:443–447
41. Shelton LM, Mukherjee P, Huysentruyt LC, Urits I, Rosenberg JA, Seyfried TN (2010) A novel pre-clinical in vivo mouse model for malignant brain tumor growth and invasion. J Neurooncol 99:165–176
42. Jacobs VL, Valdes PA, Hickey WF, De Leo JA (2011) Current review of in vivo GBM rodent models: emphasis on the CNS-1 tumor model. ASN Neuro. doi:10.1042/AN20110014
43. Ecsedy JA, Holthaus KA, Yohe HC, Seyfried TN (1999) Expression of mouse sialic acid on

gangliosides of a human glioma grown as a xenograft in SCID mice. J Neurochem 73:254–259

44. Chou HH, Takematsu H, Diaz S, Iber J, Nickerson E, Wright KL, Muchmore EA, Nelson DL, Warren ST, Varki A (1998) A mutation in human CMP-sialic acid hydroxylase occurred after the Homo-Pan divergence. Proc Natl Acad Sci USA 95:11751–11756
45. Martin MJ, Muotri A, Gage F, Varki A (2005) Human embryonic stem cells express an immunogenic nonhuman sialic acid. Nat Med 11:228–232
46. Mahoney LB, Denny CA, Seyfried TN (2006) Caloric restriction in C57BL/6 J mice mimics therapeutic fasting in humans. Lipids Health Dis 5:13
47. Mukherjee P, Abate LE, Seyfried TN (2004) Antiangiogenic and proapoptotic effects of dietary restriction on experimental mouse and human brain tumors. Clin Cancer Res 10:5622–5629
48. Kalaany NY, Sabatini DM (2009) Tumours with PI3K activation are resistant to dietary restriction. Nature 458:725–731
49. Chaparro RJ, Konigshofer Y, Beilhack GF, Shizuru JA, McDevitt HO, Chien YH (2006) Nonobese diabetic mice express aspects of both type 1 and type 2 diabetes. Proc Natl Acad Sci USA 103:12475–12480
50. Swenberg JA (1982) Neoplasms of the nervous system. In: Foster HL, Small JD, Fox JG (eds) The mouse in biomedical research. Academic Press, New York, pp 527–529
51. Fraser H (1971) Astrocytomas in an inbred mouse strain. J Pathol 103:266–270
52. Fraser H (1986) Brain tumours in mice, with particular reference to astrocytoma. Food Chem Toxicol 24:105–111
53. Kiebish MA, Han X, Cheng H, Chuang JH, Seyfried TN (2008) Brain mitochondrial lipid abnormalities in mice susceptible to spontaneous gliomas. Lipids 43:951–959
54. Ehtesham M, Winston JA, Kabos P, Thompson RC (2006) CXCR4 expression mediates glioma cell invasiveness. Oncogene 25:2801–2806
55. Esencay M, Newcomb EW, Zagzag D (2010) HGF upregulates CXCR4 expression in gliomas via NF-kappaB: implications for glioma cell migration. J Neurooncol 99:33–40
56. Ehtesham M, Mapara KY, Stevenson CB, Thompson RC (2009) CXCR4 mediates the proliferation of glioblastoma progenitor cells. Cancer Lett 274:305–312
57. Bian XW, Yang SX, Chen JH, Ping YF, Zhou XD, Wang QL, Jiang XF, Gong W, Xiao HL, Du LL, Chen ZQ, Zhao W, Shi JQ, Wang JM (2007) Preferential expression of chemokine receptor CXCR4 by highly malignant human gliomas and its association with poor patient survival. Neurosurgery 61:570–578, discussion 578–579
58. Ranes MK, El-Abbadi M, Manfredi MG, Mukherjee P, Platt FM, Seyfried TN (2001) N -butyldeoxynojirimycin reduces growth and ganglioside content of experimental mouse brain tumours. Br J Cancer 84:1107–1114
59. Kleinschmidt-Demasters BK (1996) Diffuse bone marrow metastases from glioblastoma multiforme: the role of dural invasion. Hum Pathol 27:197–201
60. Rajagopalan V, El Kamar FG, Thayaparan R, Grossbard ML (2005) Bone marrow metastases from glioblastoma multiforme–a case report and review of the literature. J Neurooncol 72:157–161
61. Rous P (1914) The influence of diet on transplanted and spontaneous mouse tumors. J Exp Med 20:433–451
62. Mukherjee P, El-Abbadi MM, Kasperzyk JL, Ranes MK, Seyfried TN (2002) Dietary restriction reduces angiogenesis and growth in an orthotopic mouse brain tumour model. Br J Cancer 86:1615–1621
63. Seyfried TN, Sanderson TM, El-Abbadi MM, McGowan R, Mukherjee P (2003) Role of glucose and ketone bodies in the metabolic control of experimental brain cancer. Br J Cancer 89:1375–1382
64. Seyfried TN, Mukherjee P (2005) Targeting energy metabolism in brain cancer: review and hypothesis. Nutr Metab 2:30
65. Seyfried TN, Shelton LM (2010) Cancer as a metabolic disease. Nutr Metab 7:7
66. Maurer GD, Brucker DP, Baehr O, Harter PN, Hattingen E, Walenta S, Mueller-Klieser W, Steinbach JP, Rieger J (2011) Differential utilization of ketone bodies by neurons and glioma cell lines: a rationale for ketogenic diet as experimental glioma therapy. BMC Cancer 11:315
67. Marsh J, Mukherjee P, Seyfried TN (2008) Akt-dependent proapoptotic effects of dietary restriction on late-stage management of a phosphatase and tensin homologue/tuberous sclerosis complex 2-deficient mouse astrocytoma. Clin Cancer Res 14:7751–7762
68. Mulrooney TJ, Marsh J, Urits I, Seyfried TN, Mukherjee P (2011) Influence of caloric

restriction on constitutive expression of NF-kappaB in an experimental mouse astrocytoma. PLoS one 6:e18085

69. Shelton LM, Huysentruyt LC, Mukherjee P, Seyfried TN (2010) Calorie restriction as an anti-invasive therapy for malignant brain cancer in the VM mouse. ASN Neuro 2:e00038
70. Nebeling LC, Miraldi F, Shurin SB, Lerner E (1995) Effects of a ketogenic diet on tumor metabolism and nutritional status in pediatric oncology patients: two case reports. J Am Coll Nutr 14:202–208
71. Zuccoli G, Marcello N, Pisanello A, Servadei F, Vaccaro S, Mukherjee P, Seyfried TN (2010) Metabolic management of glioblastoma multiforme using standard therapy together with a restricted ketogenic diet: case report. Nutr Metab 7:33
72. Seyfried TN (2012) Cancer as a metabolic disease. John Wiley Press

Neuromethods (2013) 77: 57–82
DOI 10.1007/7657_2012_35

Published online: 13 March 2012

Oligodendroglioma Models

Nanna Lindberg and Lene Uhrbom

Abstract

Oligodendroglial tumors are primary tumors of the central nervous system that largely affect adults. The cell of origin is undefined, but the tumors display many features reminiscent of oligodendrocytes or oligodendrocyte progenitor cells. Here, we briefly recapitulate the history of oligodendroglial tumor research, discuss the current knowledge concerning the biology of oligodendroglial tumors, and thoroughly review the various mouse models that have been used and that are currently in use to study oligodendroglial tumor development.

Key words: Oligodendroglioma, Oligodendrocyte, Oligodendrocyte progenitor cell (OPC), Platelet-derived growth factor receptor alpha (PDGFRA), Epidermal growth factor receptor (EGFR), CNPase, OLIG2, NG2, Mouse model

1. Introduction

Oligodendrogliomas display features of oligodendrocytes in their histological appearance. Oligodendrocytes are cells of the central nervous system (CNS) with main purpose to myelinate neurons. In this chapter, experimental models of oligodendroglioma are in focus. Histopathology and molecular genetics of human oligodendroglioma are described, and brief accounts of early oligodendroglioma models and more detailed descriptions of current models presented. Oligodendroglial tumors include according to the current WHO classification (Fig. 1a–d): oligodendroglioma (WHO grade II), anaplastic oligodendroglioma (WHO grade III), oligoastrocytoma (WHO grade II), and anaplastic oligoastrocytoma (WHO grade III). For simplicity, we refer collectively to all types of oligodendroglial tumors as oligodendroglioma. In those instances where we specifically mean the clinical subtype oligodendroglioma we add grade II where applicable.

2. Classification and Biology of Oligodendroglioma

2.1. Historical Notes

In the early twentieth century, neurosurgeons and pathologists began describing glial tumors with oligodendroglial histology establishing them as a separate tumor subtype. Neuroglia ("nerve glue," the cells that glue together the nerves) were at this time described as supportive cells of the CNS with elaborate fibrillae which today are known as astrocytes. Also neuroglial cells lacking fibrillae had been observed and were first described by W. F. Robertsson in 1899 as a different subset of neuroglia (1). These were portrayed as small process-bearing cells present throughout the gray and white matter that could be selectively stained using a platinum impregnation technique. They were believed to be derived from the mesoderm and were called mesoglia when they in fact were the first oligodendrocytes to be described. In 1921, del Río-Hortega identified oligodendroglia to be of ectodermal origin and renamed these cells oligodendrocytes. In 1924, Bailey and Cushing discussed how cell-specific chemical color reactions, such as the platinum impregnation technique used by Robertsson et al., could be used to identify the corresponding normal cell type that would most resemble that of a particular cancer cell (2). The current belief was that the supporting tissue of the CNS constituted neuroglia (astrocytes), microglia, and oligodendroglia (also called oligocytes), and that tumors of the CNS were composed of and putatively originated from these supporting cells rather than the nervous tissue itself. Tumors with oligodendroglial-like cells were either called oligocytomas or oligodendrogliomas. In 1929, Bailey and Bucy described a set of 13 oligodendrogliomas comparing the tumor cells to oligodendroglia (3).

The comparisons of morphological characteristics of oligocytes and astrocytes in normal tissue and cerebral tumors were the subject for intense investigations by Eugenia R.A. Cooper in the 1930s (4). Cooper described an oligocytoma as containing closely packed round "empty" cells, a tissue processing artifact later referred to as perinuclear halo. Other characteristics such as proliferation and migration of endothelial cells and presence of enlarged astrocytes (later termed "reactive astrocytes") infiltrating the tumor area were described by Bailey and Cooper and others. When reading Cooper's report it is evident that he did not believe that an oligocyte could change into an astrocyte or develop into an astrocytoma and vice versa. Rather, he thought that oligocytomas and astrocytomas strictly arose from their normal cellular counterparts. The possible cell of origin for gliomas has been heavily debated during the years and today we know that transitions between different cellular lineages can in fact occur, at least in experimental glioma (5–9).

The relationship between precursor cells of oligocytes and astrocytes and the cellular origin of oligodendroglioma was also discussed by Penfield who stated that the characteristics of an oligocytoma cell was that of an oligodendroblast, a less differentiated cell (10). He was supported by del Río-Hortega, reviewed in (11), who described the tumor cells of oligodendroglioma as incompletely differentiated, since myelinated fibers were absent. Many of these early observations and theories would later prove to be correct.

2.2. Histological Characteristics

Oligodendrogliomas comprises cells with oligodendrocyte appearance having small to slightly enlarged nuclei, with compact chromatin surrounded by a small amount of eosinophilic cytoplasm. As mentioned above, due to a tissue processing artifact oligodendroglioma cells typically display a clear cytoplasm around a central spherical nucleus making them look like "fried eggs" (Fig. 1a). This occurs when the tissue is formalin-fixed which causes degeneration due to acute swelling leaving an empty space between the nucleus and the cellular membrane, a feature called perinuclear halo, giving the tissue a honeycomb appearance. Tumor cells are round in shape and relatively homogenous within the tumor. As for all gliomas, they infiltrate the normal brain parenchyma irrespective of malignancy grade. A hallmark of oligodendrogliomas is the endothelial cell proliferation and migration seen as "chicken-wire"-like small, branching capillaries (Fig. 1a).

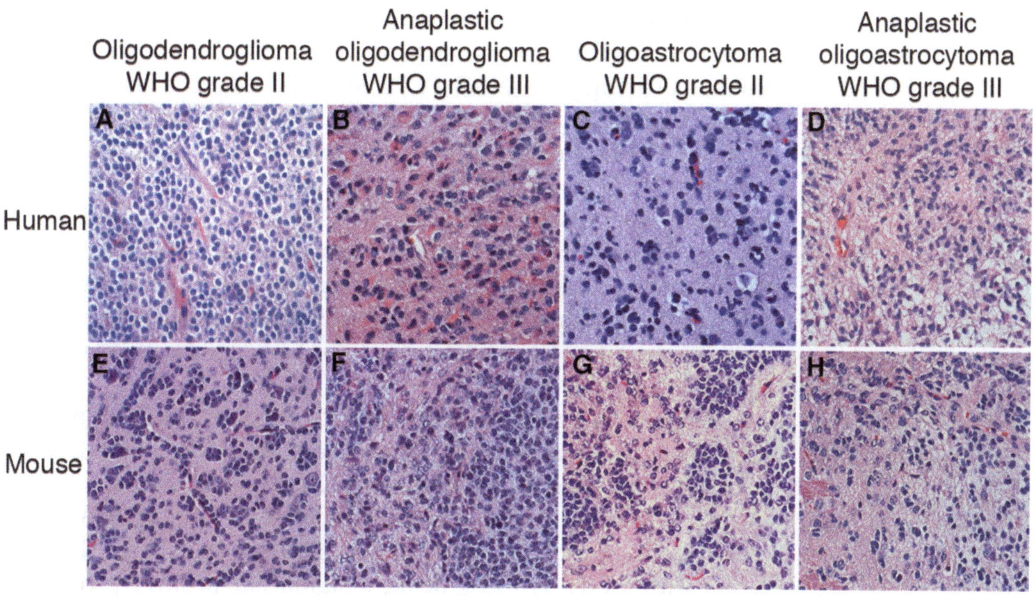

Fig. 1. H&E stained sections of the various types (oligodendroglioma, anaplastic oligodendroglioma, oligoastrocytoma, anaplastic oligoastrocytoma) and grades (II and III) of (**a–d**) human oligodendroglial tumors and (**e**, **f**) PDGF-B-induced mouse oligodendroglial tumors from the Ctv-a model (142).

Neoplastic oligodendrocytes commonly cluster around neurons, "perineuronal satellitosis," or around vessels forming "pseudorosettes." They preferentially migrate along white matter tracts or vessels spreading diffusely within the brain and across the hemispheres.

2.3. Grading, Markers, and Prognosis

Oligodendrogliomas are commonly classified using guidelines from the World Health Organization (WHO), most recently updated in 2007 (12). Currently, oligodendroglial tumors are classified into grade II (oligodendroglioma, oligoastrocytoma) and grade III tumors (anaplastic oligodendroglioma, anaplastic oligoastrocytoma) (Fig. 1a–d). Grade II tumors are composed of homogenous, diffusely infiltrating, round cells with moderate cell density. Nuclear atypia and occasional mitotic figures may be present, but significant mitotic activity, considerable vascular proliferation, and necrosis are signs of tumor progression into a grade III tumor. Grade III tumors are classified by significant hypercellularity and pleomorphism. Unlike astrocytic tumors, there is no grade IV classification for oligodendroglial tumors. Tumors with characteristics of grade IV glioma [glioblastoma (GB)] arising from a previously diagnosed oligodendroglial tumor are currently graded as "glioblastoma with an oligodendroglial component." However, it is not clear if these tumors represent a subset of GBs with differential origin, prognosis, and treatment requirements.

There are no cellular markers that are exclusive for oligodendrogliomas. Many oligodendroglial markers are expressed by astrocytic tumors and vice versa. Some markers expressed by normal oligodendrocytes are however more predominantly expressed in oligodendroglial tumors such as OLIG2 (13–16), chondroitin sulfate proteoglycan NG2 (17), SOX10 (18–20), and NKX2-2 (20, 21). Also, 1p/19q loss is rare in astrocytomas and is used in the clinic to aid in the diagnosis of oligodendrogliomas. Loss of 1p or 1p/19q is associated with better prognosis for patients with grade II and possibly also grade III oligodendrogliomas (22–29). This is likely due to an increased sensitivity to chemotherapy of these tumors (26–28), a notion further supported by several gene expression profiling analyses that have identified oligodendrogliomas carrying 1p/19q codeletions to be associated with a favorable prognosis (30, 31).

Oligodendroglioma and oligoastrocytoma respond well to combined treatment using procarbazine, lomustine (CCNU), and vincristine, in short PCV treatment (32–35). They are also sensitive to the alkylating agent temozolomide (36). The prognosis for patients diagnosed with oligodendroglioma varies from 1 to 10 years with LOH 1p/19q being a significant positive prognostic factor (37, 38).

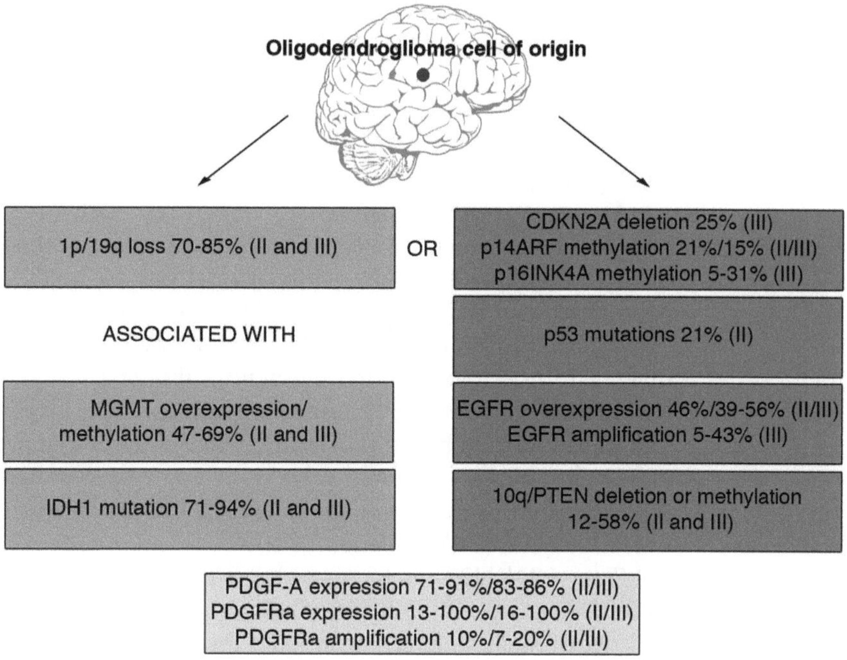

Fig. 2. The frequency of core mutations in human oligodendroglial tumors of grade II and III. There appear to exist at least two different paths to develop oligodendroglioma. One displays loss of 1p/19q (37, 39–43), MGMT overexpression/ promoter methylation (49, 50) and IDH1 mutation (55–57). The other path exhibits CDKN2A deletion/methylation (50, 61), p53 mutation (49, 64), EGFR amplification (25, 68, 70), and loss of 10q (25, 68). PDGF-A/PDGFRa overexpression or PDGFRa amplification seem to occur independent of 1p/19q status (82–84).

2.4. Core Signaling Pathways

The spectrum of mutations in oligodendroglial tumors overlap to large extent with those found in astrocytic tumors but there are also some distinct differences. Figure 2 shows a summary of the most frequently mutated genes in oligodendroglioma that are discussed below.

Combined loss of chromosomes 1p and 19q is the most common genetic alteration and is considered to be a hallmark of oligodendroglioma. While rare in astrocytic tumors single or codeletion of chromosomes 1p and 19q can be found in 70–85% of oligodendrogliomas (37, 39–43). As mentioned above, the loss of 1p/19q can be used both as a diagnostic and prognostic tool (37, 38). Since 1p/19q deletions can be found in both grade II and grade III tumors, they are believed to be early events.

DNA repair protein O6-methylguanine-DNA methyltransferase (MGMT) is often overexpressed in oligodendrogliomas causing treatment resistance to alkylating agents such as temozolomide, procarbazine, and CCNU (44, 45). Hypermethylation of *MGMT* leads to lower expression of the protein and thereby a better response to chemotherapy and a favorable prognosis for patients with oligodendroglioma grade II (46–50) and anaplastic oligodendroglioma grade III (51). In vitro, MGMT expression levels are lower in oligodendrocytes than in astrocytes which leads to higher chemosensitivity of oligodendrocytes to BCNU (52), suggesting that oligodendrocytes express low levels of MGMT and that overexpression is associated with tumorigenesis. Interestingly, hypermethylation of MGMT is associated with the loss of 19q and 1p/19q (50, 53, 54).

Heterozygous mutations of isocitrate dehydrogenase 1 (IDH1) at amino acid 132 (R132) result in the loss of the enzyme's ability to catalyze the conversion of isocitrate to α-ketoglutarate. Mutations at this residue is an early event in oligodendroglioma (55), since it is frequently found in both grade II and III oligodendroglial tumors (56, 57). In oligodendroglioma grade II, there is a correlation between IDH1 mutations and 1p/19q loss (55). This link was supported by another study, which showed that presence of IDH1 mutations correlated with both 1p/19q loss and MGMT promoter methylation, and could be used as a prognostic factor for overall survival (58).

The CDKN2a locus encodes tumor suppressor proteins $p16^{INK4a}$ and $p14^{ARF}$, both of which are frequently methylated and occasionally deleted in oligodendrogliomas (25, 50, 59–61). Alterations of $p14^{ARF}$ are more common in anaplastic oligodendroglioma than in oligodendroglioma grade II (62), but no correlations to survival have been reported. Decreased expression of $p16^{INK4a}$ predicts poor survival (63), and loss of 1p/19q and *CDKN2A* alterations appear to be mutually exclusive (37).

While common in astrocytic tumors mutations of TP53 are infrequent in oligodendroglial tumors (49, 64). Similar as for *CDKN2A* alterations, oligodendroglial tumors rarely have concomitant TP53 mutation and 1p/19q loss (42, 65).

Epidermal growth factor receptor (EGFR) overexpression, mutation, and amplification occurs less frequently in oligodendroglial than in astrocytic tumors (66, 67), and these alterations can be associated with poor prognosis (25, 68, 69). Many grade III tumors exhibit EGFR overexpression that can be coupled to EGFR amplification (66, 70), but there are also oligodendroglial tumors of both grade II and III that lack EGFR amplification but still exhibit overexpression of EGFRs (71). EGFR amplification has been found to be mutual exclusive to 1p/19q deletions (25, 68, 70).

Alterations affecting PTEN are very common in astrocytic tumors and occur in 74–87% of glioblastomas (72–74) and in 37–75% of anaplastic astrocytomas (73–75). PTEN alterations are also common in oligodendrogliomas where they occur in up to 48% of the tumors (37, 42, 76, 77). Such alterations involve mutations, promoter methylations, and 10q deletions causing loss of PTEN expression/function (25, 46, 77–81). Loss of 10q seems to be mutual exclusive to 1p/19q deletions (25, 68).

Platelet-derived growth factor-A (PDGFA) and PDGF receptor-A (PDGFRA) are commonly expressed in oligodendroglial tumors and appear to be an early event (82, 83). A subset of tumors also shows amplifications in the PDFGRA gene (83, 84).

High-grade gliomas have been grouped into three subclasses based on their molecular expression signatures; proneural, mesenchymal and proliferative (31), where all oligodendroglioma grade III tumors were stratified to the prognostically favorable proneural subclass. In a different investigation, gliomas of all grades and histologies were classified based on activation of PDGFRA, EGFR, and NF1 (85). In this study, two oligodendrogliomas grade II and two oligoastrocytomas were included, and the result indicated that oligodendroglial tumors could be overrepresented in the group of tumors driven by PDGF signaling. The Cancer Genome Atlas Research Network (TCGA) presented a gene expression signature classification of glioblastomas (grade IV astrocytomas) that described four different subtypes of glioblastoma (neural, proneural, classical, and mesenchymal) (86) that displayed reasonable similarities with the molecular subtypes described by Phillips et al. This subclass classification was used in a very recent study that included 43 oligodendrogliomas which showed that the majority of oligodendroglial tumors belonged to the proneural group as defined by TCGA. These were also associated with a significantly better prognosis than oligodendrogliomas stratified to the classical or mesenchymal subgroups (87). Loss of 1p/19q was common in the oligodendrogliomas and was overrepresented in those tumors belonging to the proneural group. Taken together, we can conclude that oligodendrogliomas appear to be driven by somewhat different signaling pathways than astrocytomas, and that there seem to be at least two independent routes leading to oligodendroglioma development.

3. Cellular Origin of Oligodendroglioma

As mentioned above, oligodendrogliomas have been named after the cell type that they share most histological and morphological resemblance to. Notably, the cell of origin for oligodendroglioma is still unknown. As described previously, already in the 1920s and 1930s neuropathologists and neurosurgeons debated whether the histology of glial tumor cells reflected their cellular origin, the relationship between oligodendrocytes and astrocytes, and the existence of a common or shared progenitor cell (4, 10, 11). During the late twentieth century, many studies were focused on characterizing markers expressed by human oligodendrogliomas, and the result showed that they indeed shared many antigens with normal oligodendrocytes and oligodendrocyte progenitor cells (OPCs). Some markers of OPCs that are commonly expressed in oligodendrogliomas are NG2 (17, 88),

PDGFRA (17, 88), OLIG2 (13, 14, 16, 89), GalC in mouse and human, respectively (90, 91), MBP (92), A2B5 (91, 93), and Leu-7 in rat oligodendrogliomas (94). Antigens of mature oligodendrocytes, for example proteins associated with myelination such as proteolipid protein (PLP) and 2′,3′-cyclic nucleotide 3′-phosphodiesterase (CNPase), are only rarely expressed in oligodendrogliomas (95), confirming the observation by del Río-Hortega (11) that oligodendrogliomas exhibit OPC characteristics. Thus, the OPC with its high capacity for self-renewal seem to be in a favorable state for promoting neoplastic transformation, which does not necessarily mean that OPCs are the (sole) cell of origin for oligodendroglioma.

It should be noted that both astrocytic and/or neuronal markers are at times expressed in oligodendrogliomas (96, 97). This could point to that oligodendrogliomas and astrocytomas in some instances may have a shared progenitor cell or merely illustrate the disturbed differentiation potential of glioma cells. Genetic analyses suggest that oligodendrogliomas and oligoastrocytomas have a common cellular origin due to their similar genetic profiles such as loss of 1p and 19q (98), and that the astrocytic component represents the skewed differentiation potential of the tumor cells.

There are studies supporting that GBs with a significant oligodendroglial contribution might have developed from an oligodendroglioma grade II or III rather than from an astrocytoma. GBs without TP53 mutations or EGFR gene amplification could be oligodendroglial GBs, since oligodendrogliomas rarely carry these genetic alterations as suggested by (99). Consequently, the histopathology of oligodendroglioma does not formally prove the cellular origin of these tumors.

4. Animal Models of Oligodendroglioma

Gliomas cannot be diagnosed until a tumor is present, and most gliomas are diagnosed at a late stage of the disease, which makes it difficult to study the mechanisms of tumor initiation in human biopsy material. Furthermore, due to ethical considerations clinical trials of novel drugs have to be preceded by preclinical trials in experimental models. Animal models of oligodendroglioma can help improve our knowledge of genetic, molecular, cellular and systemic mechanisms that contribute to tumor development, and be used in preclinical trials to identify promising candidate drugs.

4.1. Properties of a High Quality Oligodendroglioma Model

In our view, there are three basic features that need to be fulfilled in order to be regarded as a relevant and adequate animal model for oligodendroglial tumors: (1) The histopathology of the experimental tumor should recapitulate the histological characteristics of the corresponding type and grade of human oligodendroglioma. (2) The genetic alteration(s) should mimic signaling pathways activated in human oligodendroglial tumors. (3) The tumor should be orthotopic so that it develops in the proper microenvironment. Also, there are several additional properties of an experimental model that are highly desirable: (4) A defined cell of origin. (5) Temporal control of tumor initiation. (6) High tumor incidence and penetrance and short tumor latency. (7) Possibility to model different grades of oligodendroglial tumors.

The experimental models presented here have all generated tumors that have been classified as oligodendroglioma-like by their authors at the time of publication. Many of these models have produced tumors of various histopathologies, i.e., oligodendroglioma, oligoastrocytoma, and astrocytoma. This could be due to that astrocytic and oligodendroglial tumors share many genetic alterations but could also reflect that these different tumor types may in some instances arise from the same cell type. Also of note is that there have been continuous revisions of the WHO guidelines for classification of gliomas; hence, some tumors that were earlier classified as oligodendroglial tumors might not meet the criteria of today.

4.2. Early Oligodendroglioma Models

Early studies in the beginning of the twentieth century were focused on describing spontaneous oligodendroglial tumors in various animal species such as dog (100–102), cat (100, 103), rat (104–107), horse (100), and cow (100, 108). The first experimentally induced oligodendrogliomas were published in the 1960s where the Rous Sarcoma Virus (RSV) was used as the initiating agent.

4.2.1. RSV-Induced Oligodendrogliomas

RSV was the first oncovirus to be discovered by Peyton Rous in 1911, which awarded him the Nobel Prize in 1966. RSV contains the oncogene v-src (viral-sarcoma) that is derived from the proto-oncogene c-Src encoding an intracellular tyrosine kinase protein (109). v-Src lacks the C-terminal inhibitory phosphorylation of Tyr-527, present in c-Src, leading to constitutive activation and transformation of cells (110). In the 1960s and 1970s RSV was extensively used to induce brain tumors in dogs (111–113), mice (114), rabbits (115), and other mammals. Some of these tumors resembled oligodendroglioma while most were histologically similar to other types of glioma. This could be due to that RSV could not infect the appropriate target cells for oligodendroglioma and/or that v-Src signaling did not activate the pathways required for oligodendroglioma development. The low frequency of oligodendroglioma made this a poor model.

4.2.2. Chemically Induced Oligodendrogliomas

Various chemical agents have been used to induce experimental gliomas. The first and most commonly used chemical agent to induce gliomas was nitrosourea, a DNA alkylating agent that has cytotoxic, mutagenic, and carcinogenic effects (62). Nitrosourea can pass the blood–brain barrier, and glioma-like brain tumors were induced in both rats (116) (117, 118) and rabbits (119) by intravenous, intraperitoneal, or oral administration, e.g., the widely used C6 rat glioma cell line was established from a glioma induced by ethylnitrosurea (116). The CNS tumors induced resembled histologically human oligodendrogliomas, astrocytomas, neurinomas, and ependymomas, sometimes with intermixed cell types. For many years these models were used to study the histological and morphological composition of glial tumors (120–122). Oligodendroglial-like tumors (in minority to astrocytic tumors) were also induced in rats by the combination of dietary lead subacetate and 2-acetylaminofluorene (123), and by embryonal distribution of methanesulfonate (124). None of the chemically induced models generated a high frequency of oligodendrogliomas; the majority of induced tumors were in fact astrocytoma-like. Other major disadvantages were the random and unspecific genetic alterations caused by the agents and the undefined target cells making tumor induction unpredictable and experiments difficult to reproduce.

4.2.3. Transplanted Oligodendrogliomas

In the 1950s, it became feasible to maintain cells in culture as the knowledge of their nutritional and environmental requirements expanded (125). The first transplantation models using cultured glioma cells were published in the 1960s (126). However, most cell lines have been established from high-grade astrocytomas, such as the C6 rat glioblastoma cell line (116), and there are very few mouse oligodendroglioma cell lines (G26-24 (127)) and human oligodendroglioma cell lines (e.g., IN/157 (128), HOG7 (129), TC620 (130), MO3.13 (131), KG-1C (132)) described. Recently, two anaplastic oligodendrogliomas with complete loss of 1p/19q have been explanted and maintained as neurospheres in culture (133). Cultured cells have shown to have characteristics of CSCs, and one of the cell lines (BT088) could initiate tumors similar to anaplastic oligodendroglioma in immunocompromised mice.

One major advantage with transplanted glioma models is that they are highly reproducible with regard to tumor incidence, latency, growth rate and location given that an established, transplantable cell line is being used. A weakness of grafted tumors has been that they often do not recapitulate the invasive growth of human oligodendroglioma. This has lately been greatly improved by the use of glioma-derived cancer stem cells for transplantation. When grafted these cells generate tumors with histopathology highly similar to the primary tumor, which make these models much more relevant for tumor biology studies and preclinical testing. However, so far very few CSC lines have been established from oligodendrogliomas (133). An additional limitation of transplanted models is that the role of cellular origin and underlying mechanisms of tumor initiation and progression cannot be studied.

5. Genetically Induced Oligodendroglioma Models

These are the experimental oligodendroglioma models most commonly used today. Table 1 presents a summary of the contemporary models generating oligodendroglioma-like tumors.

5.1. Retroviral Models of Oligodendroglioma

Retroviruses (e.g., Avian leukosis viruses such as RSV, Murine leukemia viruses, Lentiviruses) confer stable transduction of genes into the host genome and have been widely used to study the significance of specific genetic alterations in glioma and many other tumor types. Common for all retroviral models is that the target cell needs to be dividing for retroviral infection to occur.

Eric Holland and Harold Varmus established the first RCAS/TV-A tumor model system (134, 135) that has subsequently been used to model many other tumor types outside the CNS (136–139). The RCAS/TV-A model allowed for controlled and specific transduction of exogenous genes into distinctive glial cell populations of the brain. RCAS is a plasmid originally derived from RSV that was modified to harbor an exogenous gene (140). RCAS viruses are propagated in DF-1 chicken cells and can only infect cells expressing the avian tv-a receptor, required for infection of subgroup A Avian leukosis viruses, that is not present in the mammalian genome. Two transgenic mouse lines expressing tv-a in a cell type or tissue specific manner were generated using the nestin or GFAP (Glial fibrillary acidic protein) promoters, respectively (134, 135). The Nestin promoter will direct infection by RCAS to neural/glial stem cells in the Ntv-a line, and the GFAP promoter to astrocytes and neural stem cells of the subventricular zone in the Gtv-a line (141). Recently, our lab has published a third tv-a transgenic mouse line for the study of CNS tumors, where the 2′, 3′-cyclic-nucleotide 3′-phosphodiesterase (CNPase) promoter was used to control tv-a expression (142) and target it to OPCs. Directed infection can take place by intracerebral injection of RCAS viruses in neonatal or adult mice and the tumor will develop orthotopically in a normal microenvironment. Some advantages with the RCAS/TV-A model over the early models described are that the cellular origin, genetic alterations and the time of induction can be controlled. The RCAS virus is replication-incompetent in mammalian cells, which will limit the contribution of insertional mutagenesis. The different tv-a transgenic mouse lines have been extensively used to model many types and grades of glioma. By infecting the three tv-a lines with RCAS-PDGF-B, oligodendrogliomas grade II can be reproducibly induced (Fig. 1e) (6, 142, 143). Occasional oligoastrocytomas also developed in the Ctv-a mice (Fig. 1g) (142). Through cross-breeding the tv-a transgenic mouse lines with mice carrying targeted deletions of the Cdkn2a

Table 1
A list of the most common contemporary experimental models of oligodendroglioma development

Model system	Animal species	Target cell	Oncogene	Tumor suppressor gene	Histopathology	References
Transplanted CSCs derived from OG	Mouse	NA	NA	1p19q deletion	OIII	Kelly (2010)
RCAS/TV-A	Mouse	GFAP	PyV-mt	wt	OII-III, AIII	Holland (2000)
RCAS/TV-A	Mouse	GFAP or Nestin	PDGF-B	wt, Cdkn2a$^{-/-}$	OII-III, AII-IV	Dai (2001)
RCAS/TV-A	Mouse	GFAP or Nestin	PDGF-B	wt, Ink4a$^{-/-}$, Arf$^{-/-}$, Cdkn2a$^{-/-}$	OII-III	Tchougounova (2007)
RCAS/TV-A	Mouse	CNPase	PDGF-B	wt	OII-III, OAIII	Lindberg (2009)
MoMuLV	Mouse	Unknown	PDGF-B	wt	OII-III, AII-IV	Uhrbom (1998)
MoMuLV	Mouse	Unknown	PDGF-B	p53$^{-/-}$, Cdkn2a$^{-/-}$	OII-III, AII-IV	Hesselager (2003)
transgene	Mouse	GFAP	V12Ha-Ras + EGFRvIII	wt	OII, OAII	Ding (2003)
transgene	Mouse	S100β	v-erbB	wt, Cdkn2a$^{+/-}$, p53$^{+/-}$	OII-III	Weiss (2003)
transgene	Rat	S100β	v-erbB	wt	OII-III, AIV	Ohgaki (2006)

NA not applicable, *wt* wild type, *OII* oligodendroglioma grade II, *OIII* anaplastic oligodendroglioma grade III, *OAII* oligoastrocytoma grade II, *OAIII* oligoastrocytoma grade III, *AII* astrocytoma grade II, *AIII* anaplastic astrocytoma grade III, *AIV* astrocytoma grade IV locus, the separate roles of tumor suppressors $p16^{Ink4a}$ and $p19^{Arf}$ in PDGF-B could be studied. Loss of $p19^{Arf}$ have been found to be the major suppressor causing enhanced incidence and malignancy, producing a high frequency of grade III oligodendroglioma (Fig. 1f, h) and glioblastoma in all three mouse lines (143, 144). Thus, by using tv-a transgenic mice cross-bred with Arf−/− mice high-grade oligodendrogliomas can be robustly generated. The loss of other cell cycle inhibitors such as $p21^{Cip1}$ (145) and $p27^{Kip1}$ (146, 147) have also been studied in PDGF-B-driven oligodendroglioma development, as well as the role of many putative glioma-causing genes/pathways such as SOX10 (19), IGFBP2 (148, 149), SOX5 (150), nitric oxide-induced signaling (151), and DNA damage response genes (152). The PDGF-induced Ntv-a model has also been used in preclinical investigations, e.g., to analyze the effects of the AKT inhibitor perifosine (153), and the antiangiogenic/immune-modulating histidine-rich glycoprotein (HRG) (154).

Moloney Murine Leukemia Virus (MMLV)-based vectors have also been widely used to model many different types of tumors, including oligodendroglioma, in mice. MMLV will deliver the exogenous gene in an undefined manner infecting most murine cell types. The first oligodendroglioma-like tumor generated by MMLV infection came from infection of fetal rat brain cells by a replication-incompetent MMLV encoding ras and myc oncogenes that were subsequently transplanted into the brains of syngeneic hosts (155). A few years later, our lab used a combination of replication-competent and replication-incompetent MMLV carrying the human PDGF-B cDNA that were injected into the brains of neonatal mice (156, 157). The mice developed tumors with a high incidence and many showed a histopathology similar to human oligodendroglioma. The model was made with the purpose to be used as a tool to identify retrovirally tagged cancer-causing genes complementing PDGF-B in glioma development and several novel and known brain tumor loci were subsequently identified (19, 150, 158–161). MMLV vectors have been used to transduce mouse neural stem cells and astrocytes in vitro with a wild type or constitutively active EGFR mutant, EGFRvIII (in-frame deletion of exons 2 to 7 spanning the extracellular ligand-binding domain that will produce a truncated, ligand-independent, constitutively phosphorylated receptor), which upon transplantation formed tumors with some oligodendroglial-like features (162). Some concerns with MMLV models have been the undefined cellular origin and the unknown contribution of retroviral insertional mutagenesis, the latter being applicable to retroviral models in general.

5.2. Transgenic Models of Oligodendroglioma

Transgenic mouse models have been extensively used to evaluate the significance of overexpression of oncogenes and targeted deletions of tumor suppressor genes in glioma. In addition, as exemplified above, many mouse lines carrying targeted deletions have been used in combination with retroviral models. When discussing transgenic models one needs to bear in mind that, contrary to viral models, the transgene analyzed will only be expressed as long as the promoter controlling its expression is active, which may explain why transgene models often are less efficient in generating tumors.

In 2003, Weiss et al. published a transgenic mouse model where a retroviral transforming allele of the EGF receptor, v-erbB, the viral homolog of the human EGFR that is lacking a large portion of the ligand-binding domain and has ligand-independent, constitutive tyrosine kinase activity much like the EGFRvIII, was expressed under the S100β promoter (7). Albeit S100β is expressed in many glial and neuronal cells of the CNS induced tumors were diagnosed as oligodendroglioma grade II. When cross-breeding the S100β-v-erbB mice with mice carrying targeted deletions of the p53 pathway, tumor development was accelerated and anaplastic oligodendroglioma developed. In a subsequent investigation, the cellular origin of tumors from the S100β-verbB model was deduced to be OPCs of the white matter (163). Similarly, Ding et al. described the generation of oligodendroglial-like tumors from combined transgenic expression of the EGFRvIII and ^{V12}Ha-Ras (RasB8), both under the control of the GFAP promoter (8). Also, transgenic expression of v-erbB in S100β expressing cells in rats produced oligodendrogliomas, anaplastic oligodendrogliomas, and glioblastomas (9, 164). Worth mentioning is that pCNPase-hEGFR transgenic mice did not develop any tumors (165). However, in this study the normal human EGFR was being used.

Data from human tumors suggests that aberrant PDGF signaling is more common than EGFR signaling in oligodendroglioma. However, transgenic models of PDGF expression have not been very potent in inducing gliomas. PDGF-B expressed under the control of the GFAP promoter had to be crossed with TP53 null mice to develop oligodendrogliomas and glioblastomas (166). Likewise, transgenic mice expressing PDGF-B from the nestin promoter did not produce any gliomas in the wild-type background (167). This could be due to inadequate expression levels and/or inaccurate spatiotemporal expression of PDGF-B for oligodendroglioma to develop.

5.3. Conclusions from Oligodendroglioma Modeling

Expression and mutation analyses of human oligodendroglioma have encouraged extensive studies on the role of PDGFR and EGFR signaling pathways using experimental glioma models. These studies have convincingly showed a causative role of these pathways in glioma development. Activation of PDGF signaling has been more potent than EGFR signaling in inducing experimental oligodendrogliomas in the viral models. When the EGFR mutant denoted EGFR* (168) was used intracranially in the

RCAS/TV-A model, it resulted in the development of small glioma-like lesions, but only in combination with other genetic alterations (134). The higher oncogenic potency of PDGF activation may seem rational, since PDGFR overexpression is much more common in human tumors, and PDGFRα signaling is necessary and sufficient to maintain normal OPCs in a proliferative state (169–171). However, in transgenic models EGFR signaling appears to be more potent than PDGFR signaling, although definitive comparisons cannot be made since different promoters have been used. In normal development, EGFR signaling can stimulate oligodendrocyte production by expanding subventricular zone type-C progenitor cells (SVZ OPCs) that in turn will generate oligodendrocytes (172, 173). These cells were the putative target cells in the S100β-v-erbB mice (7). However, it cannot be excluded that PDGF signaling also contributed to tumor development in these mice since PDGFα-receptors were expressed in the tumor tissue. It has recently been shown that PDGFRα can be activated in a ligand-independent manner by activation of other receptor tyrosine kinases. These in turn will activate reactive oxygen species (ROS) that will phosphorylate PDGFRα (174), and in glioma cells it has been shown that ROS levels are clearly increased in response to EGFR activation (175).

The viral and transgenic models have made it possible to study the implication of cellular origin/differentiation status for glioma development. As discussed above, OPCs and/or oligodendrocytes have long been suggested to be the most likely cellular origin for oligodendroglioma. However, the results from the cell type-specific experimental models are proposing that experimental oligodendroglioma may develop from many different types of glial cells and that the oncogenic signaling pathways dictate the glioma subtype that develops. In the RCAS/TV-A model, PDGF signaling clearly favors oligodendroglioma development, while AKT + KRAS signaling will promote astrocytic tumor development (144, 176, 177). The origin of human oligodendroglioma remains undefined but the recent molecular-based stratification of human gliomas into different subclasses could in part reflect that they have originated from different cell types. Thus, taking the current knowledge into account we suggest that there is a common signaling pathway specific to oligodendroglioma that drives the phenotype of these tumors irrespective of cell of origin. This pathway appears to be related to normal OPC development and maintenance and tumors induced from neural/glial progenitor cells will acquire the histology of OPCs through a differentiation-like process, while tumors induced in astrocytes and CNPase positive OPCs will undergo a dedifferentiation-like process. Since these are not normal but neoplastic cells, this process may be more appropriately referred to as oncogene-induced differentiation.

6. Future Perspectives

Although oligodendrogliomas have a better prognosis than astrocytomas of the same malignancy grade they are still largely incurable due to recurrence, often coupled to malignant progression. The single most important prognostic factor for oligodendroglioma is the 1p/19q codeletion associated with longer survival in both grade II and III patients.

The problem with current therapy is that it fails to incapacitate all tumor cells. Gliomas in general are diffusely growing tumors that spread throughout the brain already early in tumor development. To cure patients one need to identify tumor-specific antigens that can be used to target these dispersed tumor cells in the normal parenchyma and hit them with tailor-made (each patient's specific genetic signature) drugs that induce terminal differentiation, cellular senescence or apoptosis. Models that may help us to reach these goals are of principal importance.

Models that allow for imaging of single cells in vivo to understand the migration of tumor cells and follow the tumorigenic process in both time and space are warranted. To be able to image glioma development the Ef-luc transgenic mouse was generated that expresses firefly luciferase under the control of the E2F1 promoter. When these mice were crossed with Ntv-a mice development of PDGF-B driven oligodendrogliomas and anaplastic oligodendrogliomas could be followed in vivo (178). However, the resolution of firefly luciferase imaging is too low to be able to follow single cells and more sensitive reporters and imaging systems need be developed.

To fully understand the biology of a tumor requires knowledge of its cellular origin. Also, by targeting the right cell type the most relevant animal models are generated. Several different glial promoters have been used when trying to pinpoint the cellular origin of oligodendroglioma. To continue to analyze more CNS cell types, one could obviously generate additional tv-a transgenic lines. An alternative strategy could be to use the conditional RCAS/TV-A model that employs a Cre-loxP-based strategy to target expression of tv-a in the Rosa26 locus (179). This mouse line has been used to model pancreatic tumors through cross-breeding with the Ptf1a-cre transgenic mouse, but could potentially be crossed with new or already available Cre transgenic mouse lines to extend the analyses of cellular origin.

Oligodendrogliomas have not been modeled using inducible systems as yet. Such models may be useful for example for the study of tumor initiation and maintenance. In the RCAS/TV-A mouse model, inducible RCAS vectors have been developed where oncogenes or tumor suppressor genes can be expressed under the control of the reverse tet transcriptional transactivator (rtTA) (180). Expression from RCAS can be controlled by adding doxycycline (tetracycline analogue) to the food or drinking water, and withdrawal will result in decreased gene expression. The advantage of this system is that the change in gene expression is limited to a subset of RCAS infected cells that are surrounded by a normal cellular environment.

The role of epigenetic mechanisms in oligodendroglioma development has been poorly studied although it has been known for a long time that promoter methylation that causes gene silencing is common in oligodendrogliomas. However, few experimental animal models have addressed this issue. Hong et al. have identified SLC5A8 to be frequently aberrantly methylated in human astrocytomas and oligodendrogliomas. This aberrant methylation was confirmed in 9 of 10 murine oligodendroglial tumors from p53+/− or Ink4a-Arf +/− mice transgenic for S100b-v-erbB (7).

In addition to developing new models, it would also be important to truly validate the currently most used models and compare transcriptome signatures with those of human oligodendroglioma to identify the most relevant and useful models for different types and grades of oligodendrogliomas. Such information could be important and counseling when designing both basal tumor biology studies and preclinical investigations, and would see to that the generated results are as relevant and applicable as possible to the corresponding human situation.

Acknowledgments

This work has been supported by grants from the Swedish Cancer Society, the Swedish Childhood Cancer Foundation, the Swedish Research Council, and Ragnar Söderberg's foundation. We thank Fredrik J Swartling for his insightful comments.

References

1. Robertson WF (1899) On a new method of obtaining a black reaction in certain tissue-elements of the central nervous system (platinum method). Scott Med Surg J 4:23–30
2. Bailey P, Cushing H (1925) Microchemical color reactions as an aid to the identification and classification of brain tumors. Proc Natl Acad Sci USA 11:82–84
3. Bailey P, Bucy PC (1929) Oligodendrogliomas of the brain. J Pathol Bacteriol 32:735–751
4. Cooper ERA (1935) The relation of oligocytes and astrocytes in cerebral tumours. J Pathol Bacteriol 41:259–266
5. Holland EC, Li Y, Celestino J, Dai C, Schaefer L, Sawaya RA, Fuller GN (2000) Astrocytes give rise to oligodendrogliomas and astrocytomas after gene transfer of polyoma virus middle T antigen in vivo. Am J Pathol 157:1031–1037
6. Dai C, Celestino JC, Okada Y, Louis DN, Fuller GN, Holland EC (2001) PDGF autocrine stimulation dedifferentiates cultured astrocytes and induces oligodendrogliomas and oligoastrocytomas from neural progenitors and astrocytes in vivo. Genes Dev 15:1913–1925

7. Weiss WA, Burns MJ, Hackett C, Aldape K, Hill JR, Kuriyama H, Kuriyama N, Milshteyn N, Roberts T, Wendland MF, DePinho R, Israel MA (2003) Genetic determinants of malignancy in a mouse model for oligodendroglioma. Cancer Res 63:1589–1595
8. Ding H, Shannon P, Lau N, Wu X, Roncari L, Baldwin RL, Takebayashi H, Nagy A, Gutmann DH, Guha A (2003) Oligodendrogliomas result from the expression of an activated mutant epidermal growth factor receptor in a RAS transgenic mouse astrocytoma model. Cancer Res 63:1106–1113
9. Ohgaki H, Kita D, Favereaux A, Huang H, Homma T, Dessen P, Weiss WA, Kleihues P, Heppner FL (2006) Brain tumors in S100beta-v-erbB transgenic rats. J Neuropathol Exp Neurol 65:1111–1117
10. Penfield W (1931) The classification of gliomas and neuroglia cell types. Arch Neurol Psychiatry 26:745–753
11. del Rio-Hortega P (1934) Estructura y sistematizacion de los gliomas y paragliomas. Arch Neurol Psychiatry 32:1117–1120
12. Cavenee WK, Weller M, Furnari FB, Berens ME, Nagane M, Plate KH, Huang H-JS, Israel MA, Newcomb EW, Noble MD, Bigner DD, Kleihues P (2007) World Health Organization classification of tumors: pathology and genetics of tumors of the nervous system. International Agency for Research on Cancer Press, Lyon
13. Mikami S, Hirose Y, Yoshida K, Kawase T, Ohnishi A, Nagashima K, Mukai M, Okada Y, Ikeda E (2007) Predominant expression of OLIG2 over ID2 in oligodendroglial tumors. Virchows Arch 450:575–584
14. Ligon KL, Alberta JA, Kho AT, Weiss J, Kwaan MR, Nutt CL, Louis DN, Stiles CD, Rowitch DH (2004) The oligodendroglial lineage marker OLIG2 is universally expressed in diffuse gliomas. J Neuropathol Exp Neurol 63:499–509
15. Marie Y, Sanson M, Mokhtari K, Leuraud P, Kujas M, Delattre JY, Poirier J, Zalc B, Hoang-Xuan K (2001) OLIG2 as a specific marker of oligodendroglial tumour cells. Lancet 358:298–300
16. Lu QR, Park JK, Noll E, Chan JA, Alberta J, Yuk D, Alzamora MG, Louis DN, Stiles CD, Rowitch DH, Black PM (2001) Oligodendrocyte lineage genes (OLIG) as molecular markers for human glial brain tumors. Proc Natl Acad Sci USA 98:10851–10856
17. Nishiyama A, Chang A, Trapp BD (1999) NG2+ glial cells: a novel glial cell population in the adult brain. J Neuropathol Exp Neurol 58:1113–1124
18. Bannykh SI, Stolt CC, Kim J, Perry A, Wegner M (2006) Oligodendroglial-specific transcriptional factor SOX10 is ubiquitously expressed in human gliomas. J Neurooncol 76:115–127
19. Ferletta M, Uhrbom L, Olofsson T, Ponten F, Westermark B (2007) Sox10 has a broad expression pattern in gliomas and enhances platelet-derived growth factor-B-induced gliomagenesis. Mol Cancer Res 5:891–897
20. Rousseau A, Nutt CL, Betensky RA, Iafrate AJ, Han M, Ligon KL, Rowitch DH, Louis DN (2006) Expression of oligodendroglial and astrocytic lineage markers in diffuse gliomas: use of YKL-40, ApoE, ASCL1, and NKX2-2. J Neuropathol Exp Neurol 65: 1149–1156
21. Riemenschneider MJ, Koy TH, Reifenberger G (2004) Expression of oligodendrocyte lineage genes in oligodendroglial and astrocytic gliomas. Acta Neuropathol 107:277–282
22. Fallon KB, Palmer CA, Roth KA, Nabors LB, Wang W, Carpenter M, Banerjee R, Forsyth P, Rich K, Perry A (2004) Prognostic value of 1p, 19q, 9p, 10q, and EGFR-FISH analyses in recurrent oligodendrogliomas. J Neuropathol Exp Neurol 63:314–322
23. Iwamoto FM, Nicolardi L, Demopoulos A, Barbashina V, Salazar P, Rosenblum M, Hormigo A (2008) Clinical relevance of 1p and 19q deletion for patients with WHO grade 2 and 3 gliomas. J Neurooncol 88:293–298
24. Kujas M, Lejeune J, Benouaich-Amiel A, Criniere E, Laigle-Donadey F, Marie Y, Mokhtari K, Polivka M, Bernier M, Chretien F, Couvelard A, Capelle L, Duffau H, Cornu P, Broet P, Thillet J, Carpentier AF, Sanson M, Hoang-Xuan K, Delattre JY (2005) Chromosome 1p loss: a favorable prognostic factor in low-grade gliomas. Ann Neurol 58:322–326
25. Hoang-Xuan K, He J, Huguet S, Mokhtari K, Marie Y, Kujas M, Leuraud P, Capelle L, Delattre JY, Poirier J, Broet P, Sanson M (2001) Molecular heterogeneity of oligodendrogliomas suggests alternative pathways in tumor progression. Neurology 57:1278–1281
26. Thiessen B, Maguire JA, McNeil K, Huntsman D, Martin MA, Horsman D (2003) Loss of heterozygosity for loci on chromosome arms 1p and 10q in oligodendroglial tumors: relationship to outcome and chemosensitivity. J Neurooncol 64:271–278
27. van den Bent MJ, Looijenga LH, Langenberg K, Dinjens W, Graveland W, Uytdewilligen L, Sillevis Smitt PA, Jenkins RB, Kros JM (2003) Chromosomal anomalies in oligodendroglial tumors are correlated with clinical features. Cancer 97:1276–1284

28. Walker C, Haylock B, Husband D, Joyce KA, Fildes D, Jenkinson MD, Smith T, Broome J, du Plessis DG, Warnke PC (2006) Clinical use of genotype to predict chemosensitivity in oligodendroglial tumors. Neurology 66:1661–1667
29. Smith JS, Perry A, Borell TJ, Lee HK, O'Fallon J, Hosek SM, Kimmel D, Yates A, Burger PC, Scheithauer BW, Jenkins RB (2000) Alterations of chromosome arms 1p and 19q as predictors of survival in oligodendrogliomas, astrocytomas, and mixed oligoastrocytomas. J Clin Oncol 18:636–645
30. Freije WA, Castro-Vargas FE, Fang Z, Horvath S, Cloughesy T, Liau LM, Mischel PS, Nelson SF (2004) Gene expression profiling of gliomas strongly predicts survival. Cancer Res 64:6503–6510
31. Phillips HS, Kharbanda S, Chen R, Forrest WF, Soriano RH, Wu TD, Misra A, Nigro JM, Colman H, Soroceanu L, Williams PM, Modrusan Z, Feuerstein BG, Aldape K (2006) Molecular subclasses of high-grade glioma predict prognosis, delineate a pattern of disease progression, and resemble stages in neurogenesis. Cancer Cell 9:157–173
32. Cairncross JG, Macdonald DR (1988) Successful chemotherapy for recurrent malignant oligodendroglioma. Ann Neurol 23:360–364
33. Macdonald DR, Gaspar LE, Cairncross JG (1990) Successful chemotherapy for newly diagnosed aggressive oligodendroglioma. Ann Neurol 27:573–574
34. Bauman GS, Cairncross JG (2001) Multidisciplinary management of adult anaplastic oligodendrogliomas and anaplastic mixed oligoastrocytomas. Semin Radiat Oncol 11:170–180
35. Kim L, Hochberg FH, Thornton AF, Harsh GRT, Patel H, Finkelstein D, Louis DN (1996) Procarbazine, lomustine, and vincristine (PCV) chemotherapy for grade III and grade IV oligoastrocytomas. J Neurosurg 85:602–607
36. Chinot O (2001) Chemotherapy for the treatment of oligodendroglial tumors. Semin Oncol 28:13–18
37. Cairncross JG, Ueki K, Zlatescu MC, Lisle DK, Finkelstein DM, Hammond RR, Silver JS, Stark PC, Macdonald DR, Ino Y, Ramsay DA, Louis DN (1998) Specific genetic predictors of chemotherapeutic response and survival in patients with anaplastic oligodendrogliomas. J Natl Cancer Inst 90:1473–1479
38. Ino Y, Betensky RA, Zlatescu MC, Sasaki H, Macdonald DR, Stemmer-Rachamimov AO, Ramsay DA, Cairncross JG, Louis DN (2001) Molecular subtypes of anaplastic oligodendroglioma: implications for patient management at diagnosis. Clin Cancer Res 7:839–845
39. Reifenberger J, Reifenberger G, Liu L, James CD, Wechsler W, Collins VP (1994) Molecular genetic analysis of oligodendroglial tumors shows preferential allelic deletions on 19q and 1p. Am J Pathol 145:1175–1190
40. Barbashina V, Salazar P, Holland EC, Rosenblum MK, Ladanyi M (2005) Allelic losses at 1p36 and 19q13 in gliomas: correlation with histologic classification, definition of a 150-kb minimal deleted region on 1p36, and evaluation of CAMTA1 as a candidate tumor suppressor gene. Clin Cancer Res 11:1119–1128
41. Smith JS, Alderete B, Minn Y, Borell TJ, Perry A, Mohapatra G, Hosek SM, Kimmel D, O'Fallon J, Yates A, Feuerstein BG, Burger PC, Scheithauer BW, Jenkins RB (1999) Localization of common deletion regions on 1p and 19q in human gliomas and their association with histological subtype. Oncogene 18:4144–4152
42. Bigner SH, Matthews MR, Rasheed BK, Wiltshire RN, Friedman HS, Friedman AH, Stenzel TT, Dawes DM, McLendon RE, Bigner DD (1999) Molecular genetic aspects of oligodendrogliomas including analysis by comparative genomic hybridization. Am J Pathol 155:375–386
43. Gelpi E, Ambros IM, Birner P, Luegmayr A, Drlicek M, Fischer I, Kleinert R, Maier H, Huemer M, Gatterbauer B, Anton J, Rossler K, Budka H, Ambros PF, Hainfellner JA (2003) Fluorescent in situ hybridization on isolated tumor cell nuclei: a sensitive method for 1p and 19q deletion analysis in paraffin-embedded oligodendroglial tumor specimens. Mod Pathol 16:708–715
44. Silber JR, Bobola MS, Ghatan S, Blank A, Kolstoe DD, Berger MS (1998) O6-methylguanine-DNA methyltransferase activity in adult gliomas: relation to patient and tumor characteristics. Cancer Res 58:1068–1073
45. Buccoliero AM, Arganini L, Ammannati F, Gallina P, Di Lorenzo N, Mennonna P, Taddei GL (2005) Oligodendrogliomas lacking O6-methylguanine-DNA-methyltransferase expression. J Chemother 17:321–326
46. Kuo LT, Kuo KT, Lee MJ, Wei CC, Scaravilli F, Tsai JC, Tseng HM, Kuo MF, Tu YK (2009) Correlation among pathology, genetic and epigenetic profiles, and clinical outcome in oligodendroglial tumors. Int J Cancer 124:2872–2879

47. Brandes AA, Tosoni A, Cavallo G, Reni M, Franceschi E, Bonaldi L, Bertorelle R, Gardiman M, Ghimenton C, Iuzzolino P, Pession A, Blatt V, Ermani M (2006) Correlations between O6-methylguanine DNA methyltransferase promoter methylation status, 1p and 19q deletions, and response to temozolomide in anaplastic and recurrent oligodendroglioma: a prospective GICNO study. J Clin Oncol 24:4746–4753
48. Levin N, Lavon I, Zelikovitsh B, Fuchs D, Bokstein F, Fellig Y, Siegal T (2006) Progressive low-grade oligodendrogliomas: response to temozolomide and correlation between genetic profile and O6-methylguanine DNA methyltransferase protein expression. Cancer 106:1759–1765
49. Watanabe T, Nakamura M, Kros JM, Burkhard C, Yonekawa Y, Kleihues P, Ohgaki H (2002) Phenotype versus genotype correlation in oligodendrogliomas and low-grade diffuse astrocytomas. Acta Neuropathol 103:267–275
50. Dong SM, Pang JC, Poon WS, Hu J, To KF, Chang AR, Ng HK (2001) Concurrent hypermethylation of multiple genes is associated with grade of oligodendroglial tumors. J Neuropathol Exp Neurol 60:808–816
51. Vogelbaum MA, Berkey B, Peereboom D, Macdonald D, Giannini C, Suh JH, Jenkins R, Herman J, Brown P, Blumenthal DT, Biggs C, Schultz C, Mehta M (2009) Phase II trial of preirradiation and concurrent temozolomide in patients with newly diagnosed anaplastic oligodendrogliomas and mixed anaplastic oligoastrocytomas: RTOG BR0131. Neurooncology 11:167–175
52. Nutt CL, Noble M, Chambers AF, Cairncross JG (2000) Differential expression of drug resistance genes and chemosensitivity in glial cell lineages correlate with differential response of oligodendrogliomas and astrocytomas to chemotherapy. Cancer Res 60:4812–4818
53. Mollemann M, Wolter M, Felsberg J, Collins VP, Reifenberger G (2005) Frequent promoter hypermethylation and low expression of the MGMT gene in oligodendroglial tumors. Int J Cancer 113:379–385
54. van den Bent MJ, Dubbink HJ, Sanson M, van der Lee-Haarloo CR, Hegi M, Jeuken JW, Ibdaih A, Brandes AA, Taphoorn MJ, Frenay M, Lacombe D, Gorlia T, Dinjens WN, Kros JM (2009) MGMT promoter methylation is prognostic but not predictive for outcome to adjuvant PCV chemotherapy in anaplastic oligodendroglial tumors: a report from EORTC Brain Tumor Group Study 26951. J Clin Oncol 27:5881–5886
55. Watanabe T, Nobusawa S, Kleihues P, Ohgaki H (2009) IDH1 mutations are early events in the development of astrocytomas and oligodendrogliomas. Am J Pathol 174:1149–1153
56. Balss J, Meyer J, Mueller W, Korshunov A, Hartmann C, von Deimling A (2008) Analysis of the IDH1 codon 132 mutation in brain tumors. Acta Neuropathol 116:597–602
57. Yan H, Parsons DW, Jin G, McLendon R, Rasheed BA, Yuan W, Kos I, Batinic-Haberle I, Jones S, Riggins GJ, Friedman H, Friedman A, Reardon D, Herndon J, Kinzler KW, Velculescu VE, Vogelstein B, Bigner DD (2009) IDH1 and IDH2 mutations in gliomas. N Engl J Med 360:765–773
58. van den Bent MJ, Dubbink HJ, Marie Y, Brandes AA, Taphoorn MJ, Wesseling P, Frenay M, Tijssen CC, Lacombe D, Idbaih A, van Marion R, Kros JM, Dinjens WN, Gorlia T, Sanson M (2010) IDH1 and IDH2 mutations are prognostic but not predictive for outcome in anaplastic oligodendroglial tumors: a report of the European Organization for Research and Treatment of Cancer Brain Tumor Group. Clin Cancer Res 16:1597–1604
59. Bortolotto S, Chiado-Piat L, Cavalla P, Bosone I, Chio A, Mauro A, Schiffer D (2000) CDKN2A/p16 inactivation in the prognosis of oligodendrogliomas. Int J Cancer 88:554–557
60. Wolter M, Reifenberger J, Blaschke B, Ichimura K, Schmidt EE, Collins VP, Reifenberger G (2001) Oligodendroglial tumors frequently demonstrate hypermethylation of the CDKN2A (MTS1, p16INK4a), p14ARF, and CDKN2B (MTS2, p15INK4b) tumor suppressor genes. J Neuropathol Exp Neurol 60:1170–1180
61. Watanabe T, Yokoo H, Yokoo M, Yonekawa Y, Kleihues P, Ohgaki H (2001) Concurrent inactivation of RB1 and TP53 pathways in anaplastic oligodendrogliomas. J Neuropathol Exp Neurol 60:1181–1189
62. Watanabe T, Nakamura M, Yonekawa Y, Kleihues P, Ohgaki H (2001) Promoter hypermethylation and homozygous deletion of the p14ARF and p16INK4a genes in oligodendrogliomas. Acta Neuropathol 101:185–189
63. Miettinen H, Kononen J, Sallinen P, Alho H, Helen P, Helin H, Kalimo H, Paljarvi L, Isola J, Haapasalo H (1999) CDKN2/p16 predicts survival in oligodendrogliomas: comparison with astrocytomas. J Neurooncol 41:205–211

64. Ohgaki H, Eibl RH, Wiestler OD, Yasargil MG, Newcomb EW, Kleihues P (1991) p53 mutations in nonastrocytic human brain tumors. Cancer Res 51:6202–6205
65. Buckner JC, Gesme D Jr, O'Fallon JR, Hammack JE, Stafford S, Brown PD, Hawkins R, Scheithauer BW, Erickson BJ, Levitt R, Shaw EG, Jenkins R (2003) Phase II trial of procarbazine, lomustine, and vincristine as initial therapy for patients with low-grade oligodendroglioma or oligoastrocytoma: efficacy and associations with chromosomal abnormalities. J Clin Oncol 21:251–255
66. Viana-Pereira M, Lopes JM, Little S, Milanezi F, Basto D, Pardal F, Jones C, Reis RM (2008) Analysis of EGFR overexpression. EGFR gene amplification and the EGFRvIII mutation in Portuguese high-grade gliomas. Anticancer Res 28:913–920
67. Huang H, Okamoto Y, Yokoo H, Heppner FL, Vital A, Fevre-Montange M, Jouvet A, Yonekawa Y, Lazaridis EN, Kleihues P, Ohgaki H (2004) Gene expression profiling and subgroup identification of oligodendrogliomas. Oncogene 23:6012–6022
68. Dehais C, Laigle-Donadey F, Marie Y, Kujas M, Lejeune J, Benouaich-Amiel A, Pedretti M, Polivka M, Xuan KH, Thillet J, Delattre JY, Sanson M (2006) Prognostic stratification of patients with anaplastic gliomas according to genetic profile. Cancer 107:1891–1897
69. Ducray F, Idbaih A, de Reynies A, Bieche I, Thillet J, Mokhtari K, Lair S, Marie Y, Paris S, Vidaud M, Hoang-Xuan K, Delattre O, Delattre JY, Sanson M (2008) Anaplastic oligodendrogliomas with 1p19q codeletion have a proneural gene expression profile. Mol Cancer 7:41
70. Idbaih A, Criniere E, Marie Y, Rousseau A, Mokhtari K, Kujas M, El Houfi Y, Carpentier C, Paris S, Boisselier B, Laigle-Donadey F, Thillet J, Sanson M, Hoang-Xuan K, Delattre JY (2008) Gene amplification is a poor prognostic factor in anaplastic oligodendrogliomas. Neurooncology 10:540–547
71. Reifenberger J, Reifenberger G, Ichimura K, Schmidt EE, Wechsler W, Collins VP (1996) Epidermal growth factor receptor expression in oligodendroglial tumors. Am J Pathol 149:29–35
72. Wang SI, Puc J, Li J, Bruce JN, Cairns P, Sidransky D, Parsons R (1997) Somatic mutations of PTEN in glioblastoma multiforme. Cancer Res 57:4183–4186
73. Rasheed BK, Stenzel TT, McLendon RE, Parsons R, Friedman AH, Friedman HS, Bigner DD, Bigner SH (1997) PTEN gene mutations are seen in high-grade but not in low-grade gliomas. Cancer Res 57:4187–4190
74. Schmidt EE, Ichimura K, Goike HM, Moshref A, Liu L, Collins VP (1999) Mutational profile of the PTEN gene in primary human astrocytic tumors and cultivated xenografts. J Neuropathol Exp Neurol 58:1170–1183
75. Ichimura K, Schmidt EE, Miyakawa A, Goike HM, Collins VP (1998) Distinct patterns of deletion on 10p and 10q suggest involvement of multiple tumor suppressor genes in the development of astrocytic gliomas of different malignancy grades. Genes Chromosomes Cancer 22:9–15
76. Lin H, Bondy ML, Langford LA, Hess KR, Delclos GL, Wu X, Chan W, Pershouse MA, Yung WK, Steck PA (1998) Allelic deletion analyses of MMAC/PTEN and DMBT1 loci in gliomas: relationship to prognostic significance. Clin Cancer Res 4:2447–2454
77. Sanson M, Leuraud P, Aguirre-Cruz L, He J, Marie Y, Cartalat-Carel S, Mokhtari K, Duffau H, Delattre JY, Hoang-Xuan K (2002) Analysis of loss of chromosome 10q, DMBT1 homozygous deletions, and PTEN mutations in oligodendrogliomas. J Neurosurg 97:1397–1401
78. Fuller CE, Schmidt RE, Roth KA, Burger PC, Scheithauer BW, Banerjee R, Trinkaus K, Lytle R, Perry A (2003) Clinical utility of fluorescence in situ hybridization (FISH) in morphologically ambiguous gliomas with hybrid oligodendroglial/astrocytic features. J Neuropathol Exp Neurol 62:1118–1128
79. Reifenberger G, Louis DN (2003) Oligodendroglioma: toward molecular definitions in diagnostic neuro-oncology. J Neuropathol Exp Neurol 62:111–126
80. Sasaki H, Zlatescu MC, Betensky RA, Ino Y, Cairncross JG, Louis DN (2001) PTEN is a target of chromosome 10q loss in anaplastic oligodendrogliomas and PTEN alterations are associated with poor prognosis. Am J Pathol 159:359–367
81. Ohgaki H, Kleihues P (2005) Population-based studies on incidence, survival rates, and genetic alterations in astrocytic and oligodendroglial gliomas. J Neuropathol Exp Neurol 64:479–489
82. Di Rocco F, Carroll RS, Zhang J, Black PM (1998) Platelet-derived growth factor and its receptor expression in human oligodendrogliomas. Neurosurgery 42:341–346
83. Martinho O, Longatto-Filho A, Lambros MB, Martins A, Pinheiro C, Silva A, Pardal F, Amorim J, Mackay A, Milanezi F, Tamber N, Fenwick K, Ashworth A, Reis-Filho JS, Lopes JM, Reis RM (2009) Expression, mutation and copy number analysis of platelet-derived growth factor receptor A (PDGFRA)

and its ligand PDGFA in gliomas. Br J Cancer 101:973–982

84. Smith JS, Wang XY, Qian J, Hosek SM, Scheithauer BW, Jenkins RB, James CD (2000) Amplification of the platelet-derived growth factor receptor-A (PDGFRA) gene occurs in oligodendrogliomas with grade IV anaplastic features. J Neuropathol Exp Neurol 59:495–503
85. Brennan C, Momota H, Hambardzumyan D, Ozawa T, Tandon A, Pedraza A, Holland E (2009) Glioblastoma subclasses can be defined by activity among signal transduction pathways and associated genomic alterations. PLoS One 4:e7752
86. Verhaak RG, Hoadley KA, Purdom E, Wang V, Qi Y, Wilkerson MD, Miller CR, Ding L, Golub T, Mesirov JP, Alexe G, Lawrence M, O'Kelly M, Tamayo P, Weir BA, Gabriel S, Winckler W, Gupta S, Jakkula L, Feiler HS, Hodgson JG, James CD, Sarkaria JN, Brennan C, Kahn A, Spellman PT, Wilson RK, Speed TP, Gray JW, Meyerson M, Getz G, Perou CM, Hayes DN (2010) Integrated genomic analysis identifies clinically relevant subtypes of glioblastoma characterized by abnormalities in PDGFRA, IDH1, EGFR, and NF1. Cancer Cell 17:98–110
87. Cooper LA, Gutman DA, Long Q, Johnson BA, Cholleti SR, Kurc T, Saltz JH, Brat DJ, Moreno CS (2010) The proneural molecular signature is enriched in oligodendrogliomas and predicts improved survival among diffuse gliomas. PLoS One 5:e12548
88. Shoshan Y, Nishiyama A, Chang A, Mork S, Barnett GH, Cowell JK, Trapp BD, Staugaitis SM (1999) Expression of oligodendrocyte progenitor cell antigens by gliomas: implications for the histogenesis of brain tumors. Proc Natl Acad Sci USA 96:10361–10366
89. Mokhtari K, Paris S, Aguirre-Cruz L, Privat N, Criniere E, Marie Y, Hauw JJ, Kujas M, Rowitch D, Hoang-Xuan K, Delattre JY, Sanson M (2005) Olig2 expression, GFAP, p53 and 1p loss analysis contribute to glioma subclassification. Neuropathol Appl Neurobiol 31:62–69
90. Steck AJ, Perruisseau G (1980) Characterization of membrane markers of isolated oligodendrocytes and clonal lines of the nervous system. J Neurol Sci 47:135–144
91. de la Monte SM (1989) Uniform lineage of oligodendrogliomas. Am J Pathol 135:529–540
92. Tanaka J, Hokama Y, Nakamura H (1988) Myelin basic protein as a possible marker for oligodendroglioma. Acta Pathol Jpn 38:1297–1303
93. Colin C, Baeza N, Tong S, Bouvier C, Quilichini B, Durbec P, Figarella-Branger D (2006) In vitro identification and functional characterization of glial precursor cells in human gliomas. Neuropathol Appl Neurobiol 32:189–202
94. Galloway PG, Likavec MJ, Perry G (1990) Immunohistochemical recognition of ethylnitrosourea induced rat brain microtumors by anti-Leu 7 monoclonal antibody. Cancer Lett 49:243–248
95. Sung CC, Collins R, Li J, Pearl DK, Coons SW, Scheithauer BW, Johnson PC, Yates AJ (1996) Glycolipids and myelin proteins in human oligodendrogliomas. Glycoconjug J 13:433–443
96. Katsetos CD, Del Valle L, Geddes JF, Aldape K, Boyd JC, Legido A, Khalili K, Perentes E, Mork SJ (2002) Localization of the neuronal class III beta-tubulin in oligodendrogliomas: comparison with Ki-67 proliferative index and 1p/19q status. J Neuropathol Exp Neurol 61:307–320
97. Blumcke I, Becker AJ, Normann S, Hans V, Riederer BM, Krajewski S, Wiestler OD, Reifenberger G (2001) Distinct expression pattern of microtubule-associated protein-2 in human oligodendrogliomas and glial precursor cells. J Neuropathol Exp Neurol 60:984–993
98. Kraus JA, Koopmann J, Kaskel P, Maintz D, Brandner S, Schramm J, Louis DN, Wiestler OD, von Deimling A (1995) Shared allelic losses on chromosomes 1p and 19q suggest a common origin of oligodendroglioma and oligoastrocytoma. J Neuropathol Exp Neurol 54:91–95
99. Louis DN, Gusella JF (1995) A tiger behind many doors: multiple genetic pathways to malignant glioma. Trends Genet 11:412–415
100. Hayes HM, Priester WA Jr, Pendergrass TW (1975) Occurrence of nervous-tissue tumors in cattle, horses, cats and dogs. Int J Cancer 15:39–47
101. Holland CT, Charles JA, Smith SH, Cortaville PE (2000) Hemihyperaesthesia and hyperresponsiveness resembling central pain syndrome in a dog with a forebrain oligodendroglioma. Aust Vet J 78:676–680
102. Zaki FA (1977) Spontaneous central nervous system tumors in the dog. Vet Clin North Am 7:153–163
103. Dickinson PJ, Keel MK, Higgins RJ, Koblik PD, LeCouteur RA, Naydan DK, Bollen AW, Vernau W (2000) Clinical and pathologic features of oligodendrogliomas in two cats. Vet Pathol 37:160–167

104. Sumi N, Stavrou D, Frohberg H, Jochmann G (1976) The incidence of spontaneous tumors of the central nervous system of Wistar rats. Arch Toxicol 35:1–13
105. Fitzgerald JE, Schardein JL, Kurtz SM (1974) Spontaneous tumors of the nervous system in albino rats. J Natl Cancer Inst 52:265–273
106. Newman AJ, Mawdesley-Thomas LE (1974) Spontaneous tumours of the central nervous system of laboratory rats. J Comp Pathol 84:39–50
107. Bots GT, Kroes R, Feron VJ (1968) Spontaneous tumors of the brain in rats. A report of 3 cases. Pathol Vet 5:290–296
108. Baker JR, Kippax IS (1980) An oligodendroglioma in a bull. Vet Rec 107:42
109. Swanstrom R, Parker RC, Varmus HE, Bishop JM (1983) Transduction of a cellular oncogene: the genesis of Rous sarcoma virus. Proc Natl Acad Sci USA 80:2519–2523
110. Garcia P, Shoelson SE, Drew JS, Miller WT (1994) Phosphopeptide occupancy and photoaffinity cross-linking of the v-Src SH2 domain attenuates tyrosine kinase activity. J Biol Chem 269:30574–30579
111. Rabotti GF, Grove AS Jr, Sellers RL, Anderson WR (1966) Induction of multiple brain tumours (gliomata and leptomeningeal sarcomata) in dogs by Rous sarcoma virus. Nature 209:884–886
112. Grove AS Jr, Di Chiro G, Rabotti GF (1967) Experimental brain tumors, with a report of those induced in dogs by Rous sarcoma virus. J Neurosurg 26:465–477
113. Bigner DD, Odom GL, Mahaley MS Jr, Day ED (1969) Brain tumors induced in dogs by the Schmidt-Ruppin strain of Rous sarcoma virus. Neuropathological and immunological observations. J Neuropathol Exp Neurol 28: 648–680
114. Kumanishi T, Ikuta F, Yamamoto T (1973) Brain tumors induced by Rous sarcoma virus, Schmidt-Ruppin strain. 3. Morphology of brain tumors induced in adult mice. J Natl Cancer Inst 50:95–109
115. Rabotti GF, Sellers RL, Anderson WA (1966) Leptomeningeal sarcomata and gliomata induced in rabbits by Rous sarcoma virus. Nature 209:524–526
116. Benda P, Lightbody J, Sato G, Levine L, Sweet W (1968) Differentiated rat glial cell strain in tissue culture. Science 161:370–371
117. Takizawa S, Nishihara H (1971) Induction of tumors in the brain, kidney, and other extramammary gland organs by a continuous oral administration of N-nitrosobutylurea in Wistar-Furth rats. Gann 62:495–503
118. Searle CE, Jones EL, Smith WT (1972) Induction of brain tumours in high yield by administration of N-ethyl-N-nitrosourea to newborn rats. Experientia 28:1452–1453
119. Kleihues P, Zulch KJ, Matsumoto S, Radke U (1970) Morphology of malignant gliomas induced in rabbits by systemic application of N-methyl-N-nitrosourea. Z Neurol 198:65–78
120. Stavrou D, Haglid KG, Weidenbach W (1971) The brain specific proteins S 100 and 14.3.2 in experimental brain tumors of the rat. Z Gesamte Exp Med 156:237–242
121. Grossi-Paoletti E, Paoletti P, Pezzotta S, Schiffer D, Fabiani A (1972) Tumors of the nervous system induced by ethylnitrosourea administered either intracerebrally of subcutaneously to newborn rats. Morphological and biochemical characteristics. J Neurosurg 37:580–590
122. Stavrou D, Haglid KG, Ronnback L (1973) The S-100 protein in rat brain and in methylnitrosourea-induced tumors of the rat nervous system. A quantitative study. Eur Neurol 10:168–178
123. Oyasu R, Battifora HA, Clasen RA, McDonald JH, Hass GM (1970) Induction of cerebral gliomas in rats with dietary lead subacetate and 2-acetylaminofluorene. Cancer Res 30:1248–1261
124. Kleihues P, Mende C, Reucher W (1972) Tumours of the peripheral and central nervous system induced in BD-rats by prenatal application of methyl methanesulfonate. Eur J Cancer 8:641–645
125. Eagle H (1955) Nutrition needs of mammalian cells in tissue culture. Science 122: 501–514
126. Norrell HA Jr, Wilson CB (1966) Heterologous transplantation of human glial tumors. Surg Forum 17:431–432
127. Philipson LH, Schwartz NB (1984) Subcellular localization of hyaluronate synthetase in oligodendroglioma cells. J Biol Chem 259: 5017–5023
128. Kennedy PG, Watkins BA, LaThangue NB, Clements GB, Thomas DG (1987) A cultured human oligodendroglioma cell line and herpes simplex virus-infected cells share antigenic determinants. J Neurooncol 4:389–396
129. Post GR, Dawson G (1992) Characterization of a cell line derived from a human oligodendroglioma. Mol Chem Neuropathol 16: 303–317
130. Manuelidis L, Yu RK, Manuelidis EE (1977) Ganglioside content and pattern in human

gliomas in culture. Correlation of morphological changes with altered gangliosides. Acta Neuropathol 38:129–135

131. McLaurin J, Trudel GC, Shaw IT, Antel JP, Cashman NR (1995) A human glial hybrid cell line differentially expressing genes subserving oligodendrocyte and astrocyte phenotype. J Neurobiol 26:283–293

132. Miyake E (1979) Establishment of a human oligodendroglial cell line. Acta Neuropathol 46:51–55

133. Kelly JJ, Blough MD, Stechishin OD, Chan JA, Beauchamp D, Perizzolo M, Demetrick DJ, Steele L, Auer RN, Hader WJ, Westgate M, Parney IF, Jenkins R, Cairncross JG, Weiss S (2010) Oligodendroglioma cell lines containing t(1;19)(q10;p10). Neuro Oncol 12 (7):745–755

134. Holland EC, Hively WP, DePinho RA, Varmus HE (1998) A constitutively active epidermal growth factor receptor cooperates with disruption of G1 cell-cycle arrest pathways to induce glioma-like lesions in mice. Genes Dev 12:3675–3685

135. Holland EC, Varmus HE (1998) Basic fibroblast growth factor induces cell migration and proliferation after glia-specific gene transfer in mice. Proc Natl Acad Sci USA 95:1218–1223

136. Orsulic S, Li Y, Soslow RA, Vitale-Cross LA, Gutkind JS, Varmus HE (2002) Induction of ovarian cancer by defined multiple genetic changes in a mouse model system. Cancer Cell 1:53–62

137. Lewis BC, Klimstra DS, Varmus HE (2003) The c-myc and PyMT oncogenes induce different tumor types in a somatic mouse model for pancreatic cancer. Genes Dev 17: 3127–3138

138. Lewis BC, Klimstra DS, Socci ND, Xu S, Koutcher JA, Varmus HE (2005) The absence of p53 promotes metastasis in a novel somatic mouse model for hepatocellular carcinoma. Mol Cell Biol 25:1228–1237

139. Du Z, Podsypanina K, Huang S, McGrath A, Toneff MJ, Bogoslovskaia E, Zhang X, Moraes RC, Fluck M, Allred DC, Lewis MT, Varmus HE, Li Y (2006) Introduction of oncogenes into mammary glands in vivo with an avian retroviral vector initiates and promotes carcinogenesis in mouse models. Proc Natl Acad Sci USA 103:17396–17401

140. Hughes SH, Greenhouse JJ, Petropoulos CJ, Sutrave P (1987) Adaptor plasmids simplify the insertion of foreign DNA into helper-independent retroviral vectors. J Virol 61:3004–3012

141. Doetsch F, Caille I, Lim DA, Garcia-Verdugo JM, Alvarez-Buylla A (1999) Subventricular zone astrocytes are neural stem cells in the adult mammalian brain. Cell 97:703–716

142. Lindberg N, Kastemar M, Olofsson T, Smits A, Uhrbom L (2009) Oligodendrocyte progenitor cells can act as cell of origin for experimental glioma. Oncogene 28:2266–2275

143. Tchougounova E, Kastemar M, Brasater D, Holland EC, Westermark B, Uhrbom L (2007) Loss of Arf causes tumor progression of PDGFB-induced oligodendroglioma. Oncogene 26(43):6289–96

144. Lindberg N, Uhrbom L (2012) Unpublished data

145. Liu Y, Yeh N, Zhu XH, Leversha M, Cordon-Cardo C, Ghossein R, Singh B, Holland E, Koff A (2007) Somatic cell type specific gene transfer reveals a tumor-promoting function for p21(Waf1/Cip1). EMBO J 26: 4683–4693

146. See WL, Heinberg AR, Holland EC, Resh MD (2010) p27 deficiency is associated with migration defects in PDGF-expressing gliomas in vivo. Cell Cycle 9(8):1562–7

147. See WL, Miller JP, Squatrito M, Holland E, Resh MD, Koff A (2010) Defective DNA double-strand break repair underlies enhanced tumorigenesis and chromosomal instability in p27-deficient mice with growth factor-induced oligodendrogliomas. Oncogene 29:1720–1731

148. Moore LM, Holmes KM, Smith SM, Wu Y, Tchougounova E, Uhrbom L, Sawaya R, Bruner JM, Fuller GN, Zhang W (2009) IGFBP2 is a candidate biomarker for Ink4a-Arf status and a therapeutic target for high-grade gliomas. Proc Natl Acad Sci USA 106: 16675–16679

149. Dunlap SM, Celestino J, Wang H, Jiang R, Holland EC, Fuller GN, Zhang W (2007) Insulin-like growth factor binding protein 2 promotes glioma development and progression. Proc Natl Acad Sci USA 104: 11736–11741

150. Tchougounova E, Jiang Y, Brasater D, Lindberg N, Kastemar M, Asplund A, Westermark B, Uhrbom L (2009) Sox5 can suppress platelet-derived growth factor B-induced glioma development in Ink4a-deficient mice through induction of acute cellular senescence. Oncogene 28:1537–1548

151. Charles N, Ozawa T, Squatrito M, Bleau AM, Brennan CW, Hambardzumyan D, Holland EC (2010) Perivascular nitric oxide activates notch signaling and promotes stem-like

character in PDGF-induced glioma cells. Cell Stem Cell 6:141–152

152. Squatrito M, Brennan CW, Helmy K, Huse JT, Petrini JH, Holland EC (2010) Loss of ATM/Chk2/p53 pathway components accelerates tumor development and contributes to radiation resistance in gliomas. Cancer Cell 18:619–629

153. Momota H, Nerio E, Holland EC (2005) Perifosine inhibits multiple signaling pathways in glial progenitors and cooperates with temozolomide to arrest cell proliferation in gliomas in vivo. Cancer Res 65:7429–7435

154. Karrlander M, Lindberg N, Olofsson T, Kastemar M, Olsson AK, Uhrbom L (2009) Histidine-rich glycoprotein can prevent development of mouse experimental glioblastoma. PLoS One 4:e8536

155. Radner H, el-Shabrawi Y, Eibl RH, Brustle O, Kenner L, Kleihues P, Wiestler OD (1993) Tumor induction by ras and myc oncogenes in fetal and neonatal brain: modulating effects of developmental stage and retroviral dose. Acta Neuropathol 86:456–465

156. Uhrbom L, Hesselager G, Nister M, Westermark B (1998) Induction of brain tumors in mice using a recombinant platelet-derived growth factor B-chain retrovirus. Cancer Res 58:5275–5279

157. Hesselager G, Uhrbom L, Westermark B, Nister M (2003) Complementary effects of platelet-derived growth factor autocrine stimulation and p53 or Ink4a-Arf deletion in a mouse glioma model. Cancer Res 63: 4305–4309

158. Wolf RM, Draghi N, Liang X, Dai C, Uhrbom L, Eklof C, Westermark B, Holland EC, Resh MD (2003) p190RhoGAP can act to inhibit PDGF-induced gliomas in mice: a putative tumor suppressor encoded on human chromosome 19q13.3. Genes Dev 17:476–487

159. Swartling FJ, Ferletta M, Kastemar M, Weiss WA, Westermark B (2009) Cyclic GMP-dependent protein kinase II inhibits cell proliferation, Sox9 expression and Akt phosphorylation in human glioma cell lines. Oncogene 28:3121–3131

160. Johansson FK, Brodd J, Eklof C, Ferletta M, Hesselager G, Tiger CF, Uhrbom L, Westermark B (2004) Identification of candidate cancer-causing genes in mouse brain tumors by retroviral tagging. Proc Natl Acad Sci USA 101:11334–11337

161. Johansson FK, Goransson H, Westermark B (2005) Expression analysis of genes involved in brain tumor progression driven by retroviral insertional mutagenesis in mice. Oncogene 24:3896–3905

162. Bachoo RM, Maher EA, Ligon KL, Sharpless NE, Chan SS, You MJ, Tang Y, DeFrances J, Stover E, Weissleder R, Rowitch DH, Louis DN, DePinho RA (2002) Epidermal growth factor receptor and Ink4a/Arf: convergent mechanisms governing terminal differentiation and transformation along the neural stem cell to astrocyte axis. Cancer Cell 1: 269–277

163. Persson AI, Petritsch C, Swartling FJ, Itsara M, Sim FJ, Auvergne R, Goldenberg DD, Vandenberg SR, Nguyen KN, Yakovenko S, Ayers-Ringler J, Nishiyama A, Stallcup WB, Berger MS, Bergers G, McKnight TR, Goldman SA, Weiss WA (2010) Non-stem cell origin for oligodendroglioma. Cancer Cell 18:669–682

164. Yokoo H, Tanaka Y, Nobusawa S, Nakazato Y, Ohgaki H (2008) Immunohistochemical and ultrastructural characterization of brain tumors in S100beta-v-erbB transgenic rats. Neuropathology 28:591–598

165. Ling BC, Wu J, Miller SJ, Monk KR, Shamekh R, Rizvi TA, Decourten-Myers G, Vogel KS, DeClue JE, Ratner N (2005) Role for the epidermal growth factor receptor in neurofibromatosis-related peripheral nerve tumorigenesis. Cancer Cell 7:65–75

166. Hede SM, Hansson I, Afink GB, Eriksson A, Nazarenko I, Andrae J, Genove G, Westermark B, Nister M (2009) GFAP promoter driven transgenic expression of PDGFB in the mouse brain leads to glioblastoma in a Trp53 null background. Glia 57:1143–1153

167. Niklasson M, Bergstrom T, Zhang XQ, Gustafsdottir SM, Sjogren M, Edqvist PH, Vennstrom B, Forsberg M, Forsberg-Nilsson K (2010) Enlarged lateral ventricles and aberrant behavior in mice overexpressing PDGF-B in embryonic neural stem cells. Exp Cell Res 316:2779–2789

168. Ekstrand AJ, Sugawa N, James CD, Collins VP (1992) Amplified and rearranged epidermal growth factor receptor genes in human glioblastomas reveal deletions of sequences encoding portions of the N- and/or C-terminal tails. Proc Natl Acad Sci USA 89: 4309–4313

169. Bogler O, Wren D, Barnett SC, Land H, Noble M (1990) Cooperation between two growth factors promotes extended self-renewal and inhibits differentiation of oligodendrocyte-type-2 astrocyte (O-2A) progenitor cells. Proc Natl Acad Sci USA 87: 6368–6372

170. Redwine JM, Armstrong RC (1998) In vivo proliferation of oligodendrocyte progenitors expressing PDGFalphaR during early remyelination. J Neurobiol 37:413–428
171. Redwine JM, Blinder KL, Armstrong RC (1997) In situ expression of fibroblast growth factor receptors by oligodendrocyte progenitors and oligodendrocytes in adult mouse central nervous system. J Neurosci Res 50:229–237
172. Aguirre A, Gallo V (2007) Reduced EGFR signaling in progenitor cells of the adult subventricular zone attenuates oligodendrogenesis after demyelination. Neuron Glia Biol 3: 209–220
173. Aguirre A, Dupree JL, Mangin JM, Gallo V (2007) A functional role for EGFR signaling in myelination and remyelination. Nat Neurosci 10:990–1002
174. Lei H, Kazlauskas A (2009) Growth factors outside of the platelet-derived growth factor (PDGF) family employ reactive oxygen species/Src family kinases to activate PDGF receptor alpha and thereby promote proliferation and survival of cells. J Biol Chem 284:6329–6336
175. Nitta M, Kozono D, Kennedy R, Stommel J, Ng K, Zinn PO, Kushwaha D, Kesari S, Furnari F, Hoadley KA, Chin L, DePinho RA, Cavenee WK, D'Andrea A, Chen CC (2010) Targeting EGFR induced oxidative stress by PARP1 inhibition in glioblastoma therapy. PLoS One 5:e10767
176. Holland EC, Celestino J, Dai C, Schaefer L, Sawaya RE, Fuller GN (2000) Combined activation of Ras and Akt in neural progenitors induces glioblastoma formation in mice. Nat Genet 25:55–57
177. Uhrbom L, Dai C, Celestino JC, Rosenblum MK, Fuller GN, Holland EC (2002) Ink4a-Arf loss cooperates with KRas activation in astrocytes and neural progenitors to generate glioblastomas of various morphologies depending on activated Akt. Cancer Res 62:5551–5558
178. Uhrbom L, Nerio E, Holland EC (2004) Dissecting tumor maintenance requirements using bioluminescence imaging of cell proliferation in a mouse glioma model. Nat Med 10:1257–1260
179. Seidler B, Schmidt A, Mayr U, Nakhai H, Schmid RM, Schneider G, Saur D (2008) A Cre-loxP-based mouse model for conditional somatic gene expression and knockdown in vivo by using avian retroviral vectors. Proc Natl Acad Sci USA 105:10137–10142
180. Pao W, Klimstra DS, Fisher GH, Varmus HE (2003) Use of avian retroviral vectors to introduce transcriptional regulators into mammalian cells for analyses of tumor maintenance. Proc Natl Acad Sci USA 100:8764–8769

Neuromethods (2013) 77: 83–140
DOI 10.1007/7657_2012_37

Published online: 5 April 2012

Medulloblastoma Model

Jasmine Lau and William A. Weiss

Abstract

Medulloblastoma, the largest group of embryonal brain tumors, is highly aggressive, with a dismal prognosis for high-risk patients. This disease has historically been classified into five variants based on histopathology. Recent genetic, epigenetic, and transcriptional analyses have categorized tumors into 4–6 subgroups, which are often inconsistent with histopathology subclasses. Mouse models of medulloblastoma are key to improving our understanding of disease progression and developing targeted therapies for patients. In this chapter, we describe different genetically engineered mouse models, and how they fit in the molecular and histopathological subclasses of human medulloblastoma. We discuss the strengths and weaknesses with which these mouse models mirror human disease, highlighting how insights from these models have informed initiation and progression of disease. Lastly, we review targeted therapeutic agent testing in some of these models and discuss future steps and goals in medulloblastoma modeling and research.

Key words: Medulloblastoma, Genetically engineered mouse models, Molecular subgroups, Targeted therapies, Preclinical testing

1. Introduction

Medulloblastoma, a highly aggressive embryonal neuroepithelial tumor of the cerebellum (WHO grade IV tumor), is the most common malignant brain tumor in children (1, 2). Multimodal treatment—surgical resection, radiation, and chemotherapy—has improved overall survival rates significantly over the last decade. Five-year survival rates are now as high as 70–80% in standard-risk patients (3–5). However, this survival benefit is achieved at the cost of long-term treatment-induced morbidities, principally neurocognitive and endocrinological (6–9). Moreover, there is continued risk of secondary tumors, relapse, and metastasis (10). The prognoses for high-risk patients, especially patients less than 3 years of age with leptomeningeal dissemination at presentation, remain dismal at only 25–40% 5-year event-free survival (11, 12).

1.1. Clinical Presentation

Medulloblastoma accounts for about 20% of all pediatric central nervous system tumors (13), with an incidence of about 0.6 per 100,000 children in patients of 0–19 years (14–16). The peak age is about 7 years and is 1.6 times more common in males than females (14, 17). The incidence of medulloblastoma decreases with age: about 70% of cases occur in children younger than 10 years. Incidence progressively decreases below 0.07 per 100,000 persons above 35 years of age (16). Pediatric medulloblastoma commonly arises as a tumor of the midline cerebellar vermis; while in older patients, tumors tend to occur within the cerebellar hemispheres (18, 19).

Current risk stratifications are based broadly on age, degree of surgical resection, and disease spread. About 30% of cases display leptomeningeal dissemination at presentation, the strongest predictor of poor prognosis (20, 21). Further, the delicate proximity of medulloblastoma to the brainstem affects the extent of surgical resection which, together with the degree of tumor spread, is highly predictive of patient outcome (20, 22). Completely resected tumors in patients older than 3 years with no leptomeningeal spread at time of diagnosis are classified as standard risk, while all others are considered high risk (1, 9). Leptomeningeal spread is graded M0–M4, where M0 represents no evidence of metastasis, M1 shows some microscopic cells in the CSF, M2–M4 represents increasing metastasis from the cerebellar space to the spine and outside of the cerebrospinal axis, respectively (2).

Medulloblastoma is a heterogeneous disease comprising of four molecular and five histopathological variants. Improved stratification is required to better predict prognoses and to direct rational treatment strategies that maximize cure and minimize adverse effects. Identification of risk groups also enables development of targeted therapy based on distinct molecular subtypes and should improve clinical trial design and enrollment.

In this chapter, we will first describe the classification of different medulloblastoma subtypes based on their histopathology and molecular differences, and the genetically engineered mouse models that have been developed. We then discuss how well these mouse models mirror human disease, and how these models have contribute to the developmental therapeutics.

2. Classification of Medulloblastoma

2.1. Histopathological Features

Medulloblastoma is classified as a World Health Organization (WHO) grade 4 tumor, defined as a neoplasm that arises from the posterior fossa (23). Medulloblastoma tumor cells are aggressive in nature: they can invade into the cortex and white matter and spread via the cerebrospinal fluid in the leptomeninges and to the spine. The majority of medulloblastoma tumor cells have an undifferentiated stem/progenitor cell appearance. The 2007 WHO classification separates medulloblastoma into five variants based on their histopathological features: (i) classic, (ii) desmoplastic/nodular

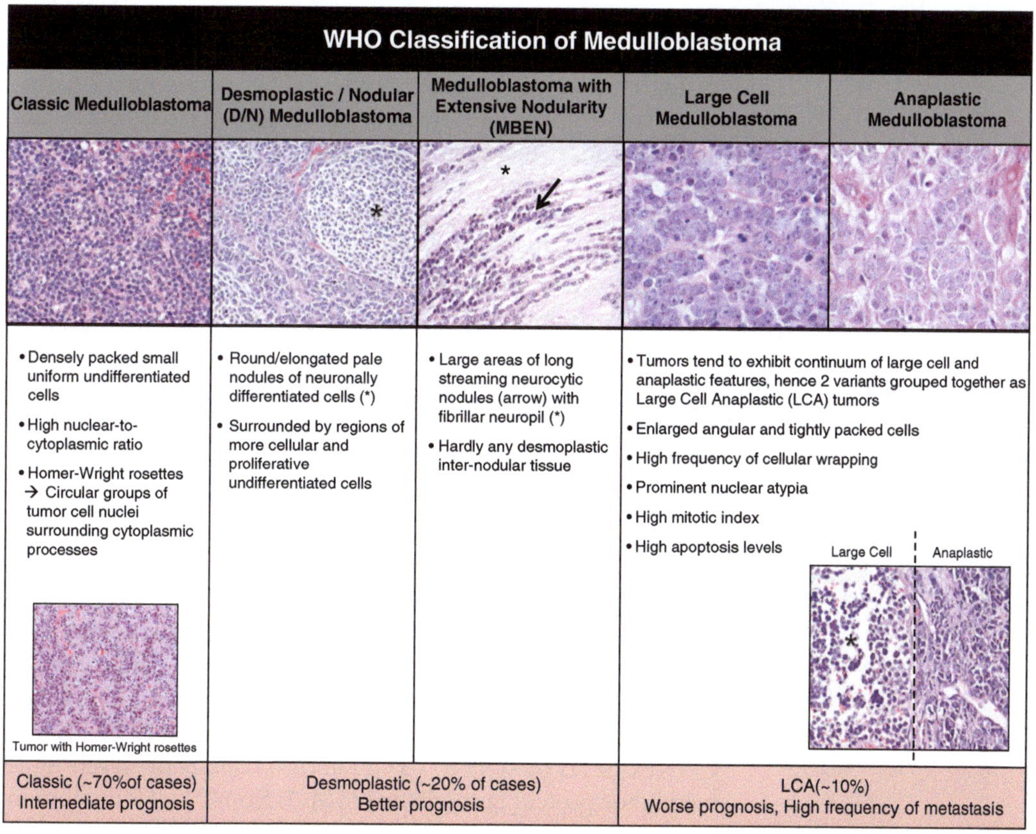

Fig. 1. World Health Organization (WHO) classification of medulloblastoma. Medulloblastoma is divided into five variants based on their histopathological features (from *left*): (i) classic, (ii) desmoplastic/nodular (D/N), (iii) medulloblastoma with extensive nodularity (MBEN), (iv) large-cell medulloblastoma, and (v) anaplastic medulloblastoma. Images adapted from Eberhart, Brain Pathology (2011) and Eliison, Acta Neuropathol (2010).

(D/N), (iii) medulloblastoma with extensive nodularity (MBEN), (iv) large-cell medulloblastoma, and (v) anaplastic medulloblastoma (24) (Fig. 1). D/N and MBEN tumors have better prognoses than classic tumors, while large cell and anaplastic tumors show the worst prognoses with higher frequency of metastasis. Historically, atypical teratoid and rhabdoid tumors (AT/RT) were grouped together with medulloblastomas, as these tumors were also small round blue cell tumors by histology that often develop in the cerebellum. However, AT/RTs are now felt to represent a distinct subgroup of small round blue cell tumors that lack expression of the *hSNF5/INI1* tumor suppressor gene and are distinctly aggressive (25).

Classic subtype tumors (70% of cases) are made of densely packed small uniform undifferentiated cells with high nuclear-to-cytoplasmic ratio. Homer–Wright rosettes, circular groups of tumor cell nuclei surrounding cytoplasmic processes, observed in a subset of classic tumors. Large cell and anaplastic features tend to form a continuum and are thus often grouped together. Large cell anaplastic (LCA) medulloblastoma (10%) have enlarged angular and tightly packed cells, prominent nuclear atypia, a high mitotic index, high levels of apoptosis and cellular wrapping (24, 26). Desmoplastic tumors, comprised of closely related D/N and MBEN subgroups, make up about 20% of all cases. D/N tumors appear as round or elongated pale nodules of neuronally differentiated cells surrounded by regions of more cellular and proliferative undifferentiated cells. MBEN tumors are made up of many large areas of long streaming neurocytic nodules with fibrillar neuropil and barely any desmoplastic internodular tissue (27).

2.2. Chromosomal and Molecular Biomarkers

Several biomarkers showing prognostic significance have been identified for specific clusters of tumors with some also validated in clinical trials.

1. SHH pathway activation

The Sonic Hedgehog (SHH) signaling pathway is required for normal cerebellar development (24). During cerebellar development, a progenitor population, cerebellar granule neuronal precursors (CGNPs), proliferates and migrates across the top of the cerebellar anlage to form the external granule layer (EGL) (Fig. 2a). Large Purkinje neurons lying deep in the EGL secrete SHH, a ligand that binds its receptor Patched (PTCH1) on CGNPs. SHH binding to PTCH in turn releases the seven-transmembrane protein Smoothened (SMO) from PTCH1 inhibition. SMO subsequently activates a cytoplasmic intermediate protein complex, which results in release of the GLI family of transcription factors from their suppressor complexes that include suppressor of fused (SUFU) (13, 24). GLI transcription factors initiate downstream target gene expression and activity, stimulating the CGNPs to proliferate in the EGL (Fig. 2b). The CGNPs then terminally differentiate and migrate inwards, leaving behind a cell-sparse molecular layer, and granule-neuron dense internal granular layer (IGL). The EGL will thus cease to exist by postnatal day 15 (P15) in mice and by 1 year in humans (28–30).

SHH pathway was first implicated in medulloblastoma when germline inactivating mutations of *PTCH1* were found to cause Gorlin's Syndrome (nevoid basal cell carcinoma) (31). These patients also showed increased incidences of rhabdomyosarcoma and medulloblastoma, suggesting that inappropriate regulation of SHH signaling in cerebellar cells contributed to medulloblastoma formation (32–34).

In about 8% of sporadic medulloblastoma cases, SHH signaling is activated due to mutational inactivation of *PTCH1* (34–37). Subsequently, mutations in downstream SHH pathway components *SMO* and *SUFU* were also identified in sporadic medulloblastoma (38). The majority of SHH-dysregulated medulloblastomas present as desmoplastic or classic histopathology. Altogether, these tumors contribute to 25–30% of all cases, with desmoplastic tumors in particular associated with better prognoses (24, 39). Recently, large-scale molecular profiling has recently identified high-level amplification of additional SHH downstream effector transcription factors in a subset of medulloblastomas. This includes *GLI1* and *GLI2*, miR17-92 complex, and *YAP1*, further cementing the importance of SHH signaling in medulloblastoma tumorigenesis (40–43).

2. WNT pathway activation

   The canonical WNT pathway is activated upon the binding of WNT ligands to Frizzled receptors, activating Disheveled. Activated Disheveled inhibits an AXIN/GSK-3β/APC protein complex that, in its basal state, promotes degradation of β-catenin. WNT-driven inhibition of this complex therefore stabilizes β-catenin, enabling its transport from the cytoplasm into the nucleus to activate downstream signaling (13) (Fig. 2c).

   The WNT signaling pathway was first implicated in medulloblastoma from observations that a subset of patients with germline mutations in tumor suppressor *APC* (Turcot syndrome) developed medulloblastomas (44, 45). Subsequently, 5–10% of patients with sporadic medulloblastoma were also shown to harbor activating point mutations in the β-catenin gene *CTNNB1*, resulting in a degradation-resistant form of β-catenin that aberrantly activated downstream WNT signaling (46, 47). Other abnormalities found in some WNT tumors include promoter methylation (and hence gene silencing) of the secreted frizzled-related protein 1 (SFRP) family of WNT inhibitors and monosomy 6 (14, 17). SFRP1, SFRP2, and SFRP3 inhibit WNT signaling by binding to WNT ligand and sequestering WNT from its receptor Frizzled, hence epigenetic silencing of *SFRP* genes releases the inhibition of WNT signaling (48). Monosomy 6 is a cytogenetic marker that identifies almost all WNT tumors and tumors with monosomy 6 and β-catenin mutations associated with favorable prognosis (17, 24, 49). Comparisons between a β-catenin mouse model and human data identified a common deletion in syntenic region at human 6q25.3 encoding *TULP4*, which may be the candidate gene commonly deleted in the WNT tumors with monosomy 6 that is important for tumorigenesis (50). *TULP4* is related to the tubby gene family and is associated with neuronal cell apoptosis. Altogether, the WNT-associated tumors account for about 18–25% of all cases and are usually of classic histology (47, 51). Patients with tumors displaying nuclear β-catenin positivity have significantly better survival (92.3%) compared to patients with tumors that are immuno-negative (65.3%), suggesting that activation of WNT, like activation of SHH (which collectively denote a minority of medulloblastoma tumors), correlates with a relatively favorable outcome (47, 51).

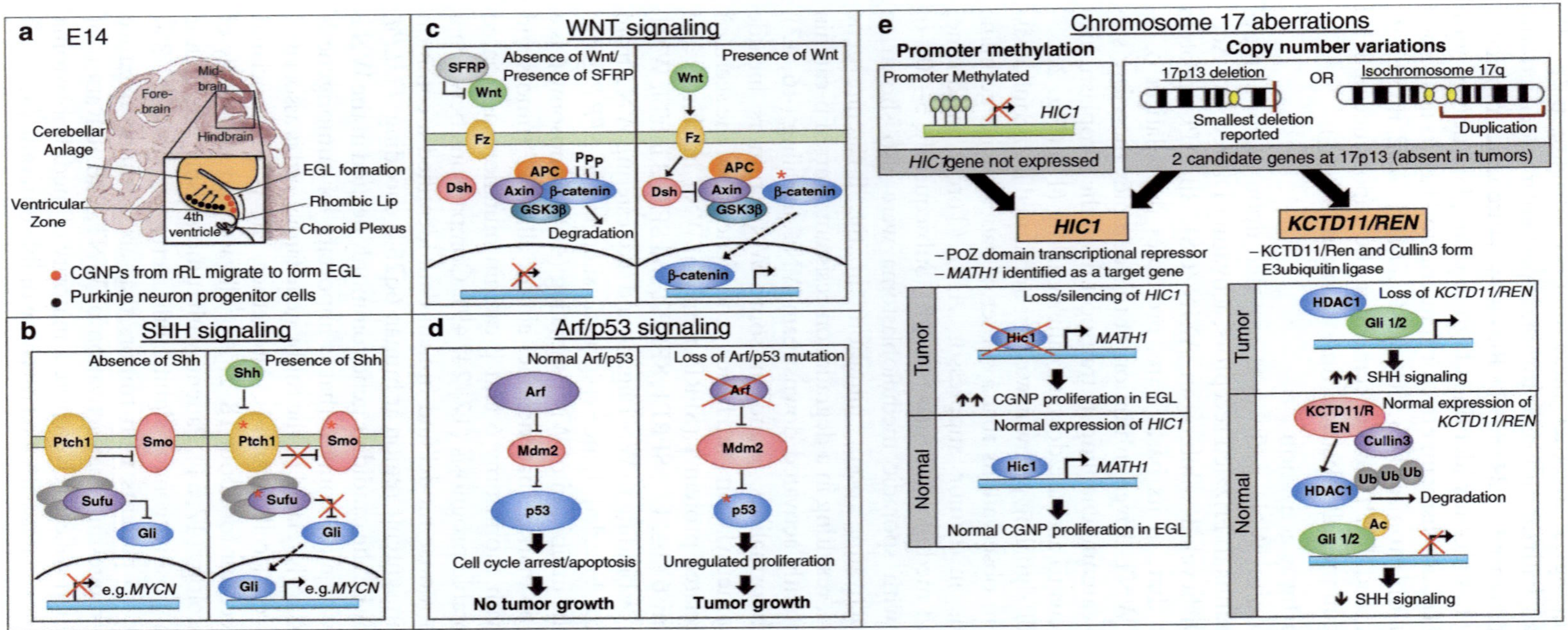

Image from (a) adapted from Schüller et al., Cancer Cell (2008)

Fig. 2. Signaling pathways and chromosomal aberrations in medulloblastomagenesis. (**a**) Development of the murine cerebellar anlage at E14. Cerebellar granule neuron precursors (CGNPs, in *red*) arise from the rhombic lip at the posterior edge of the cerebellar anlage and begin to migrate over the top surface of the anlage to form the external granule layer (EGL). Purkinje neuron precursors (in *black*) arise from the ventricular zone adjacent to the fourth ventricle, and migrate radially upwards through the anlage towards the EGL. Purkinje neurons are the source of Sonic Hedgehog (SHH) ligand, which activates SHH signaling in CGNPs. Image adapted from Schüller et al., Cancer Cell (2008). (**b**) SHH signaling pathway. In the absence of SHH ligand, Patched (PTCH1) inhibits Smoothened (SMO) to prevent downstream signaling. Suppressor of Fused (SUFU) is a negative regulator of GLI transcription factor, binding to and preventing the translocation of GLI to the nucleus, where it activates downstream target genes. SHH binding to PTCH releases inhibition of SMO, enabling GLI-driven transcription. Activation of SHH signaling in CGNPs stimulates their proliferation in the EGL. *Red asterisks* indicate members of the pathway that have been found mutated in human medulloblastomas. (**c**) WNT signaling pathway. In the absence of WNT ligand, or when SFRP family of WNT inhibitors sequesters WNT from its receptor Frizzled (Fz), the AXIN/GSK-3β/APC complex phosphorylates β-catenin, resulting in its degradation. Hence, downstream target genes are not activated. WNT ligand binds to Fz, leading to the activation of Disheveled (Dsh). Activated Dsh inhibits the AXIN/GSK-3β/APC complex, allowing β-catenin to enter the nucleus to activate downstream signaling. Activating (stabilizing) mutations of β-catenin are often found in medulloblastoma patients with activated WNT signaling [as indicated by *red asterisk*] (**d**) p53 pathway. In response to stress such as DNA damage, the transcription factor p53 can activate target genes that induce cell cycle arrest or apoptosis of the damaged cell. In the nucleolus, the Arf tumor suppressor binds to and inhibits Mdm2, which is a negative regulator of p53. Mdm2 binds to p53 to directly inhibit p53 transcriptional activity, and also has E3 ubiquitin ligase activity, targeting p53 for export out of the nucleolus into the cytoplasm for degradation. However, in tumor cells with Arf loss, Arf loss releases inhibition on Mdm2, which negatively regulates p53. This results in unchecked proliferation that leads to genetic instability in tumor cells. Mutations in p53 (*asterisk*) are also commonly found to result in failure to induce cell cycle arrest or apoptosis in these tumor cells. (**e**) Chromosome 17 aberrations. Copy number variations in chromosome 17 are present in a significant number of medulloblastoma, typically due to the formation of isochromosome 17q from the duplication of the 17q arm. This results in the gain of 17q, with a corresponding loss of 17p. The smallest deletion reported is at 17p13. Two candidate genes have been identified: Hypermethylated in Cancer 1 (*HIC1*) and *KCTD11/REN*. In addition to being deleted, *HIC1* is also frequently hypermethylated at its promoter and its expression is hence silenced in tumors. HIC1 is a transcriptional repressor and *MATH1* has been identified as a target gene. Hence, absence of HIC1 permits *MATH1* expression and induces proliferation of the CGNPs. KCTD11/REN cooperates with Cullin3 target HDAC1 for degradation. HDAC1 deacetylates Gli1 to permit SHH signaling.

If the association of SHH and WNT with outcome holds up, such patients may benefit from milder therapies that will reduce debilitating treatment-associated side effects.

3. *MYC/MYCN* overexpression and amplification

   Amplification of *MYC* genes (*c-MYC* and *MYCN*) correlate strongly with poor prognosis and are often found in tumors with large cell anaplastic histopathology (52). Amplification of *c-MYC* has been reported in 5–15% of all cases, with a small subset reported to have high-level amplification of *c-MYC* of >4 copies (52–54). Similarly, *MYCN* amplification has been found in about 10% of cases (52–54).

   MYCN is a downstream target of the SHH pathway and *MYCN* expression levels are high in SHH-driven tumors (55–57). *MYCN* is the principle *MYC* gene that drives proliferation during cerebellar development, with little *MYCN* detectable after development is complete (57). MYCN levels are elevated in the majority of medulloblastoma tumors, irrespective of subtype, and overexpressing *Nmyc* in premalignant CGNP cells that harbor partial *PTCH1* loss result in robust tumor formation (58). Further, overexpression of a *MYCN* transgene to the hindbrain of mice result in formation of medulloblastoma with classic and LCA histology. Over 90% of tumors that arose were demonstrated to be SHH-independent, except one mouse that harbored *PTCH1* loss (59). This suggests that MYCN overexpression is sufficient to produce medulloblastoma that is both SHH dependent and independent.

4. ARF/TP53 pathway

   Li-Fraumeni patients with germline mutations in *TP53* have increased risk of cancers including medulloblastoma (60). *TP53* is found on chromosome 17p and was initially the prime candidate involved in chromosome 17 aberrations. However, *TP53* loss or mutation contributes only up to 10–15% of all medulloblastoma (61, 62). Other members of the ARF/TP53 tumor suppressor pathway have also been implicated in medulloblastoma. Loss of *ARF* due to homozygous deletion or promoter hypermethylation (with wild-type TP53) has been reported in about 10% of tumors (61). An analysis of a small subset of human tumors found that *TP53* mutations, methylation, and deletions of *ARF* are common in more aggressive LCA tumors, demonstrating that ARF/TP53 is important in promoting tumor malignancy (61).

5. Chromosome 17 aberrations

   The only frequent chromosome aberration found in medulloblastoma is isochromosome 17q, present in 30–50% of cases (4, 17). Isochromosome 17q is formed when two copies of 17q arms and two copies of centromeric 17p are fused together to form an especially long chromosome with two centromeres in place of a normal chromosome 17. This results in one copy of 17p and three copies of 17q, leading to the proposition that genes encoding for oncogenes may be present on 17q while genes encoding tumor suppressors are present on 17p. 17p losses or 17q gains have been reported as well, with loss of 17p present in up to 30–40% of cases (63–65). Altogether, copy number variations in chromosome 17, including 17q gains, isochromosome i17q, and 17p deletions, have been associated with poor prognosis (52).

The smallest region of deletion reported at 17p is at 17p13.3. Two candidate tumor suppressors have been identified at 17p13 that likely play a role in medulloblastoma tumorigenesis. These are namely Hypermethylated in Cancer 1 (HIC1) and REN^{KCTD11} (66–69) (Fig. 2e). *HIC1* maps exactly at 17p13.3, the region of deletion most commonly found in medulloblastoma. HIC1 is a POZ domain transcriptional repressor that is frequently hypermethylated at its promoter and thus silenced in medulloblastoma tumors (68, 70, 71). REN^{KCTD11}, working with Cullin3, forms an E3 ubiquitin ligase complex that degrades HDAC1, which is responsible for deacetylating GLI1 and GLI2 to permit SHH-downstream signaling (72). Thus, loss of REN^{KCTD11} during 17p deletion can result in increased deacetylated GLI1/2 to enhance SHH signaling in SHH-driven tumors.

6. Other epigenetic mechanisms

Hypermethylation of promoter regions of other tumor suppressor genes have been identified. *RASSF1A* (ras association domain family protein 1, isoform A) is a tumor suppressor gene whose promoter is hypermethylated in a number of adult cancers leading to epigenetic gene silencing (73, 74). *RASSF1A* is involved in cell cycle control, microtubule stabilization, cell adhesion, motility, and apoptosis (75). In medulloblastomas, promoter-associated CpG island of *RASSF1A* was found to be extensively hypermethylated in both alleles, resulting in epigenetic silencing of *RASSF1A* (76). Biallelic promoter hypermethylation of tumor samples were of all histological variants and of both adult and pediatric patients, while non-tumor samples demonstrated no methylation evidence.

SPINT2 (serine protease inhibitor kunitz-type 2) is another tumor suppressor gene that acts as an inhibitor of HGF/MET signaling. *SPINT2* expression was found to be silenced via promoter methylation in 34% of primary tumor samples examined (77). In addition, some tumors harbored hemizygous deletions in the *SPINT2* locus and gains in regions of *HGF* and *MET* loci. Altogether, *SPINT2* expression was downregulated in 73% of tumors, while *MET* expression was upregulated in 45% of tumors. Re-expression of *SPINT2* in medulloblastoma cell lines resulted in reduced cell proliferation and motility, and improved survival in vivo in a xenograft system, supporting the potential role of SPINT2 in medulloblastoma tumorigenesis (77).

Other genes reported to be genetically and epigenetically altered include Kruppel-like Factor 4 (*KLF4*) and Dickkopf-1 (*DKK1*). *KLF4* expression is absent in 40% of tumor samples; CpG methylation of *KLF4* promoter occurs in about 16% of tumors, while recurrent homozygous deletions were found in rare cases (78). Accordingly, introducing demethylating agent 5-azacytidine (5-azaC) to medulloblastoma cell lines results in the re-expression of *KLF4*. *DKK1* is a WNT antagonist whose expression is downregulated in primary samples compared to normal cerebellum (79). Treatment of medulloblastoma cells with histone deacetylase (HDAC) inhibitor results in *DKK1* upregulation, suggesting that histone acetylation may down-regulate *DKK1* expression, reducing its tumor suppressor functions in medulloblastoma.

Caspase-8 is involved in mediating TRAIL receptor-induced apoptosis, and it was reported that caspase-8 (*CASP8*) promoter region is hypermethylated in 55% of primitive neuroectodermal tumors/medulloblastomas (73). However, while *CASP8* expression was reduced in half of tumor samples, expression of related members in the pathway like TRAIL receptor *DR4*, the adapter protein *FADD* and caspase-3 expression was unchanged in tumor samples compared to normal brain samples. This result suggests that downregulation of caspase-8 is particularly critical in permitting medulloblastoma tumor cell growth.

Investigations also suggest a link between histone lysine methylation and medulloblastoma tumorigenesis. In one study, SNP analysis revealed aberrant genetic events that target the methylation of repressive chromatin mark on histone H3, lysine 9 (H3K9) (80). Homozygous deletions in histone lysine methyltransferase *EHMT1* and *SMYD4* and hemizygous or homozygous deletions of H3K9-interacting Polycomb genes *L3MBTL3*, *L3MBTL2*, and *SCML2* also occur. On the other hand, amplification and gains of histone lysine demethylases *JMJD2C*, *JMJD2B* and histone lysine acetyltransferase *MYST3* that regulate H3K9 methylation were found in a subset of tumors. Thus, the authors suggest that these gene aberrations may lead to H3K9 hypomethylation, contributing to the transformation in medulloblastoma. Recently, in another sequencing effort, inactivating mutations of the histone-lysine *N*-methyltransferase genes *MLL2* or *MLL3* were also identified in a small subset (16%) of patients, suggesting role of these putative tumor suppressor genes in medulloblastoma tumorigenesis (81).

7. Other pathways implicated in medulloblastomagenesis
Notch signaling is activated in a subset of medulloblastoma, via overexpression or amplification of ligands NOTCH1 and NOTCH2, as well as receptors HES1 and HES5 (82, 83), or via silencing of miR199b-5p, a negative regulator of NOTCH (84). PI3K pathway has also been implicated in medulloblastoma tumorigenesis. Mutations in the catalytic subunit *PIK3CA* are found in about 5% of cases (85). Deletions or mutations in the opposing phosphatase *PTEN* (a negative regulator of PI3K) occurs in 30–35% of cases (86).

Developmental genes *OTX2* and *TRKC* have been associated with medulloblastoma as well (87). OTX2, a transcription factor expressed in the posterior cerebellum in the EGL and IGL and important in normal cerebellar development, is commonly amplified particularly in non-SHH, non-WNT tumor cases and those of anaplastic histology (88–90). TRKC, a receptor for neurotrophic factor neurotrophin-3 that is widely expressed by granule cells in the cerebellum during development, is expressed in a subset of tumors and correlated with favorable prognosis, as TRKC activation promotes apoptosis and hence inhibits tumor growth (87, 91, 92). Cell cycle factor CDK6 is overexpressed in a subset and is correlated significantly with poor prognosis (93).

2.3. Molecular Profiling

Several analyses of medulloblastoma tumor samples have attempted to subclassify this disease based on transcriptomal profiles (17, 94–96). These unbiased analyses from human tumor samples reveal some parallels that can be drawn between these subclassification and the molecular lesions discussed above.

Two subclasses of tumors, with dysregulated WNT or SHH signaling, have been identified consistently (17, 94–96) (Fig. 3). The clustering of WNT and SHH tumors away from other medulloblastomas suggests that these tumors arise and develop differently from other subtypes. WNT subtype of tumors demonstrates overexpression of WNT signaling pathway members, and these tumors typically harbor monosomy 6 and activating mutations in β-catenin gene *CTNNB1*. WNT tumors tend to show classic pathology and patients are associated with favorable prognosis. Demographic analyses show that patients are usually older children more than 3 years (with median of about 9–10 years)(17, 94). High MYC levels and upregulated NOTCH and PDGF signaling have also been associated with this tumor subclass (17, 94).

SHH tumor subclass demonstrates overexpression of genes in SHH signaling pathway such as *GLI1* and *MYCN*, and *PTCH1* mutations are exclusively found in these tumors (17, 94). In addition, 9q deletions are also exclusively found in these tumors, with corresponding 9p gains (forming i9p isochromosome)(17, 94, 96). Demographic analyses reveal that patients tend to be very young

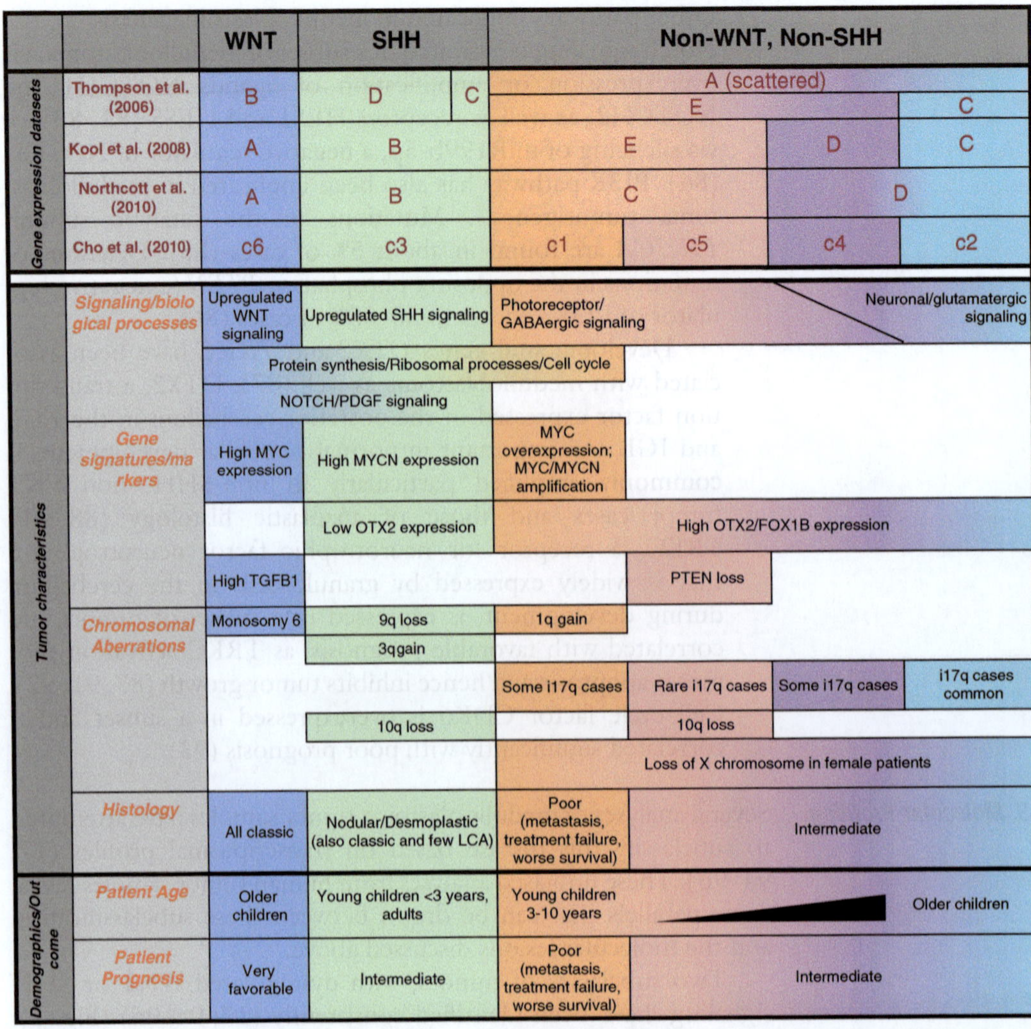

Fig. 3. Subclassification of medulloblastomas based on transcriptional profiling. Expression profiling from four independent groups has identified WNT and SHH tumors as distinct subgroups, while subclassifying non-WNT, non-SHH tumors into 2–4 subgroups. Listed are some common features of the subgroups based on tumor characteristics and patient demographics and outcome.

children (<3 years) or adults, and these patients tend to show favorable prognoses (17). Desmoplastic tumors largely segregate in SHH subclass, although some desmoplastic tumors are found in the other subclasses (17). Similar to WNT tumors, SHH tumors have been associated with upregulated NOTCH and PDGF signaling (94). Interestingly, oncogene *OTX2*, which is overexpressed in a large majority of medulloblastomas, is lowly or not expressed in SHH tumors (94).

Independent groups have classified non-WNT, non-SHH tumors into 2–4 subgroups (Fig. 3). These tumors are hard to subclassify as they are more closely related. Gene expression studies have generally identified neuronal and photoreceptor markers as cluster determinants. Thompson and colleagues were the first to conduct gene expression arrays on medulloblastoma patient samples, and they classified the tumors into five subclasses A–E, of which groups B and D were the WNT and SHH subclasses, respectively (95). Of the other groups, they observed that deletion of 17p (hence i17q isochromosome) was found primarily in group C tumors, and anaplastic cases were most common in group E tumors.

A subsequent study by Kool et al. classified WNT and SHH tumors as groups A and B respectively, and divided the non-WNT, non-SHH tumors into three other clusters (94). Cluster determinants used include expression of neuronal differentiation (groups C and D) or retinal photoreceptor genes (groups D and E). Notably, metastatic disease was strongly associated with these non-WNT, non-SHH tumors, and most strongly with group E tumors. Additionally, female cases within these non-WNT, non-SHH tumors had a high incidence of loss of inactivated X chromosome. Other observations were that groups C and D weakly expressed protein synthesis and cell cycle-related genes compared to all other groups, and chromosome 17 aberrations were strongly associated with groups C or D tumors. Lastly, patients with group D or E tumors were mostly young children (<3 years), similar to SHH tumors.

A later study by Northcott et al. identified four subclasses of medulloblastomas, of which the WNT and SHH subclasses also segregated distinctly (17). The non-WNT, non-SHH tumor subgroups, groups C and D, showed expression of axon guidance and neuronal developmental genes, and also had high expression of *OTX2* and *FOXG1B*. Group C/D tumors also harbored i17q chromosomal aberrations that were not found in WNT or SHH tumors. Notably, several of these observations mirrored findings by Kool et al. in their classification of subgroups C to E (94).

Analysis by Northcott et al. also revealed that group C tumors often amplified and/or overexpressed c-MYC, had the poorest prognosis, presented with frequent metastatic dissemination, and were represented only in young children (17). This group was also associated with genes involved in phototransduction and glutamate signaling. On the other hand, Group D tumors with no expression of *MYC* or *MYCN*, and the patients were more distributed across all age groups, particularly in adolescents. Group D tumors were associated with neuronal semaphorin signaling, G-protein coupled receptors and cyclic AMP signaling.

Lastly, the most recent study by Cho et al. analyzed the largest number of tumor samples, characterizing the mRNA and miRNA expression profiles, as well as conducted single nucleotide polymorphism (SNP) analysis for chromosomal changes (96). Six subgroups of medulloblastomas were identified, of which the WNT (c6) and SHH (c3) subgroups clustered similarly to the other published reports. Similar to previous studies, chromosome 17 aberrations were not found in the WNT or SHH subgroups.

Of the non-WNT, non-SHH subgroups, c1 and c5 were similar to each other, displaying photoreceptor/GABAergic signatures, while c2 and c4 showed neuronal/glutamatergic signatures. Comparing between c1 and c5 subclasses, upregulation/amplification of c-MYC and translational/ribosomal signatures were specific to c1. Importantly, c1 subgroup, but not c5, was associated with poor clinical outcome. Some of the signatures that distinguish c2 and c4 subclasses are as below: c2 subgroup was observed to upregulate immune-associated markers such as tumor necrosis factor α, interferon-β, and nuclear factor-κB, while c4 subgroup tumors were more enriched in *MYC* and *MYCN*. c4 subgroup was also striking in that although the tumors expressed neuronal/glutamatergic markers, they also expressed some photoreceptor markers. Recall that this result mirrors Kool's classification of group D tumors (94).

A comparison of the broad histopathological WHO classification of classic, desmoplastic and large cell/anaplastic tumors with the molecular/transcriptional subclasses draws some parallels, but also highlights some inconsistencies in these subclassification schemes. Classic tumors are the most common subtype, but histology is not informative as these tumors span across all molecular subtypes. Desmoplastic tumors are indeed most commonly found in the SHH tumor group, but importantly, are also found in non-WNT, non-SHH tumors. LCA histopathology in tumor samples does not offer much insight, as LCA tumors can occur in all molecular subgroups; interestingly, the WNT LCA tumors have been reported to have good prognosis while a subset of non-WNT, non-SHH LCA tumors has worse prognosis. This is in line with the correlation of WNT tumors (regardless of histopathology) with good prognosis, and the correlation of high levels of *MYC/MYCN* (mostly group C tumors in Northcott's analysis) with poor prognosis. Altogether, we need to employ a combination of various prognostic markers from both histopathological and molecular classifications to better stratify patients into risk groups.

3. Modeling Medulloblastoma

Research groups and pharmaceutical companies are continually developing a wide spectrum of compounds that inhibit potential targets in tumor biology. However, a big challenge is finding the most appropriate platform to screen for effective drugs against different subclasses of medulloblastoma. The majority of preclinical testing is done in tumor-derived cell lines, which are easy to use in screening assays. However, cells grown in plastic or matrix-coated plates lack cellular heterogeneity and tumor–stromal interactions and thus do not faithfully recapitulate in vivo settings.

Transplantation of human or murine cells into immunodeficient or immunocompetent mice is a popular in vivo tumor model system as highly penetrant and reproducible tumors can be induced. However, the tumors that develop do not recapitulate the sequential acquisition of changes in both tumor and tumor microenvironment changes that both moderate aspects of malignant progression. Within this context, orthotopic transplantations into immunocompetent mice are closer representations of the tumor setting over subcutaneous injections or use of immunodeficient mice, as tumors are formed in the appropriate tissue context and allow for antitumor adaptive immune responses. Also, direct implantation of fresh tumors or minimally passaged cells has been used to reduce shortcomings of the xenograft/allograft system. Regardless, results generated from these models should still be evaluated with caution.

The use of genetically engineered mouse (GEM) models provides the most physiologically relevant insight into human tumor settings, albeit not without its own limitations. GEM models initiate tumorigenesis via introduced perturbations in molecular pathways relevant to the disease. Evolving with stromal cells in the microenvironment, the tumor is more likely to mimic genetic and molecular changes and exhibit angiogenic and invasive characteristics representative of human tumors.

Historically, the first models of medulloblastoma were derived in the 1970–1980s by inoculating viral agents such as adenoviruses and polyoma viruses into rodents early postnatally during cerebellar development. One of the early successes was the use of Simian adenovirus SA7 inoculated intracerebrally in newborn hamsters that resulted in brain tumors in 22% of animals in 1–2 months, most of them resembling human medulloblastomas (97, 98). Similarly, JC human polyoma virus introduced in hamsters resulted in medulloblastomas in 3–6 months in 80–95% of animals (99, 100). The tumors were found to originate from the EGL which then migrated to the IGL (100), providing one of the first evidence in animal models of the origin of medulloblastoma. Transgenic mice engineered to express JC virus T antigen (viral early regulatory protein) also resulted in medulloblastomas in 9–13 months (101). These tumors were resembled human tumors of classic histology and presented occasional Homer–Weight rosettes, demonstrating the importance of the early viral T antigen in medulloblastoma formation. These models will not be elaborated upon in this chapter. Instead, we will focus on the more recent mouse models that have been developed using introduction or deletion of specific genes to better mimic the molecular heterogeneity of the human tumors (Table 1).

Table 1
Mouse models of medulloblastoma

Mouse model	Tissue specific	MB tumor incidence; Latency (Lat)	Histology/tumor features	Immunoreactivity/ gene expression	Other tumors or defects observed	Molecular/chromosomal changes, other characteristics	Reference(s)
SHH-driven models							
SHH pathway—germline-modified models							
Ptch1$^{+/-}$	Ubiquitous	14–19%; Lat: 10 months (tumors seen as early as 5 weeks, peaks at 16–24 weeks)	Classic EGL expansion and compression of normal cerebellum Circumscribed, no infiltration of parenchyma/white matter	Neurofilament, synaptophysin, GFAP, reelin, trkC	Sarcomas, hydrocephalus, hindlimb polydactyly or syndactyly, larger body size	Contradicting reports in retention of remaining WT *Ptch1* or LOH or mutations of remaining *Ptch1* High Gli1 expression	(102–106)
Single dose 3 Gy X-ray	Ubiquitous	P1 has decreased incidence from 81% to 3% compared to P10 pups (compared to 6.7% non-irradiated *Ptch1*$^{+/-}$ mice); Lat: 7–20 weeks P4 pups: 51% compared to 7.7% non-irradiated *Ptch1*$^{+/-}$ mice; Lat: 38 weeks 3-month mice: no increase in tumor incidence	Similar to *Ptch1*$^{+/-}$ non-irradiated tumors	N.D.	Sarcomas, lymphomas, BCCs	P1 pups have lower apoptosis, increased pro-survival Akt and decreased p53 with IR, compared to P10 LOH of remaining *Ptch1*	(107, 108)
Ptch1$^{+/-}$; *Ptch2*$^{\pm/-}$	Ubiquitous	15.4–17.4% compared to 6.4% in *Ptch*$^{+/-}$ mice; Lat: 1–10 months	Similar to *Ptch1*$^{+/-}$ tumors	GFAP, Tuj1, synaptophysin	Sarcomas, BCCs, angiectasis	Overexpression of *Gli1/2*, *Math1*, *Sfrp1* LOH of remaining *Ptch1*	(109)

Ptch$^{+/-}$; *p53*$^{-/-}$	Ubiquitous	95–100% compared to 14% in *Ptch*$^{+/-}$ mice; Lat: 4–16 weeks	Similar to *Ptch1*$^{+/-}$ tumors	N.D.	Sarcomas, lymphomas, BCCs	WT *Ptch1* retained in tumors Elevated *Gli1* expression Random chromosomal losses	(110, 111)
Conditional *Ptch1* homozygous deletion	*hGFAP-Cre*, *Math1-Cre*	100%; Lat: 1–3 months (*hGFAP*: 3–4 weeks; *Math1*: 2–3 months)	Similar to conventional *Ptch1*$^{+/-}$ tumors, disrupted normal cerebellar architecture	Math1, NeuN, synaptophysin, GABRA6	None	N.D.	(119)
Conditional *Ptch1* homozygous inactivation; conditional *p53* deletion	*Pax7-Cre*	30%; (p53 WT: Lat: 88–100 days; p53 partial/complete deletion: 32–65 days)	Similar to Ptch1$^{+/-}$ tumors	Tuj1, NeuN, synaptophysin	None	Invaded subarachnoid space, 50% of mice show spread to fourth ventricle	(122)
CAGGS-CreER; R26-SmoM2 (ubiquitous)	Ubiquitous	40%; Lat:18 weeks	Classic	NeuN, Zic1	Rhabdomyosarcomas, BCCs, pancreatic and gastrointestinal lesions	Upregulated Shh pathway members, like *Ptch1/2*, *Gli1 and Cyclin D1*. Upregulated PDGFRα, IGF and MAPK pathways, *c-Myc*, *N-myc*, *Otx2*, *Ezh2*, *Sox18*, *Foxm1*	(117)
Conditional *SmoM2* activation	*hGFAP-*, *Math1-*, *Olig2-*, *Tlx3-Cre*	100%; Lat: 1–2 months (*hGFAP*: shorter survival)	Classic *hGFAP* and *Math1-Cre*: diffuse EGL tumors *Olig2* and *Tlx3-Cre*: focal tumors in posterior lateral EGL lobes	GFAP, Tuj1, synaptophysin, Math1, Zic1, Pax6, Olig2	None	N.D.	(28)
Conditional *SmoM2* activation	*Gli1-CreERT2*	40% Incidence (tamoxifen administration at P10)	Classic, Focal tumors	N.D.	None	N.D.	
SmoA1	*NeuroD2* (*ND2*)	Hemizygotes—48%; Lat: 26 weeks Homozygotes—94%; Lat: 2 months	Classic, EGL hyperproliferation Homozygotes: Disruption of normal cerebellar architecture. Polygonal tumor cells, rosette	Nestin, NeuN Homozygotes: loss of NeuN expression representing loss	None	Upregulation of *Gli1/2*, *N-myc and cyclin D1*. Induction of NOTCH signaling genes such as *Notch2* and *HES5*, and increased expression of	(83, 118)

(continued)

Table 1 (continued)

Mouse model	Tissue specific	MB tumor incidence; Latency (Lat)	Histology/tumor features	Immunoreactivity/ gene expression	Other tumors or defects observed	Molecular/chromosomal changes, other characteristics	Reference(s)
			formation. Necrosis, neovascularization, invasion of fourth ventricle/molecular layer, leptomeningeal spread to spinal cord and distant brain sites	of neuronal differentiation		cell cycle regulatory genes and ribosomal proteins	
Sufu$^{+/-}$; *p53*$^{-/-}$	Ubiquitous	58%; Lat: 10 months	Classic	Synaptophysin, Tuj1, GFAP	Rhabdomyosarcomas, lymphoma, thymoma	LOH of *Sufu* in tumors Upregulation of *Math1*, *N-myc*, *Ptch2*, *Sfrp1*, *cyclin D1* *Gli1/2* not upregulated	(123)
SHH pathway—retroviral somatic gene transfer models							
SHH-MMLV, injected E13.5	Ubiquitous	76%; Lat: 2–3 weeks	Classic	GFAP, NeuN	None	Upregulation of *Math1*	(125)
RCAS-SHH in *Ntv-a* mice	*Nestin tv-a*	9–34%; Lat: 3 months	Classic	Tuj1, NeuN, synaptophysin, trkC, Math1	None	N.D.	(128–132)
SHH + IGF2		39% (SHH alone—15%); Lat: 12 weeks	Classic, some infiltration to brain stem			Comparable proliferation and reduced apoptosis compared to SHH alone (SHH + IGF2: Upregulated IGF pathway, phospho-AKT and N-MYC)	(131)
SHH + activated AKT-Myr-Δ11-60		48% (SHH alone—15%)	Classic			(SHH + AKT: high levels of N-MYC)	

SHH + wild-type NMYC		47%; Lat: 47 days (SHH alone—15%; Lat: 48 days)	Classic			Increased proliferation and reduced apoptosis compared to SHH alone, high levels of cyclin D2	(130)
SHH + NMYC T50A		78%; Lat: 22 days (SHH alone—15%; Lat: 48 days)	Classic				
SHH + BCL-2		78%; Lat: 47 days (SHH alone—34%; Lat: 48 days)	Classic			Comparable proliferation and reduced apoptosis compared to Shh alone	(129)
SHH + PTEN deletion (via PTEN-floxed)		N.D.	70% of tumors: 100% extensive nodularity 30% of tumors: extensive nodularity with 10–30% areas with high proliferation			N.D.	(135)
SHH + MYC		23% (SHH alone—9%)	Classic, invasion into brainstem and fourth ventricle			N.D.	(132)
SHH + HGF		78% (SHH alone—39%)	Classic, some tumors with desmoplastic regions (D/N and MBEN)			Increased proliferation and reduced apoptosis compared to SHH alone	(128)
Models showing upregulation of SHH pathway							
Cell cycle-associated genes							
Conditional *Rb* deletion in *p53* floxed/null mice	*mGFAP-Cre*	*Rb* complete loss: 85–100%; Lat: 3–4 months *Rb* partial loss, floxed *p53*: 75%; Lat: 3–7 months	LCA tumors, EGL expansion from Math1^{+} CGNPs	Tuj1, MAP-2, GFAP, Math1 Negative for neuron-specific enolast, NeuN and synaptophysin	Lymphomas, sarcomas in *p53*-null background	*N-myc*, *Ptch2* and *Gli2* mutations	(144, 145)
Ink4c$^{-/-}$; *p53*$^{-/-}$	Ubiquitous	8%; Lat: 5 months	Classic, some LCA regions Failure of CGNP migration into IGL, hyperproliferation.	GFAP, synaptophysin, nestin	Vascular tumors such as angiomas and hemangiosarcomas, pituitary ademona,	Overexpression of *Gli1*, *N-myc*, *c-Myc*, *cyclin D1*	(142)

(continued)

Table 1 (continued)

Mouse model	Tissue specific	MB tumor incidence; Latency (Lat)	Histology/tumor features	Immunoreactivity/ gene expression	Other tumors or defects observed	Molecular/chromosomal changes, other characteristics	Reference(s)
			Some tumors infiltrate through molecular layer into granule layer and white matter, into brain stem		B and T cell lymphoma, etc.		
Single dose 4 Gy γ-irradiation of *Ink4c*$^{\pm/-}$; *p53*$^{-/-}$ at P5-6	Ubiquitous	75–88% (compared to 68% in *p53*$^{-/-}$ mice and 20% in *Ink4c*$^{+/-}$; *p53*$^{+/-}$ mice)	Not mentioned, presumably similar to *Ink4c*$^{-/-}$; *p53*$^{-/-}$ mice	Not mentioned, presumably similar to *Ink4c*$^{-/-}$; *p53*$^{-/-}$ mice	Other tumors (not mentioned)	Not mentioned, presumably similar to *Ink4c*$^{-/-}$; *p53*$^{-/-}$ mice	(147)
Conditional p53 deletion in *Ink4c*$^{-/-}$ mice + 4 Gy γ-irradiation at P5	*Nestin-Cre*	100%; Lat: 5 months			None		
Ink4c$^{\pm/-}$; *Ptch1*$^{+/-}$	Ubiquitous	37–46% (compared to 7% in *Ptch1*$^{+/-}$ mice); Lat: 40 weeks	Similar to tumors from *Ptch1*$^{+/-}$ and *Ink4c*$^{-/-}$; *p53*$^{-/-}$ mice	Not mentioned, presumably • similar to tumors from *Ptch1*$^{+/-}$ and *Ink4c*$^{-/-}$; *p53*$^{-/-}$ mice	N.D.	LOH of *Ptch1*, but no LOH of *Ink4c* in *Ink4c* heterozygotes Similar to tumors from *Ptch1*$^{+/-}$, *Ptch1*$^{+/-}$; *p53*$^{-/-}$ or *Ink4c*$^{-/-}$; *p53*$^{-/-}$ mice: Upregulation of SHH signaling pathway members *Gli1/2*, *Foxm1*, *Ptch2* High levels of *Math1*, *cyclin D1/2*, *N-myc* expression	

DNA repair enzymes							
Lig4$^{-/-}$; *p53*$^{-/-}$	Ubiquitous	100%; Lat: 9 months (tumors seen as early as 3 weeks)	Classic, from EGL expansion	Synaptophysin, MAP2, Tuj1, GFAP Negative for NF	Pro B-cell lymphoma	High levels of *Math1*, *Gli1*	(150)
Ku80$^{-/-}$; *p53*$^{-/-}$; *Rag1*$^{-/-}$	Ubiquitous	66.7%; Lat: 4–5 months	Classic, from EGL expansion	GFAP, synaptophysin	T-cell lymphoma	N.D.	(151)
Conditional *Xrcc4* deletion in *p53*$^{-/-}$ mice	*Nestin-Cre*	100%; Lat: 25 weeks (tumors seen as early as 4 weeks)	Classic, from EGL expansion Occasional Homer-Wright rosettes 2 mice have tumors with LCA regions, 1 mouse with spinal metastasis (out of 18 mice)	GFAP, NeuN	Thymic lymphoma, solid tumors	*N-myc* and *cyclin D2* amplification, chromosomal translocations, single-copy deletion of *Ptch1*	(153)
Conditional *Brca2* deletion in *p53*$^{\pm/-}$ mice	*Nestin-Cre*	In *p53*$^{+/-}$ mice: 83%; Lat: 13 weeks In *p53*$^{-/-}$ mice: 72%; Lat: 21 weeks	Classic, from EGL expansion	N.D.	Lymphoid tumors	*N-myc* amplification, Ptch1 loss LOH of *p53* in *p53*$^{+/-}$ mice	(154)
Parp-1$^{-/-}$; *p53*$^{-/-}$	Ubiquitous	49%; Lat: 16 weeks (tumors seen as early as 8 weeks)	Classic, from EGL expansion, invasive Some tumors with Homer-Wright rosettes, larger tumors invaded fourth ventricle, spread into surrounding structures	NeuN, Nestin, MAP2 Weak synaptophysin Negative for GFAP	Lymphomas, angiosarcomas	High expression of *Math1* Upregulation of SHH pathway via reduction in *Ptch1* and increase in *Gli1* Aneuploid tumor cells Chromosomal aberrations such as fusons, chromosome/chromatid breaks, fragmentation	(152)
Hypermethylated in Cancer 1 (Hic1)							
Ptch1$^{+/-}$; *Hic1*$^{+/-}$	Ubiquitous	40%; Lat: 25 weeks (compared to 10%; similar latency in *Ptch1*$^{+/-}$ mice)	Classic, similar to tumors from *Ptch1*$^{+/-}$ mice	Nestin, synaptophysin, GFAP	None	High expression of *Smo*, *Gli1/2*, *Sufu* Wild-type *Ptch1* allele retained but not expressed *Hic1* expression low or absent, Hic1 protein absent	(69)

(continued)

Table 1 (continued)

Mouse model	Tissue specific	MB tumor incidence; Latency (Lat)	Histology/tumor features	Immunoreactivity/ gene expression	Other tumors or defects observed	Molecular/chromosomal changes, other characteristics	Reference(s)
Overexpression of Ifng							
hGFAP-tTA; TRE-Ifng	*hGFAP-tTA*	E16 early induction of *Ifng* expression: 80%; Lat: 3–4 weeks (Ifng expression required before P16)	Classic, similar to tumors from *Ptch1*$^{+/-}$ mice Tumors in molecular layer and EGL, also invaded into white matter and brainstem regions	Synaptophysin, Math1	None	N.D.	(155)
Non-SHH models							
WNT pathway models							
RCAS (c-MYC and S37A β-catenin) in *Gtv-a; p53*$^{-/-}$ mice	*GFAP tv-a*	83%; Lat: 6 weeks	LCA tumors	N.D.	None	N.D.	(156)
Conditional β-catenin activation in *p53*$^{\pm/-}$ mice	*Blbp-Cre*	*p53*$^{+/-}$ mice: 4%; Lat: 10 months *p53*$^{-/-}$ mice: 15%; Lat: 10 months	Classic, non-EGL tumor origin Tumor formed from proliferation of *Zic1*$^{+}$ mossy-fiber neuron precursors in dorsal brainstem/lower rhombic lip	*Zic1*	None	N.D.	(50)
MYCN overexpression							
Glt-tTA; TRE-MYCN/Luc (GLT/TML)	*Glt1*	TML hemizygotes: 40%; Lat: 7 months TML homozygotes: 75%; Lat: 7 months	Classic and LCA tumors, rare metastasis to spine Non-EGL tumor origin: EGL cells in GTML mice identical to wild-type	GFAP, moderate synaptophysin, negative for NeuN and Math1	None	No SHH pathway activation except one case with *Ptch1* loss Elevated levels of *Otx2* Chromosomal aberrations, particularly syntenic to human chromosome 17	(59)

IR irradiation, *LOH* loss of heterozygosity, *BCC* basal cell carcinoma, *MB* medulloblastoma, *N.D.* not determined

Various GEM models have been developed using 2 main strategies: (A) germline modifications via transgenes, genetic knock-ins or knockouts, and (B) somatic cell gene transfer methods using Moloney murine leukemia virus (MMLV) or avian retrovirus RCAS/tv-a (replication-competent ASLV long terminal repeat with splice acceptor/tumor virus-A) systems. Because of the molecular and genetic complexity of medulloblastoma, it is unlikely that a single mouse model can perfectly encompass all variants of the disease. Multiple mouse models, each driven by specific but relevant genetic modifications, will together provide avenues to understand the disease.

4. Mouse Models of Medulloblastoma

4.1. SHH-Driven

Many investigations in the past have centered on the SHH signaling pathway because of its importance in normal cerebellar development and its implication in Gorlin's Syndrome. The established link of SHH pathway with medulloblastoma has led to the several mouse models that modulate members of the pathway. These includes (a) germline-modified mouse models with partial deletion of *Ptch1*, activation of *Smo* and partial deletion of *Sufu*; as well as models using somatic gene transfer in which SHH and other cooperative proteins are ectopically expressed using retroviruses. Each of these will be elaborated in respective sections below.

1. Germline-modified mouse models
   (a) Deletion of *Ptch1*

   While homozygous deletion of *Ptch1* results in early embryonic lethality due to cardiac and neural tube defects, heterozygous mice are viable. Medulloblastoma tumors developed in 14–19% of heterozygous mice by 10 months, with peak occurrence at 16–25 weeks of age (102–104). Due to systemic heterozygosity for *Ptch1*, these mice also develop other tumors, notably sarcomas, as well as other developmental abnormalities like hydrocephalus and polydactyly. Medulloblastoma tumors arise from the external germinal layer of the posterior fossa in these mice, form well-circumscribed tumors of classic histology that compress normal cerebellar tissue but did not infiltrate the normal parenchyma or migrate along the white matter tracks. Immunoreactivity of these tumors is similar to human tumors and displays positive neurofilament, synaptophysin and glial fibrillary acidic protein (GFAP) reactivity, which are consistent with tendency of neuronal and glial differentiation of medulloblastoma tumors (104).

While decreased SHH signaling is likely to be sufficient in promoting tumor initiation, the sufficiency of heterozygous *Ptch1* to promote tumor progression remains controversial. Some groups have observed the retention of wild-type *Ptch1* allele in these *Ptch1*$^{+/-}$ murine tumors (102, 104), while others report loss of heterozygosity (LOH) or mutation of the remaining *Ptch1* allele (105, 106). Of note, preneoplastic proliferative lesions developed early in these mice, but most did not progress to develop full-blown tumors, suggesting that additional hits are required to achieve transformation (106).

Subsequently, several approaches were employed to increase the incidence of tumors in this model. Pazzaglia et al. demonstrated that X-ray irradiation of newborn *Ptch1*$^{+/-}$ markedly increased incidence of tumors from 7.7% to 81% by 5–10 months (107, 108). However, tumor susceptibility is dependent on age of irradiation. In comparison to mice irradiation at P1, mice irradiated at P10 had much reduced susceptibility to tumors while mice irradiated at 3 months of age had unaltered tumor incidence (107, 108). CGNPs from P1 pups had reduced apoptosis in response to radiation via increased pro-survival AKT signaling and reduced TP53 levels compared to those from P10 pups, likely accounting for their increased resistance to irradiation (108). Further, characterization of medulloblastomas that developed from irradiated *Ptch1*$^{+/-}$ mice revealed the LOH of wild-type *Ptch1* allele, suggesting that complete loss of *Ptch1* was required for increased malignancy.

Ptch2 is highly similar to *Ptch1*, sharing 73% amino acid similarity. Although *Ptch2*-deficient mice demonstrated no obvious developmental defects or tumor susceptibility, partial or complete loss of *Ptch2* in a *Ptch1*$^{+/-}$ background modestly reduced tumor latency for tumor formation and increased incidence of tumors from 6.4% to 16% compared to *Ptch1*$^{+/-}$ mice (109). Similar to *Ptch1*$^{+/-}$ mice, a wide range of tumors formed including sarcomas and basal cell carcinoma and there was also LOH of *Ptch1* in these tumors. Altogether, these results suggest that loss of *Ptch2* can cooperate with *Ptch1* loss to promote medulloblastoma tumor progression.

Complete loss of *p53* in *Ptch1*$^{+/-}$ mice increased incidence of tumors (95–100%) and reduced latency to 4–16 weeks (110, 111). Combined loss of *p53* and loss or haploinsufficiency for *Ptch1* is proposed to (1) Deregulate the G2/M cell cycle checkpoint while also increasing genomic instability and (2) Allow cells that harbor secondary mutations to proliferate unchecked and promote tumorigenesis. Interestingly, tumors that arise from *Ptch1*$^{+/-}$ mice did not harbor *p53* mutations or deletions, suggesting that medulloblastoma formation in these mice occur independently of TP53. Tumors from *Ptch1*$^{+/-}$; *p53*$^{-/-}$ mice did demonstrate increased random chromosome loss compared to the *Ptch1*$^{+/-}$; *p53*$^{+/+}$ mice. The tumor suppressor p19^{Arf} indirectly activates TP53, but interestingly, combination of *Arf* loss with *Ptch1* heterozygosity did not phenocopy *Ptch1*$^{+/-}$; *p53*$^{-/-}$ mice. Loss of *Arf* did not drive increased incidence of tumors in mice mutant for *Ptch1*, hinting at a synergistic effect between TP53 and PTCH1 in medulloblastoma tumorigenesis (110).

Loss of *Ptch1* results in increased activity of downstream genes *Gli1* and *Cyclin D1* in CGNPs. Accordingly, the complete inactivation of *Gli1* in *Ptch1*$^{+/-}$ mice resulted in reduced incidence of tumor (3.4% of observed mice), demonstrating that GLI1 contributes significantly to SHH-driven medulloblastoma. These compound mutants upregulated GLI2, suggesting that GLI2 can partly compensate for the loss of GLI1 (112). Similar to *Gli1*, a complete loss of *Cyclin D1* in *Ptch1*$^{+/-}$ mice also suppressed formation of medulloblastoma (113). Lastly, *Ptch1*$^{+/-}$ mice were crossed into an *Igf2*-null background, because rhabdomyosarcomas in *Ptch1*$^{+/-}$ mice had consistently increased IGF2 levels (114). No tumors developed in the double mutant mice, demonstrating that IGF2 is essential for SHH-driven medulloblastoma (115). It was revealed that these double mutants harbored preneoplastic lesions, but these lesions failed to progress to form tumors. Further analysis showed that *Igf2* transcription is not initiated until late tumor stages, demonstrating that unlike SHH signaling, IGF2 is not necessary for initial lesion formation, but is necessary for malignant progression (116).

(b) Activation of *Smo*

A number of transgenic mouse models have been generated using activated SMO, either with mutation W535L (*SmoM2*) in human *SMO* or the corresponding mutation W539L in mouse *Smo* (*SmoA1*). The mutation resides in the transmembrane domain of SMO and results in SHH ligand-independent activation of downstream signaling.

The *SmoM2* mouse model used a ubiquitously expressed *CreER* transgene to allow for tamoxifen-mediated activation of mutant *SmoM2* and induction of downstream SHH signaling (117). Sporadic leakiness of Cre activity resulted in all mice developing rhabdomyosarcomas and 27% of mice developing medulloblastoma. Acute activation of *SmoM2* via tamoxifen administration at post-natal day P10 increased rhabdomyosarcoma numbers per mouse, induced basal cell carcinoma-like lesions and increased medulloblastoma incidence to 40%. Medulloblastoma tumors had typical features of classic tumors, were highly proliferative and exhibited neuronal differentiation markers like ZIC1 and NeuN. These tumors also upregulated PDGFRα, components of the IGF2 and MAPK pathways, as well as markers previously implicated in medulloblastoma like c-MYC, NMYC, and OTX2.

The *SmoA1* mouse model uses the NeuroD2 (ND2) promoter to drive *SmoA1* expression in CGNPs (83, 118). Hemizygous mice developed medulloblastoma at a median age of 26 weeks with 48% incidence, while 94% of homozygous mice developed tumors at 2 months. Initial analyses on tumors from hemizygous mice demonstrated upregulation of SHH targets *Gli1* and *Nmyc*, as well as concomitant induction of NOTCH signaling, suggesting that SHH and NOTCH pathways cooperate in promoting survival of tumor cells in SHH-driven tumors (83). Like other SHH-driven tumors, these tumors developed from CGNPs in the EGL. In particular, tumors from homozygous mice displayed areas of necrosis, neovascularization, and invasion into the fourth ventricle and molecular layer. Notably, these tumors often displayed leptomeningeal spread that is characteristic of high-risk patients with poor prognoses. Tumor cells were found in the leptomeninges of the brain, at distant sites from the tumor and within the gray matter of the spinal cord. Hemizygous mice also had limited frequency of leptomeningeal spread (118).

(c) Other tissue-specific *Ptch1* and *Smo* models

Other tissue-specific knockouts have been generated to restrict activated SHH signaling to the cerebellar neural stem cells (NSCs) or CGNPs. Utilizing *hGFAP-Cre* or *Math1-Cre* drivers to deleted *Ptch1* conditionally in mice, Yang et al. observed that incidence of tumors increased dramatically to 100% by 1–3 months of age, compared to *Ptch1*$^{+/-}$ mice (119). *hGFAP-Cre* mice employed the human regulatory sequences for GFAP that were reported to drive Cre in Nestin-positive radial glia cells, which generate most cerebellar cell types including CGNPs. *Math1-Cre* expression is restricted to unipotent CGNPs of the rostral rhombic lip (rRL), a transiently lived structure from which CGNPs are generated and migrate to form the EGL. Similar tumors developed using either driver; however, tumors initiated from NSCs with *hGFAP-Cre* arose at higher incidence than those initiated from CGNPs using *Math1-Cre*, resulting in death by 3–4 weeks compared to 2–3 months. Similar to the systemic deletion of *Ptch1* and activation of *Smo* in mouse models described earlier, CGNPs in the EGL of *Math1-Cre; Ptch1$^{flox/flox}$* mice were highly proliferative at 6 weeks, with an otherwise normal cerebellar organization. This result suggests that *Ptch1* loss in GCNPs induced increased proliferation and self-renewal capacity, while still permitting neuronal differentiation. By 10–11 weeks, however, full-blown tumors were present that disrupted normal cerebellar structure (119).

In a similar fashion, Schüller et al. expressed activated *SmoM2* in a tissue-specific fashion using *hGFAP-*, *Math1-*, *Olig2-*, and *Tlx3-Cre* drivers (28). *Olig2-Cre* and *Tlx3-Cre* are specifically expressed in neuronal progenitors in the posterior EGL and remain expressed in IGL. Medulloblastoma developed robustly in 100% of mice using all 4 drivers, with an average survival of about 1–2 months. Like the conditional *Ptch1* model, the different Cre drivers gave rise to tumors with similar classic histopathology and expression profiles. Tumors varied in extent of tumor spread, reflective of Cre expression. Medulloblastomas driven by *hGFAP-Cre:SmoM2* and *Math1-Cre:SmoM2* were diffuse, while *Olig2-Cre* and *Tlx3-Cre* drivers generated focal tumors restricted to the posterior lateral lobes. Like the *Ptch1* model, *hGFAP-Cre:SmoM2* mice had significantly shortened survival compared to the other Cre drivers. The authors also targeted *SmoM2* using the Gli1 promoter. However, incidence of tumors was only at 40%, and the animals had prolonged survival, suggesting that driving tumors using promoters of genes downstream from SHH less potently promotes oncogenesis (28).

Both studies highlight the usefulness of tissue-specific genetic manipulations in mice, as restricting oncogenic signals to NSCs or CGNPs greatly increased penetrance for medulloblastoma and eliminated sarcomas and skin carcinomas that have confounded earlier studies conducted on systemic genetic knockouts. These two studies also revealed that activating SHH signaling in NSCs manifested only as medulloblastomas was conditional upon their commitment towards the CGNP lineage. Activated SHH signaling results in increased proliferation of NSCs and this explains the reduced tumor latency in *hGFAP-Cre* mice. Only the CGNP lineage expanded to form medulloblastomas. Although NSCs are also capable of differentiating into astrocytes, oligodendrocytes, and neurons in the forebrain and hindbrain, astrocytomas, oligodendrogliomas, and non-CGNP medulloblastomas were not found in these mice (28, 119). Curiously, while unipotent CGNPs appear to drive SHH-pathway medulloblastoma, SHH-driven tumor cells themselves show some evidence of multipotency, as they are capable of differentiating into multiple lineages (120, 121). Further investigations are required to examine the potential reacquisition of multipotency in medulloblastoma tumor cells.

In another study, *Pax7-Cre* was used to inactivate *Ptch1* in mice and 30% of double homozygous mice developed medulloblastomas in 88–100 days (122). *Pax7* is expressed in the developing cerebellum and mid-brain, as well as muscle satellite cells, but *Ptch1* inactivation resulted in only medulloblastomas and not rhabdomyosarcomas in these mice. These medulloblastoma tumors were similar to the other SHH tumors: they were of classic histology, likely originated from the CGNPs in the EGL and were immunopositive for TUJ1, NeuN, and synaptophysin. These tumors also invaded the subarachnoid space and in half the mice, showed spread to the fourth ventricle. Conditional *p53* allele was introduced into the model and partial or complete *p53* deletion decreases tumor latency to 32–65 days although tumor histology or aggressiveness was unaltered.

(d) Deletion of *Sufu*

Mice deleted for *Sufu* and for *Ptch1* exhibit many parallels. Complete loss of *Sufu* is embryonic lethal due to cardiac and neural tube defects. Partial loss of *Sufu* did not make mice prone to medulloblastoma, suggesting that *Sufu* is not as potent as *Ptch1* in inducing tumorigenesis. $Sufu^{+/-}$; $p53^{-/-}$ mice developed medulloblastomas at 58% incidence and rhabdomyosarcomas at 9% incidence by 10 months (123). The reduced tumorigenic potential of *Sufu* was also reflected in $Sufu^{+/-}$; $p53^{-/-}$ mice having a longer median survival of 4 months compared to 2 months in $Ptch1^{+/-}$; $p53^{-/-}$ mice. Medulloblastoma tumors exhibited LOH of *Sufu*, and were positive for neuronal and glial markers like synaptophysin, TUJ1, and GFAP. Interestingly, in contrast to $Ptch^{+/-}$ mice, GLI1/2 expression was not upregulated, although MATH1, NMYC, and PTCH2 were upregulated in these tumors, suggesting that downstream GLI is not a crucial effector of SHH pathway in this model.

2. Somatic cell gene transfer models

(a) MMLV retroviral system

Somatic cell gene transfer has also been used to investigate the role of SHH pathway in medulloblastoma. Early studies used SHH-expressing Moloney murine leukemia viruses (MMLV) injected into embryonic day E13.5 mouse cerebellum via in utero ultrasound biomicroscopy guidance. Medulloblastoma tumors arose in 76% of animals by P14-21, proving that direct overexpression of SHH alone is sufficient to initiate tumor formation (124, 125). In line with germline-modified SHH-driven models, persistence and subsequent expansion of the EGL resulted in the growth of tumors that compressed the cerebellum.

SHH-expressing retroviruses were also introduced in the cerebellum of *Gli1*$^{-/-}$ mice and medulloblastoma tumors still formed, aligned with germline results using *Sufu* (123), and showing that *Gli1* is not critical in some models of SHH-induced medulloblastoma. *Gli2*, in contrast, was expressed in the tumors induced in both wild-type and *Gli1*$^{-/-}$ mice, suggesting that *Gli2* may compensate for *Gli1* loss. These results parallel observations made from *Ptch1* knockout mice, which showed similar upregulation of *Gli2* levels in the tumors that formed, even though incidence of tumors was drastically lower in *Ptch1*$^{+/-}$; *Gli1*$^{-/-}$ compared to *Ptch1*$^{+/-}$ mice (112).

(b) RCAS/tv—a retroviral system

A disadvantage of the MMLV system is that the gene transfer is not cell-type specific. To enable cell-type-specific infection, the RCAS/tv-a system is now commonly used in the modeling of a number of tumors including medulloblastoma (126, 127). The avian retrovirus RCAS (replication-competent ASLV long terminal repeat with splice acceptor) only infects murine cells that express the avian RCAS receptor tv-a (tumor virus-A), which is not expressed in mammalian cells. Transgenic mice that expresses tv-a from cell-type-specific promoters, including GFAP or nestin (Gtv-a or Ntv-a) have been generated, and can be used to mediate gene transfer to NSCs or progenitor populations, opening up opportunities to conduct lineage tracing studies and overexpress multiple gene combinations within the same population of cells.

Investigations primarily by the Fults laboratory have used the RCAS/tv-a system to generate SHH pathway medulloblastoma. Virus-producing DF-1 chicken fibroblasts were injected into the cerebella of newborn mice. Injected DF-1 cells persist for less than 3 days, producing viruses that transduce the surrounding tv-a-expressing cells (127). RCAS-SHH retroviruses injected into Ntv-a mice induced medulloblastoma formation at 9–39% incidence within 3 months (128–132). This system was subsequently used to test genes for cooperation with SHH, as elaborated below.

- IGF2/PI3K pathway in combination with SHH.
  The insulin-like growth factor (IGF)2 signaling pathway is a key mitogenic and cell survival pathway implicated in medulloblastoma. Expression of IGF2 is increased in desmoplastic medulloblastomas, most commonly driven by SHH (133). Medulloblastomas

fail to form in *Ptch1*$^{+/-}$; *IGF2-null* mice, arguing for the importance of IGF2 in SHH-driven medulloblastoma pathogenesis (115). IGF2 ligand binds to IGF1 receptor, triggering the phosphorylation of insulin receptor substrate 1 (IRS-1). This leads to the recruitment of phosphatidylinositol 3-kinase (PI3K), which activates AKT and associated downstream signaling. NMYC serves as a link between IGF2/PI3K and SHH signaling pathways in regulating CGNP proliferation. Activated AKT phosphorylates GSK-3β, stabilizing NMYC by preventing its phosphorylation at threonine-58, a mark for proteolytic degradation (134).

Several genes from the IGF2/PI3K pathway were each introduced concurrently with SHH using the RCAS/tv-a system to test for their ability to enhance medulloblastoma tumorigenesis. These are: (i) IGF2, (ii) activated transforming form of AKT (Akt-Myr-Δ11-60), (iii) wild-type NMYC, and (iv) stabilized, non-degradable T58A mutant NMYC (130, 131). Combinations with all genes tested showed at least a doubling in incidence of tumors. Mice infected with Shh + IGF2 or Shh + Akt-Myr-Δ11-60 had medulloblastoma incidence of 39% and 49% respectively (compared to 15% with SHH alone), supporting the cooperation of IGF2/PI3K signaling with SHH in medulloblastoma tumorigenesis. As expected, tumors displayed an increase in levels of phosphorylated-IGF1 and -AKT, IRS1, and NMYC (130, 131). While co-injecting SHH with wild-type NMYC resulted in similar latency but higher incidence of medulloblastoma (47%) compared to SHH alone (15%), combining SHH with stabilized NMYC T58A was more potent. Incidence of tumors increased to 78% with decreased tumor latency of 22 days rather than 48 days (130). RCAS/tv-a-derived tumors were immunopositive for EGL marker MATH1 and neuronal markers TUJ1, NeuN, and synaptophysin, reflecting markers observed in germline-modified SHH models, and further supporting CGNPs as the cells of origin for SHH-driven tumors.

Tumors that developed from the combination of SHH with IGF2, activated AKT, or wild-type NMYC had reduced levels of apoptosis, with comparable proliferation compared to SHH alone. Together, these results suggest that IGF2/PI3K pathway reduces apoptosis and enhances survival of CGNPs, without affecting their proliferation. In contrast, NMYC T58A increased proliferation concomitant with enhancing pro-survival effects. To uncouple apoptosis from proliferation, anti-apoptotic protein BCL-2 was combined with

SHH. Medulloblastoma arose at high incidence (78% compared to 34%). Thus, the ability to evade intrinsic mechanisms driving cell death represents a critical step towards increased malignancy (129).

Of note, merely adding IGF2, activated AKT, wild-type or mutant NMYC, or BCL-2 alone in all the above studies did not drive medulloblastoma, neither did combining wild-type NMYC with IGF2/PI3K signaling elements (GLI1, AKT, and IGF2), nor combining NMYC T58A with anti-apoptotic BCL-2. Together, these experiments suggest that the pro-tumorigenic effect of SHH is not solely mediated via NMYC, and that other downstream effectors of SHH cooperate with IGF2/PI3K signaling to drive tumorigenesis.

Further, introducing RCAS-SHH with RCAS-Cre in Ntv-a; PTEN-floxed mice drive concurrent overexpression of SHH with PTEN loss and resulted in medulloblastoma formation as well (135). Interestingly, the majority of tumors that developed showed extensive nodularity similar to the human MBEN histology. These nodular regions made up the tumor bulk, were NeuN positive and showed no proliferation, while a small proportion of cells that were perivascular were Nestin-positive, NeuN-negative, and highly proliferative.

Although much was learned from these RCAS/tv-a models, the resulting pathology of tumors was mostly classic rather than desmoplastic, with the exception of the SHH/PTEN model. Just like the germline-modified mouse models discussed in the previous section, these SHH-driven models were unable to drive desmoplastic features characteristic of many SHH-driven human medulloblastoma tumors.

- c-MYC and HGF in combination with SHH

  Hepatocyte growth factor (HGF) has been shown to regulate expression of c-MYC and can cooperate with c-MYC to drive cell proliferation and apoptosis (136, 137). Accordingly, HGF and c-MYC have each been demonstrated to promote proliferation of CGNPs (138–140). High levels of HGF/c-MET receptor and amplification of c-MYC in human medulloblastoma are independently associated with LCA tumors and poor prognosis (137). When expressed together using the RCAS/Ntv-a system, c-MYC and SHH cooperated to drive aggressive tumors that invaded the brainstem and fourth ventricle, with incidence of 23%, compared to 9% with SHH alone and 0% with c-MYC alone (132). Like previous RCAS/tv-a-derived tumors, these showed classic pathological features and expressed early neuronal markers. Combining HGF with SHH yielded increased incidence of tumors to 78% (compared to 39%) due to enhanced proliferation and reduced apoptosis. Interestingly, some tumors displayed areas of neurocytic differentiation similar to D/N and MBEN tumors (128).

Although both c-MYC and HGF cooperated with SHH to drive increased penetrance, aggressiveness, and even generated regional desmoplasia in murine medulloblastoma, it is not clear that this combination of genes mirrors cooperative expression in human tumors. As previously mentioned, high levels of HGF and c-MYC are associated with LCA tumors and poor prognosis in patients. However, human tumors with activated SHH pathways have transcriptional profiles and characteristics that are distinct from tumors with upregulated HGF/c-MYC (17, 136, 137): SHH tumors tend to be desmoplastic or classic and are associated with better prognosis. In fact, molecular profiling classifies c-MYC-driven tumors primarily into group C and WNT tumors, subclasses that segregate distinctly from the SHH tumors.

3. Remarks

Both germline-modified and somatic cell gene transfer models that activate the SHH pathway are aligned in that targeting this pathway in CGNPs results in their persistence and accumulation of transformed CGNPs in the EGL, leading to medulloblastomas. In human cases, SHH-driven tumors are often of classic or desmoplastic histology, with desmoplastic tumors almost exclusively in the SHH subclass. An outstanding question remains as to why the vast majority of SHH-driven mouse models generate classic rather than desmoplastic histopathology (the exception being the SHH/PTEN and SHH + HGF RCAS/Ntv-a models). The inability of most mouse models to model desmoplasia highlights either inherent differences between the murine and human brain or a need to better understand the genetic and mechanistic basis for desmoplasia in medulloblastoma.

4.2. Models with Other Perturbations Showing Upregulated SHH Pathway

Other mouse models have been developed by disrupting of cell cycle-associated genes, DNA repair enzymes, and the candidate gene on chromosome 17 *Hic1*, as well as overexpression of pro-inflammatory cytokine interferon-gamma. Although the introduced aberrations are not known to be associated with SHH signaling, the tumors that formed showed upregulated SHH signaling as well, and displayed similar histopathological features and expression profiles as the SHH models elaborated in the previous section (141). In many of these models discussed in the next section, investigators have introduced mutation in *p53* to drive tumorigenesis. It should be noted that *p53* dysregulation occurs in 15% of medulloblastoma and, therefore, does not play a particularly important role in medulloblastoma tumorigenesis.

1. Deletion of cell cycle-associated genes—*Rb* and *Ink4c*
   RB is a tumor suppressor that inhibits cell cycle progression. Hyperphosphorylation of RB by a cyclin-dependent kinase (CDK) relieves this inhibition, permitting cell cycle progression. CDKs in turn are regulated by inhibitors, including INK4C (*IN*hibitor of CD*K4*). Hence, loss of CDK inhibitors, like loss of RB, leads to cell cycle progression. *INK4C* is transiently expressed in proliferating CGNPs in the EGL when they start to differentiate into granule neurons (142, 143). This time period corresponds to the window in which dysregulation of CGNP proliferation or blockade of differentiation can lead to medulloblastoma.

   In one model, GFAP-Cre mediated deletion of *Rb* in NSCs under a *p53*-null or -floxed background resulted in 85–100% incidence of tumors at a latency of 3–4 months (144). Partial loss of *Rb* resulted in slightly lower frequency and longer latency for tumors, suggesting that LOH of the remaining *Rb* allele or other genetic aberrations may drive malignancy.

   These *Rb/p53* compound mutants highlight important factors regulating tumor formation, with parallels in SHH models. First, CGNPs in the EGL were the only population that proliferated in response to *Rb* deletion, while other GFAP-derived lineages were unaffected (144). Recall that similar observations were made in SHH models that conditionally deleted *Ptch1* or activated *Smo* in NSCs (28, 119), suggesting that the proliferating CGNP lineage is particularly susceptible to transformation. Second, synergy between coordinate loss of *p53* and *Rb* permits tumor formation. Loss of *p53* cannot be substituted by loss of *Arf*, *p21* or cell cycle checkpoint gene *Atm*, as no tumors formed with these combinations (144). The requirement for loss of *p53* was also observed in the $Ptch1^{+/-}$ model, where loss of *p53* but not *Arf* increased incidence of tumors (110), demonstrating that loss of *p53* is essential in promoting tumorigenesis in these mouse models. Finally, although the *Rb/p53* model was engineered to combine loss of *Rb* and *p53*, preneoplastic lesions were found to harbor mutations in *N-myc*, *Ptch2*, and *Gli2*, activating SHH signaling, demonstrating the tight link between SHH pathway and CGNP proliferation (145). Strikingly, tumors from *Rb/p53* mutants were all of LCA histology (144), in contrast to the classic tumors generated by SHH models (28, 119).

The *Rb/p53* conditional knockout model using the GFAP-Cre driver demonstrated that CGNP population gave rise to tumors in the model. In a follow-up study, postnatal P7 cerebellar NSCs that were CD133-positive and Math1-negative (non-CGNPs) were isolated from mice with floxed *Rb* and *p53* alleles. *Rb* and *p53* were deleted in vitro with adenoviral-Cre and injected subcutaneously into NOD/SCID mice (146). Interestingly, a small 15% of mice developed medulloblastomas that were strongly immunopositive for stem cell markers *Sox2* and *Sox9* but had low expression of SHH pathway members like *N-myc*, *Gli1*, and *Ptch2* (146). Conversely, the CGNP-derived *Rb/p53* medulloblastomas were weakly positive for stem cell markers but exhibited SHH pathway activation as expected. These results suggest *Rb* and *p53* deletion in postnatal cerebellar NSCs could also serve as medulloblastoma initiating cells, in addition to CGNPs that are widely accepted as a population important in contributing to medulloblastoma formation.

In another model, 8% of *Ink4c/p53* double knockout mice developed medulloblastoma tumors by 5 months of age (142). Loss of *Ink4c* is proposed to block the ability of CGNPs to migrate into the IGL, resulting in accumulation of CGNPs in the molecular layer. Together with added genomic instability from loss of *p53*, these foci of cells proliferate and form medulloblastomas. Treatment of $Ink4c^{+/-}$; $p53^{-/-}$ or $Ink4c^{-/-}$; $p53^{-/-}$ compound mutant mice with 4 Gy total body γ-irradiation at P5-6 modestly enhanced incidence of medulloblastoma to 75–88%, compared to 68% incidence in $p53^{-/-}$ mice (147). In contrast, only 20% of irradiated $Ink4c^{+/-}$; $p53^{+/-}$ mice developed medulloblastoma, suggesting that complete loss of *p53* is important in permitting radiation-induced tumorigenesis. Lastly, use of Nestin-Cre in $Ink4c^{-/-}$ mice, restricting loss of *p53* to EGL CGNPs and irradiation of these mice with 4 Gy at P5 resulted in 100% incidence of tumors. Similar results were seen in *Ptch1* and *Smo* models in which tissue-specific genetic knockouts are more efficient in generating tumors. Thus, *Ink4c* loss cooperates with *p53* loss to promote tumorigenesis (147), and tissue-specific knockouts generally have higher penetrance for medulloblastoma than systemic knockouts.

Concurrent loss of *Ink4c* and *Ptch1* also enhances medulloblastoma tumorigenesis. While $Ptch1^{+/-}$ mice developed tumors at only 7% incidence, combination with $Ink4c^{+/-}$ or $Ink4c^{-/-}$ increased incidence to 37–46% (147). Similar to other $Ptch1^{+/-}$ studies, these *Ptch1/Ink4c* tumors were found to have LOH of *Ptch1*. Interestingly, tumors from *Ptch1/Ink4c* or *Ink4c/p53* compound mutant mice had similar expression profiles of upregulated SHH pathway members such as *Gli1/2*, *Ptch2*, and *N-myc* and EGL marker *Math1*, similar to the *Rb/p53* tumor model.

Some genetic aberrations engineered in these models are of uncertain relevance to human cases. First, loss of *TP53* is rarely seen in human medulloblastoma tumors; however, loss of *Rb* or *Ink4c* alone, without introducing deletion of *p53*, was not sufficient to induce tumors in either model. Second, the *Rb/p53* mouse model developed LCA tumors with activated SHH signaling, although the SHH subclass of tumors rarely show LCA histology. Besides, concurrent loss of *RB* and *TP53* is even more rare among patients. Lastly, although decreased levels of INK4C protein were observed in a small subset of human medulloblastoma cases, it is not known whether loss of INK4C clusters with activated SHH signaling, as in the *Ptch1/Ink4c* and *Ink4c/p53* mouse models.

2. Deletion of DNA repair enzymes genes—*Lig4*, *Xrcc4*, *Ku80*, *Brca2*, *Parp1*

   Double-stranded DNA breaks are repaired via two main mechanisms: Non-homologous end-joining (NHEJ) or homologous recombination repair (HRR). DNA ligase IV (LIG4), KU80, and XRCC4 are involved in NHEJ-mediated DNA repair, while BRCA2 plays a key role in HRR. BRCA2 has also been previously identified as a tumor suppressor, as loss of BRCA2 is linked to increased susceptibility to breast, ovarian, and brain cancer (148, 149). The integrity of the DNA repair machinery is critical in maintaining proper regulation of cell proliferation and cell cycle arrest, and knockouts of the above-mentioned NHEJ or HRR members lead to early embryonic lethality. Compound knockouts also deleted for *p53* rescues the lethality; however, many of these mice develop lymphomas and medulloblastomas.

   A conventional knockout strategy was used to delete *Lig4* and *Ku80* in *p53*-null mice. All *Lig4*$^{-/-}$; *p53*$^{-/-}$ mice developed B-cell lymphoma and medulloblastoma by 9 weeks, starting as early as 3 weeks of age (150). Two-thirds of *Ku80*$^{-/-}$; *p53*$^{-/-}$; *Rag-1*$^{-/-}$ triple mutant mice developed medulloblastoma and T-cell lymphoma, with other cancers developing in small numbers of animals (151). Further, deletion of poly ADP-ribose polymerase (*Parp-1*), an early DNA damage response gene that senses and binds to DNA breaks to suppress recombination, in conjunction with deletion of *p53*, led to medulloblastomas with 49% incidence and latency of 16 weeks (152). Lymphomas and angiosarcomas were observed in some mice as well (152).

   The tumors from the *Lig4*, *Ku80*, or *Parp-1* models were of classic histology and resulted from expansion of the EGL, as in SHH models. Tumor cells also generally expressed the EGL fate marker MATH1, showed upregulation of the SHH downstream effector *Gli1*, and expressed neuronal and glial markers like synaptophysin, TUJ1, NeuN, and GFAP. In particular, some tumors from *Parp-1*$^{-/-}$; *p53*$^{-/-}$ mice had neuroblastic Homer–Wright rosettes, a characteristic of classic tumors, with larger tumors also invading the fourth ventricle and spreading into the surrounding structures.

Conditional inactivation of *Xrcc4* using a Nestin-Cre driver in a *p53*-null background led to medulloblastoma without concurrent lymphoma development. Foci of tumors started developing in the cerebella at 4 weeks, achieving 100% incidence of medulloblastoma by 25 weeks (153). Like tumors in conventional knockout models of DNA repair-associated genes, tumors from the *Xrcc4* model showed mostly classic histopathology with occasional Homer–Wright rosettes, developed in the EGL and were immunopositive for GFAP and NeuN. Interestingly, two of the 18 tumors had regions with LCA histology, and one mouse had spinal metastases. Frequent recurrent abnormalities were also found in these tumors, including high levels of *N-myc* amplification, chromosomal translocations, and single-copy deletion of *Ptch1*, highlighting the importance of DNA repair mechanisms in preventing genomic instability and tumor formation. Similar to the *Xrcc4* model, the Nestin-Cre driver was used to conditionally inactivate *Brca2* in a *p53*-null background. 83% of mice developed medulloblastoma with a tumor latency of ~13 weeks (154). These tumors were also found to harbor *Ptch1* loss and *N-myc* amplifications.

3. Deletion of chromosome 17 candidate gene—*Hic1*

Hic1, a putative tumor suppressor gene at chromosome 17, is frequently silenced epigenetically in medulloblastoma tumors. Although *Hic1*$^{+/-}$ mice do not spontaneously develop medulloblastoma, the tumor suppressor function of HIC1 was revealed in compound *Ptch1*$^{+/-}$; *Hic1*$^{+/-}$ mice, which exhibited a fourfold increase in medulloblastoma incidence compared to *Ptch1*$^{+/-}$ mice (69). In these tumors, both of the remaining wild-type *Ptch1* and *Hic1* alleles were silenced. HIC1 was found to transcriptionally repress *Math1*, whose expression is restricted to CGNPs and is critical for their development. Hence, low levels or loss of HIC1 expression likely relieves suppression of *Math1*, promoting proliferation of CGNPs and leading to medulloblastoma. In addition, tumors from both *Ptch1*$^{+/-}$ and *Ptch1*$^{+/-}$; *Hic1*$^{+/-}$ mice showed silencing of *Hic1* via promoter methylation. Loss of *Hic1* expression was tumor-cell specific, as surrounding blood vessels and fibroblasts showed normal expression of *Hic1*, suggesting that efficient silencing of *Hic1* is important in tumor formation.

Even though HIC1 appears to be downstream of the SHH pathway by repressing *Math1*, chromosome 17 aberrations like i17q, 17p losses and 17q gains are usually found in groups C and D medulloblastoma, and not in the SHH tumors (17), raising the question of whether HIC1 may function independently of SHH, or whether other loci on chromosome 17 contribute to tumor formation in group C and D tumors. Investigations of *Hic1* loss in mouse models that develop group C and D tumors will be necessary to shed light on the role of *Hic1* in medulloblastoma tumorigenesis.

4. Overexpression of *Ifng*
   Interferon-gamma was overexpressed in astrocytes via the hGFAP promoter, using the tetracycline-controllable system of the tetracycline transcriptional activator (tTA) and tetracycline response element (TRE) in *hGFAP-tTA; TRE-Ifng* mice (155). Early expression of *Ifng* starting from E16 resulted in more than 80% of mice developing medulloblastomas by 3–4 weeks. These tumors were found in the molecular layer and EGL of the cerebellum, and also invaded into the white matter and brainstem regions. Tumors displayed activation of SHH signaling including induction of SHH expression itself and had expression of synaptophysin, EGL marker MATH1. Doxycycline was used to shut down expression of *Ifng* at various developmental time points, and *Ifng* was found to be required for early tumor initiation before P16 when the EGL is actively proliferating. Shutting down transgene expression at P12 resulted in no tumor formation, while shutting down transgene at P16 still resulted in tumor formation and progression that led to mice lethality. However, tumors that developed with continued *Ifng* expression displayed more necrosis, apoptosis, and immune infiltration that were absent in the tumors that progressed without *Ifng* expression. This result is consistent with interferon-gamma being a proinflammatory cytokine and likely plays a role in inducing host response from tumor formation.

4.3. Non-SHH Driven

Although most mouse models of medulloblastoma have focused on the SHH pathway or resulted in tumors which upregulated SHH signaling, SHH-driven tumors only make up about 25–30% of all cases, and these patients tend to have better prognosis. Below, we will discuss mouse models of non-SHH tumors that have been developed by activating the WNT pathway and overexpressing *MYCN*.

1. Activation of WNT pathway—activated β-catenin
   WNT drives a molecularly distinct class of tumors associated with *APC* or *CTNNB1* mutations and high expression of c-MYC. The RCAS/Gtv-a system was used to overexpress human *c-MYC* and constitutively active S37A mutant β-catenin in $p53^{-/-}$ mice. Tumors with LCA histopathology developed in five of six mice within 6 weeks (156). The aggressive histology

of these tumors may in part be due the introduction of genes into a larger population of GFAP-positive cells instead of Nestin-positive CGNPs (Ntv-a mice) used in the SHH studies. However, the fidelity of this model to human cases is questionable: although the dysregulation of c-MYC and β-catenin is characteristic of WNT tumors, human tumors are almost exclusively classic and rarely LCA tumors.

In another study, Gibson et al. demonstrated that medulloblastomas could be initiated from *Blbp*-expressing cells in the lower rhombic lip/dorsal brainstem via activated β-catenin combined with *p53* loss (50). The *Blbp-Cre* driver was used to drive mutant β-catenin in all cerebellar progenitor cell populations in the ventricular zone, EGL (CGNP population), and lower rhombic lip. Cells in the ventricular zone and EGL were unaffected by mutant β-catenin. In contrast, Zic1-positive postmitotic mossy-fiber neuron precursor cells failed to migrate from the dorsal brainstem to form pontine gray nuclei in the ventral brainstem, resulting in their accumulation in the dorsal brainstem. However, for this hyperplasia to form, classic medulloblastoma tumors required partial or complete loss of *p53* (4% and 15% incidence respectively, with a median time of 10 months). The resulting tumors were in contact with the brainstem and had a transcriptome similar to human WNT tumors. The authors propose that WNT tumors originate from the lower rhombic lip/dorsal brainstem region and are distinct from SHH tumors that originate from CGNPs in the upper rhombic lip/EGL.

The murine data matches MRI and clinical reports of human medulloblastomas: WNT tumors tend to infiltrate the dorsal brainstem while SHH tumors tend to grow within the cerebellar parenchyma (50). Altogether, this study demonstrates that the WNT subclass of tumors likely originates from a distinct population in the lower rhombic lip/dorsal brainstem, and opens up questions about whether specific transcriptomal subclasses of medulloblastomas arise from distinct cellular origins or regions.

Note also that in these two models, deletion of *p53* was also introduced into the model to drive tumorigenesis.

2. *MYCN* overexpression

Expression profiling of human SHH tumors and SHH-driven mouse models consistently reveals high levels of *MYCN* (*NMYC* in mouse). However, RCAS/tv-a studies have demonstrated that wild-type or mutationally stabilized NMYC T58A alone is insufficient to initiate medulloblastoma formation, suggesting that other effectors are required to cooperate in tumorigenesis (130). Consistent with these observations, medulloblastoma tumors can be initiated by a *MYCN* transgene, generating a model with recurrent genomic aberrations, suggesting that additional events cooperate with *MYCN* in this model. In this model, the *G*lutamate transporter (Glt1) promoter and a bidirectional *T*et-operator were used to drive bidirectional and simultaneous expression of human *MYCN* and *l*uciferase in the cerebellum (GTML), resulting in primarily SHH-independent tumors with classic or LCA histopathology (59). Notably, a rare number of mice show spinal metastases, making this model one of the few that have documented leptomeningeal spread of the malignant primary brain tumor.

GTML tumors demonstrated genomic instability, with frequent gains in mouse chromosome 11. These regions are syntenic to regions of human chromosome 17q, where copy number aberrations are frequently observed in human tumors and are associated with poor prognosis (4, 17). Thus, while *MYCN* is able to initiate medulloblastoma tumors, additional genetic aberrations are subsequently required. The use of Glt1 promoter to drive *MYCN* in this transgenic model likely targets a larger pool of cerebellar cells compared to the RCAS/tv-a retroviral system, increasing the number of cells that could acquire a secondary mutation. *MYC/MYCN* amplification is highly correlated with poor prognosis and aggressive LCA subtype in human tumors (52), thus supporting the use of this model in pursuing investigations of high-risk patients with poor prognosis.

- Remarks

  In the following section, we highlight some of the main concerns of current medulloblastoma models and some important lessons that we have gathered from investigations thus far.

  1. Use of *p53* loss/relevant genetic changes

     It is striking that only (most of) SHH-driven models and the *MYCN*-driven GTML model have driven tumor formation without the deletion of *p53*, while other models have all relied on additional loss of *p53* to drive tumorigenesis. As highlighted before, *p53* loss is not a prominent feature in most human cases, only accounting for 10–15% of all cases. Hence, although medulloblastomas develop in these mouse models, the relevance and use of drug screens and preclinical testing using these models should proceed with caution.

     Additionally, in the light of new information about genetic aberrations associated with each molecular subclass of medulloblastoma, we should evaluate the relevance of genetic changes engineered in mouse models. Some of the genetic changes not commonly seen in human cases include combined c-MYC and SHH overexpression, combined *Rb* and *p53* loss and deletion of DNA repair enzymes. In order for investigations to translate to the clinical setting, the use of relevant mouse models that mimic closely to the human disease is critical to achieve maximum benefit for patients.

2. Appropriate histological representations of molecular subclasses

   Mouse models not only should mimic genetic events of molecular subtypes but also should faithfully recapitulate the corresponding histologies found in these subtypes. It is unfortunate that current mouse models largely model the SHH tumor subclass. In addition, these models report mainly tumors of classic and not desmoplastic histology, which represent a typical histopathology of SHH-driven tumors. There is a dire lack of mouse models of non-SHH tumors with relevant genetic aberrations and histology. In particular, more LCA tumor models modeling non-WNT, non-SHH tumors are needed, especially since these patients represent the group with the poorest prognoses, with only 25–40% patients achieving 5-year event-free survival.

3. Different cell of origins in subtypes

   Results from two particular mouse models highlight that different medulloblastoma variants may originate from different anatomical regions and cellular populations. Firstly, β-catenin loss mediated by *Blbp-Cre* in *p53*-null mice resulted in an expansion of *Zic1*-positive mossy fiber neuronal precursor cells in the lower rhombic lip/dorsal brainstem region, suggesting that the WNT subclass of tumors may originate from this region, and not the EGL as with SHH tumors. The authors further confirmed anatomical differences in human WNT and SHH tumors. Secondly, in *MYCN*-driven GTML medulloblastoma tumors, levels of EGL marker *Math1* and SHH downstream target *Gli1* were found to be low, suggesting that GTML tumors had a non-EGL origin. There were no noticeable lesions or thickening in the EGL layer in P7 GTML mice and isolated EGL cells from GTML or wild-type mice had comparable proliferation and survival. This suggests that non-SHH, non-WNT tumors may also have a distinct cell of origin from the EGL CGNPs that form SHH-driven tumors. Altogether, these two studies highlight the need for further investigations to better define the heterogeneity of medulloblastoma subclasses and understand the biology of these tumors.

   Additionally, recall that *Rb* and *p53* deletion in both postnatal cerebellar NSCs and CGNPs can serve as medulloblastoma initiating cells and result in tumors with different profiles. The *Rb/p53* conditional knockout model using the GFAP-Cre driver demonstrated that CGNP population gave rise to tumors with upregulated SHH pathway that were weakly positive for stem cell markers, while *Rb/p53*-deleted cerebellar NSCs that were CD133-positive and Math1-negative (non-CGNPs) gave rise to tumors with low SHH activation but were positive for stem cell markers. (146).

4. Metastasis modeling

One critical factor that influences patient prognoses is leptomeningeal spread (metastasis) of the primary tumor. About 30% of patients present with evidence of disease dissemination at initial diagnosis, and this is one of the clear indicators of poor prognosis. However, of the engineered mouse models, few have successfully modeled leptomeningeal spread of the tumor. This greatly hampers our ability to identify genes that drive metastasis of the tumor, as well as to develop and screen treatment strategies that are effective for these patients with poor prognoses. Identifying genetic factors promoting tumor spread is critical for future efforts to model tumor leptomeningeal spread in mice.

5. Current and Future Use of Medulloblastoma Models

5.1. Inhibitor Screening and Preclinical Testing

In the light of the molecular subclassification of medulloblastomas, it is clear that development of treatment strategies should take into consideration the molecular and histological differences. Inhibitors targeting activated WNT pathway, and SHH/NOTCH pathways, will likely be effective against WNT and SHH tumors respectively. Further, inhibitors that curb *MYC/MYCN*-mediated proliferation and malignancy are promising targets to treat *MYC/MYCN*-amplified aggressive LCA tumors that currently have dismal prognoses. In designing targeted treatments regimens, we should keep in mind the need to maximize disease control, without compromising the long-term quality of life of patients, especially for the patients who have better prognoses.

A number of inhibitors have been tested in vitro and/or in vivo for their efficacy in inhibiting medulloblastoma tumor growth. Unfortunately, for ease of investigations, only some of these inhibitors have been tested on transgenic mouse models, while many inhibitors have only been tested in allograft systems using immunocompromised mice. With more GEM models of medulloblastoma being developed, it will be favorable that inhibitors and treatment protocols are tested and validated in these models before their enrollment into clinical trials for patients.

1. SMO antagonists

The early implication of SHH pathway in medulloblastoma and the vast efforts of SHH tumor modeling have led to the preclinical testing of a number of SHH inhibitors. Cyclopamine is a steroidal alkaloid isolated from the corn lily *Veratrum californicum* that was discovered to inhibit tissue response to SHH signaling (157, 158), and was effective in blocking growth of tumor cells with activated SHH signaling such as medulloblastoma (105, 159). Cyclopamine acts by binding directly to SMO, inhibiting its function of activating downstream SHH signaling (160, 161). Treatment of mice with subcutaneous allograft tumors derived from *Ptch*$^{+/-}$; *p53*$^{+/-}$ mice with cyclopamine at 50 mg/kg daily decreased in volume due to inhibited cell proliferation (105).

However, cyclopamine has a relatively low affinity for SMO. Hence, later drug screens identified more attractive SMO inhibitors (based on effectiveness, stability, solubility, complexity, ease of manufacturing, etc.) as therapeutic agents (162, 163). HhAntag-691 (HhAntag) is a benzimidazole derivative that has a higher affinity for SMO, is effective at tenfold lower concentrations than cyclopamine, and also penetrated the blood–brain barrier via oral delivery just like cyclopamine (163). A 2-week HhAntag treatment of *Ptch1*$^{+/-}$; *p53*$^{-/-}$ mice at 20 mg/kg reduced tumor volume, while high dose 100 mg/kg treatment completely eliminated tumors (163). Analysis of tumors displayed suppression of SHH downstream gene expression, such as *Gli1/2*, *Ptch2*, *Sfrp1*, and *Math1*, inhibition of cell proliferation and induction of cell death. Long-term treatment at 100 mg/kg led to prolonged disease-free survival from 39 days (vehicle) to 147 days, with no observed weight loss or deleterious side effects, hence supporting the use of SHH antagonists for patients. However, the potential success of this drug was put into question when a further study demonstrated that a short-term 2-day drug exposure in young mice resulted in permanent bone growth defects although SHH signaling was restored with drug removal (164). This highlights the potential danger of affecting bone development in pediatric patients who make up the majority of medulloblastoma tumor patients.

GDC-0449 was identified from another screen of benzimidazole derivatives that inhibits SMO at a higher efficacy than cyclopamine. Treatment of allograft models of *Ptch1*$^{+/-}$ tumor cells at doses of 12.5–100 mg/kg resulted in successful tumor regression (162, 165), with corresponding downregulation of *Gli1* expression (162). A 26-year-old medulloblastoma patient with metastatic disease who harbored an inactivating mutation of *Ptch1* (W844C) was treated with GDC-0449 and this treatment effectively regressed tumors (165). Unfortunately, the treatment was transient and most of the metastatic tumors recurred. Analysis of one of his biopsies revealed the acquisition of a missense mutation in *Smo* resulting in amino acid change D473H and that disrupted drug binding to SMO and conferred tumor cell resistance to GDC-0449 treatment

(166). Strikingly, one of the GDC-0449-resistant strains of *Ptch1*$^{+/-}$; *p53*$^{-/-}$ tumor cells that was derived from allograft mice also harbored a missense mutation that exactly corresponded to the mutation in the patient. Further screens of inhibitors have since identified bis-amide compound 5 that could inhibit tumor growth of this resistant tumor strain (167).

Similar to the previous drug screens, a semisynthetic analogue of cyclopamine IPI-926 was selected for its improved potency and pharmacokinetic properties compared to cyclopamine. Treatment of subcutaneous allograft tumors of *Ptch1*$^{+/-}$; *Hic1*$^{+/-}$ cells with IPI-926 at a dose of 40 mg/kg led to a regression of tumors in a median time of 9 days with no observed tumor recurrence 21 days post-treatment (168). Lastly, NVP-LDE225 is a biphenyl carboxamide compound that was identified as another SMO antagonist that displaces the binding of SHH ligand (169). Daily NVP-LDE225 treatment of mice with subcutaneous *Ptch1*$^{+/-}$; *p53*$^{-/-}$ tumors at 20 mg/kg resulted in near-complete tumor regression for the initial 2 weeks (170). However, all tumors developed resistance and regrew, with accompanying re-expression of SHH pathway signaling molecules like *Gli1* and *Cyclin D1*. Similar results were obtained from allografts of tumor cells from *Ptch1*$^{+/-}$; *Hic1*$^{+/-}$; and *Ptch1*$^{+/-}$ mice. Further analysis of resistant tumors revealed *Gli2* amplification, some rare missense mutations in *Smo* and upregulation of IGF/PI3K signaling. Accordingly, concurrent treatment with PI3K inhibitors and NVP-LDE225 could delay resistant tumor regrowth after initial tumors had subsided, but addition of PI3K inhibitors after resistant tumors were established could not inhibit their growth. This suggests that the resistant tumors were dependent on PI3K signaling for tumor initiation, but not for maintenance.

2. Other SHH pathway inhibitors
   In addition to the SMO antagonists, a few other drug screens have identified some compounds not previously known to inhibit the SHH signaling pathway. Arsenic trioxide (ATO) is an FDA-approved drug for use in acute promyelocytic leukemia, believed to degrade promyelocytic leukemia-retinoic acid receptor α (PML-RAR) fusion protein. ATO was identified as an effective GLI inhibitor by directly binding to GLI1 (and GLI2) protein to inhibit transcriptional activity and decreasing expression of downstream target genes (171). ATO was administered to tumor-bearing *ND2:SmoA1* homozygous mice 0.15 mg/kg three times a week and this resulted in significant increase in survival from 4.8 months to 7.4 months, with corresponding decrease in GLI target genes like *Ptch1* and *N-myc*. ATO is an FDA-approved drug for use in acute promyelocytic leukemia patients, hence the authors suggest that ATO can be incorporated in medulloblastoma treatment as an inhibitor of SHH signaling targeting GLI.

In another study, a systemic antifungal, itraconazole, was identified as an antagonist of SHH signaling (172). The SHH pathway effects were separate from its inhibitory effects on fungal sterol biosynthesis, as other azole antifungal drugs did not exhibit SHH pathway inhibitory abilities. Interestingly, although itraconazole blocked SHH signaling at the level of SMO, its mechanism was distinct from cyclopamine in that itraconazole prevented ciliary accumulation of SMO (172), while cyclopamine allows the accumulation of SMO at the primary cilium but blocks full activation of SMO to activate downstream signaling (173, 174). Daily itraconazole treatment of mice with subcutaneous allografts of *Ptch1*$^{+/-}$; *p53*$^{-/-}$ tumor cells at a dose of 75 or 100 mg/kg inhibited tumor growth, and tumor inhibition was enhanced with combination treatment with cyclopamine. Tumors revealed a reduction in SHH pathway activity reflected by reduced *Gli1* expression.

Clinically approved glucocorticoid dexamethasone was administered at a dose of 3 μg/g to *Math1-Cre:SmoM2* tumor-prone P10 mice and this increased survival of these mice modestly from 16 to 25 days (175). Dexamethasone-treated mice were found to have lower NMYC protein levels, while GLI1 protein levels were unchanged. Further analysis of dexamethasone treatment on tumor cells suggest that glucocorticoid signaling activated GSK-3β, thus leading to increased phosphorylation of NMYC at T58 which promotes its subsequent degradation. When conditional *N-myc* deletion was introduced in *SmoM2* mice, 15% of mice that successfully knocked out *N-myc* survived and were disease-free, while the remaining mice failed to recombine and delete *N-myc*, resulting in medulloblastoma formation. Altogether, this suggests that *N-myc* downstream of SHH signaling plays a pivotal role in promoting medulloblastoma formation and maintenance, and that glucocorticoids may be effective in targeting SHH signaling in patients with SHH-driven tumors.

Lastly, in the *Pax7-Cre*-mediated *Ptch1* and *p53* compound mutants, further analysis revealed that *Ptch1* mRNA expression was unaltered but protein expression was absent, suggesting that *Ptch1* was regulated post-transcriptionally. A proteasome inhibitor, bortezomib/PS-341 was administered to tumor-bearing mice when they showed symptoms of tumor onset (122). This increased survival of mice from 1 day to 10 days post-treatment and restored PTCH1 protein levels in tumor cells, demonstrating that proteasome-mediated PTCH1 degradation may be important in medulloblastoma tumor progression.

3. Epigenetic therapy targeting SHH pathway

Agents targeting epigenetic regulation such as histone deacetylase inhibitors have shown success in treatment of hematologic and solid tumors by permitting transcription of genes such as tumor suppressors, and those regulating cell cycle and apoptosis. DNA methylating agent 5-azaC restored *Ptch1* expression in tumor cells derived from *Ptch*$^{+/-}$; *p53*$^{+/-}$ mice (105), suggesting that the remaining *Ptch1* allele in these mice is epigenetically silenced via promoter methylation. Subsequently, other studies have investigated the effectiveness of inhibiting epigenetic silencing mechanisms in treating medulloblastoma in mouse models.

Suberoylanilide hydroxamic acid (SAHA) is a histone deacetylase inhibitor that is known to cross the blood–brain barrier and has shown some promise in treatment of glioma and other tumors. *ND2:SmoA1* mice were administered SAHA at a dose of 200 mg/kg daily for 3 days and tumors had significantly higher levels of apoptosis compared to controls (176). These results were mirrored in human medulloblastoma cell line xenografts in athymic mice and cell culture studies (176). In a follow-up study, SAHA was combined with 13-cis retinoic acid (RA), a drug that has been used in treatment of cancers leukemia and neuroblastoma (177). RA has been shown to cause apoptotic cell death in medulloblastoma tumor cells by inducing *Bmp2* transcription that activates p38 MAPK signaling and SAHA was found to cooperate with RA to induce *Bmp2* transcription and boost apoptotic effects in a p38-independent manner (177, 178). Combination therapy of SAHA and RA in conjunction with a very low dose of chemotherapy drug cisplatin was tested on the *ND2:SmoA1* model, and this combination achieved the highest tumor cell apoptosis levels, compared to single or double drug therapy, while causing no toxicity in the normal brain regions (177).

DNA methyltransferase1 (DNMT1) is one of the Dnmt enzymes involved in DNA methylation at gene promoters that results in transcriptional silencing of these genes. Genetic reduction of *Dnmt1* has been shown in mice to reduce tumor foci numbers with lung and intestinal tumors, and *Dnmt* inactivation has been demonstrated to lead to increased histone acetylation. Compound *Dnmt1* and *Ptch1* double heterozygotes had decreased medulloblastoma tumor incidence of 47% compared to the *Ptch1*$^{+/-}$ mice, suggesting that reduced *Dnmt1* activity delayed tumor growth (179). In addition, combination treatment of DNMT inhibitor 5-aza-2′deoxycytidine (5-aza-dC) and histone deacetylase (HDAC) inhibitor valproic acid (VPA) resulted in reduced tumor incidence of 17% compared to 44% in untreated control mice (179). Separately, compound *Dnmt1* and *Ptch1* mutants in a different genetic background develop rhabdomyosarcomas, and 5-aza-dC/VPA combination therapy also reduced tumor incidence in the mice. Further analysis of these rhabdomyosarcomas revealed that the tumors from treated mice displayed an upregulation of *Ptch1* expression compared to the controls, suggesting the *Ptch1* is usually silenced in the tumors and its expression was reactivated with the 5-aza-dC/VPA treatment. Together, this suggests that both pharmacologic and genetic intervention with epigenetic mechanisms of methylation and histone deacetylation likely results in the reversal of *Ptch1* silencing that inhibits formation of SHH-driven tumors.

4. Other non-SHH therapy strategies

   Other classes of inhibitors have been reported to have potential therapeutic benefits for patients. In the *Pax7-Cre*-mediated *Ptch1* and *p53* compound mutants, IGF signaling was found to be upregulated, with increased mRNA and protein levels of IGF1R receptor (180). Picropodophyllin, an inhibitor of IGF1R tyrosine phosphorylation, effectively inhibited proliferation of tumor cells isolated from the murine tumors and cultured in vitro (180). Accordingly, phosphorylation of downstream targets AKT and MAPK were reduced and apoptosis was induced in tumor cells. From these studies, it appears that inhibitors that target other pathways such as the IGF signaling or proteasome function may serve as possible alternative strategies to curb SHH tumor growth, instead of the direct inhibitors against the SHH pathway members.

   In the *SmoA1* mouse model, recall that tumors were found to concurrently upregulate NOTCH pathway with SHH pathway. Consequently, inhibition of NOTCH pathway using soluble Delta ligand or gamma-secretase inhibitors could decrease proliferation and induce apoptosis in vitro. This suggests that SHH tumors that exhibit concurrent NOTCH activation may be effectively treated by targeting the NOTCH pathway.

   Concurrent introduction of SHH and HGF using the RCAS/tv-a system induced a high incidence of medulloblastoma. To test if these tumors were responsive to SHH and/or HGF-based therapy, the authors tested single or combination therapy using HGF-neutralizing antibody L2G7 and SHH pathway inhibitors such as cyclopamine and SHH-neutralizing antibody 5E1 (128, 181). Agents were administered to mice 2 weeks after RCAS virus introduction to induce tumor formation. Single therapy using either HGF- or SHH-targeted agents all successfully improved survival, although HGF neutralizing antibody L2G7 was more effective than either SHH agents, increasing median survival time to 111 days (L2G7) compared to 62 (cyclopamine) and 102 days (5E1). Survival of control mice treated with non-specific antibody was only 23 days. Histochemical analysis of the harvested tumors showed increased apoptosis with very slight decreased proliferation in the tumors of mice that had better survival. Interestingly, combination of

L2G7 with SHH agents achieved similar or even worse results than L2G7 single agent therapy. Assess of HGF signaling of these tumors revealed that addition of SHH agents may have partially reversed the suppression of c-MET phosphorylation achieved by L2G7 single therapy although the mechanism is still unclear. HGF has been shown to play an important role in medulloblastoma tumorigenesis in a subset of patients, although these tumors tend to be of LCA histology and have poor prognosis, segregating away from the SHH-driven tumors. Regardless, the study hints at the potential use of HGF-targeted therapy for patients with elevated HGF levels.

5. Remarks

Altogether, various studies have shown some success in identifying candidate compounds from inhibitor screens and treating medulloblastoma in GEM and allograft models. Follow-up studies are important to examine the effects of these compounds, the likelihood of acquiring resistance against these drugs and identifying newer, more effective drugs that can overcome or eliminate these potential drug resistances.

On a separate note, some of the models described attain only a low tumor incidence with a variable tumor latency, which does not permit investigators to predict the age and probability of tumor onset. Hence, this complicates efforts to evaluate the effectiveness of inhibitors or regimens in preclinical testing. Our hopes are that future mouse models that correctly identified major drivers of the disease in the different subclasses will induce tumor formation faithfully to facilitate future research.

5.2. Cancer Stem Cells and Tumor Resistance

The identification of the subset of tumor cell populations that is responsible for propagating tumor growth from continued proliferation is important so that we can identify treatment strategies that not only target the tumor bulk but also eradicate the "cancer stem cell" population that serves as the tumor cell source. One of the concerns in cancer patients is the risk of tumor relapse, and cancer stem cells have been shown in other tumors to be also the cells that are resistant to treatments like chemotherapy and radiation that will rebound and lead to tumor regrowth. In human neural tumors such as medulloblastomas, astrocytomas, and ependymomas, CD133 has been identified as a marker of the "cancer stem cell" population that have tumor-propagating properties. However, in the *Ptch1*$^{+/-}$ mouse model, it was discovered that the tumor-propagating cells were defined not by CD133 marker, as CD133-negative sorted cells resulted in tumors from an orthotopic transplant and not the CD133-positive cells (182). Instead, cells positive for CD15/SSEA-1 within MATH1-positive CGNP cells resulted in secondary tumors upon transplant (182, 183). These CD15-positive cells upregulated genes associated with proliferation, self-renewal, and anti-apoptosis with corresponding downregulation of pro-apoptotic genes and differentiation. Further, CD15-positive cells had increased expression of SHH pathway genes such as *Cyclin D1* and *Gli1* that pointed to increased SHH pathway activation. Lastly, human medulloblastomas which had an expression profile more similar to that of CD15-positive cells in the mouse model were associated with poorer prognosis, suggesting that CD15-positive cells define the "cancer stem cell" population in medulloblastomas that will associate with more aggressive phenotype.

5.3. Treatment Response

Current standard therapy for medulloblastoma patients includes radiation and chemotherapy, but much of the treatment-induced effects remain unclear. Patients also have continued risk of tumor relapse from the resistant tumor cells that proliferate and repopulate the primary tumor site, or even invade to secondary sites. As we develop more GEM models that faithfully recapitulate aspects of the disease, we can then try to understand how the tumor and surrounding tissue react to these treatment regimens. This will open up avenues to better refine our treatment strategies to select combination therapy that target treatment-resistant tumor populations and modulate the tumor microenvironment to increase survival prognosis and prevent relapse.

The effect of radiation on tumor growth and resistance was investigated by using the $Ptch1^{+/-}$ mouse model and the RCAS/Ntv-a mouse models using co-introduction of SHH with AKT or NMYC (135). Radiation induced a widespread apoptotic cell death in the tumor bulk in all the models, but there were some resistant Nestin-positive stem cell-like tumor cells that remain post-irradiation in the perivascular regions. These radiation-resistant cells showed activation of PI3K/AKT/mTOR signaling and elevation of p53 levels and underwent a brief cell cycle arrest before resuming proliferation after 72 h. When the RCAS/Ntv-a models were crossed to a *p53*-null background, radiation no longer had effects on the proliferating tumor bulk that showed little apoptosis, and the perivascular stem-like cells did not arrest like in the *p53* wild-type background (135). Hence, radiation-induced apoptotic cell death and cell cycle arrest of the tumor cells were *p53*-dependent. Since the radiation-resistant cells showed an activation of PI3K/AKT signaling, the mice were pre-treated with AKT inhibitor perifosine 3 days before irradiation and this combination therapy resulted in decreased survival of this resistant cell, suggesting the potential use of this combination therapy to prevent tumor relapse in patients (135). In contrast, concurrent SHH overexpression with PTEN loss in mice resulted in nodular tumors and these cells were largely non-proliferating and not affected by radiation (135). In addition, the perivascular Nestin-positive cells in these tumors did not arrest and remained highly proliferative with no associated *p53* elevation. These results demonstrate that post-irradiation cell cycle arrest and *p53* increase was PTEN-dependent.

5.4. The Future of Medulloblastoma Modeling and Research

Several of the mouse models have only been developed in the recent few years. Hence, we look forward to more studies using these models to identify drug targets and relevant inhibitors that have potential to serve as novel therapies for medulloblastoma patients. This will include targeted therapy for different molecular variants and relapse patients. Further, we are now more equipped with profiling information from human medulloblastoma tumor samples that have come out of various research groups. Hence, future mouse models can be engineered to mimic as closely the aberrations that exist in the distinct tumor subclasses for more relevant studies.

References

1. Rossi A, Caracciolo V, Russo G, Reiss K, Giordano A (2008) Medulloblastoma: from molecular pathology to therapy. Clin Cancer Res 14:971–976
2. Gilbertson RJ (2004) Medulloblastoma: signalling a change in treatment. Lancet Oncol 5:209–218
3. Schmidt AL, Brunetto AL, Schwartsmann G, Roesler R, Abujamra AL (2010) Recent therapeutic advances for treating medulloblastoma: focus on new molecular targets. CNS Neurol Disord Drug Targets 9:335–348
4. Huse JT, Holland EC (2009) Genetically engineered mouse models of brain cancer and the promise of preclinical testing. Brain Pathol 19:132–143
5. Packer RJ, Vezina G (2008) Management of and prognosis with medulloblastoma: therapy at a crossroads. Arch Neurol 65:1419–1424
6. Eberhart CG (2011) Molecular diagnostics in embryonal brain tumors. Brain Pathol 21:96–104
7. Boman KK, Hovén E, Anclair M, Lannering B, Gustafsson G (2009) Health and persistent functional late effects in adult survivors of childhood CNS tumours: a population-based cohort study. Eur J Cancer 45:2552–2561
8. Mulhern RK, Palmer SL, Merchant TE, Wallace D, Kocak M, Brouwers P, Krull K, Chintagumpala M, Stargatt R, Ashley DM, Tyc VL, Kun L, Boyett J, Gajjar A (2005) Neurocognitive consequences of risk-adapted therapy for childhood medulloblastoma. J Clin Oncol 23:5511–5519
9. Frange P, Alapetite C, Gaboriaud G, Bours D, Zucker JM, Zerah M, Brisse H, Chevignard M, Mosseri V, Bouffet E, Doz F (2009) From childhood to adulthood: long-term outcome of medulloblastoma patients. The Institut Curie experience (1980–2000). J Neurooncol 95:271–279
10. Oyharcabal-Bourden V, Kalifa C, Gentet JC, Frappaz D, Edan C, Chastagner P, Sariban E, Pagnier A, Babin A, Pichon F, Neuenschwander S, Vinchon M, Bours D, Mosseri V, Le Gales C, Ruchoux M, Carrie C, Doz F (2005) Standard-risk medulloblastoma treated by adjuvant chemotherapy followed by reduced-dose craniospinal radiation therapy: a French Society of Pediatric Oncology Study. J Clin Oncol 23:4726–4734
11. Pizer BL, Clifford SC (2009) The potential impact of tumour biology on improved clinical practice for medulloblastoma: progress towards biologically driven clinical trials. Br J Neurosurg 23:364–375
12. Taylor RE, Bailey CC, Robinson KJ, Weston CL, Walker DA, Ellison D, Ironside J, Pizer BL, Lashford LS (2005) Outcome for patients with metastatic (M2-3) medulloblastoma treated with SIOP/UKCCSG PNET-3 chemotherapy. Eur J Cancer 41:727–734
13. Saran A (2009) Medulloblastoma: role of developmental pathways. DNA repair signaling, and other players. Curr Mol Med 9:1046–1057
14. Pfister SM, Korshunov A, Kool M, Hasselblatt M, Eberhart C, Taylor MD (2010) Molecular diagnostics of CNS embryonal tumors. Acta Neuropathol 120:553–566
15. Momota H, Holland EC (2009) Mouse models of CNS embryonal tumors. Brain Tumor Pathol 26:43–50
16. CBTRUS (2010). CBTRUS statistical report: primary brain and central nervous system

tumors diagnosed in the United States in 2004–2006. Central Brain Tumor Registry of the United States, Hinsdale, IL. www.cbtrus.org. pp 1–61.

17. Northcott PA, Korshunov A, Witt H, Hielscher T, Eberhart CG, Mack S, Bouffet E, Clifford SC, Hawkins CE, French P, Rutka JT, Pfister S, Taylor MD (2010) Medulloblastoma comprises four distinct molecular variants. J Clin Oncol 29(11):1408–1414
18. Padovani L, Sunyach M-P, Perol D, Mercier C, Alapetite C, Haie-Meder C, Hoffstetter S, Muracciole X, Kerr C, Wagner J-P, Lagrange J-L, Maire J-P, Cowen D, Frappaz D, Carrie C (2007) Common strategy for adult and pediatric medulloblastoma: a multicenter series of 253 adults. Int J Radiat Oncol Biol Phys 68:433–440
19. Kunschner LJ (2002) Harvey Cushing and medulloblastoma. Arch Neurol 59:642–645
20. von Hoff K, Hinkes B, Gerber NU, Deinlein F, Mittler U, Urban C, Benesch M, Warmuth-Metz M, Soerensen N, Zwiener I, Goette H, Schlegel PG, Pietsch T, Kortmann RD, Kuehl J, Rutkowski S (2009) Long-term outcome and clinical prognostic factors in children with medulloblastoma treated in the prospective randomised multicentre trial HIT'91. Eur J Cancer 45:1209–1217
21. Zeltzer PM, Boyett JM, Finlay JL, Albright AL, Rorke LB, Milstein JM, Allen JC, Stevens KR, Stanley P, Li H, Wisoff JH, Geyer JR, McGuire-Cullen P, Stehbens JA, Shurin SB, Packer RJ (1999) Metastasis stage, adjuvant treatment, and residual tumor are prognostic factors for medulloblastoma in children: conclusions from the Children's Cancer Group 921 randomized phase III study. J Clin Oncol 17:832–845
22. Albright AL, Wisoff JH, Zeltzer PM, Boyett JM, Rorke LB, Stanley P (1996) Effects of medulloblastoma resections on outcome in children: a report from the Children's Cancer Group. Neurosurgery 38:265–271
23. Louis DN, Ohgaki H, Wiestler OD, Cavenee WK, Burger PC, Jouvet A, Scheithauer BW, Kleihues P (2007) The 2007 WHO classification of tumours of the central nervous system. Acta Neuropathol 114:97–109
24. Gilbertson RJ, Ellison DW (2008) The origins of medulloblastoma subtypes. Annu Rev Pathol 3:341–365
25. Biegel JA, Zhou JY, Rorke LB, Stenstrom C, Wainwright LM, Fogelgren B (1999) Germline and acquired mutations of INI1 in atypical teratoid and rhabdoid tumors. Cancer Res 59:74–79
26. Ellison D (2002) Classifying the medulloblastoma: insights from morphology and molecular genetics. Neuropathol Appl Neurobiol 28:257–282
27. Rutkowski S, von Hoff K, Emser A, Zwiener I, Pietsch T, Figarella-Branger D, Giangaspero F, Ellison DW, Garre M-L, Biassoni V, Grundy RG, Finlay JL, Dhall G, Raquin M-A, Grill J (2010) Survival and prognostic factors of early childhood medulloblastoma: an international meta-analysis. J Clin Oncol 28:4961–4968
28. Schüller U, Heine VM, Mao J, Kho AT, Dillon AK, Han Y-G, Huillard E, Sun T, Ligon AH, Qian Y, Ma Q, Alvarez-Buylla A, McMahon AP, Rowitch DH, Ligon KL (2008) Acquisition of granule neuron precursor identity is a critical determinant of progenitor cell competence to form Shh-induced medulloblastoma. Cancer Cell 14:123–134
29. Borghesani PR, Peyrin JM, Klein R, Rubin J, Carter AR, Schwartz PM, Luster A, Corfas G, Segal RA (2002) BDNF stimulates migration of cerebellar granule cells. Development 129:1435–1442
30. Abrahám H, Tornóczky T, Kosztolányi G, Seress L (2001) Cell formation in the cortical layers of the developing human cerebellum. Int J Dev Neurosci 19:53–62
31. Evans DG, Farndon PA, Burnell LD, Gattamaneni HR, Birch JM (1991) The incidence of Gorlin syndrome in 173 consecutive cases of medulloblastoma. Br J Cancer 64:959–961
32. Johnson RL, Rothman AL, Xie J, Goodrich LV, Bare JW, Bonifas JM, Quinn AG, Myers RM, Cox DR, Epstein EH, Scott MP (1996) Human homolog of patched, a candidate gene for the basal cell nevus syndrome. Science 272:1668–1671
33. Hahn H, Wicking C, Zaphiropoulous PG, Gailani MR, Shanley S, Chidambaram A, Vorechovsky I, Holmberg E, Unden AB, Gillies S, Negus K, Smyth I, Pressman C, Leffell DJ, Gerrard B, Goldstein AM, Dean M, Toftgard R, Chenevix-Trench G, Wainwright B, Bale AE (1996) Mutations of the human homolog of Drosophila patched in the nevoid basal cell carcinoma syndrome. Cell 85:841–851
34. Raffel C, Jenkins RB, Frederick L, Hebrink D, Alderete B, Fults DW, James CD (1997) Sporadic medulloblastomas contain PTCH mutations. Cancer Res 57:842–845
35. Dong J, Gailani MR, Pomeroy SL, Reardon D, Bale AE (2000) Identification of PATCHED mutations in medulloblastomas by direct sequencing. Hum Mutat 16:89–90

36. Vorechovský I, Tingby O, Hartman M, Strömberg B, Nister M, Collins VP, Toftgård R (1997) Somatic mutations in the human homologue of Drosophila patched in primitive neuroectodermal tumours. Oncogene 15:361–366
37. Pietsch T, Waha A, Koch A, Kraus J, Albrecht S, Tonn J, Sörensen N, Berthold F, Henk B, Schmandt N, Wolf HK, von Deimling A, Wainwright B, Chenevix-Trench G, Wiestler OD, Wicking C (1997) Medulloblastomas of the desmoplastic variant carry mutations of the human homologue of Drosophila patched. Cancer Res 57:2085–2088
38. Taylor MD, Liu L, Raffel C, Hui C-C, Mainprize TG, Zhang X, Agatep R, Chiappa S, Gao L, Lowrance A, Hao A, Goldstein AM, Stavrou T, Scherer SW, Dura WT, Wainwright B, Squire JA, Rutka JT, Hogg D (2002) Mutations in SUFU predispose to medulloblastoma. Nat Genet 31:306–310
39. Ellison DW (2010) Childhood medulloblastoma: novel approaches to the classification of a heterogeneous disease. Acta Neuropathol 120:305–316
40. Zurawel RH, Allen C, Chiappa S, Cato W, Biegel J, Cogen P, de Sauvage F, Raffel C (2000) Analysis of PTCH/SMO/SHH pathway genes in medulloblastoma. Genes Chromosomes Cancer 27:44–51
41. Uziel T, Karginov FV, Xie S, Parker JS, Wang Y-D, Gajjar A, He L, Ellison D, Gilbertson RJ, Hannon G, Roussel MF (2009) The miR-17 92 cluster collaborates with the Sonic Hedgehog pathway in medulloblastoma. Proc Natl Acad Sci U S A 106:2812–2817
42. Northcott PA, Fernandez-L A, Hagan JP, Ellison DW, Grajkowska W, Gillespie Y, Grundy R, Van Meter T, Rutka JT, Croce CM, Kenney AM, Taylor MD (2009) The miR-17/92 polycistron is up-regulated in sonic hedgehog-driven medulloblastomas and induced by N-myc in sonic hedgehog-treated cerebellar neural precursors. Cancer Res 69:3249–3255
43. Fernandez-L A, Northcott PA, Dalton J, Fraga C, Ellison D, Angers S, Taylor MD, Kenney AM (2009) YAP1 is amplified and up-regulated in hedgehog-associated medulloblastomas and mediates Sonic hedgehog-driven neural precursor proliferation. Genes Dev 23:2729–2741
44. Huang H, Mahler-Araujo BM, Sankila A, Chimelli L, Yonekawa Y, Kleihues P, Ohgaki H (2000) APC mutations in sporadic medulloblastomas. Am J Pathol 156:433–437
45. Paraf F, Jothy S, Van Meir EG (1997) Brain tumor-polyposis syndrome: two genetic diseases? J Clin Oncol 15:2744–2758
46. Zurawel RH, Chiappa SA, Allen C, Raffel C (1998) Sporadic medulloblastomas contain oncogenic beta-catenin mutations. Cancer Res 58:896–899
47. Eberhart CG, Tihan T, Burger PC (2000) Nuclear localization and mutation of beta-catenin in medulloblastomas. J Neuropathol Exp Neurol 59:333–337
48. Kongkham PN, Northcott PA, Croul SE, Smith CA, Taylor MD, Rutka JT (2010) The SFRP family of WNT inhibitors function as novel tumor suppressor genes epigenetically silenced in medulloblastoma. Oncogene 29:3017–3024
49. Clifford SC, Lusher ME, Lindsey JC, Langdon JA, Gilbertson RJ, Straughton D, Ellison DW (2006) Wnt/Wingless pathway activation and chromosome 6 loss characterize a distinct molecular sub-group of medulloblastomas associated with a favorable prognosis. Cell Cycle 5:2666–2670
50. Gibson P, Tong Y, Robinson G, Thompson MC, Currle DS, Eden C, Kranenburg TA, Hogg T, Poppleton H, Martin J, Finkelstein D, Pounds S, Weiss A, Patay Z, Scoggins M, Ogg R, Pei Y, Yang Z-J, Brun S, Lee Y, Zindy F, Lindsey JC, Taketo MM, Boop FA, Sanford RA, Gajjar A, Clifford SC, Roussel MF, McKinnon PJ, Gutmann DH, Ellison DW, Wechsler-Reya R, Gilbertson RJ (2010) Subtypes of medulloblastoma have distinct developmental origins. Nature 468(7327): 1095–1099
51. Ellison DW, Onilude OE, Lindsey JC, Lusher ME, Weston CL, Taylor RE, Pearson AD, Clifford SC, United Kingdom Children's Cancer Study Group Brain Tumour Committee (2005) beta-Catenin status predicts a favorable outcome in childhood medulloblastoma: the United Kingdom Children's Cancer Study Group Brain Tumour Committee. J Clin Oncol 23:7951–7957
52. Pfister S, Remke M, Benner A, Mendrzyk F, Toedt G, Felsberg J, Wittmann A, Devens F, Gerber NU, Joos S, Kulozik A, Reifenberger G, Rutkowski S, Wiestler OD, Radlwimmer B, Scheurlen W, Lichter P, Korshunov A (2009) Outcome prediction in pediatric medulloblastoma based on DNA copy-number aberrations of chromosomes 6q and 17q and the MYC and MYCN loci. J Clin Oncol 27:1627–1636

53. Eberhart CG, Kratz JE, Schuster A, Goldthwaite P, Cohen KJ, Perlman EJ, Burger PC (2002) Comparative genomic hybridization detects an increased number of chromosomal alterations in large cell/anaplastic medulloblastomas. Brain Pathol 12:36–44

54. Aldosari N, Bigner SH, Burger PC, Becker L, Kepner JL, Friedman HS, McLendon RE (2002) MYCC and MYCN oncogene amplification in medulloblastoma. A fluorescence in situ hybridization study on paraffin sections from the Children's Oncology Group. Arch Pathol Lab Med 126:540–544

55. Hatton BA, Knoepfler PS, Kenney AM, Rowitch DH, de Alborán IM, Olson JM, Eisenman RN (2006) N-myc is an essential downstream effector of Shh signaling during both normal and neoplastic cerebellar growth. Cancer Res 66:8655–8661

56. Thomas WD, Chen J, Gao YR, Cheung B, Koach J, Sekyere E, Norris MD, Haber M, Ellis T, Wainwright B, Marshall GM (2009) Patched1 deletion increases N-Myc protein stability as a mechanism of medulloblastoma initiation and progression. Oncogene 28: 1605–1615

57. Kenney AM, Cole MD, Rowitch DH (2003) Nmyc upregulation by sonic hedgehog signaling promotes proliferation in developing cerebellar granule neuron precursors. Development 130:15–28

58. Kessler JD, Hasegawa H, Brun SN, Emmenegger BA, Yang Z-J, Dutton JW, Wang F, Wechsler-Reya RJ (2009) N-myc alters the fate of preneoplastic cells in a mouse model of medulloblastoma. Genes Dev 23:157–170

59. Swartling FJ, Grimmer MR, Hackett CS, Northcott PA, Fan Q-W, Goldenberg DD, Lau J, Masic S, Nguyen K, Yakovenko S, Zhe X-N, Gilmer HCF, Collins R, Nagaoka M, Phillips JJ, Jenkins RB, Tihan T, Vandenberg SR, James CD, Tanaka K, Taylor MD, Weiss WA, Chesler L (2010) Pleiotropic role for MYCN in medulloblastoma. Genes Dev 24:1059–1072

60. Kleihues P, Schäuble B, zur Hausen A, Estève J, Ohgaki H (1997) Tumors associated with p53 germline mutations: a synopsis of 91 families. Am J Pathol 150:1–13

61. Frank AJ, Hernan R, Hollander A, Lindsey JC, Lusher ME, Fuller CE, Clifford SC, Gilbertson RJ (2004) The TP53-ARF tumor suppressor pathway is frequently disrupted in large/cell anaplastic medulloblastoma. Brain Res Mol Brain Res 121:137–140

62. Tabori U, Baskin B, Shago M, Alon N, Taylor MD, Ray PN, Bouffet E, Malkin D, Hawkins C (2010) Universal poor survival in children with medulloblastoma harboring somatic TP53 mutations. J Clin Oncol 28:1345–1350

63. Reardon DA, Michalkiewicz E, Boyett JM, Sublett JE, Entrekin RE, Ragsdale ST, Valentine MB, Behm FG, Li H, Heideman RL, Kun LE, Shapiro DN, Look AT (1997) Extensive genomic abnormalities in childhood medulloblastoma by comparative genomic hybridization. Cancer Res 57:4042–4047

64. Gilbertson R, Wickramasinghe C, Hernan R, Balaji V, Hunt D, Jones-Wallace D, Crolla J, Perry R, Lunec J, Pearson A, Ellison D (2001) Clinical and molecular stratification of disease risk in medulloblastoma. Br J Cancer 85:705–712

65. Lamont JM, McManamy CS, Pearson AD, Clifford SC, Ellison DW (2004) Combined histopathological and molecular cytogenetic stratification of medulloblastoma patients. Clin Cancer Res 10:5482–5493

66. Mendrzyk F, Korshunov A, Toedt G, Schwarz F, Korn B, Joos S, Hochhaus A, Schoch C, Lichter P, Radlwimmer B (2006) Isochromosome breakpoints on 17p in medulloblastoma are flanked by different classes of DNA sequence repeats. Genes Chromosomes Cancer 45:401–410

67. Di Marcotullio L, Ferretti E, De Smaele E, Argenti B, Mincione C, Zazzeroni F, Gallo R, Masuelli L, Napolitano M, Maroder M, Modesti A, Giangaspero F, Screpanti I, Alesse E, Gulino A (2004) REN(KCTD11) is a suppressor of Hedgehog signaling and is deleted in human medulloblastoma. Proc Natl Acad Sci USA 101:10833–10838

68. Rood BR, Zhang H, Weitman DM, Cogen PH (2002) Hypermethylation of HIC-1 and 17p allelic loss in medulloblastoma. Cancer Res 62:3794–3797

69. Briggs KJ, Corcoran-Schwartz IM, Zhang W, Harcke T, Devereux WL, Baylin SB, Eberhart CG, Watkins DN (2008) Cooperation between the Hic1 and Ptch1 tumor suppressors in medulloblastoma. Genes Dev 22:770–785

70. Waha A, Waha A, Koch A, Meyer-Puttlitz B, Weggen S, Sörensen N, Tonn JC, Albrecht S, Goodyer CG, Berthold F, Wiestler OD, Pietsch T (2003) Epigenetic silencing of the HIC-1 gene in human medulloblastomas. J Neuropathol Exp Neurol 62:1192–1201

71. Lindsey JC, Anderton JA, Lusher ME, Clifford SC (2005) Epigenetic events in medulloblastoma development. Neurosurg Focus 19:E10

72. Canettieri G, Di Marcotullio L, Greco A, Coni S, Antonucci L, Infante P, Pietrosanti

L, De Smaele E, Ferretti E, Miele E, Pelloni M, De Simone G, Pedone EM, Gallinari P, Giorgi A, Steinkühler C, Vitagliano L, Pedone C, Schinin ME, Screpanti I, Gulino A (2010) Histone deacetylase and Cullin3-REN (KCTD11) ubiquitin ligase interplay regulates Hedgehog signalling through Gli acetylation. Nat Cell Biol 12:132–142

73. Peters I, Rehmet K, Wilke N, Kuczyk MA, Hennenlotter J, Eilers T, Machtens S, Jonas U, Serth J (2007) RASSF1A promoter methylation and expression analysis in normal and neoplastic kidney indicates a role in early tumorigenesis. Mol Cancer 6:49

74. Lindsey JC, Lusher ME, Anderton JA, Bailey S, Gilbertson RJ, Pearson ADJ, Ellison DW, Clifford SC (2004) Identification of tumour-specific epigenetic events in medulloblastoma development by hypermethylation profiling. Carcinogenesis 25:661–668

75. Agathanggelou A, Cooper WN, Latif F (2005) Role of the Ras-association domain family 1 tumor suppressor gene in human cancers. Cancer Res 65:3497–3508

76. Lusher ME, Lindsey JC, Latif F, Pearson ADJ, Ellison DW, Clifford SC (2002) Biallelic epigenetic inactivation of the RASSF1A tumor suppressor gene in medulloblastoma development. Cancer Res 62:5906–5911

77. Kongkham PN, Northcott PA, Ra YS, Nakahara Y, Mainprize TG, Croul SE, Smith CA, Taylor MD, Rutka JT (2008) An epigenetic genome-wide screen identifies SPINT2 as a novel tumor suppressor gene in pediatric medulloblastoma. Cancer Res 68:9945–9953

78. Nakahara Y, Northcott PA, Li M, Kongkham PN, Smith C, Yan H, Croul S, Ra Y-S, Eberhart C, Huang A, Bigner D, Grajkowska W, Van Meter T, Rutka JT, Taylor MD (2010) Genetic and epigenetic inactivation of Kruppel-like factor 4 in medulloblastoma. Neoplasia 12:20–27

79. Vibhakar R, Foltz G, Yoon J-G, Field L, Lee H, Ryu G-Y, Pierson J, Davidson B, Madan A (2007) Dickkopf-1 is an epigenetically silenced candidate tumor suppressor gene in medulloblastoma. Neuro Oncol 9:135–144

80. Northcott PA, Nakahara Y, Wu X, Feuk L, Ellison DW, Croul S, Mack S, Kongkham PN, Peacock J, Dubuc A, Ra Y-S, Zilberberg K, McLeod J, Scherer SW, Sunil Rao J, Eberhart CG, Grajkowska W, Gillespie Y, Lach B, Grundy R, Pollack IF, Hamilton RL, Van Meter T, Carlotti CG, Boop F, Bigner D, Gilbertson RJ, Rutka JT, Taylor MD (2009) Multiple recurrent genetic events converge on control of histone lysine methylation in medulloblastoma. Nat Genet 41:465–472

81. Parsons DW, Li M, Zhang X, Jones S, Leary RJ, Lin JC-H, Boca SM, Carter H, Samayoa J, Bettegowda C, Gallia GL, Jallo GI, Binder ZA, Nikolsky Y, Hartigan J, Smith DR, Gerhard DS, Fults DW, Vandenberg S, Berger MS, Marie SKN, Shinjo SMO, Clara C, Phillips PC, Minturn JE, Biegel JA, Judkins AR, Resnick AC, Storm PB, Curran T, He Y, Rasheed BA, Friedman HS, Keir ST, McLendon R, Northcott PA, Taylor MD, Burger PC, Riggins GJ, Karchin R, Parmigiani G, Bigner DD, Yan H, Papadopoulos N, Vogelstein B, Kinzler KW, Velculescu VE (2011) The genetic landscape of the childhood cancer medulloblastoma. Science 331(6016):435–439

82. Fan X, Mikolaenko I, Elhassan I, Ni X, Wang Y, Ball D, Brat DJ, Perry A, Eberhart CG (2004) Notch1 and notch2 have opposite effects on embryonal brain tumor growth. Cancer Res 64:7787–7793

83. Hallahan AR, Pritchard JI, Hansen S, Benson M, Stoeck J, Hatton BA, Russell TL, Ellenbogen RG, Bernstein ID, Beachy PA, Olson JM (2004) The SmoA1 mouse model reveals that notch signaling is critical for the growth and survival of sonic hedgehog-induced medulloblastomas. Cancer Res 64:7794–7800

84. Garzia L, Andolfo I, Cusanelli E, Marino N, Petrosino G, De Martino D, Esposito V, Galeone A, Navas L, Esposito S, Gargiulo S, Fattet S, Donofrio V, Cinalli G, Brunetti A, Vecchio LD, Northcott PA, Delattre O, Taylor MD, Iolascon A, Zollo M (2009) MicroRNA-199b-5p impairs cancer stem cells through negative regulation of HES1 in medulloblastoma. PLoS One 4:e4998

85. Broderick DK, Di C, Parrett TJ, Samuels YR, Cummins JM, McLendon RE, Fults DW, Velculescu VE, Bigner DD, Yan H (2004) Mutations of PIK3CA in anaplastic oligodendrogliomas, high-grade astrocytomas, and medulloblastomas. Cancer Res 64:5048–5050

86. Inda MM, Mercapide J, Muñoz J, Coullin P, Danglot G, Tuñon T, Martínez-Peñuela JM, Rivera JM, Burgos JJ, Bernheim A, Castresana JS (2004) PTEN and DMBT1 homozygous deletion and expression in medulloblastomas and supratentorial primitive neuroectodermal tumors. Oncol Rep 12:1341–1347

87. Eberhart CG, Kratz J, Wang Y, Summers K, Stearns D, Cohen K, Dang CV, Burger PC (2004) Histopathological and molecular prognostic markers in medulloblastoma: c-myc, N-myc, TrkC, and anaplasia. J Neuropathol Exp Neurol 63:441–449

88. de Haas T, Oussoren E, Grajkowska W, Perek-Polnik M, Popovic M, Zadravec-Zaletel L, Perera M, Corte G, Wirths O, van Sluis P,

Pietsch T, Troost D, Baas F, Versteeg R, Kool M (2006) OTX1 and OTX2 expression correlates with the clinicopathologic classification of medulloblastomas. J Neuropathol Exp Neurol 65:176–186

89. Di C, Liao S, Adamson DC, Parrett TJ, Broderick DK, Shi Q, Lengauer C, Cummins JM, Velculescu VE, Fults DW, McLendon RE, Bigner DD, Yan H (2005) Identification of OTX2 as a medulloblastoma oncogene whose product can be targeted by all-trans retinoic acid. Cancer Res 65:919–924
90. Northcott PA, Rutka JT, Taylor MD (2010) Genomics of medulloblastoma: from Giemsa-banding to next-generation sequencing in 20 years. Neurosurg Focus 28:E6
91. Segal RA, Goumnerova LC, Kwon YK, Stiles CD, Pomeroy SL (1994) Expression of the neurotrophin receptor TrkC is linked to a favorable outcome in medulloblastoma. Proc Natl Acad Sci USA 91:12867–12871
92. Kim JY, Sutton ME, Lu DJ, Cho TA, Goumnerova LC, Goritchenko L, Kaufman JR, Lam KK, Billet AL, Tarbell NJ, Wu J, Allen JC, Stiles CD, Segal RA, Pomeroy SL (1999) Activation of neurotrophin-3 receptor TrkC induces apoptosis in medulloblastomas. Cancer Res 59:711–719
93. Mendrzyk F, Radlwimmer B, Joos S, Kokocinski F, Benner A, Stange DE, Neben K, Fiegler H, Carter NP, Reifenberger G, Korshunov A, Lichter P (2005) Genomic and protein expression profiling identifies CDK6 as novel independent prognostic marker in medulloblastoma. J Clin Oncol 23:8853–8862
94. Kool M, Koster J, Bunt J, Hasselt NE, Lakeman A, van Sluis P, Troost D, Meeteren NS-V, Caron HN, Cloos J, Mrsić A, Ylstra B, Grajkowska W, Hartmann W, Pietsch T, Ellison D, Clifford SC, Versteeg R (2008) Integrated genomics identifies five medulloblastoma subtypes with distinct genetic profiles, pathway signatures and clinicopathological features. PLoS One 3:e3088
95. Thompson MC, Fuller C, Hogg TL, Dalton J, Finkelstein D, Lau CC, Chintagumpala M, Adesina A, Ashley DM, Kellie SJ, Taylor MD, Curran T, Gajjar A, Gilbertson RJ (2006) Genomics identifies medulloblastoma subgroups that are enriched for specific genetic alterations. J Clin Oncol 24:1924–1931
96. Cho Y-J, Tsherniak A, Tamayo P, Santagata S, Ligon A, Greulich H, Berhoukim R, Amani V, Goumnerova L, Eberhart CG, Lau CC, Olson JM, Gilbertson RJ, Gajjar A, Delattre O, Kool M, Ligon K, Meyerson M, Mesirov JP, Pomeroy SL (2011) Integrative genomic analysis of medulloblastoma identifies a molecular subgroup that drives poor clinical outcome. J Clin Oncol 29 (11):1424–1430
97. Chen TT, Mora EC, Mealey J (1975) Cultivation of medulloblastoma cells derived from simian adenovirus SA7-induced hamster brain tumor. Cancer Res 35:3566–3570
98. Rapp F, Pauluzzi S, Waltz TA, Burdine JA, Matsen FA, Levy B (1969) Induction of brain tumors in newborn hamsters by simian adenovirus SA7. Cancer Res 29:1173–1178
99. Zu Rhein GM, Varakis JN (1979) Perinatal induction of medulloblastomas in Syrian golden hamsters by a human polyoma virus (JC). Natl Cancer Inst Monogr 51:205–208
100. Nagashima K, Yasui K, Kimura J, Washizu M, Yamaguchi K, Mori W (1984) Induction of brain tumors by a newly isolated JC virus (Tokyo-1 strain). Am J Pathol 116:455–463
101. Krynska B, Otte J, Franks R, Khalili K, Croul S (1999) Human ubiquitous JCV(CY) T-antigen gene induces brain tumors in experimental animals. Oncogene 18:39–46
102. Zurawel RH, Allen C, Wechsler-Reya R, Scott MP, Raffel C (2000) Evidence that haploinsufficiency of Ptch leads to medulloblastoma in mice. Genes Chromosomes Cancer 28:77–81
103. Goodrich LV, Milenković L, Higgins KM, Scott MP (1997) Altered neural cell fates and medulloblastoma in mouse patched mutants. Science 277:1109–1113
104. Wetmore C, Eberhart DE, Curran T (2000) The normal patched allele is expressed in medulloblastomas from mice with heterozygous germ-line mutation of patched. Cancer Res 60:2239–2246
105. Berman DM, Karhadkar SS, Hallahan AR, Pritchard JI, Eberhart CG, Watkins DN, Chen JK, Cooper MK, Taipale J, Olson JM, Beachy PA (2002) Medulloblastoma growth inhibition by hedgehog pathway blockade. Science 297:1559–1561
106. Oliver TG, Read TA, Kessler JD, Mehmeti A, Wells JF, Huynh TTT, Lin SM, Wechsler-Reya RJ (2005) Loss of patched and disruption of granule cell development in a pre-neoplastic stage of medulloblastoma. Development 132:2425–2439
107. Pazzaglia S, Mancuso M, Atkinson MJ, Tanori M, Rebessi S, Majo VD, Covelli V, Hahn H, Saran A (2002) High incidence of medulloblastoma following X-ray-irradiation of newborn Ptc1 heterozygous mice. Oncogene 21:7580–7584
108. Pazzaglia S, Tanori M, Mancuso M, Rebessi S, Leonardi S, Di Majo V, Covelli V, Atkinson MJ, Hahn H, Saran A (2006) Linking DNA

damage to medulloblastoma tumorigenesis in patched heterozygous knockout mice. Oncogene 25:1165–1173

109. Lee Y, Miller HL, Russell HR, Boyd K, Curran T, McKinnon PJ (2006) Patched2 modulates tumorigenesis in patched1 heterozygous mice. Cancer Res 66:6964–6971
110. Wetmore C, Eberhart DE, Curran T (2001) Loss of p53 but not ARF accelerates medulloblastoma in mice heterozygous for patched. Cancer Res 61:513–516
111. Romer J, Curran T (2005) Targeting medulloblastoma: small-molecule inhibitors of the Sonic Hedgehog pathway as potential cancer therapeutics. Cancer Res 65:4975–4978
112. Kimura H, Stephen D, Joyner A, Curran T (2005) Gli1 is important for medulloblastoma formation in Ptc1+/− mice. Oncogene 24:4026–4036
113. Pogoriler J, Millen K, Utset M, Du W (2006) Loss of cyclin D1 impairs cerebellar development and suppresses medulloblastoma formation. Development 133:3929–3937
114. Hahn H, Wojnowski L, Zimmer AM, Hall J, Miller G, Zimmer A (1998) Rhabdomyosarcomas and radiation hypersensitivity in a mouse model of Gorlin syndrome. Nat Med 4:619–622
115. Hahn H, Wojnowski L, Specht K, Kappler R, Calzada-Wack J, Potter D, Zimmer A, Müller U, Samson E, Quintanilla-Martinez L, Zimmer A (2000) Patched target Igf2 is indispensable for the formation of medulloblastoma and rhabdomyosarcoma. J Biol Chem 275:28341–28344
116. Corcoran RB, Bachar Raveh T, Barakat MT, Lee EY, Scott MP (2008) Insulin-like growth factor 2 is required for progression to advanced medulloblastoma in patched1 heterozygous mice. Cancer Res 68:8788–8795
117. Mao J, Ligon KL, Rakhlin EY, Thayer SP, Bronson RT, Rowitch D, McMahon AP (2006) A novel somatic mouse model to survey tumorigenic potential applied to the Hedgehog pathway. Cancer Res 66:10171–10178
118. Hatton BA, Villavicencio EH, Tsuchiya KD, Pritchard JI, Ditzler S, Pullar B, Hansen S, Knoblaugh SE, Lee D, Eberhart CG, Hallahan AR, Olson JM (2008) The Smo/Smo model: hedgehog-induced medulloblastoma with 90% incidence and leptomeningeal spread. Cancer Res 68:1768–1776
119. Yang Z-J, Ellis T, Markant SL, Read T-A, Kessler JD, Bourboulas M, Schüller U, Machold R, Fishell G, Rowitch DH, Wainwright BJ, Wechsler-Reya RJ (2008) Medulloblastoma can be initiated by deletion of Patched in lineage-restricted progenitors or stem cells. Cancer Cell 14:135–145
120. Singh SK, Clarke ID, Terasaki M, Bonn VE, Hawkins C, Squire J, Dirks PB (2003) Identification of a cancer stem cell in human brain tumors. Cancer Res 63:5821–5828
121. Hemmati HD, Nakano I, Lazareff JA, Masterman-Smith M, Geschwind DH, Bronner-Fraser M, Kornblum HI (2003) Cancerous stem cells can arise from pediatric brain tumors. Proc Natl Acad Sci USA 100:15178–15183
122. Taniguchi E, Cho MJ, Arenkiel BR, Hansen MS, Rivera OJ, McCleish AT, Qualman SJ, Guttridge DC, Scott MP, Capecchi MR, Keller C (2009) Bortezomib reverses a post-translational mechanism of tumorigenesis for patched1 haploinsufficiency in medulloblastoma. Pediatr Blood Cancer 53:136–144
123. Lee Y, Kawagoe R, Sasai K, Li Y, Russell HR, Curran T, McKinnon PJ (2007) Loss of suppressor-of-fused function promotes tumorigenesis. Oncogene 26:6442–6447
124. Gaiano N, Kohtz JD, Turnbull DH, Fishell G (1999) A method for rapid gain-of-function studies in the mouse embryonic nervous system. Nat Neurosci 2:812–819
125. Weiner HL, Bakst R, Hurlbert MS, Ruggiero J, Ahn E, Lee WS, Stephen D, Zagzag D, Joyner AL, Turnbull DH (2002) Induction of medulloblastomas in mice by sonic hedgehog, independent of Gli1. Cancer Res 62:6385–6389
126. Orsulic S (2002) An RCAS-TVA-based approach to designer mouse models. Mamm Genome 13:543–547
127. Holland EC, Varmus HE (1998) Basic fibroblast growth factor induces cell migration and proliferation after glia-specific gene transfer in mice. Proc Natl Acad Sci USA 95:1218–1223
128. Binning MJ, Niazi T, Pedone CA, Lal B, Eberhart CG, Kim KJ, Laterra J, Fults DW (2008) Hepatocyte growth factor and sonic Hedgehog expression in cerebellar neural progenitor cells costimulate medulloblastoma initiation and growth. Cancer Res 68:7838–7845
129. McCall TD, Pedone CA, Fults DW (2007) Apoptosis suppression by somatic cell transfer of Bcl-2 promotes Sonic hedgehog-dependent medulloblastoma formation in mice. Cancer Res 67:5179–5185
130. Browd SR, Kenney AM, Gottfried ON, Yoon JW, Walterhouse D, Pedone CA, Fults DW (2006) N-myc can substitute for insulin-like growth factor signaling in a mouse model of

sonic hedgehog-induced medulloblastoma. Cancer Res 66:2666–2672

131. Rao G, Pedone CA, Del Valle L, Reiss K, Holland EC, Fults DW (2004) Sonic hedgehog and insulin-like growth factor signaling synergize to induce medulloblastoma formation from nestin-expressing neural progenitors in mice. Oncogene 23:6156–6162

132. Rao G, Pedone CA, Coffin CM, Holland EC, Fults DW (2003) c-Myc enhances sonic hedgehog-induced medulloblastoma formation from nestin-expressing neural progenitors in mice. Neoplasia 5:198–204

133. Pomeroy SL, Tamayo P, Gaasenbeek M, Sturla LM, Angelo M, McLaughlin ME, Kim JYH, Goumnerova LC, Black PM, Lau C, Allen JC, Zagzag D, Olson JM, Curran T, Wetmore C, Biegel JA, Poggio T, Mukherjee S, Rifkin R, Califano A, Stolovitzky G, Louis DN, Mesirov JP, Lander ES, Golub TR (2002) Prediction of central nervous system embryonal tumour outcome based on gene expression. Nature 415:436–442

134. Kenney AM, Widlund HR, Rowitch DH (2004) Hedgehog and PI-3 kinase signaling converge on Nmyc1 to promote cell cycle progression in cerebellar neuronal precursors. Development 131:217–228

135. Hambardzumyan D, Becher OJ, Rosenblum MK, Pandolfi PP, Manova-Todorova K, Holland EC (2008) PI3K pathway regulates survival of cancer stem cells residing in the perivascular niche following radiation in medulloblastoma in vivo. Genes Dev 22:436–448

136. Li Y, Lal B, Kwon S, Fan X, Saldanha U, Reznik TE, Kuchner EB, Eberhart C, Laterra J, Abounader R (2005) The scatter factor/hepatocyte growth factor: c-met pathway in human embryonal central nervous system tumor malignancy. Cancer Res 65:9355–9362

137. Li Y, Guessous F, Johnson EB, Eberhart CG, Li X-N, Shu Q, Fan S, Lal B, Laterra J, Schiff D, Abounader R (2008) Functional and molecular interactions between the HGF/c-Met pathway and c-Myc in large-cell medulloblastoma. Lab Invest 88:98–111

138. Fults D, Pedone C, Dai C, Holland EC (2002) MYC expression promotes the proliferation of neural progenitor cells in culture and in vivo. Neoplasia 4:32–39

139. Hossain MA, Russell JC, Gomez R, Laterra J, Gomes R (2002) Neuroprotection by scatter factor/hepatocyte growth factor and FGF-1 in cerebellar granule neurons is phosphatidylinositol 3-kinase/akt-dependent and MAPK/CREB-independent. J Neurochem 81:365–378

140. Ieraci A, Forni PE, Ponzetto C (2002) Viable hypomorphic signaling mutant of the Met receptor reveals a role for hepatocyte growth factor in postnatal cerebellar development. Proc Natl Acad Sci USA 99:15200–15205

141. Lee Y, Miller HL, Jensen P, Hernan R, Connelly M, Wetmore C, Zindy F, Roussel MF, Curran T, Gilbertson RJ, McKinnon PJ (2003) A molecular fingerprint for medulloblastoma. Cancer Res 63:5428–5437

142. Zindy F, Nilsson LM, Nguyen L, Meunier C, Smeyne RJ, Rehg JE, Eberhart C, Sherr CJ, Roussel MF (2003) Hemangiosarcomas, medulloblastomas, and other tumors in Ink4c/p53-null mice. Cancer Res 63:5420–5427

143. Zindy F, Cunningham JJ, Sherr CJ, Jogal S, Smeyne RJ, Roussel MF (1999) Postnatal neuronal proliferation in mice lacking Ink4d and Kip1 inhibitors of cyclin-dependent kinases. Proc Natl Acad Sci USA 96:13462–13467

144. Marino S, Vooijs M, van Der Gulden H, Jonkers J, Berns A (2000) Induction of medulloblastomas in p53-null mutant mice by somatic inactivation of Rb in the external granular layer cells of the cerebellum. Genes Dev 14:994–1004

145. Shakhova O, Leung C, van Montfort E, Berns A, Marino S (2006) Lack of Rb and p53 delays cerebellar development and predisposes to large cell anaplastic medulloblastoma through amplification of N-Myc and Ptch2. Cancer Res 66:5190–5200

146. Sutter R, Shakhova O, Bhagat H, Behesti H, Sutter C, Penkar S, Santuccione A, Bernays R, Heppner FL, Schüller U, Grotzer M, Moch H, Schraml P, Marino S (2010) Cerebellar stem cells act as medulloblastoma-initiating cells in a mouse model and a neural stem cell signature characterizes a subset of human medulloblastomas. Oncogene 29:1845–1856

147. Uziel T, Zindy F, Xie S, Lee Y, Forget A, Magdaleno S, Rehg JE, Calabrese C, Solecki D, Eberhart CG, Sherr SE, Plimmer S, Clifford SC, Hatten ME, McKinnon PJ, Gilbertson RJ, Curran T, Sherr CJ, Roussel MF (2005) The tumor suppressors Ink4c and p53 collaborate independently with patched to suppress medulloblastoma formation. Genes Dev 19:2656–2667

148. Hughes-Davies L, Huntsman D, Ruas M, Fuks F, Bye J, Chin S-F, Milner J, Brown LA, Hsu F, Gilks B, Nielsen T, Schulzer M, Chia S, Ragaz J, Cahn A, Linger L, Ozdag H, Cattaneo E, Jordanova ES, Schuuring E, Yu DS, Venkitaraman A, Ponder B, Doherty A,

Aparicio S, Bentley D, Theillet C, Ponting CP, Caldas C, Kouzarides T (2003) EMSY links the BRCA2 pathway to sporadic breast and ovarian cancer. Cell 115:523–535

149. Offit K, Levran O, Mullaney B, Mah K, Nafa K, Batish SD, Diotti R, Schneider H, Deffenbaugh A, Scholl T, Proud VK, Robson M, Norton L, Ellis N, Hanenberg H, Auerbach AD (2003) Shared genetic susceptibility to breast cancer, brain tumors, and Fanconi anemia. J Natl Cancer Inst 95:1548–1551
150. Lee Y, McKinnon PJ (2002) DNA ligase IV suppresses medulloblastoma formation. Cancer Res 62:6395–6399
151. Holcomb VB, Vogel H, Marple T, Kornegay RW, Hasty P (2006) Ku80 and p53 suppress medulloblastoma that arise independent of Rag-1-induced DSBs. Oncogene 25:7159–7165
152. Tong W-M, Ohgaki H, Huang H, Granier C, Kleihues P, Wang Z-Q (2003) Null mutation of DNA strand break-binding molecule poly (ADP-ribose) polymerase causes medulloblastomas in p53(−/−) mice. Am J Pathol 162:343–352
153. Yan CT, Kaushal D, Murphy M, Zhang Y, Datta A, Chen C, Monroe B, Mostoslavsky G, Coakley K, Gao Y, Mills KD, Fazeli AP, Tepsuporn S, Hall G, Mulligan R, Fox E, Bronson R, De Girolami U, Lee C, Alt FW (2006) XRCC4 suppresses medulloblastomas with recurrent translocations in p53-deficient mice. Proc Natl Acad Sci USA 103:7378–7383
154. Frappart P-O, Lee Y, Lamont J, McKinnon PJ (2007) BRCA2 is required for neurogenesis and suppression of medulloblastoma. EMBO J 26:2732–2742
155. Lin W, Kemper A, McCarthy KD, Pytel P, Wang J-P, Campbell IL, Utset MF, Popko B (2004) Interferon-gamma induced medulloblastoma in the developing cerebellum. J Neurosci 24:10074–10083
156. Momota H, Shih AH, Edgar MA, Holland EC (2008) c-Myc and beta-catenin cooperate with loss of p53 to generate multiple members of the primitive neuroectodermal tumor family in mice. Oncogene 27:4392–4401
157. Cooper MK, Porter JA, Young KE, Beachy PA (1998) Teratogen-mediated inhibition of target tissue response to Shh signaling. Science 280:1603–1607
158. Incardona JP, Gaffield W, Kapur RP, Roelink H (1998) The teratogenic Veratrum alkaloid cyclopamine inhibits sonic hedgehog signal transduction. Development 125:3553–3562
159. Taipale J, Chen JK, Cooper MK, Wang B, Mann RK, Milenkovic L, Scott MP, Beachy PA (2000) Effects of oncogenic mutations in Smoothened and Patched can be reversed by cyclopamine. Nature 406:1005–1009
160. Chen JK, Taipale J, Cooper MK, Beachy PA (2002) Inhibition of Hedgehog signaling by direct binding of cyclopamine to Smoothened. Genes Dev 16:2743–2748
161. Chen JK, Taipale J, Young KE, Maiti T, Beachy PA (2002) Small molecule modulation of Smoothened activity. Proc Natl Acad Sci U S A 99:14071–14076
162. Robarge KD, Brunton SA, Castanedo GM, Cui Y, Dina MS, Goldsmith R, Gould SE, Guichert O, Gunzner JL, Halladay J, Jia W, Khojasteh C, Koehler MFT, Kotkow K, La H, Lalonde RL, Lau K, Lee L, Marshall D, Marsters JC, Murray LJ, Qian C, Rubin LL, Salphati L, Stanley MS, Stibbard JHA, Sutherlin DP, Ubhayaker S, Wang S, Wong S, Xie M (2009) GDC-0449—a potent inhibitor of the hedgehog pathway. Bioorg Med Chem Lett 19:5576–5581
163. Romer JT, Kimura H, Magdaleno S, Sasai K, Fuller C, Baines H, Connelly M, Stewart CF, Gould S, Rubin LL, Curran T (2004) Suppression of the Shh pathway using a small molecule inhibitor eliminates medulloblastoma in Ptc1(+/−)p53(−/−) mice. Cancer Cell 6:229–240
164. Kimura H, Ng JMY, Curran T (2008) Transient inhibition of the Hedgehog pathway in young mice causes permanent defects in bone structure. Cancer Cell 13:249–260
165. Rudin CM, Hann CL, Laterra J, Yauch RL, Callahan CA, Fu L, Holcomb T, Stinson J, Gould SE, Coleman B, LoRusso PM, Von Hoff DD, de Sauvage FJ, Low JA (2009) Treatment of medulloblastoma with hedgehog pathway inhibitor GDC-0449. N Engl J Med 361:1173–1178
166. Yauch RL, Dijkgraaf GJP, Alicke B, Januario T, Ahn CP, Holcomb T, Pujara K, Stinson J, Callahan CA, Tang T, Bazan JF, Kan Z, Seshagiri S, Hann CL, Gould SE, Low JA, Rudin CM, de Sauvage FJ (2009) Smoothened mutation confers resistance to a Hedgehog pathway inhibitor in medulloblastoma. Science 326:572–574
167. Dijkgraaf GJP, Alicke B, Weinmann L, Januario T, West K, Modrusan Z, Burdick D, Goldsmith R, Robarge K, Sutherlin D, Scales SJ, Gould SE, Yauch RL, de Sauvage FJ (2011) Small molecule inhibition of GDC-0449 refractory smoothened mutants and downstream mechanisms of drug resistance. Cancer Res 71:435–444
168. Tremblay MR, Lescarbeau A, Grogan MJ, Tan E, Lin G, Austad BC, Yu L-C, Behnke

ML, Nair SJ, Hagel M, White K, Conley J, Manna JD, Alvarez-Diez TM, Hoyt J, Woodward CN, Sydor JR, Pink M, MacDougall J, Campbell MJ, Cushing J, Ferguson J, Curtis MS, McGovern K, Read MA, Palombella VJ, Adams J, Castro AC (2009) Discovery of a potent and orally active hedgehog pathway antagonist (IPI-926). J Med Chem 52:4400–4418

169. Pan S, Wu X, Jiang J, Gao W, Wan Y, Cheng D, Han D, Liu J, Englund NP, Wang Y, Peukert S, Miller-Moslin K, Yuan J, Guo R, Matsumoto M, Vattay A, Jiang Y, Tsao J, Sun F, Pferdekamper AC, Dodd S, Tuntland T, Maniara W, Kelleher JF, Yao Y-M, Warmuth M, Williams J, Dorsch M (2010) Discovery of NVP-LDE225, a potent and selective smoothened antagonist. ACS Med Chem Lett 1:130–134

170. Buonamici S, Williams J, Morrissey M, Wang A, Guo R, Vattay A, Hsiao K, Yuan J, Green J, Ospina B, Yu Q, Ostrom L, Fordjour P, Anderson DL, Monahan JE, Kelleher JF, Peukert S, Pan S, Wu X, Maira S-M, García-Echeverría C, Briggs KJ, Watkins DN, Yao Y-M, Lengauer C, Warmuth M, Sellers WR, Dorsch M (2010) Interfering with resistance to smoothened antagonists by inhibition of the PI3K pathway in medulloblastoma. Sci Transl Med 2:51–70

171. Beauchamp EM, Ringer L, Bulut G, Sajwan KP, Hall MD, Lee Y-C, Peaceman D, Ozdemirli M, Rodriguez O, MacDonald TJ, Albanese C, Toretsky JA, Uren A (2011) Arsenic trioxide inhibits human cancer cell growth and tumor development in mice by blocking Hedgehog/GLI pathway. J Clin Invest 121:148–160

172. Kim J, Tang JY, Gong R, Kim J, Lee JJ, Clemons KV, Chong CR, Chang KS, Fereshteh M, Gardner D, Reya T, Liu JO, Epstein EH, Stevens DA, Beachy PA (2010) Itraconazole, a commonly used antifungal that inhibits Hedgehog pathway activity and cancer growth. Cancer Cell 17:388–399

173. Wang Y, Zhou Z, Walsh CT, McMahon AP (2009) Selective translocation of intracellular Smoothened to the primary cilium in response to Hedgehog pathway modulation. Proc Natl Acad Sci U S A 106:2623–2628

174. Rohatgi R, Milenkovic L, Corcoran RB, Scott MP (2009) Hedgehog signal transduction by Smoothened: pharmacologic evidence for a 2-step activation process. Proc Natl Acad Sci USA 106:3196–3201

175. Heine VM, Priller M, Ling J, Rowitch DH, Schüller U (2010) Dexamethasone destabilizes Nmyc to inhibit the growth of hedgehog-associated medulloblastoma. Cancer Res 70:5220–5225

176. Spiller SE, Ravanpay AC, Hahn AW, Olson JM (2006) Suberoylanilide hydroxamic acid is effective in preclinical studies of medulloblastoma. J Neurooncol 79:259–270

177. Spiller SE, Ditzler SH, Pullar BJ, Olson JM (2008) Response of preclinical medulloblastoma models to combination therapy with 13-cis retinoic acid and suberoylanilide hydroxamic acid (SAHA). J Neurooncol 87:133–141

178. Hallahan AR, Pritchard JI, Chandraratna RAS, Ellenbogen RG, Geyer JR, Overland RP, Strand AD, Tapscott SJ, Olson JM (2003) BMP-2 mediates retinoid-induced apoptosis in medulloblastoma cells through a paracrine effect. Nat Med 9:1033–1038

179. Ecke I, Petry F, Rosenberger A, Tauber S, Mönkemeyer S, Hess I, Dullin C, Kimmina S, Pirngruber J, Johnsen SA, Uhmann A, Nitzki F, Wojnowski L, Schulz-Schaeffer W, Witt O, Hahn H (2009) Antitumor effects of a combined 5-aza-2′deoxycytidine and valproic acid treatment on rhabdomyosarcoma and medulloblastoma in Ptch mutant mice. Cancer Res 69:887–895

180. Ohshima-Hosoyama S, Hosoyama T, Nelon LD, Keller C (2010) IGF-1 receptor inhibition by picropodophyllin in medulloblastoma. Biochem Biophys Res Commun 399:727–732

181. Coon V, Laukert T, Pedone CA, Laterra J, Kim KJ, Fults DW (2010) Molecular therapy targeting Sonic hedgehog and hepatocyte growth factor signaling in a mouse model of medulloblastoma. Mol Cancer Ther 9:2627–2636

182. Read T-A, Fogarty MP, Markant SL, McLendon RE, Wei Z, Ellison DW, Febbo PG, Wechsler-Reya RJ (2009) Identification of CD15 as a marker for tumor-propagating cells in a mouse model of medulloblastoma. Cancer Cell 15:135–147

183. Ward RJ, Lee L, Graham K, Satkunendran T, Yoshikawa K, Ling E, Harper L, Austin R, Nieuwenhuis E, Clarke ID, Hui C-C, Dirks PB (2009) Multipotent CD15+ cancer stem cells in patched-1-deficient mouse medulloblastoma. Cancer Res 69:4682–4690

Neuromethods (2013) 77: 141–152
DOI 10.1007/7657_2011_25

Published online: 13 March 2012

Learning About Retinoblastoma from Mouse Models That Missed

David Cobrinik

Abstract

Retinoblastoma has been an attractive cancer for mouse modeling, as it is thought to develop in response to the inactivation of a single gene (*RB*) and thus provides an extraordinary opportunity to dissect mechanisms of tumor initiation. However, difficulties have repeatedly arisen in efforts to model this malignancy. For example, mice with inactivating *Rb* mutations fail to develop retinoblastoma, and instead develop pituitary and thyroid cancers. Moreover, mice with combined mutations in *Rb* and *Rb*-related genes develop retinal tumors that differ from human retinoblastomas in important aspects, including the mutations necessary to form the tumors, the retinal cell phenotypes of the tumors, and potentially the tumor cell of origin. Despite these discrepancies, mouse models may provide insight into human retinoblastoma by revealing features that contribute to retinal tumors in both species. Nevertheless, the mouse's inexact representation of retinoblastoma challenges us to develop more accurate models and to decipher how human-specific circuitry sensitizes to the loss of Rb and enables retinoblastoma initiation.

Key words: Retinoblastoma, RB, p107, p130, Mouse models, Cancer, Retina

1. Introduction: Mouse Modeling to Define Mechanisms of Human Tumorigenesis

Genetically altered mouse models are exceedingly important tools in cancer biology, as they can simulate tumor initiation, progression, and response to therapy. Prior to their development, in vivo studies of tumorigenesis were largely restricted to xenografts and chemically induced tumors. Xenografts are comparatively poor models for characterizing early stages of tumorigenesis, as they are usually initiated by fully transformed cells, often fail to develop in a relevant microenvironment, and can have vastly different histology and behavior compared to their originating tumors. Chemically induced cancers are also problematic, as they commonly involve unknown originating mutations and fail to model the full spectrum of human malignancies. Genetically engineered models circumvent these concerns, by enabling tumors to develop autochthonously (i.e., from a single cell within a normal tissue), from precisely defined cell types, and in response to precisely defined genomic changes. Accordingly, mouse models that accurately mimic the histopathological features, originating mutations, and cell of origin of a human cancer provide powerful platforms with which to explore numerous aspects of tumorigenesis and targeted therapeutics.

Notwithstanding their value, accurate cancer models can be challenging to develop. This is not only because of the technical challenges but also because the originating mutations and cells of origin of many cancers are unknown. Thus, rather than modeling events that are unequivocally known to occur in human tumors, many mouse models examine whether introduction of a *candidate* originating mutation into a *candidate* cell of origin can induce a particular neoplasm. When a tumor with appropriate histopathological features results, this approach can provide support for a proposed oncogenic mechanism. However, in the absence of corroborating data, this approach also leaves open the possibility that the corresponding human tumors develop with different originating mutations or from different cell types, which could preclude the model's depiction of essential disease features.

2. Retinoblastoma Is a Human Cancer with a Defined Initiating Event: The Biallelic Inactivation of the RB Gene

The thorny problem of identifying authentic originating mutations, as needed for accurate modeling, applies to nearly all human cancers. However, an exception to this concern is the childhood eye cancer, retinoblastoma, in which there is unusually strong evidence that the tumors initiate in response to the inactivation of the *retinoblastoma* gene (known as *RB* or *RB1* in humans, and as Rb or Rb1 in mice) in the developing human retina. The notion that retinoblastomas result from a defined initiating event has propelled tremendous interest in this tumor, as it provides an opportunity to decipher the complex process of tumor initiation by focusing on the consequences of a single genetic aberration. Retinoblastoma also provides special insight into brain tumor biology (the focus of this volume), as the retina is considered to be an extension of the brain based on its embryologic origin and cellular organization (1).

The concept that retinoblastomas develop in response to the mutation of a single gene derives from clinical and molecular observations stretching back more than 100 years (2). For example, retinoblastomas have long been known to occur both in a hereditary form, associated with the development of multiple independent tumors in the two eyes, and in a sporadic nonhereditary form that manifests as a single, unilateral tumor (3). Importantly, the two forms of retinoblastoma have indistinguishable histologies and behaviors, except that sporadic retinoblastomas appear with a later and kinetically distinct age distribution. These distinct kinetics led Knudson to propose that retinoblastomas develop in response to two mutations or "hits," and with one of the hits pre-existing in the germ line in familial cases (4). Later studies suggested that the two hits represent the inactivation the two alleles of a single *RB* gene at chromosome 13q14 (5). Molecular studies confirmed this view, by showing that the two *RB* alleles are inactivated through a combination of germ line and somatic mutations in hereditary retinoblastoma, and through two somatic mutations in nonhereditary tumors (6, 7).

While the inactivation of both *RB* alleles is thought to initiate retinoblastoma, most tumors carry additional genomic changes that may also contribute to tumorigenesis (8). Indeed, recent evidence suggests that *RB* inactivation first enables the development of premalignant retinoma lesions and that additional changes are needed for tumor progression (9). However, unlike other cancers, retinoblastomas may be unusually prone to progress. For example, germ line *RB* mutations predispose to an average of five retinoblastomas distributed between the two eyes (10) and elicit at least one retinoblastoma with ~95% penetrance, but predispose to retinomas that fail to progress in only ~2–10% of $RB^{+/-}$ individuals (11, 12). The rarity of retinomas that fail to progress suggests that *RB* loss routinely triggers an inexorable oncogenic process.

Despite the compelling evidence that *RB* mutations initiate retinoblastoma, the cell of origin in which *RB* suppresses tumorigenesis has been subject to long-standing debate. Early analyses revealed that many retinoblastomas contain cells with photoreceptor-like morphology, raising the possibility that they derive from a photoreceptor-related cell type (13, 14). More specifically, most of the tumor cells were found to express cone photoreceptor markers, both in situ (15–17) and in early passage cultures (18). However, the tumors also contained rare cells that expressed markers of other retinal cells, and retinoblastoma cultures were reported to have evidence of glial as well as neuronal differentiation (19), suggestive of an origin from a multipotential cell type. Notably, the evidence of glial differentiation in retinoblastoma cultures was later refuted (20). However, the presence of diverse cellular phenotypes in human retinoblastomas in situ continued to suggest an origin from a multipotential cell, and cast doubt on the relevance of the predominant cone phenotype to the cell-of-origin issue.

3. Mice with Rb Mutations Fail to Develop Retinoblastoma

Because of the overwhelming evidence that retinoblastoma is initiated by *RB* mutations, this tumor proved to be an irresistible subject for cancer modeling and represented the first attempt to produce a cancer using targeted mutagenesis. Clearly, the objectives were to test Knudson's two hit hypothesis as it pertains to retinoblastoma and to provide a platform for further investigation of the cell of origin and the tumor initiation mechanism. To this end, three groups produced mice that were heterozygous for null *Rb* alleles ($Rb^{+/-}$), to test whether somatic inactivation of the wild-type *Rb* allele elicits $Rb^{-/-}$ retinal tumors similar to those in humans. The results of these first "retinoblastoma" models were as unambiguous as they were surprising: $Rb^{+/-}$ mice failed to develop retinal tumors (21–23), but instead formed pituitary tumors and thyroid C-cell carcinomas with nearly complete penetrance, along with other lesions that developed at lower frequencies (24, 25).

The surprising absence of retinoblastomas in $Rb^{+/-}$ mice represented a failure to confirm Knudson's two-hit hypothesis as it pertains to retinoblastoma, yet the mice nevertheless confirmed the hypothesis in more general terms. Specifically, the pituitary and thyroid tumors in these mice consisted of $Rb^{-/-}$ cells that originated from $Rb^{+/-}$ cells after inactivation of the wild-type *Rb* allele (24, 25). Subsequent studies using chimeric mice derived from $Rb^{-/-}$ embryonic stem cells confirmed that homozygous *Rb* loss triggers oncogenic transformation of mouse pituitary cells but not mouse retinal cells (26, 27). Thus, these models showed that *Rb* can function as a tumor suppressor whose inactivation suffices to initiate tumorigenesis, yet also revealed that the cell types that are sensitive to *Rb* loss differ according to species.

The failure of $Rb^{+/-}$ mice to develop retinoblastoma shone a harsh light on the assumption that mice should recapitulate signaling events that are central to human oncogenesis. However, *Rb* mutant mice also gave an opportunity to learn about human cancer by providing a platform with which to identify human-specific oncogenic features. Specifically, they showed that mice lack circuitry (either through losses or gains of function) that is critical to the development of retinoblastoma, and they instigated efforts to identify circuitry that underlies the human retina's unique sensitivity to *RB* mutations.

4. Chimeric Mice with Inactivation of Multiple Rb Family Genes Produce Retinal Tumors with Interneuron Properties

Soon after the realization that $Rb^{+/-}$ mice are not predisposed to retinoblastoma, *RB* was found to be a member of a gene family that also includes *p107* and *p130* (also known as *RBL1* and *RBL2*) (28). Like the *RB*-encoded p105 protein, p107 and p130 were originally identified as proteins bound by the adenovirus E1A, SV40 T antigen, and HPV E7 DNA tumor virus oncoproteins, and appeared to be targeted by the oncoproteins' transforming activity (reviewed in (28)). Moreover, p107 and p130 were found to directly inhibit CDK2 and regulated "repressor" E2Fs in apparent complementarity to Rb's regulation of "activator" E2Fs. These observations stimulated efforts to determine whether redundant or compensatory actions of p107 and p130 in mouse (but not human) retinas underlie the lack of retinoblastoma tumorigenesis in $Rb^{+/-}$ mice.

To address this issue, mice with combined mutations in *Rb* and either *p107* or *p130* were produced and examined for retinal tumors. Initial studies employed $Rb^{+/-}$; $p107^{-/-}$ mice and $Rb^{+/-}$; $p130^{-/-}$ mice, which relied on spontaneous loss of the wild-type *Rb* allele to produce $Rb^{-/-}$; $p107^{-/-}$ or $Rb^{-/-}$; $p130^{-/-}$ cells. However, among many $Rb^{+/-}$; $p107^{-/-}$ and $Rb^{+/-}$; $p130^{-/-}$ mice examined, none developed malignancies other than the pituitary and thyroid tumors already evident in *Rb* heterozygotes (29). These findings suggested that mouse $Rb^{-/-}$; $p107^{-/-}$ and $Rb^{-/-}$; $p130^{-/-}$ retinal cells generally fail to initiate retinoblastoma tumorigenesis.

A more direct test of the tumorigenic potential of *Rb/p107*-deficient and *Rb/p130*-deficient cells was provided by chimeric mice derived from $Rb^{-/-}$; $p107^{-/-}$ or $Rb^{-/-}$; $p130^{-/-}$ embryonic stem cells. Importantly, the resulting *Rb/p107*-deficient and *Rb/p130*-deficient chimeras did develop retinal tumors, with six such tumors occurring among 14 eyes of *Rb/p107* chimeras, and eight tumors occurring among the 22 eyes of *Rb/p130* chimeras (30, 31). Gratifying as the production of retinal tumors was, it is important to note that the chimeric mice had likely produced thousands if not millions of $Rb^{-/-}$; $p107^{-/-}$ or $Rb^{-/-}$; $p130^{-/-}$ retinal progenitor cells. Thus, the incomplete penetrance of retinal tumors again implied that the vast majority of $Rb^{-/-}$; $p107^{-/-}$ or $Rb^{-/-}$; $p130^{-/-}$ retinal cells not form tumors. This contrasts with the notion that rare, stochastically produced $RB^{-/-}$ retinal cells routinely give rise to retinoblastomas in humans, and implied that further genetic changes were needed to mimic the human retina's sensitivity to the loss of Rb function.

The incomplete penetrance of retinal tumors in *Rb/p107* and *Rb/p130* chimeras was not the only concern brought on by the chimeric models. Indeed, the retinal tumors in the chimeras had histopathological features significantly different from the human disease. For example, the vast majority of the tumor cells expressed markers of amacrine interneurons rather than the photoreceptor markers routinely detected in human retinoblastomas (30, 31). The mouse tumors also lacked Flexner–Wintersteiner rosettes, which are characteristic of retinoblastoma but not other malignancies. The mouse tumors did form Homer–Wright rosettes, but these structures are common to most neuronal tumors and have no specificity for retinoblastoma.

The histopathological differences between the retinal tumors in *Rb/p107* and *Rb/p130* chimeric mice, as compared to human retinoblastomas, suggested that the mouse tumors might develop through a different process. However, it was unclear whether the histologic discrepancies reflected the use of a fundamentally distinct tumorigenesis mechanism or a less significant change in how the tumorigenic process plays out. For example, it remained possible that both the mouse and the human tumors originate from a multipotential cell, but undergo predominant interneuron rather than photoreceptor differentiation in the mouse versus human ocular environments. It was also possible that the mouse and human tumors arise from different retinal cell types, yet employ a similar oncogenic mechanism. Thus, *Rb/p107* and *Rb/p130* chimeric mice identified genetic contexts for the production of mouse retinal tumors, yet their anomalous histopathological features necessitated additional studies to assess how their tumor initiation mechanism related to that of the human malignancy.

5. Retina-Specific Inactivation of Multiple Rb Family Genes Elicits Tumors That Appear to Originate from Retinal Interneurons

Rb/p107 and *Rb/p130* chimeric mice were poorly suited to studying tumor initiation, because of the arduous task of producing chimeras and their low tumor penetrance. To circumvent these limitations, new models were designed to selectively inactivate *Rb* in retinas in combination with germ line *p107* or *p130* loss. In these models, homozygous *Rb* deletion was targeted to the retina using Rb^{lox} alleles and either an α-*Cre* transgene active in retinal progenitor cells (32, 33), a *NesCre* transgene active mosaically in retinal and neural progenitor cells (33), or a *Chx10-Cre* transgene active in a smaller subset of retinal progenitors (34). The Cre-mediated *Rb* inactivation in each of the models was thought to produce *Rb/p107*- or *Rb/p130*-deficiency in retinal progenitor cells as well as in their ganglion, amacrine, horizontal, bipolar, rod, cone, and Müller cell progeny. As expected, each of the models developed tumors similar to those in *Rb/p107* and *Rb/p130* chimeras, with interneuron but not photoreceptor differentiation, and Homer– Wright but not Flexner–Wintersteiner rosettes.

Importantly, mice with *Cre*-mediated *Rb* loss enabled analyses of the step-by-step response to the loss of Rb family proteins. In contrast to expectations, α-*Cre*-mediated Rb loss, or combined Rb/p107 loss, had little or no effect on the proliferation of retinal progenitor cells, identified by their Chx10 expression (prior to Chx10 expression in other retinal cells) and their position at the ventricular surface during mitosis (32). A claim that retinal progenitor cell proliferation was increased in one model (34) might be inaccurate because the progenitors were identified by staining for Pax6, which labels amacrine and horizontal cells in addition to progenitors. Interestingly, loss of Rb or Rb/p107 did not affect specification or production of the seven major types of retinal precursors (i.e., ganglion, amacrine, horizontal, bipolar, rod, cone, and Müller cells). Instead, Rb or Rb/p107 loss impaired the precursors' ability to exit the cell cycle, and in some cases their ability to terminally differentiate, and induced apoptosis in some of the aberrantly proliferating retinal precursors (32).

Specifically, retina-specific Rb loss elicited aberrant proliferation and nearly total ablation of ganglion and bipolar cells, as well as ectopic proliferation and reduced survival of rods (32, 33). Rb ablation with the *Chx10-Cre* also appeared to impair rod differentiation (35). However, in studies with *α-Cre*, all aberrant neuronal proliferation and rod abnormalities were reversed by co-deletion of E2F1, implying that Rb's effects on rod differentiation might indirectly relate to the aberrant rod proliferation. On the other hand, Rb had a proliferation-independent effect on the differentiation of a particular amacrine cell subtype (36).

Notably, retina-specific loss of Rb plus p107 elicited aberrant proliferation and ablation of both rods and cones, in addition to the loss of ganglion and bipolar cells in retinas lacking only Rb. Moreover, Rb/p107-deficient horizontal and amacrine cells proliferated, but did not undergo extensive apoptosis. Rather, many of these cells eventually exited the cell cycle, and their proliferation in most of the retina ceased by the 30th postnatal day (32). In addition, microtumors were detected in some Rb/p107-deficient retinas by postnatal day 8, and tumors were eventually detected in 60% of the Rb/p107-deficient eyes. As the tumors mainly had amacrine features, they were deemed likely to derive from the pool of aberrantly proliferating amacrine cells present in the postnatal period (32).

A comparison of *α-Cre*-mediated *Rb/p107* and *Rb/p130* knockouts showed that *Rb/p130* mutant retinas develop tumors with histopathology similar to that of *Rb/p107* knockouts, yet with higher penetrance (37). This study also showed that (1) *Rb/p107*-deficient and *Rb/p130*-deficient tumors arise from a region in the retinal periphery in which many cells failed to exit the cell cycle, (2) early hyperplasias and tumors include distinct regions with predominant horizontal or amacrine cell differentiation, (3) some proliferating amacrine or horizontal cells die rather than arrest, and (4) tumor progression is associated with genomic changes that in some instances mimic those in human retinoblastomas. While the amacrine and horizontal cell heterogeneity was noted to be consistent with an origin from a multipotential progenitor cell (37), it could also be explained by a multifocal origin from amacrine and horizontal cells or by instability of the highly related amacrine and horizontal cell phenotypes.

Finally, a study using a *Chx10-Cre* examined effects of homozygous inactivation of Rb in retinal progenitor cells in combination with constitutive homozygous loss of p130 and heterozygous loss of p107, in *Chx10-Cre*; $Rb^{lox/lox}$; $p130^{-/-}$; $p107^{+/-}$ mice (38). These mice are expected to produce retinal cells with the genotype $Rb^{-/-}$; $p130^{-/-}$; $p107^{+/-}$, and were thus referred to as "*p107-*

single." In *p107-single* retinas, horizontal cell precursors appeared to undergo terminal differentiation, yet retained the ability to proliferate. Notably, the *p107 single* tumors appeared to develop more rapidly than those in *α-Cre*-mediated *Rb/p107*-deficient and *Rb/p130*-deficient retinas. These findings provided support for a horizontal cell origin of retinal tumors in *p107-single* mice, and by extension indirectly supported an amacrine and/or horizontal cell origin of the similar *Rb/p107*-deficient and *Rb/p130*-deficient tumors.

In sum, mice with retina-specific inactivation of the *Rb* family provided no evidence of tumor initiation (or indeed any abnormality) in retinal progenitor cells. Rather, they suggested that the tumors arise from amacrine or horizontal cell interneurons, which are cell types that seem unlikely to give rise to the photoreceptor-like tumors in the eyes of human retinoblastoma patients.

6. Clues to the Origin of Human Retinoblastoma, and an Analogous Origin of Mouse Retinal Tumors

Had mouse models faithfully replicated human retinoblastoma histopathology, they would have provided definitive insight into the context in which Rb (or the Rb family) suppresses retinoblastoma initiation. Thus, their failure to mimic human retinoblastoma represented a missed opportunity for the retinoblastoma field, which could be attributed to species-specific signaling differences. On the other hand, the mouse models could provide insights into human retinoblastoma if their tumors originate through an analogous process, albeit from a different type of retinal neuron. Indeed, recent evidence suggests that an analogous oncogenic process might occur.

As noted earlier, the high proportion of cone-like cells in human retinoblastomas was suggestive of a cone-related cell of origin, whereas the presence of cells with other retinal phenotypes appeared consistent with an origin from a multipotential cell type. However, it remained possible that cells lacking a cone phenotype were non-neoplastic, and thus irrelevant to the cell-of-origin issue. To address this possibility, Xu et al. used Rb protein expression and *RB* genotyping to distinguish neoplastic from nonneoplastic cells in retinoblastoma tumors (39). They found that the vast majority of cells that lacked cone markers, or that expressed markers of other retinal cell types, also expressed wild-type Rb protein and were thus nonneoplastic. They also found that rare cells that lacked cone markers *and* Rb immunoreactivity nevertheless retained both *RB* alleles, and were thus nonneoplastic as well. Moreover, tumor cells expressing the progenitor and glial marker Sox2 were found to retain the two *RB* alleles and had properties of nonneoplastic retinal astrocytes (40). Overall, these analyses implied that retinoblastomas consist of cone-like tumor cells that have inactivation of both *RB* alleles, intermixed with diverse nonneoplastic retinal cells that retain wild-type *RB* and derive from the tumor's retinal environment.

To gain insight into the possible basis for the cone phenotype, Xu et al. examined the expression of potential collaborators of *RB* mutations in the developing human retina. They found that human cone precursors expressed exceptionally high levels of the MDM2 and N-Myc oncoproteins, along with high levels of Rb. This finding is notable because MDM2 and N-Myc collaborate with *RB* mutations to elicit tumors in other contexts (41–43). Moreover, human retinoblastoma cells prominently expressed MDM2 and N-Myc, and depended on these oncoproteins for their proliferation and survival. The high levels of MDM2 and N-Myc expression generally occurred without gene amplification, and MDM2 expression depended upon a cone-specific transcription factor. These observations appeared consistent with a mechanism in which Rb loss collaborates with intrinsically high levels of N-Myc and MDM2 in normal cone precursors to initiate retinoblastoma tumorigenesis (39).

In contrast to human retinas, mouse retinas lacked prominent cone precursor Mdm2 and N-Myc expression. On the other hand, Mdm2 was especially prominent in mouse horizontal cells and in a subset of amacrine cells (39)—which are candidate cells of origin of the mouse retinal tumors as described above. The especially prominent MDM2 in the human cone cells that resemble human retinoblastoma, and in the mouse amacrine and horizontal cells that resemble the mouse retinal tumors, supports the notion that intrinsically high MDM2 expression might sensitize distinct retinal cells to the loss of Rb function, both in humans and in mouse models. If confirmed, this notion would demonstrate that species-specific features can dictate the response of specific cell types to specific oncogenic insults.

7. Humanizing Mouse Signaling Circuitry to Improve Cancer Modeling and to Define Human-Specific Aspects of Tumorigenesis

It has long been recognized that cellular signaling circuitry evolves in concert with gross morphological features of different species. However, we know little about these species-specific aspects of signaling circuitries and their effects on cancer predisposition. Understanding such differences is relevant to mouse cancer models, as the mouse's failure to recapitulate human signaling mechanisms could preclude its depiction of essential disease features. Indeed, a number of important human versus mouse differences have been identified, and steps have been taken to alter or "humanize" the mouse germ line to more accurately mimic human cancer biology (44, 45).

Mouse retinoblastoma models tell us that much more needs to be learned about the roles of human-specific signaling circuitry in cancer pathogenesis. For example, they oblige us to determine how cell type-specific circuitry differs among species and sensitizes different cells to different oncogenic mutations. Mouse models will undoubtedly have a key role in this endeavor, as engineering a human-specific feature into the mouse germ line could prove the feature's involvement in human cancer, while improving the model's authenticity and its ability to predict human therapeutic responses. An important addition to mouse retinoblastoma modeling was recently published (Sangwan M, McCurdy SR, Livne-bar I, Ahmad M, Wrana JL, Chen D, and Bremner R (2012) Established and new mouse models reveal E2F1 and CDK2 dependency of retinoblastoma, and expose effective strategies to block tumor initiation. Oncogene, advance online publication). This study suggests that mouse retinal tumors require combined Rb inactivation and Cdk2 deregulation, and that p107 and p130 mutations are needed to deregulate Cdk2, rather than to impair a redundant Rb function, in Rb/p107-deficient and Rb/p130-deficient models. It will be of interest to assess how this idea, derived from a mouse retinal tumor model, relates to human retinoblastoma genesis.

Acknowledgments

The author acknowledges support from the Fund for Ophthalmic Knowledge, the Elsa U. Pardee Foundation, and the NIH (1R01CA137124) during the preparation of this review, and thanks Rod Bremner for critical reading of the manuscript.

References

1. Chow RL, Lang RA (2001) Early eye development in vertebrates. Annu Rev Cell Dev Biol 17:255–296
2. Albert DM (1987) Historic review of retinoblastoma. Ophthalmology 94:654–662
3. Weller CV (1941) The inheritance of retinoblastoma and its relationship to practical eugenics. Cancer Res 1:517–535
4. Knudson AG Jr (1971) Mutation and cancer: statistical study of retinoblastoma. Proc Natl Acad Sci USA 68:820–823
5. Cavenee WK, Dryja TP, Phillips RA, Benedict WF, Godbout R, Gallie BL, Murphree AL, Strong LC, White RL (1983) Expression of recessive alleles by chromosomal mechanisms in retinoblastoma. Nature 305:779–784
6. Dunn JM, Phillips RA, Becker AJ, Gallie BL (1988) Identification of germline and somatic mutations affecting the retinoblastoma gene. Science 241:1797–1800
7. Friend SH, Bernards R, Rogelj S, Weinberg RA, Rapaport JM, Albert DM, Dryja TP (1986) A human DNA segment with properties of the gene that predisposes to retinoblastoma and osteosarcoma. Nature 323:643–646
8. Corson TW, Gallie BL (2007) One hit, two hits, three hits, more? Genomic changes in

the development of retinoblastoma. Genes Chromosomes Cancer 46:617–634

9. Dimaras H, Khetan V, Halliday W, Orlic M, Prigoda NL, Piovesan B, Marrano P, Corson TW, Eagle RC Jr, Squire JA, Gallie BL (2008) Loss of RB1 induces non-proliferative retinoma; increasing genomic instability correlates with progression to retinoblastoma. Hum Mol Genet 17(10): 1363–1372
10. Abramson DH, Gombos DS (1996) The topography of bilateral retinoblastoma lesions. Retina 16:232–239
11. Balmer A, Munier F, Gailloud C (1991) Retinoma. Case studies. Ophthalmic Paediatr Genet 12:131–137
12. Singh AD, Santos CM, Shields CL, Shields JA, Eagle RC Jr (2000) Observations on 17 patients with retinocytoma. Arch Ophthalmol 118:199–205
13. Flexner S (1891) A peculiar glioma (neuroepithelioma?) of the retina. Bull Johns Hopkins Hosp Balt 2:115–119
14. Ts'o MO, Zimmerman LE, Fine BS (1970) The nature of retinoblastoma. I. Photoreceptor differentiation: a clinical and histopathologic study. Am J Ophthalmol 69:339–349
15. Gonzalez-Fernandez F, Lopes MB, Garcia-Fernandez JM, Foster RG, De Grip WJ, Rosemberg S, Newman SA, VandenBerg SR (1992) Expression of developmentally defined retinal phenotypes in the histogenesis of retinoblastoma. Am J Pathol 141:363–375
16. Nork TM, Schwartz TL, Doshi HM, Millecchia LL (1995) Retinoblastoma. Cell of origin. Arch Ophthalmol 113:791–802
17. Rodrigues MM, Rajagopalan S, Lee L, Nair CN, Advani SH, Donoso L, Chader GJ, Wiggert B (1992) Retinoblastoma: messenger RNA for interphotoreceptor retinoid binding protein. Curr Eye Res 11:425–433
18. Bogenmann E, Lochrie MA, Simon MI (1988) Cone cell-specific genes expressed in retinoblastoma. Science 240:76–78
19. Kyritsis AP, Tsokos M, Triche TJ, Chader GJ (1984) Retinoblastoma—origin from a primitive neuroectodermal cell? Nature 307:471–473
20. Virtanen I, Kivela T, Bugnoli M, Mencarelli C, Pallini V, Albert DM, Tarkkanen A (1988) Expression of intermediate filaments and synaptophysin show neuronal properties and lack of glial characteristics in Y79 retinoblastoma cells. Lab Invest 59:649–656
21. Jacks T, Fazeli A, Schmitt EM, Bronson RT, Goodell MA, Weinberg RA (1992) Effects of an Rb mutation in the mouse. Nature 359:295–300
22. Lee EY, Chang CY, Hu N, Wang YC, Lai CC, Herrup K, Lee WH, Bradley A (1992) Mice deficient for Rb are nonviable and show defects in neurogenesis and haematopoiesis. Nature 359:288–294
23. Clarke AR, Maandag ER, van Roon M, van der Lugt NM, van der Valk M, Hooper ML, Berns A, te Riele H (1992) Requirement for a functional Rb-1 gene in murine development. Nature 359:328–330
24. Hu N, Gutsmann A, Herbert DC, Bradley A, Lee WH, Lee EY (1994) Heterozygous Rb-1 delta 20/+mice are predisposed to tumors of the pituitary gland with a nearly complete penetrance. Oncogene 9:1021–1027
25. Williams BO, Remington L, Albert DM, Mukai S, Bronson RT, Jacks T (1994) Cooperative tumorigenic effects of germline mutations in Rb and p53. Nat Genet 7:480–484
26. Maandag EC, van der Valk M, Vlaar M, Feltkamp C, O'Brien J, van Roon M, van der Lugt N, Berns A, te Riele H (1994) Developmental rescue of an embryonic-lethal mutation in the retinoblastoma gene in chimeric mice. EMBO J 13:4260–4268
27. Williams BO, Schmitt EM, Remington L, Bronson RT, Albert DM, Weinberg RA, Jacks T (1994) Extensive contribution of Rb-deficient cells to adult chimeric mice with limited histopathological consequences. EMBO J 13:4251–4259
28. Classon M, Dyson N (2001) p107 and p130: versatile proteins with interesting pockets. Exp Cell Res 264:135–147
29. Lee MH, Williams BO, Mulligan G, Mukai S, Bronson RT, Dyson N, Harlow E, Jacks T (1996) Targeted disruption of p107: functional overlap between p107 and Rb. Genes Dev 10:1621–1632
30. Dannenberg JH, Schuijff L, Dekker M, van der Valk M, te Riele H (2004) Tissue-specific tumor suppressor activity of retinoblastoma gene homologs p107 and p130. Genes Dev 18:2952–2962
31. Robanus-Maandag E, Dekker M, van der Valk M, Carrozza ML, Jeanny JC, Dannenberg JH, Berns A, te Riele H (1998) p107 is a suppressor of retinoblastoma development in pRb-deficient mice. Genes Dev 12:1599–1609
32. Chen D, Livne-bar I, Vanderluit JL, Slack RS, Agochiya M, Bremner R (2004) Cell-specific effects of RB or RB/p107 loss on retinal development implicate an intrinsically death-resistant cell-of-origin in retinoblastoma. Cancer Cell 5:539–551
33. MacPherson D, Sage J, Kim T, Ho D, McLaughlin ME, Jacks T (2004) Cell type-

specific effects of Rb deletion in the murine retina. Genes Dev 18:1681–1694
34. Zhang J, Schweers B, Dyer MA (2004) The first knockout mouse model of retinoblastoma. Cell Cycle 3:952–959
35. Zhang J, Gray J, Wu L, Leone G, Rowan S, Cepko CL, Zhu X, Craft CM, Dyer MA (2004) Rb regulates proliferation and rod photoreceptor development in the mouse retina. Nat Genet 36:351–360
36. Chen D, Opavsky R, Pacal M, Tanimoto N, Wenzel P, Seeliger MW, Leone G, Bremner R (2007) Rb-mediated neuronal differentiation through cell-cycle-independent regulation of E2f3a. PLoS Biol 5:e179
37. MacPherson D, Conkrite K, Tam M, Mukai S, Mu D, Jacks T (2007) Murine bilateral retinoblastoma exhibiting rapid-onset, metastatic progression and N-myc gene amplification. EMBO J 26:784–794
38. Ajioka I, Martins RA, Bayazitov IT, Donovan S, Johnson DA, Frase S, Cicero SA, Boyd K, Zakharenko SS, Dyer MA (2007) Differentiated horizontal interneurons clonally expand to form metastatic retinoblastoma in mice. Cell 131:378–390
39. Xu XL, Fang Y, Lee TC, Forrest D, Gregory-Evans C, Almeida D, Liu A, Jhanwar SC, Abramson DH, Cobrinik D (2009) Retinoblastoma has properties of a cone precursor tumor and depends upon cone-specific MDM2 signaling. Cell 137:1018–1031
40. Xu XL, Lee TC, Offor N, Cheng C, Liu A, Fang Y, Jhanwar SC, Abramson DH, Cobrinik D (2010) Tumor-associated retinal astrocytes promote retinoblastoma cell proliferation through production of IGFBP-5. Am J Pathol 177:424–435
41. Kowalik TF, DeGregori J, Leone G, Jakoi L, Nevins JR (1998) E2F1-specific induction of apoptosis and p53 accumulation, which is blocked by Mdm2. Cell Growth Differ 9:113–118
42. Lomazzi M, Moroni MC, Jensen MR, Frittoli E, Helin K (2002) Suppression of the p53- or pRB-mediated G1 checkpoint is required for E2F-induced S-phase entry. Nat Genet 31:190–194
43. Weiss WA, Aldape K, Mohapatra G, Feuerstein BG, Bishop JM (1997) Targeted expression of MYCN causes neuroblastoma in transgenic mice. EMBO J 16:2985–2995
44. Rangarajan A, Weinberg RA (2003) Opinion: comparative biology of mouse versus human cells: modelling human cancer in mice. Nat Rev Cancer 3:952–959
45. Demetrius L (2005) Of mice and men. When it comes to studying ageing and the means to slow it down, mice are not just small humans. EMBO Rep 6(S1):S39–S44

Neuromethods (2013) 77: 153–161
DOI 10.1007/7657_2011_26

Published online: 13 March 2012

Ependymoma Models

Jesús Vaquero and Mercedes Zurita

Abstract

Experimental neurocarcinogenesis using chemical agents or certain viruses can get brain tumors; some of them have been considered classically as ependymomas according to their histological aspects. However, it is now accepted that these tumors do not meet a clear criteria to be considered as ependymoma models. Xenografts of human ependymomas in nude mouse are most useful models, but they have the disadvantage that the absence of host immune response may mask important aspects currently considered in the biology of brain tumors.

Key words: Ependymoma, Experimental brain tumors, ENU, Chemical neurocarcinogenesis

1. Introduction

Ependymomas are rare neuroepithelial tumors that arise from the ependymal cells of the cerebral ventricles, the central canal of the spinal cord, and cortical rests. They constitute 8–10% of brain tumors in children and up to 4% of brain tumors in adults (1).

Data concerning the genetic and molecular abnormalities in these tumors are sparse (2). Early studies revealed a high incidence of loss of heterozygosity (LOH) on chromosome arm 22q, and mutations in the *MEN1* gene, located at 11q13, especially in cases of tumor recurrence or malignant transformation. Molecular pathways abnormalities include overexpression of the ERBB2 and ERBB4 receptors that have been identified in over 75% of pediatric ependymomas and correlated with proliferative index and prognosis (3).

Currently, the concept that tumors develop from cancer stem cells is gaining acceptance, and in this line of thought, recent studies suggested that radial-glial cells may be the stem cells for ependymomas (4) which can facilitate obtaining experimental ependymoma models useful for testing new therapies potentially applicable to patients. However, at present, we have no experimental models that allow such strategy, and details about the experimental induction of ependymoma or ependymoma-like tumors are scarce and contradictory (5).

It is well known that experimental oncology can induce brain tumors of different histology and location, but there are many problems in establishing a clear correlation between induced brain tumors and brain tumors that we can find in humans. Many of the descriptions in the literature about the presence of ependymomas induced in animals by different methods are based on the location of tumors within the brain or are based on morphology of tumor cells. However, the location of a brain tumor in the ventricular region or occupying the ventricular cavity is not a valid argument to classify this tumor as ependymoma. The presence of a tissue structure similar to that presented human ependymomas is not a valid argument either. Moreover, there are many data supporting that experimental brain tumors should be considered mostly as undifferentiated neoplasms, regardless of their histologic appearance (6–9). Given these limitations, it is true that experimental tumors have been described in which the location or morphology of induced tumors has led to their classification as "ependymomas" or "ependymomatous tumors."

2. Genetically Engineered Animals Models of Ependymoma

In the last years, genetically manipulated animals, generally mice, have been reported to study the roles of specific genes and of multiple genes in combination, in the etiology of cancer. At present, it is known that a large number of genetically engineered mice predisposed to show specific tumors of the central and peripheral nervous system, and these models were reviewed and evaluated by a panel convened by the NCI Mouse Models of Cancer Consortium (10). An important recommendation of this panel is that tumors arising in genetically engineered mice should be classified by the same nomenclature system used for the corresponding human tumors, but preceded by the designator GEM (genetically engineered murine) (e.g., GEM-ependymoma). This convention explicitly acknowledges that the murine tumors resulting from these constructs may not be exact equivalents of the corresponding and morphologically similar human tumors, and the same consideration applies also to naturally occurring and carcinogen-induced tumors in nonhuman species, as those also may not be the exact equivalents of their human counterparts. Many GEM tumors provide good biological models of malignant gliomas, but at present, any clear GEM-ependymoma model has been described.

3. Virus-Induced Ependymomatous Tumors

The role of certain viruses to cause human brain tumors has been discussed for many years and although occasionally virus particles have been described in gliomas, there are no reliable data about an association between virus and development of brain tumors in humans. However, in experimental neuro-oncology, clear evidences support that certain oncogenic viruses can induce brain tumors in animals, including tumors whose morphology suggests their classification as malignant ependymomas. Among these viruses, the Simian adenovirus (SA7) is capable of inducing both ependymomas and malignant choroid plexus papillomas, after intracerebral inoculation to rats and Syrian hamsters (11). The chick embryo lethal orphan virus (CELO) is an avian adenovirus that can also induce intraventricular tumors in newborn Syrian hamsters (12) and although intraventricular location has led to their classification as malignant ependymomas, the ultrastructural features of these tumors are more similar to those of choroid plexus papillomas.

After intracerebral inoculation in Syriam hamsters, some papovavirus have also given rise to tumors identified as ependymomas, and even in some experiments papovaviruses isolated from the brain of humans with progressive multifocal leukoencephalopathy (PML) have been used to cause brain tumors in animals (13). Among the major types of human papovavirus capable of inducing ependymoma or ependymomatous tumors, the BK-virus can induce brain tumors with the appearance of malignant ependymomas, showing atypia, pseudorosettes, and necrosis. This virus can also induce tumors in other organs, such as pancreatic insulinomas and osteosarcomas (14, 15).

The simian vacuolating virus (SV40) induces tumors difficult to classify, after intracerebral inoculation in neonatal Syrian hamsters. However, many of them have been regarded as malignant papillary ependymomas or malignant choroid plexus tumors, according to their histological features (16).

4. Ependymomatous Tumors Induced by Chemical Agents

Intraventricular tumors in mice are easily produced by implanting crystals of carcinogenic aromatic hydrocarbons, such as 3-methylcholanthrene, into the cerebral ventricle (17), but this experimental method is far from an acceptable model of ependymoma. On the other hand, suspected ependymomas have been identified among the primary brain tumors induced in rats by transplacental dimethyl-benzo-anthracene (DBA) (18) or by ethyl-nitrosourea (ENU) and other alkylating agents (19, 20, 21–28).

These tumors occurred within the cavities of the ventricles and represent only a small fraction of all brain tumors that can be induced in these experiments (Fig. 1). Their classification as ependymomas is controversial, and generally they have been considered as such, either by their intraventricular location or due to their histological features, reminiscent of ependymomatous tumors of humans.

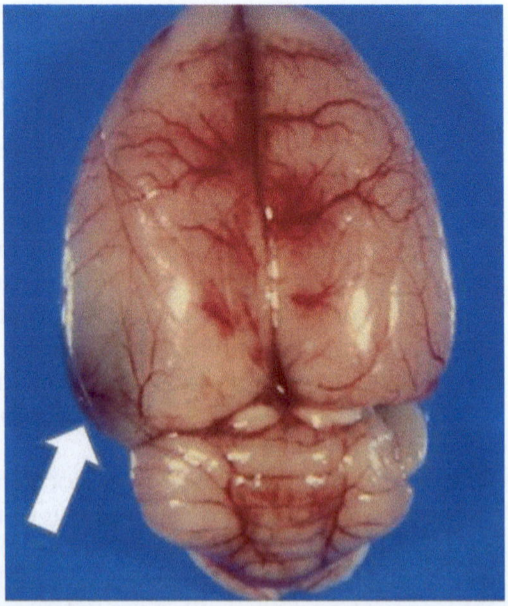

Fig. 1. Macroscopical view of the brain, in a case of ENU-induced brain tumor. The left hemisphere shows an edematous aspect due to the presence of the intracerebral tumor (*arrow*).

Unequivocal ependymomas were not seen in series of mice exposed transplacentally to ENU, but according to accepted classifications, approximately 20% of the ENU-induced brain tumors could be diagnosed as ependymomas, anaplastic ependymomas, or mixed glial tumors with ependymoma areas (24). In many classifications, ependymomatous tumors were termed as "anaplastic glioependymomas" due to the presence of ependymoma-like cells, with other glial-like cells, pleomorphic cells, and generally with rounded cells very similar to those of human oligodendrogliomas.

In 1973, Jones et al. (22) distinguished five groups of ENU-induced tumors which were, in order of frequency: periventricular subependymal plate gliomas, astrocytic and oligodendrocytic gliomas, neuronal tumors of the spinal and intracranial ganglia and nerves, neuronal tumors of the central nervous system, and meningeal tumors. Periventricular subependymal plate gliomas were further divided into ependymomas and mixed ependymo-oligodendroastrocytomas, these latter tumors being equivalent to the anaplastic glioependymomas of Koestner et al. (26). In these tumors, the neoplastic ependymomatous cells are arranged in groups, cords, and rosettes or packed in solid masses. The majority of cells are medium or large, with oval or irregular nuclei showing prominent nucleoli. The cytoplasm varied from being abundant to forming a narrow ring around the nucleus. Mitosis and vascular hyperplasia is frequent, with hemorrhagic and necrotic areas, especially in large tumors. Ultrastructurally, two cell types predominate, a smaller, undifferentiated, and another cell type, showing greater size and nuclear pleomorphism (11, 24, 25). However, in these cells, generally there are no characteristics of ependymoma cells, such as the presence of cilia or specialized cell junctions.

In any case, studies on the early stages of the tumor development suggest that most, if not all, ENU-induced brain gliomas originate from the subependymal plate (11), a layer of primitive cells which exists throughout postnatal life in mammals as a remnants of the periventricular matrix, and represents an area of neurogenesis in adults. After the interaction of the carcinogen with cells in this region, the transformed cells can migrate into the white matter, immediately beneath the cortex, or remain in situ. As a result of neoplastic transformation, neural stem cells of the subependymal plate are affected, and the predominance of cells showing differentiation toward a glial phenotype or the other determines the different morphology of each tumor, explaining the frequent heterogeneity of cell population in the ENU-induced gliomas.

In 1992, we studied the immunohistochemical expression of synaptophysin in seven experimental brain tumors previously classified as anaplastic ependymoma or mixed oligodendroglioma-ependymoma on the basis to their histologic pattern (6). The seven brain tumors were induced by transplacentary administration of ENU (50 mg/kg) to adult Wistar rats, at day 17 of gestation, and all of them were clinically manifested in the offspring by the appearance of neurologic signs after a mean latency of 10 months. Our results showed strong synaptophysin-positivity in most of the tumor cells (Fig. 2), and electron microscopy revealed rosettes of neuroblastic type and tumor cells processes with microtubules, electron-dense bodies, and dense core vesicles, without features that

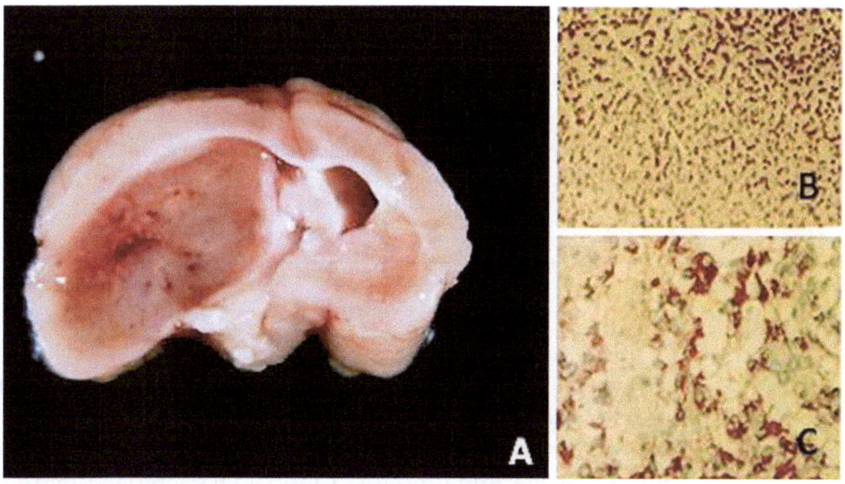

Fig. 2. (**a**) Macroscopic aspect of an ENU-induced brain tumor in the Wistar rat. The tumor was classified as a malignant ependymoma. (**b**) Histological aspect of the tumor (×100) (H.E.). (**c**) Strong synaptophysin expression in the tumor cells (×200).

characterize ependymomatous tumors. These findings support the opinion that ENU-induced tumors are undifferentiated neuro-epithelial tumors, regardless of their location or morphology (7–9).

5. The Ependymoblastoma Murine Model

Ependymoblastoma is considered in the actual WHO classification of human brain tumors as a rare, malignant, and embryonary tumor, manifestating in neonates and young children, and histologically characterized by distinctive multilayered rosettes (29). Nevertheless, in the past, the term "ependymoblastoma" has been used in human pathology for the description of malignant ependymomas or undifferentiated malignant gliomas containing areas with ependymomatous aspect. This fact explains the description as ependymoblastoma of experimental brain tumors consisting of malignant undifferentiated cells and forming the pseudorosettes that usually can be observed in ependymomatous tumors.

In 1939, Seligman and Shear induced brain tumors in mice, through the application of methyl cholantrene (11), and 2 years later, Zimmerman and Arnold (30) were able to transplant these tumors intracerebrally in other mices, obtaining growth in 90–100% of cases. From these original tumors, Ausman et al. (31) obtained several cell lines, including a line named as "murine ependymoblastoma A," with histological structure that suggested a malignant and embryonary ependymomatous tumor. This line has been used in numerous studies of experimental neuro-oncology, due to its easy transplantation on the brain of recipient mices, but it lack many of the characteristics noted in the human ependymoblastomas or malignant ependymomas, including ultrastructural ependymal features and cell kinetics (32). Furthermore, the murine ependymoblastoma contains presumed endogenous retrovirus, murine mammary tumor particles, and murine C-type viral structural antigens (33), which has led some authors to suggest that it can actually be a virus-induced tumor (34).

6. Xenografts of Ependymomatous Tumors

The possibility to obtain xenografts of human brain tumors has been considered an acceptable model for studies in experimental neuro-oncology. However, it requires the implantation of the tumor on the nude mouse, a model characterized by the absence of immunological defense, which allows the development of xenografts, but interferes with many aspects of the neurobiology of brain tumors, in which, the immune response determines the outcome and prognosis (35).

Some authors have studied the growth and survival of human ependymomas xenografted to nude mouse. As an example, in 1996, McLendon et al. (36) reported nude mouse xenografts of human ependymomas obtained from two nonrelated childhood. Xenografts were performed in both subcutaneous and intracranial locations, with a median postimplantation survival about 3 months. Both xenografts grow as well-formed masses, with no evidence of infiltration into either brain or subcutaneous tissues. While perivascular pseudopalisading was found in both xenografts, true ependymal rosette formation was absent. Ultrastructurally, neither xenografts exhibited cilia, but both produced abundant intermediate filaments. Neoplastic cells showed positivity to glial fibrillary acidic protein (GFAP), vimentin, and nestin. Thus, it is obvious that xenografts of human ependymomas can be performed on the nude mouse model, which offers the possibility of several studies about the progression and therapy of these tumors.

In this same way of research, recently, Yu et al. (37) injected a fresh ependymoma specimen from a 9-year-old patient into the brain of RAG2/severe complex immune deficiency (SCID) mice. All five mice receiving the graft developed tumor growth, which were serially subtransplanted in vivo in mouse brains for four generations and can be cryopreserved for long term. Xenografts shared identical morphological features with the original tumors, and they contained a small population of CD133+ stem cells that can form neurospheras and display multipotent capabilities in vitro. The authors obtained a new and permanent cell line (BXD-1425EPN) that can facilitate further biological studies on ependymomas.

7. Conclusions

When the possibility to obtain an experimental model of ependymoma is considered, it is obvious that none of the animal models presently used fits the criteria for an ideal model. Genetically engineered animal models have been used in recent years to the study of brain tumors, but none of them can be considered useful as an experimental model of ependymoma. Viral neurocarcinogenesis obtains occasionally tumors with ependymoma aspect, but ultrastructural studies disclosed important questions about whether they really can be classified as ependymomas or choroidal plexus tumors. Most of the experimental cerebral neoplasms identified as "ependymoma" or "ependymoma-like" tumors are, in fact, malignant neuroectodermal tumors, and their intraventricular location or their histological features are not arguments supporting that they are tumors arising from ependymal cells. The named “murine ependymoblastoma” is a tumor of uncertain classification, whose origin is also controversial, and data exist for the possibility that it represents a virus-induced tumor. Finally, the possibility of using xenografts of human ependymomatous tumors on the model of nude mouse or other animal models presenting immune deficiency presents the difficulty of working on an experimental model in which the absence of host immune response may mask many issues that currently are considered important in the biology of brain tumors.

References

1. Chamberlain MC (2003) Ependymomas. Curr Neurol Neurosci Rep 3:193–199
2. Ruda R, Gilbert M, Soffietti R (2008) Ependymomas of the adult: molecular biology and treatment. Curr Opin Neurol 21:754–761
3. Gilbertson RJ, Bentley L, Hernan R, Junttila TT, Frank AJ, Haapasalo H, Connelly M, Wetmore C, Curran T, Elenius K, Ellison DW (2002) ERBB receptor signalling promotes ependymoma cell proliferation and represents a potential novel therapeutic target for this disease. Clin Cancer Res 8:3054–3064
4. Poppleton H, Gilbertson RJ (2007) Stem cells of ependymoma. Br J Cancer 96:6–10
5. Rorke LB (1994) Experimental production of primitive neuroectodermal tumors and its relevante to human neuro-oncology. Am J Pathol 144:444–448
6. Vaquero J, Coca S, Zurita M, Oya S, Arias A, Moreno M, Morales C (1992) Synaptophysin expresión in “ependymal tumors” induced by ethyl-nitrosourea in rats. Am J Pathol 141:1037–1041
7. Vaquero J, Oya S, Coca S, Zurita M (1994) Experimental induction of primitive neuroectodermal tumours in rats: a reappraisement of the ENU-induced model of neurocarcinogenesis. Acta Neurochir (Wien) 131:294–301
8. Vaquero J, Oya S, Coca S, Zurita M (1992) Expression of neuronalmarkers in the so-called ENU-induced schwannomas. Neuroreport 3:619–621
9. Vaquero J, Zurita M, Arias A, Oya S, Coca S, Morales C (1992) Expression of glial and neuronal markers in so-called ENU-induced oligodendrogliomas of the rat. Histol Histopathol 7:647–651
10. Weiss WA, Israel M, Cobbs C, Holland E, James CD, Louis DN, Marks C, McClatchey AI, Roberts T, Van Dyke T, Wetmore C, Chiu IM, Giovannini M, Guha A, Higgins RJ, Marino S, Radovanovic I, Reilly K, Aldape K (2002) Neuropathology of genetically engineered mice: consensus report and recommendations from an international forum. Oncogene 21:7453–7463
11. Lantos PL (1980) Chemical induction of tumours in the nervous system. In: Thomas DGT, Graham DI (eds) Brain tumors. Scientific basis, clinical investigation and current therapy. Butterworth, London, pp 85–108
12. Mancini LO, Yates VJ, Jasty V, Anderson J (1969) Ependymomas induced in hamsters inoculted with an avian adenovirus (CELO). Nature 222:190–191
13. Padgett BL, Walker DL, ZuRhein GM, Varakis JN (1977) Differential neurooncogenicity of strains of JC virus, a human polyoma virus, in newborn Syrian hamsters. Cancer Res 37:718–720
14. Altavilla G, Carra L, Alberti S, Corallini A, Cavazzini L, Fabris G, Aleotti A, Barbani-Brodano G (1983) BK virus-induced tumors in hamsters: a morphological, histochemical and ultrastructural study. Oncology 40:427–441
15. Corallini A, Barbanti-Brodano G, Bortoloni W, Nenci I, Cassai E, Tampieri M, Portolani M, Borgatti M (1977) High incidence of ependymomas induced by BK virus, a human papovavirus: brief communication. J Natl Cancer Inst 59:1561–1564
16. Rabson AS, O’Connor GT, Kirschtein RL, Branigan WJ (1962) Papillary ependymomas produced in Rattus (Mastomys) natalensis inoculated with vacuolating virus (SV40). J Natl Cancer Inst 29:765–787
17. Janisch W, Screiber D (1977) Methods of induction of experimental CNS tumors. In: Bigner DD, Swenberg JA (eds) Experimental tumors of the central nervous system. Upjohn Co, Michigan
18. Rice JM, Joshi SR, Shenefelt RE, Wenk ML (1978) Transplacental carcinogenic activity of 7,12-dimethylbenz[*a*]anthracene. In: Jones

PW, Freudenthal RI (eds) Carcinogenesis polynuclear aromatic hydrocarbons, vol 3. Raven, New York, pp 413–422

19. Tomatis L, Mohr U (1973) Transplacental carcinogenesis. IARC Scientific Publications No. 4, International Agency for Research on Cancer, Lyon
20. Grossi-Paoletti E, Paoletti P, Pezzotta S, Schiffer D, Fabiani A (1972) Tumors of the nervous system induced by ethylnitrosourea administered either intracerebrally or subcutaneously to newborn rats. Morphological and biochemical characteristics. J Neurosurg 37:580–590
21. Grossi-Paoletti E, Paoletti P, Schiffer D, Fabiani A (1970) Experimental brain tumors induced in rats by nitrosoures derivatives. Part 2. Morphological aspects of nitrosoethylurea tumors obtained by transplacentalinduction. J Neurol Sci 11:573–581
22. Jones EL, Searle CE, Smith WT (1973) Tumours of the nervous system induced in rats by the neonatal administration of N-ethyl-N-nitrosourea. J Pathol 109:123–139
23. Schiffer D, Giordana MT, Pezzotta S, Lechner C, Paoletti P (1978) Cerebral tumors induced by transplacental ENU: study of the different stages, particularly of early proliferations. Acta Neuropathol 41:27–31
24. Mandybur TI, Alvira MM (1982) Ultrastructural findings in so-called ependymal tumors induced by transplacental administration of ethylnitrosourea (ENU). Acta Neuropathol 57:51–58
25. Goebel HH, Cravioto H (1972) Ultrastructure of human and experimental ependymomas. J Neuropathol Exp Neurol 31:54–71
26. Koestner A, Swenberg JA, Wechsler W (1971) Transplacental production with ethylnitrosourea of neoplamas of the nervous system in Sprague-Dawley rats. Am J Pathol 63:37–56
27. Swenberg JA (1974) Animal models of brain tumor research. In: Sano K, Ishii S (eds) Recent progress in neurological surgery. Excerpta Medica, Amsterdam
28. Swenberg JA, Koestner A, Wechsler W (1972) The induction of tumors of the nervous system with intravenous methylnitrosourea. Lab Invest 26:74–85
29. McLendon RE, Judkins AR, Eberthart CG, Fuller GN, Sarkar C, Ng HK (2007) Central nervous system primitive neuroectodermal tumours. In: Louis DN, Ohgaki H, Wiestler OD, Cavenee WK (eds) WHO classification of tumours of the central nervous system. IARC, Lyon, pp 141–146
30. Zimmerman HM, Arnold H (1941) Experimental brain tumors. I. Tumors produced with methylcholanthrene. Cancer Res 1:919–938
31. Ausman JI, Shapiro WR, Rall PD (1970) Studies on the chemotherapy of experimental brain tumors: development of an experimental model. Cancer Res 30:2394–2400
32. Rubin R, Sutton CH, Zimmerman HM (1968) Experimental ependymoblastoma (Fine structure). J Neuropathol Exp Neurol 27:421–438
33. Rubin R, Ames RP, Sutton CH, Zimmerman HM (1969) Virus-like particles in murine ependymoblastoma. J Neuropathol Exp Neurol 28:371–387
34. Ames RP, Rubin RC (1970) Morphology of virus-like particles persisting in murine ependymoblastoma in vitro. Cancer Res 30:1142–1148
35. Brooks WH, Roszman TL (1980) Cellular immune responsiveness of patients with primary intracranial tumors. In: Thomas DGT, Graham DI (eds) Brain tumours. Scientific basis, clinical investigation and current therapy. Butterworths, London, pp 121–132
36. McLendon RE, Fung KM, Bentley RC, Ahmed Rasheed BK, Trojanowski JQ, Bigner SH, Bigner DD, Friedman HS (1996) Production and characterization of two ependymoma xenografts. J Neuropathol Exp Neurol 55:540–548
37. Yu L, Baxter PA, Volcu H, Gurusiddappa S, Zhao Y, Adesina A, Man TK, Shu Q, Zhang YJ, Zhao XM, Su JM, Perlaky L, Dauser R, Chintagumpala M, Lau CC, Blaney SM, Rao PH, Leung HC, Li XN (2010) A clinical relevant orthotopic xenograft model of ependymoma that maintains the genomic signature of the primary tumor and preserves cancer stem cells in vivo. Neuro Oncol 12:580–594

Neuromethods (2013) 77: 163–173
DOI 10.1007/7657_2011_27

Published online: 20 March 2012

Meningioma Models

Jessica Rainey and Randy L. Jensen

Abstract

Intracranial xenograft meningioma models are considered best because of their biological relevance in tumor location. Traditionally, difficulty in obtaining serial measurements and high costs hindered intracranial tumor research; however, with the development of bioluminescence imaging systems, serial measurements of intracranial xenografted immortal meningioma cell lines can now be completed in a noninvasive, cost-effective, and efficient manner. This method involves transfecting immortal meningioma cells with the firefly *luciferase* gene, which then expresses the luciferase enzyme. This enzyme oxidizes luciferin in a reaction that releases photon energy that can be easily measured and recorded, even after the transfected tumor cells have been implanted intracranially in a mouse model. The transfected cells replicate key macroscopic, microscopic, immunohistochemical, and ultrastructure features of aggressive meningiomas.

Key words: Bioluminescence, CH-157-MN, CH-157-MN-LucNeo, IOMM-Lee, IOMM-Lee-LucNeo, Luciferase, Meningioma, Mouse model, Xenograft

1. Introduction

Meningiomas account for approximately 20% of all adult primary intracranial tumors, with a higher prevalence in women (2:1) and older adults (1). They arise from specialized meningothelial cells known as arachnoid cap cells and usually present within the arachnoid villi (2). The vast majority (~91%) of these tumors are graded as benign, while atypical and anaplastic/malignant tumors have an incidence of 5% and 4%, respectively (3). The location and intrinsic biology of these tumors primarily determine patient outcome, although most meningiomas can be surgically excised (1, 4, 5).

Animal models of meningiomas are important to studying and developing novel treatment options for these tumors that can be translated into clinical practice. In 1945, Greene and Arnold (6) described the first animal model, in which they transplanted operative meningioma specimens into guinea pig eyes. Through the years, animal models of xenograft transplantation have been described covering a variety of animals and numerous transplantation locations (7–10). Intracranial xenograft locations are the ideal meningioma model because this location is the most biologically relevant; however, tumors within the cranium are traditionally difficult and costly to measure serially. For ease and cost-effectiveness, we have used the bioluminescence imaging (BLI) system to measure tumor changes via luciferase-expressing meningioma cells.

This model involves transfecting the desired human immortal cell lines with luciferase and growing the transduced cells in culture. The transfected cells are then implanted intracranially into a mouse model and later imaged using the BLI system. This BLI system works via an intraperitoneal injection of firefly D-Luciferin into the xenograft mouse model, which is then oxidized by luciferase-containing cells. This reaction releases photon energy that can be detected using the Xenogen IVIS 100 imaging system. This can be supplemented with magnetic resonance (MR) imaging if further anatomical, functional (tumor blood flow or peritumoral edema), or metabolic (magnetic resonance spectroscopy) information is desired. After the animals are sacrificed, the tumors are removed for immunohistochemical, histological, and ultrastructure analysis.

The BLI and MR imaging results show excellent correlation on tumor volume between the two systems (11). The transfected xenograft tumors appear as discrete tumors insinuated along the cranial nerves and Virchow–Robin spaces, analogous to operative meningiomas. Also like clinical meningiomas, the transfected immortal cell lines stain positive for vimentin and negative for glial fibrillary acidic protein (GFAP) and epithelial membrane antigen (EMA). Finally, these tumors exhibit desmosomes and junctional complexes upon ultrastructure analysis, again consistent with clinical meningiomas (11).

The mouse intracranial meningioma bioluminescence model allows researchers to obtain serial measurements of the tumors in a noninvasive and cost-effective manner. Importantly, this model allows for biologically and anatomically relevant results as these tumors appear macroscopically, histologically, immunohistochemically, and ultrastructurally consistent with clinical aggressive meningiomas.

2. Materials

The products and suppliers used are listed below. Comparable products should also be effective.

2.1. Meningioma Cell Culture and Luciferase Gene Insertion

1. Human immortal cell lines, such as CH-157-MN and IOMM-Lee, grown at 37°C in 7.5% CO_2
2. Growth medium: Dulbecco's modified Eagle medium, supplemented with 10% fetal calf serum, L-glutamine (2 μM), penicillin (50 IU/ml), and streptomycin (50 mg/ml)
3. T-175 flasks

4. 10-cm Dishes
5. Modified pMMP retrovirus, fused with the coding sequences for luciferase and neomycin phosphotransferase (pMMP-LucNeo)
6. Opti-MEM with 8 ng/ml hexadimethrine bromide (polybrene, Aldrich)
7. 300 mg/ml G418 (Gibco)
8. 500 μl of Glo Lysis Buffer and 100 μl of Brite Glo luciferin (Promega)
9. Trypsin
10. Luminometer
11. Six-well plates

2.2. Transfected Tumor Cell Implantation

1. CH-157-MN-LucNeo and IOMM-Lee-LucNeo cell lines at ~80% confluence in T-175 flasks
2. Ice
3. Hemocytometer
4. Growth medium: Dulbecco's modified Eagle medium, supplemented with 10% fetal calf serum, L-glutamine (2 μM), penicillin (50 IU/ml), and streptomycin (50 mg/ml)
5. 3-Week-old immunodeficient mice
6. Intraperitoneal injections of 100 mg/kg ketamine and 10 mg/kg xylazine hydrochloride (Sigma) in phosphate-buffered saline
7. Mouse or small rodent stereotactic head frame
8. 21-Gauge needle
9. 10-μl Hamilton syringe (Hamilton)

2.3. BLI System Imaging

1. Xenogen IVIS 100 Imaging System (Xenogen Corporation)
2. 50 mg/kg Firefly D-luciferin (Xenogen)
3. 100 mg/kg Ketamine and 10 mg/kg xylazine hydrochloride (Sigma) in phosphate-buffered saline
4. Living Image software (Xenogen)

2.4. Analysis

2.4.1. Magnetic Resonance Imaging

1. 100-μl Gadolinium (per animal)
2. Phenobarbital
3. 3-T magnet (Siemens)

2.4.2. Immunohistochemical Staining for Vimentin, EMA, GFAP, and Ki 67 (MIB-1)

1. Paraffin-fixed tissue slides cut at 4 μm
2. Xylene
3. Graded alcohol (100, 95, and 70%)
4. Ventana ES (Ventana Medical Systems) at 40°C

5. Citrate buffer (pH 6.0)
6. Primary antibodies: vimentin (1:300, mouse monoclonal Ab, Cline Vim 3B4, Dako Cytomation); EMA (1:200, mouse monoclonal Ab, Clone E29, Dako Cytomation); GFAP (1:400, mouse monoclonal Ab, Clone 6 F2, Dako Cytomation); Ki 67 (1:160, mouse monoclonal Ab, Dako)
7. Secondary antibodies: Mouse Fab (1:200, mouse immunoglobulin G, Dako Cytomation)
8. IView DAB Detection Kit (Ventana)
9. Positive controls: staining normal skin, pancreatic, brain, and colon cancer cells for vimentin, EMA, GFAP, and Ki 67, respectively
10. Negative controls as above (without primary antibody)
11. Image-Pro Plus 5.0 graphic analysis software

2.4.3. Transmission Electron Microscopy for Desmosomes

1. Mouse tumor samples in 1-mm^3 blocks
2. 2.5% Paraformaldehyde
3. 1% Glutaraldehyde
4. 0.1 M Sodium cacodylate buffer, with sucrose and calcium chloride at pH 7.4
5. 2% Osmium tetroxide in 0.1 M sodium cacodylate buffer
6. Distilled water
7. Aqueous uranyl acetate
8. Alcohol (50–100%)
9. Spurr plastic
10. Leica UCT Ultramicrotome with Diatome diamond knife
11. Uranyl acetate and Reynolds lead citrate
12. Hitachi H7100 transmission electron microscope
13. Kodak 4489 film

2.4.4. Statistical Analysis

1. Two-tailed Spearman correlation test (GraphPad Prism 4.0)

3. Methods

3.1. Meningioma Cell Culture and Luciferase Gene Insertion

1. Grow the CH-157-MN and IOMM-Lee cell lines in Dulbecco's modified Eagle medium to confluence in T-175 flasks in an incubator at 37°C and 7.5% CO_2.
2. Plate the cells at 10^5 cells per 10-cm dish and infect with the modified pMMP retrovirus (pMMP-LucNeo) overnight in Opti-MEM with 8 ng/ml hexadimethrine bromide (see Note 1).

3. Select the transfected cells with 300 mg/ml G418.
4. To confirm stable luciferase transfection, lyse 10^5 cells in 500-μl Glo Lysis buffer for 10 min. Add 100 μl of Brite Glo Luciferin, and measure the bioluminescence with a luminometer to assess in vitro luciferase expression (see Note 2).
5. Harvest the CH-157-MN, CH-157-MN-LucNeo, IOMM-Lee, and IOMM-Lee-LucNeo cell lines with trypsin, and plate into 6-well plates at a density of 10^5 cells per well (see Note 3).
6. Cells counts are done in triplicate for 3 days by harvesting the cells with trypsin and using a bright-line hemocytometer.

3.2. Transfected Tumor Cell Implantation

1. Harvest the transfected tumor cells for transplantation when they achieve logarithmic growth (see Note 3).
2. Grow cells briefly to ~80% confluence in T-175 flasks.
3. All steps are carried out on ice.
4. Determine cell counts using a bright-light hemocytometer; centrifuge the cells at 1,000 rpm for 5 min at 4°C, and resuspend the pellet in medium to obtain final cell concentrations of 10^4–10^6 cells per mouse in 3 μl of growth medium (see Note 4).
5. Anesthetize 3-week-old immunodeficient mice with intraperitoneal injections of 100 mg/kg ketamine and 10 mg/kg xylazine hydrochloride in phosphate-buffered saline.
6. Place mouse in stereotactic head frame.
7. For
   (a) Skull base locations
      - Cannulate a burr hole in the frontal bone 3 mm anterior and 2 mm lateral to the bregma with a 21-gauge needle.
      - Resuspend the desired number of cells in 3 μl of medium and draw into a 10-μl Hamilton syringe (mounted on the stereotactic mouse frame).
      - Pass the syringe needle through the burr hole and drop it slowly to the floor of the middle fossa.
      - Inject the cell suspension over 45 s, let stay 30 s, and then withdraw the needle over 30 s (see Note 5).

   (b) Convexity locations
      - Follow the same anatomical landmarks, but infuse the cell suspension 2 mm below the frontal bone.
8. After tumor cell implantation, give the animals free access to water, and examine twice weekly (see Note 6).
9. Kill animals that appear ill or with neurological deficits in accordance with the animal protocol (see Note 7).

10. Harvest tumors from all animals at time of death for paraffin blocks and electron microscopy.

3.3. BLI System Imaging

1. Assess luciferase activity biweekly, 1 week after tumor transplantation.
2. Inject mice intraperitoneally with 50 mg/kg firefly D-luciferin, 100 mg/kg ketamine, and 10 mg/kg xylazine hydrochloride in phosphate-buffered saline.
3. Photograph mice 10 min after this injection using the Xenogen IVIS 100 Imaging System with 30 s of exposure time (see Note 8 and Fig. 1).

3.4. Analysis

3.4.1. Magnetic Resonance Imaging

1. Follow institution's animal handling protocol.
2. One day after final intracranial BLI, sacrifice mice with an overdose of intraperitoneal phenobarbital.
3. Five minutes before phenobarbital injection, administer 100 μl of gadolinium via tail vein injection for contrast-enhanced MR images.
4. Center each mouse in a radiofrequency wrist-coil and image with T1- and T2-weighted 1-mm imaging sequences in axial, coronal, and sagittal planes on a 3-T magnet.

3.4.2. Immunohistochemical Staining for Vimentin, EMA, GFAP, and Ki 67

1. Cut paraffin-fixed tissue slides at 4 μm.
2. Melt slides at 55–60°C for 30 min.
3. Deparaffinize in xylene for 5 min.
4. Rehydrate in graded alcohols, washing twice in both 100 and 95%, and once in 70% for 1 min each.
5. Using the Ventana ES at 40°C, apply heat-induced epitope retrieval in citrate buffer (pH 6.0) in microwave oven for 15 min at 50% power.
6. Allow to cool for 15 additional minutes.
7. Apply the primary antibodies for 32 min as follows: vimentin (1:300, mouse monoclonal Ab, Cline Vim 3B4, Dako Cytomation); EMA (1:200, mouse monoclonal Ab, Clone E29, Dako Cytomation); GFAP (1:400, mouse monoclonal Ab, Clone 6 F2, Dako Cytomation); Ki 67 (1:160, mouse monoclonal Ab, Dako).
8. Follow primary antibody application with the appropriate secondary antibody for 8 min.
9. Using the IView DAB Detection Kit, determine staining.
10. Counterstain with hematoxylin for 4 min.

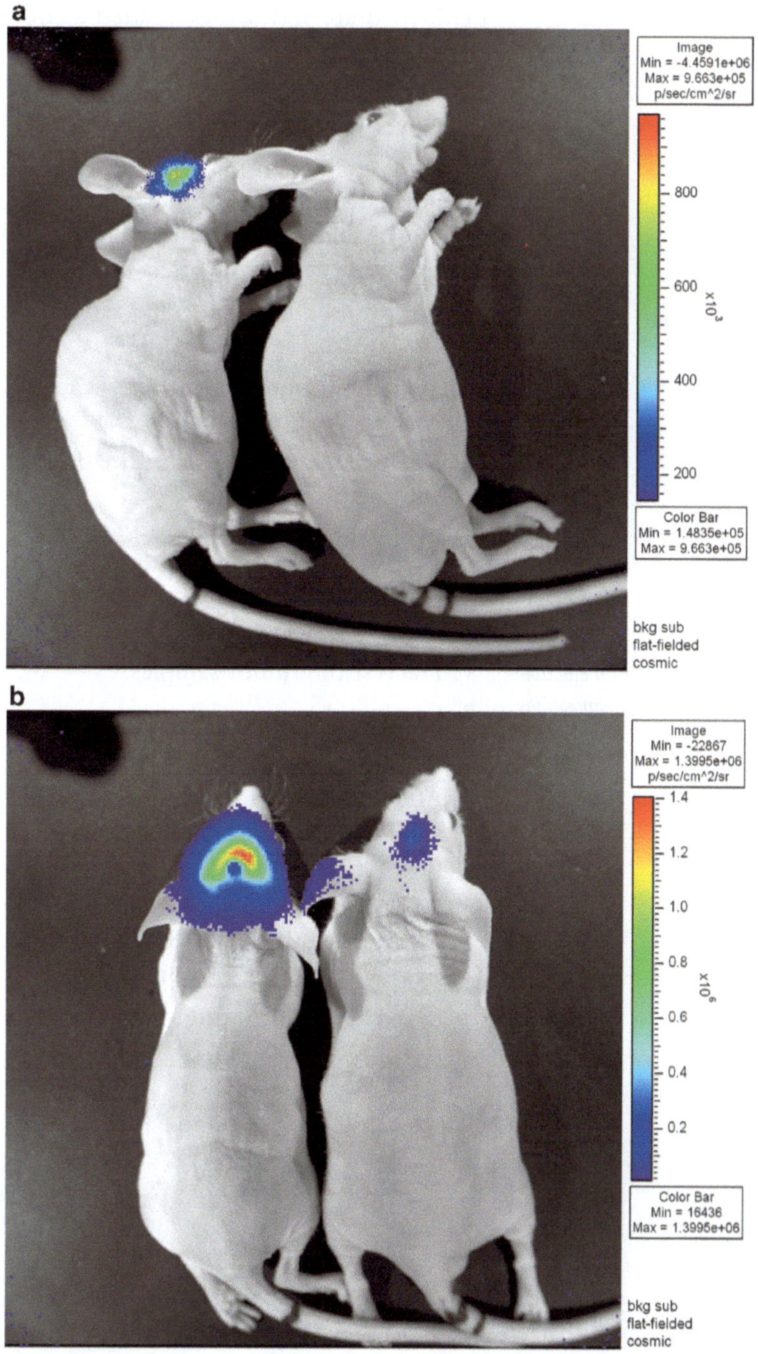

Fig. 1. (**a**) Lateral view of bioluminescence measurement on intracranial meningioma. Mouse on *left* with high activity and mouse on *right* with low activity. (**b**) Axial view of same mice, once again *left* mouse with high activity and *right* with low activity.

11. Dehydrate slides with graded alcohol, washing once in 70%, and twice each in 95 and 100% for 30 s each, dehydrate in alcohol and xylene.
12. Cover with a coverslip.
13. After staining using the above procedure, stain normal skin, normal pancreatic, brain, and colon cancer cells for EMA, GFAP, and Ki 67, respectively.
14. For the negative controls, follow the same procedure used for the positive control tissue samples, but without the primary antibody.
15. Take three random pictures of each slide at 400× magnification with an Olympus Microfire camera.
16. Analyze the images using Image-Pro Plus 5.0 graphic analysis software.
17. To calculate the MIB-1 (Ki 67) index, divide the number of MIB-1 (Ki 67)-stained cells by the total number of stained cells (Fig. 2).

3.4.3. Transmission Electron Microscopy for Desmosomes

1. Harvest the tumor samples.
2. Cut into 1-mm^3 blocks.
3. Fix in 2.5% paraformaldehyde with 1% glutaraldehyde.
4. After at least 24 h, rinse samples twice for 10 min each in 0.1 M sodium cacodylate buffer with sucrose and calcium chloride (pH 7.4).
5. Post-fix the samples in 2% osmium tetroxide in 0.1 M sodium cacodylate buffer for 45 min at room temperature on a rotator.
6. Dehydrate samples with increasing alcohol concentrations (50–100%) at room temperature on a rotator.
7. Infiltrate cells and embed in Spurr plastic.

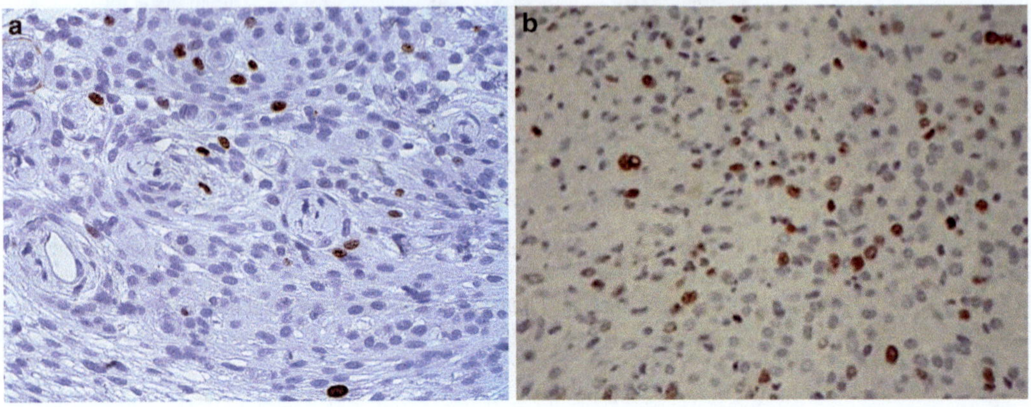

Fig. 2. MIB-1-stained meningiomas taken from intracranial mouse model. (**a**) Low MIB-1 index. (**b**) Higher MIB-1 activity.

8. Cut thin sections on a Leica UCT Ultramicrotome using a Diatome diamond knife.
9. Stain with uranyl acetate and Reynolds lead citrate.
10. Examine using a Hitachi H7100 transmission electron microscope at 75 kV.
11. Take photographs using Kodak 4489 film.

3.4.4. Statistical Analysis

1. Calculate CH-157-MN-LucNeo skull base tumor volumes from MR images (see Note 9).
2. Compare with BLI-determined values (see Note 10).
3. Analyze using 95% confidence intervals generated by the two-tailed Spearman correlation test.
4. Set statistical significance at a probability value <0.05.
5. Calculate the curve fit using a nonlinear regression analysis with exponential growth equation.
6. Present mean values ± standard deviations.

4. Notes

1. The modified pMMP retrovirus, pMMP-LucNeo, is a reporter plasmid under the control of a modified thymidine kinase promoter, which constitutively expresses luciferase. The neomycin phosphotransferase gene encodes for an enzyme that inhibits multiple aminoglycoside compounds, including G418. G418 is toxic to mammalian cells, so cells transfected with the pMMP-LucNeo construct can be selected to make the CH-157-MN-LucNeo and IOMM-Lee-LucNeo cell lines (11–13).
2. It is important to verify a stable transfection. In addition, it is possible that the transfection may alter the cell line of interest so confirming analysis of the tumor may be required.
3. Native CH-157-MN and IOMM-Lee and transduced CH-157-MN-LucNeo and IOMM-Lee-LucNeo cells showed similar doubling times of approximately 30 h (11).
4. We found 10^4 cells per animal to be ideal for skull base tumors because of longer survival times. For convexity tumors, we suggest 5×10^5 cells per animal (11).
5. The needle and Hamilton syringe are left in place for a time and removed slowly to help prevent cell leakage and brain damage along the needle track.
6. Overall, intraoperative mortality was approximately 10% and tumor induction was greater than 90%.

7. Animals were sacrificed when they appeared sickly, usually 16–24 days after xenograft. In general, the lower the number of transplanted tumor cells, the longer the survival and observation time. In addition, for both cell lines, increasing the number of tumor cells xenografted per animal resulted in parallel logarithmic graphs, which indicates constant exponential growth rates for both immortal cell lines transplanted intracranially (11).
8. Bioluminescence values for both transduced intracranial tumors increased logarithmically over time.
9. The MR image revealed ellipsoid tumors; therefore, the volumes were determined using the formula for ellipsoid volume [volume = $(4/3)\pi abc$], where a, b, and c are the radii of three orthogonal planes of the tumor (14).
10. Szentirmai et al. (15) showed there is a reduction of luminance in vivo as photons travel through tissue versus in vitro samples; therefore, comparisons between control and treatment groups should only be done from the same xenograft location.

Acknowledgment

We thank Kristin Kraus, M.Sc., for her excellent editorial assistance.

References

1. Louis DN, Scheithauer BW, Budka H, von Deimling A, Kepes JJ (2000) Meningiomas. In: Kleihues P, Cavenee WK (eds) World Health Organization Classification of Tumors: pathology and genetics of tumors of the nervous system. IARC, Lyon, France, pp 176–184
2. Al-Rodhan RF, Laws ER (1991) The history of intracranial meningiomas. In: Al-Mefty O (ed) Meningiomas. Raven, New York, pp 1–7
3. Gonzales MF (1995) Classification and pathogenesis of brain tumors. In: Laws ER, Kaye AH (eds) Brain tumors: an encyclopedic approach. Churchill Livingstone, Edinburgh, pp 31–45
4. Burger PC, Scheithauer BW, Vogel FS (2002) Surgical pathology of the nervous system and its coverings, 4th edn. Churchill Livingstone, New York
5. Simpson D (1957) The recurrence of intracranial meningiomas after surgical treatment. J Neurol Neurosurg Psychiatry 20:22–39
6. Greene HS, Arnold H (1945) The homologous and heterologous transplantation of brain and brain tumors. J Neurosurg 2:315–329
7. Medhkour A, Van Roey M, Sobel RA, Fingert HJ, Lee J, Martuza RL (1989) Implantation of human meningiomas into the subrenal capsule of the nude mouse. A model for studies of tumor growth. J Neurosurg 71:545–550
8. Rana MW, Pinkerton H, Thornton H, Nagy D (1977) Heterotransplantation of human glioblastoma multiforme and meningioma to nude mice. Proc Soc Exp Biol Med 155:85–88
9. Ueyama Y, Morita K, Ochiai C, Ohsawa N, Hata J, Tamaoki N (1978) Xenotransplantation of a human meningioma and its lung metastasis in nude mice. Br J Cancer 37:644–647
10. van Furth WR, Laughlin S, Taylor MD, Salhia B, Mainprize T, Henkelman M, Cusimano MD, Ackerley C, Rutka JT (2003) Imaging of murine brain tumors using a 1.5 Tesla clinical MRI system. Can J Neurol Sci 30:326–332
11. Ragel BT, Elam IL, Gillespie DL, Flynn JR, Kelly DA, Mabey D, Feng H, Couldwell WT,

Jensen RL (2008) A novel model of intracranial meningioma in mice using luciferase-expressing meningioma cells. Laboratory investigation. J Neurosurg 108:304–310

12. Ragel BT, Couldwell WT, Gillespie DL, Wendland MM, Whang K, Jensen RL (2008) A comparison of the cell lines used in meningioma research. Surg Neurol 70:295–307, discussion 307
13. Rubin JB, Kung AL, Klein RS, Chan JA, Sun Y, Schmidt K, Kieran MW, Luster AD, Segal RA (2003) A small-molecule antagonist of CXCR4 inhibits intracranial growth of primary brain tumors. Proc Natl Acad Sci U S A 100: 13513–13518
14. Jung HW, Yoo H, Paek SH, Choi KS (2000) Long-term outcome and growth rate of subtotally resected petroclival meningiomas: experience with 38 cases. Neurosurgery 46:567–574, discussion, pp 574–575
15. Szentirmai O, Baker CH, Lin N, Szucs S, Takahashi M, Kiryu S, Kung AL, Mulligan RC, Carter BS (2006) Noninvasive bioluminescence imaging of luciferase expressing intracranial U87 xenografts: correlation with magnetic resonance imaging determined tumor volume and longitudinal use in assessing tumor growth and antiangiogenic treatment effect. Neurosurgery 58:365–372

Neuromethods (2013) 77: 175–179
DOI 10.1007/7657_2011_28

Published online: 13 March 2012

Pituitary Models

Walavan Sivakumar and William T. Couldwell

Abstract

Pituitary tumor animal models provide researchers a microenvironment that simulates the clinical situation; however, in comparison with astrocytoma and meningioma tumor research where intracranial xenograft transplantations are increasingly being used to test various therapeutic modalities, in vivo therapeutic research on pituitary animal models focuses on direct drug therapy to the tumor because of the lack of established intracranial pituitary tumor models. The rat subcutaneous prolactin-secreting pituitary model allows investigators to noninvasively measure tumor size and the effect of direct tumor-guided therapy in a serial manner and is considered biologically relevant because it has proven to be histologically, immunocytochemically, and ultrastructurally consistent with human pituitary tumors.

Key words: Pituitary tumor, Pituitary adenoma, Adenohypophysis, Neurohypophysis

1. Introduction

Pituitary tumors are a broad group that accounts for approximately 15% of all symptomatic adult primary intracranial tumors. They are remarkably prevalent, present in approximately 20% of the population (1). Most of these tumors are small and incidental, but roughly one patient in 600–700 requires treatment (2).

Pituitary tumors can be classified by multiple different schemes, including by endocrine function, by histological staining methods, and by electron microscopic appearance (3). The mean age of affected patients is between 30 and 50 years. There is a slight female predominance with prolactin- and adrenocorticotropic hormone-secreting tumors and a male predominance in growth hormone-secreting tumors. Nonfunctioning pituitary adenomas account for 25% of these tumors and less than 1% are malignant.

Animal tumor model systems are vital to the process of cancer therapy development (4). Pituitary tumor animal models provide researchers a microenvironment that simulates the clinical situation. Since 1953 when Furth et al. (5) described transplantable pituitary tumors in the rat, numerous constructs and induction methods enlisting mouse, rat, and canine models have been described (6–9). In comparison with astrocytoma and meningioma tumor research where intracranial xenograft transplantations are increasingly being used to test various therapeutic modalities, in vivo therapeutic research on pituitary animal models focuses on direct drug therapy to the tumor because of the lack of established intracranial pituitary tumor models.

1.1. History of Pituitary Tumor Models

Early pituitary tumors in animal models were induced by exposing rats to elevated levels of estrogen (10). Other methods of inducing tumors included antithyroid procedures, carcinogen exposure, and increasing levels of ionized radiation (11). Further research with the estrogen-induced tumors showed that these tumors were completely hormone dependent and not autonomous neoplasms, making them only viable in estrogen-treated hosts (8, 12, 13). In addition, investigators found that these inducible tumors that were transplanted into other animals differed from most human pituitary tumors by their malignancy and their undifferentiated aspects, reducing their value as a human pituitary neoplasm model (14, 15).

In 1982, Trouillas et al. described spontaneous tumors that arose in the pituitary of approximately 70% of all female Wister/Furth rat strains (16, 17). Histological, immunocytochemical, and ultrastructural similarities between these spontaneous tumors and those found in the human pituitary made this entity a very promising animal tumor model for study (16). Initial difficulties using this model included the length of time required to mature enough animals with acceptable tumors, the small size of the tumors, and the unusual location, making it a difficult and costly model for pituitary study. By grafting these spontaneous tumors directly under the kidney capsule and subcutaneous region of consanguineous animals within the same strain, however, investigators developed an animal model of the pituitary tumor that was easy to transplant into experimental groups, effectively reproducible, and analogous to the human pituitary tumor. This initial cell strain was named the spontaneous mammotrophic tumor in Wistar (SMtTW) rats line (8). Since that time, other pituitary tumor cell lines that have been shown to have similar morphological and functional properties as the SMtTW line include the ArT and GH strains.

The rat subcutaneous prolactin-secreting pituitary model allows investigators to noninvasively measure tumor size and the effect of direct tumor-guided therapy in a serial manner. The results from these studies are considered biologically relevant as these animal models have been proved to be histologically, immunocytochemically, and ultrastructurally consistent with human pituitary tumors (18, 19). The majority of our work has been done with the GH pituitary tumor cell line, namely, the GH_4C_1 strain.

2. Materials

The products and suppliers used are listed below. Comparable products should also be effective.

2.1. Pituitary Tumor Cell Culture

1. Rat pituitary cell line, such as GH_4C_1, grown at 37°C in 5% CO_2
2. Growth medium: Ham's F-10 Nutrient, supplemented with 15% horse and 2.5% bovine serum in the absence of antibiotics
3. 10-cm Cell culture dishes
4. T-175 flasks

2.2. Tumor Cell Implantation and Tumor Size Measurement

1. GH_4C_1 cell line at ~80% confluence in 10-cm dishes (maximum of 9 weekly passages)
2. Ice
3. 0.025% Trypsin
4. Hemocytometer
5. Growth medium: Ham's F-10 Nutrient, supplemented with 15% horse and 2.5% bovine serum in the absence of antibiotics
6. 8-Week-old female Wistar-Furth rats
7. Intraperitoneal injections of 100 mg/kg ketamine and 10 mg/kg xylazine hydrochloride (Sigma) in phosphate-buffered saline
8. 21-Gauge needle
9. Syringe
10. Calipers

3. Methods

3.1. Pituitary Tumor Cell Culture

1. Grow the GH_4C_1 tumor cell lines in Ham's F-10 Nutrient supplemented with 15% horse and 2.5% bovine serum in the absence of antibiotics to 80% confluence in cell culture dishes at 37°C and 5% CO_2 (see Note 1).

3.2. Tumor Cell Implantation and Tumor Size Measurement

1. Grow cells to approximately 80% confluence in T-175 flasks.
2. All steps are carried out on ice.
3. Treat cells with 0.025% trypsin to detach cells from flasks.

4. Determine cell counts using a bright-light hemocytometer; centrifuge the cells at 1,000 rpm for 5 min at 4°C and resuspend the pellet in medium to obtain final cell concentrations of 10^7 cells per rat in 100 μl of growth medium (see Note 2).
5. Anesthetize 8-week-old Wistar-Furth rats with intraperitoneal injections of 100 mg/kg ketamine and 10 mg/kg xylazine hydrochloride in phosphate-buffered saline (see Note 3).
6. Inoculate approximately 1×10^7 GH4C1 cells subcutaneously between the scapulae of each rat.
7. After implantation, allow tumors to grow in vivo for 28-day period.
8. After 28-day growth period, determine tumor sizes using surface diameter measurements with calipers in three orthogonal directions.
9. Subsequently, animals can be equally divided so that each group has tumors of comparable sizes (see Note 4).
10. Give animals open access to food, water, and evaluate tumor size three times weekly.
11. Sacrifice animals that appear ill or with neurological deficits in accordance with the animal protocol at your institution (see Note 5).
12. Harvest tumors from all animals at the time of death for paraffin blocks and electron microscopy.

4. Notes

1. Stock cultures are grown on cell culture plates before being transplanted to T-175 flasks for further passaging. GH_4C_1 tumor cell line quality begins to diminish in regards to replication ability after ten passages. Therefore, we recommend only using cell lines that have undergone a maximum of nine passages.
2. For subcutaneous tumor inoculations in the rat model, we found 10^7 cells per animal to be ideal because of easier visibility of tumor and longer time before tumor-related side effects.
3. Wistar-Furth rats should be allowed to acclimatize for 1 week prior to tumor cell injection. Follow your Institutional Animal Care and Use Committee protocols regarding all animal storage and handling.
4. To minimizing fighting between rats and risk of bodily damage to animals or tumor specimens, we use only female rats in our protocol and follow institutional guidelines regarding the maximum number of animals per cage.

5. While it did not occur in our series, all animals, per institutional care protocols, that appear ill or show neurologic deficits, should be euthanized.

Acknowledgment

We would like to thank Kristin L. Kraus, M.Sc., for her incredible editorial assistance in preparing this book chapter.

References

1. Ezzat S, Asa SL, Couldwell WT, Barr CE, Dodge WE, Vance ML, McCutcheon IE (2004) The prevalence of pituitary adenomas: a systematic review. Cancer 101:613–619
2. Couldwell WT, Cannon-Albright L (2010) A heritable predisposition to pituitary tumors. Pituitary 13:130–137
3. DeLellis R, Lloyd R, Heitz P, Eng C (eds) (2004) Pathology and genetics of tumours of endocrine organs (IARC WHO classification of tumours). IARC, Lyon
4. Khleif SN (ed) (2003) Animal models in developmental therapeutics, vol. 2, 6th edn. BC Decker, Hamilton
5. Furth J, Gadsen EL, Upton AC (1953) ACTH secreting transplantable pituitary tumors. Proc Soc Exp Biol Med 84:253–254
6. Lloyd RV (1991) Ultrastructure of spontaneous and transplanted pituitary tumors in laboratory animals. J Electron Microsc Tech 19:64–79
7. Ito A (1976) Animal model of human disease: pituitary tumors. Am J Pathol 83:423–426
8. Trouillas J, Girod C, Claustrat B, Joly-Pharaboz MO, Chevallier P (1990) Spontaneous prolactin transplantable tumor in the Wistar/Furth rat (SMtTW): a new animal model of human prolactinoma. Cancer Res 50:4081–4086
9. El Etreby MF, Muller-Peddinghaus R, Bhargava AS, Trautwein G (1980) Functional morphology of spontaneous hyperplastic and neoplastic lesions in the canine pituitary gland. Vet Pathol 17:109–122
10. Lloyd RV (1983) Estrogen-induced hyperplasia and neoplasia in the rat anterior pituitary gland. An immunohistochemical study. Am J Pathol 113:198–206
11. Yokoro K, Ito A (1981) Experimental pituitary. In: Nagasawa N, Abe K (eds) Hormone related tumors, pp. 3–21. Japan Scientific Societies Press
12. Treip C (1961) Pathology of human and experimental pituitary tumours. Proc R Soc Med 54:623–627
13. Treip CS (1983) The regression of oestradiol-induced pituitary tumours in the rat. J Pathol 141:29–40
14. Ito A, Furth J, Moy P (1972) Growth hormone-secreting variants of a mammotropic tumor. Cancer Res 32:48–56
15. Lamberts SW, MacLeod RM (1979) The inability of bromocriptine to inhibit prolactin secretion by transplantable rat pituitary tumors: observations on the mechanism and dynamics of the autofeedback regulation of prolactin secretion. Endocrinology 104:65–70
16. Trouillas J, Girod C, Claustrat B, Cure M, Dubois MP (1982) Spontaneous pituitary tumors in the Wistar/Furth/Ico rat strain. An animal model of human prolactin adenoma. Am J Pathol 109:57–70
17. Prysor-Jones RA, Jenkins JS (1981) Effect of bromocriptine on DNA synthesis, growth and hormone secretion of spontaneous pituitary tumours in the rat. J Endocrinol 88: 463–469
18. Cole CD, Liu JK, Sheng X, Chin SS, Schmidt MH, Weiss MH, Couldwell WT (2008) Hypericin-mediated photodynamic therapy of pituitary tumors: preclinical study in a GH4C1 rat tumor model. J Neurooncol 87:255–261
19. Hamilton HB, Hinton DR, Law RE, Gopalakrishna R, Su YZ, Chen ZH, Weiss MH, Couldwell WT (1996) Inhibition of cellular growth and induction of apoptosis in pituitary adenoma cell lines by the protein kinase C inhibitor hypericin: potential therapeutic application. J Neurosurg 85:329–334

Neuromethods (2013) 77: 181–200
DOI 10.1007/7657_2012_40

Published online: 5 April 2012

Growth Regulation of Nervous System Tumours: Models for Assessment of Angiogenesis in Brain Tumours

Kelly Burrell, Elena Bogdanovic, Shahrzad Jalali, Abhijit Guha, and Gelareh Zadeh

Abstract

The metabolic demand of rapidly proliferating tumour cells is reliant on an adequate blood supply that allows the continual delivery of oxygen, nutrients and growth factors. The growth and progression of tumours is significantly reduced in the absence of neovascularization and often increased abnormal neoangiogenesis correlates with the increased malignancy and poor prognosis in many tumours. By far, the most studied and understood mechanism of blood vessel formation is via angiogenesis, a process that initiates the sprouting and elongation of existing vessels into the tumour. However, more recent concepts suggest that in large tumours, the process of vasculogenesis, whereby bone marrow derived progenitor cells (BMDPCs) are recruited to the tumour and differentiate into ECs and other vascular cell types, is a more important mechanism of generating *de novo* vessels.

The mechanisms underlying both processes are poorly understood and the redundancy between signalling pathways involved leads to complications in elucidating the control mechanisms involved. Using various experimental techniques to investigate the processes of tumour neovascularization is an evolving field, one which in this chapter we try to summarize and provide an overview of both the traditional and more novel experimental techniques used to study angiogenesis.

Key words: Gliomas, Tumour metabolism, Neoangiogenesis, Vasculogenesis, Endothelial cells

1. Introduction

Glioblastoma multiforme (GBM) is the most common and aggressive of all primary intracranial tumours with a median survival of 14–16 months, despite the current treatment regimen which includes maximal surgical resection followed by both radiation therapy (RTx) and chemotherapy (CHx) (1). Categorized by the World Health Organization as a Grade IV astrocytoma, GBMs are characterized by rapid cellular proliferation, migration and necrosis resulting in large diffuse tumours with centralized necrotic regions, (2, 3). GBMs form by either of two routes; most appear de novo without indication of a prior lesion or, less commonly, progress from pre-existing lower grade gliomas (Grades II and III) (1) (Fig. 1).

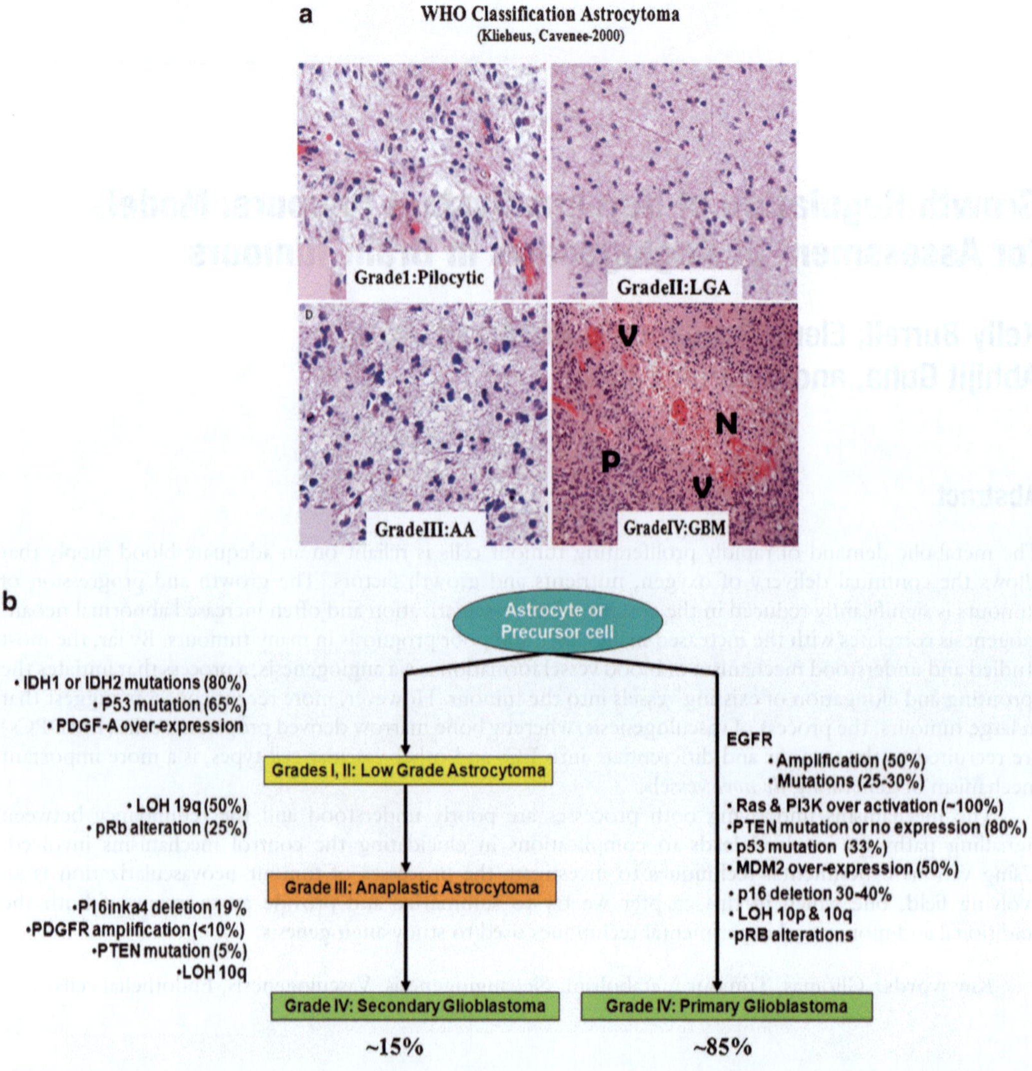

Fig. 1. WHO pathological grading and molecular characterization of adult GBMs. (**a**) Diffuse gliomas are pathologically graded by the WHO. Grade I astrocytomas, also known as pilocytic astrocytomas, are found in children and NF1 patients, and most often remain indolent. Grade II astrocytomas or low-grade astrcytomas (LGA) tend to occur in younger adults, but do progress to higher grades. Grade III astrocytomas or anaplastic astrocytomas (AA) demonstrate increased cellularity, nuclear atypia and neoangiogenesis. Grade IV astrocytomas or glioblastoma multiforme (GBM) include features of AAs plus areas of geographical necrosis (N) surrounded by pseudopalasading cells (P) and increased aberrant neoangiogenesis (V). (**b**) On a clinical and molecular, but not pathological basis, GBMs can be classified as "secondary" which occur in minority of cases, or "primary" which are most common and have a worse prognosis. Secondary GBMs progress from LGA to AA and harbour mutations in *IDH* and aberrations in *p53* and Rb cell cycle pathways, along with alterations in PDGFR signalling. Primary GBMs are characterized by amplifications and mutations of EGFR, and to a higher degree loss of PTEN. The TCGA GBM project looking at a large number of GBMs have further refined the molecular characterization to four groups, with prognostic and associated genetic alterations.

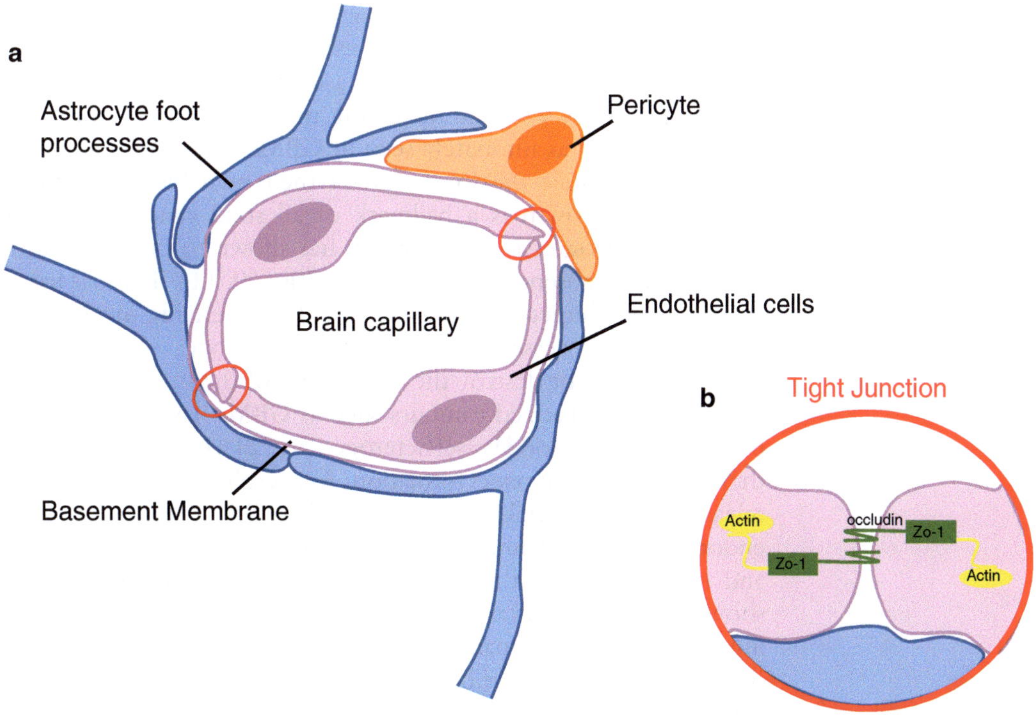

Fig. 2. Cytostructural basis of the blood–brain barrier (BBB). (**a**) The integrity of BBB is reliant on the close proximity of the capillary endothelial cells and the formation of tight junction, which are more abundant in the CNS than any other system in the body. The endothelial cells are surrounded by a thick basement membrane, which in turn is surrounded by numerous astrocytic foot-end processes. The presence of pericytes in the BBB is thought to aid the interactions between the astrocytes and the endothelial cells (50). (**b**) Tight junctions form between two cells when their membranes are in a suitably close proximity forming a nearly impenetrable barrier. The proteins underlying the junction assembly are not fully understood due to the limitations of the molecular size; however, it is known that zonula occludens (ZO-1) is the protein which attaches to the underlying cellular actin structure. From the ZO-1 protein, many occludin molecules branch and it is these strands that are thought to assemble directly in the intercellular space to firmly bind the junction.

The entire vascular system; arterial, venous, lymphatic; is lined by a single layer of endothelial cells (ECs) connected by adherens and tight junctions to allow free exchange of blood constituents with the interstitial fluid. The ECs lining the vessels of the brain and spinal cord; however, form a more specialized endothelial lining where the ECs are connected by tight junctions forming a seal-tight barrier shielding the brain and spinal cord from the blood and preventing this free exchange, thus forming the blood–brain barrier (BBB) (4, 5) (Fig. 2). ECs of the BBB are selective in the molecules that they themselves transport into the brain parenchyma, increasing the protection of the brain (6).

The metabolic demand of rapidly proliferating GBM tumour cells is reliant on an adequate blood supply that allows the continual delivery of oxygen, nutrients and growth factors. The neoangiogenic process is also involved in efficient removal of metabolic end products along with diffusion into the cerebrospinal spinal fluid (CSF), since the CNS unlike other organs does not have lymphatics. Although there are several CNS primary tumours which are highly vascularized, increased and abnormal neoangiogenesis correlates with the increasing malignancy and poor prognosis in gliomas as ultimately seen in GBMs (7, 8). The ability of tumours to induce the formation of their own blood supply to sustain their growth is a hallmark of all solid cancers (9). The growth and progression of tumours has shown to be significantly reduced in the absence of neoangiogenesis (10, 11). This has led to intense efforts aimed at understanding the underlying mechanisms driving tumour vessel formation, with the hopes of developing drugs to inhibit or destroy the vasculature of the tumour.

Highly vascularized tumours such as GBMs have increased EC proliferation and blood vessel heterogeneity (2) which could relate to their mechanisms of vessel formation. As a result of various cytokines released by glioma cells the structure of the BBB (Fig. 2) including endothelial cell tight junctions are disrupted resulting in haemorrhage and oedema (12). Although increasing EC proliferation has been shown to affect GBM vasculature, the underlying mechanisms initiating vessel formation remain poorly understood. Several mechanism(s) of neoangiogenesis have been demonstrated or proposed (Fig. 3). Initially, the tumour cells physically migrate towards blood vessels in response to signals within the tumour microenvironment and grow alongside existing vessels, a indirect mechanism of neoangiogenesis termed "vessel co-option". By far, the most studied and understood mechanism(s) of blood vessel formation is "angiogenesis," a process that initiates the sprouting and elongation of existing vessels into the tumour area (13). In response to oncogene activation and/or metabolic stress (nutrient deprivation and hypoxia) tumour cells secrete growth factors (GFs) such as VEGF and Angiopoietin-2 (Ang-2) (14–18). These GFs bind to their cognate receptors on nearby ECs on pre-existing blood vessels to first induce vessel destabilization followed by EC proliferation, migration and differentiation; thus forming a new blood vessel at the site of destabilization. Another mechanism of neoangiogenesis is "vasculogenesis," a process whereby bone marrow derived progenitor cells (BMDPCs) are recruited to the area and differentiate into ECs and other vascular cell types to produce de novo blood vessels (19–21). Until recently, the process of vasculogenesis was thought to occur primarily during embryonic development but recent studies have suggested that vasculogenesis may also occur in adulthood in response to pathological stimuli such as tumour growth (19). Vasculogenesis is more prevalent in larger tumours where an increasing level of hypoxia is seen. In these tumours, increasing HIF1α signalling initiates the release of stromal derived factor 1 (SDF-1) from tumour cells, a known chemo-attractive factor which enhances the migration of bone marrow derived cells (22). Recently, there have been suggestions that

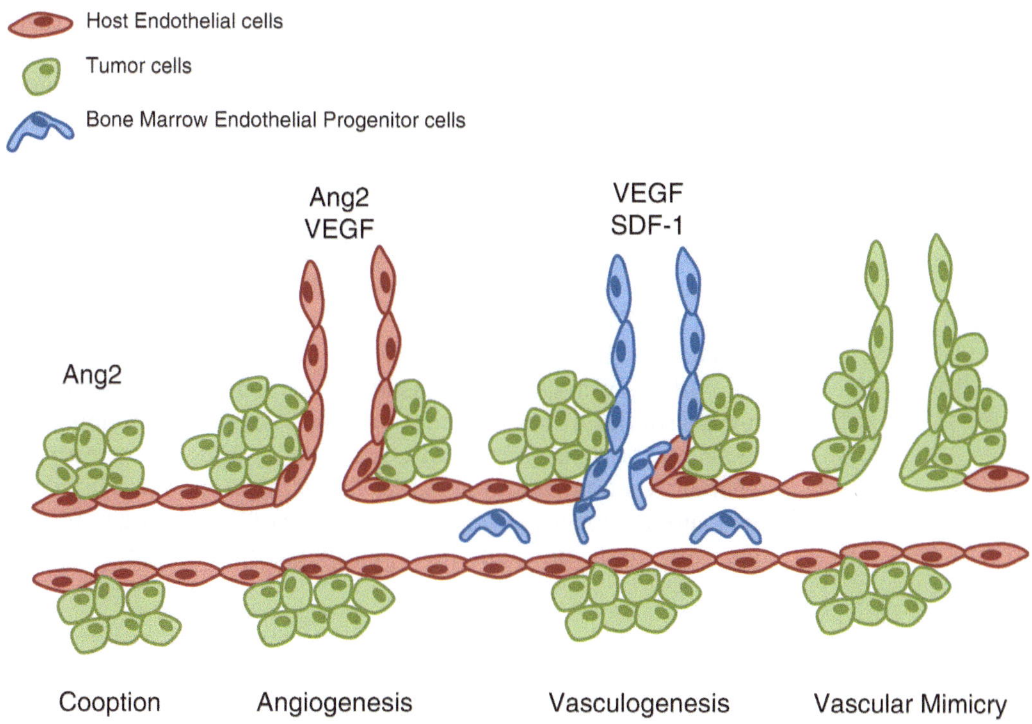

Fig. 3. Mechanisms of neoangiogenesis: co-option, angiogenesis, vasculogenesis and vascular mimicry. Tumour progression is heavily reliant on neoangiogenesis for many metabolic aspect of growth. As such there are many mechanisms underlying vessel formation. *Co-option* allows tumours to invade areas where extensive networks of vessels already exist, limiting the effort the tumour needs to go through. In larger tumours, it becomes necessary to extend existing networks further and three mechanisms exist: *Angiogenesis* is the process of direct branching of the existing vessels, initiated by a sudden increase and angiopoietin and VEGF signalling from the tumour. *Vasculogenesis* is a more involved process in which endothelial precursor cells are recruited to the tumour area and in turn differentiate into the endothelial cells required to form new vessels. This process is usually observed in larger more hypoxic tumours where increasing VEGF coincides with an increase in HIF1α signalling, which in turn increases stromal derived factor (SDF-1) release, a stimulant for bone marrow derived cell migration. *Vascular mimicry* is the most recent hypothesis for derivation of tumour vessels. Although conflicting reports exists, it is thought that particularly aggressive tumour cells are capable of trans-differentiating into endothelial cells to form their own vessels.

tumour cells themselves can form vascular channels to transport blood independently of EC lined vessels, a phenomenon referred to as "vascular mimicry" (13, 23, 24).

2. Methods for Assessing Brain Tumour Angiogenesis

Solid tumour formation and prognosis is intimately related to its vascularization, hence understanding the mechanism(s) by which the vessels are formed and how they develop will potentially lead to define new drug targets aimed at tumour neoangiogenesis. In addition to growth and progression, tumour neoangiogenesis also is a determinant of tumour sensitivity/resistance to current non-specific therapies such as radiation (RTx) and chemotherapy (CTx), perhaps related to issues around drug delivery as well as oxygen and nutrients to the tumour cells themselves (25). In addition, the tumour initiation cells, including brain tumour initiation cells (BTICs) in GBMs, have been speculated to have a special relationship with the vasculature, residing more in the "vascular niche" (26). For these reasons, angiogenesis assays are becoming a more functional and important prognostic factor in many human cancers, demonstrating a patient's potential to respond to treatment and the likelihood of tumour recurrence (27–30).

Due to the inaccessibility of tumours and their vasculature, many of the well-established assays of angiogenesis are set up for in vitro studies. It has therefore become essential to develop novel assays and systems for in vivo real-time studies to increase the reliability for human tumours. The aim of this section is to highlight the major angiogenesis assays used in the past decade and to further demonstrate the novel methods being developed for future use.

3. In Vitro Assays

A significant portion of our understanding of angiogenesis and vascular biology to date has come from in vitro assays which focus particularly on the proliferation, differentiation and migration of ECs in culture. Although these in vitro methods allow the user a greater degree of control over the variables within the experimental situation, the pseudo-environment created by the assay limits the extrapolation of the results to in vivo situations.

3.1. Tubule Formation Assay

Kubota et al. (31) developed the tubule formation assay more than 20 years ago. It is based on the ability of ECs to rapidly attach and align themselves into capillary-like tubules when exposed to the basement membrane and other extra-cellular matrix (ECM) components. The tubules formed have basic vessel characteristics including a lumen, adherens and tight junctions and luminal–abluminal cell polarity (Fig. 4).

Tubule Formation Protocol

1. Thaw Matrigel overnight at 4°C (see Note 1) (32).
2. Dilute Matrigel 1:100 in ice-cold phosphate-buffered saline (PBS).
3. Add 50 μl of diluted Matrigel solution to each well of a pre-chilled sterile 96-well plate.
4. Incubate 1 h at 37°C to allow the Matrigel solution to form a gel.

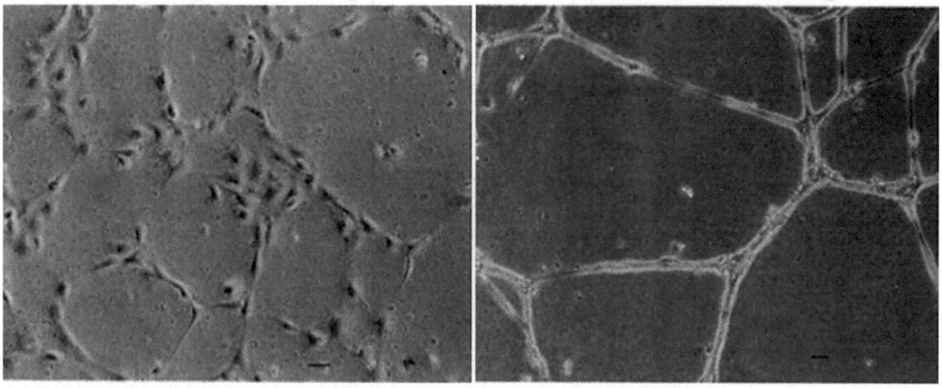

Fig. 4. Endothelial cell tubule assay. Following initial seeding of 4,800 cells/cm^2 on to matrigel substrates, HUVECs (human umbilical vascular endothelial cells) develop visible tubules within 4 h, which go onto become more defined. *Adapted from* (31).

5. Harvest endothelial cells and resuspend in desired culture medium, supplemented with desired angiogenesis factors at 1–2 × 10^5 cells/ml.
6. Add 150 μl of cell suspension (1.5–3 × 10^4 cells) per well directly onto the solidified Matrigel.
7. Incubate the assay plate at 37°C for 4–18 h.
8. Examine the endothelial tubes using a light microscope at high magnification. It is possible to visually estimate the extent of tube formation by inspecting the *overall tube length* and *branch points* (see Note 2). Staining with Haematoxylin and Eosin post-fixation can be done if needed.

Adapted from (33).

3.2. Adapting the Tubule Formation Assay to Cancer Biology

The basic assay described above can be adapted in order to examine the angiogenic potential of both the tumour microenvironment and the tumour cells themselves. First, in order to assess the angiogenic potential of the tumour microenvironment, components of the microenvironment can be included in the tubule formation assay through either the direct addition of tumour stromal fluid to the Matrigel substrate or through the use of tumour cell conditioned media obtained from tumour cell cultures. By examining the effect the addition has on endothelial tubule formation when compared to the standard protocol, it will elucidate how permissive to angiogenesis the tumour cells are. Those tumours which have an increased propensity to vascularize will have more heavily branched tubules in the assay. For example, conditioned media obtained from cancer stem cells (CSCs) have been shown to significantly increase the length of tubules formed during an 18-h culture, compared to ECs receiving non-conditioned media (34, 35).

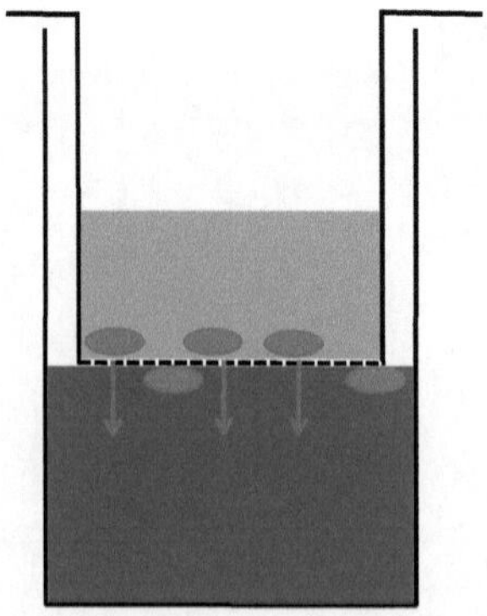

Fig. 5. Cell migration assay setup. To test the chemo-attractant properties of a substance, a cell migration assay can be set up with the cells grown on a porous membrane and the test substance underneath. The assay is quantified as the number of cells which have moved through the membrane after 48 h of culture. Different matrix can be applied to the top chamber, such as myelin, to evaluate migration through relevant organ-specific stroma such as those found in the brain.

Second, the tubule assay has been used to directly assess the potential of cells derived from the tumour to differentiate into tubules, so demonstrating the presence of endothelial precursors. GBM-derived BTICs are thought to contain an endothelial precursor subpopulation based on their expression of vascular markers such as CD144. When plated onto Matrigel, they align and differentiate into tubules of a similar phenotype to those seen in endothelial cell cultures (36). The presence of endothelial progenitors within tumours suggests that one of the vascular mechanisms arises from the tumour being able to generate its own vessels directly.

3.3. Cell Migration Assay

The migration assay is based on the motility of ECs and their ability to migrate through a porous membrane in a double layer culture system. The assay can not only look at different cells and their migratory characteristics but also look at cellular components that may stimulate or inhibit migration during a short period in culture (37). It is essential to know the cell characteristics because if left long enough all cells will migrate but not due to the stimulus. The Boyden Chamber technique is most widely used, but laboratories have tried adapting the procedure to include beads and other base substrates in the lower compartment. The technique in recent years has been improved with the commercial availability of kits containing prefabricated assay plates which house a specifically designed porous insert (Cell Biolabs) (33, 38) (Fig. 5).

Cell Migration Protocol

1. Dissolve Collagen IV gel matrix in culture media.
2. Collagen is plated over the porous membrane inserts in the culture vessels.

3. Plates are incubated at 37°C for 1 h.
4. Angio-attractants or repellents are dissolved in culture media to the correct concentration.
5. The angio-attractants or -repellents are plated in the lower culture chamber of the chosen culture vessel.
6. The collagen coated porous insert is then placed on top.
7. Cells of interest are harvested and resuspended at 1–2 × 10^5 cells/ml in media.
8. Add 150 μl of cell suspension (1.5–3 × 10^4 cells) per well directly on to the porous insert.
9. The assay plate is then incubated at 37°C for up to 24 h depending on the culture system in use.
10. Non-migrated cells that remain on the top surface of the membrane can be scraped and lysed for genetic or protein analysis.
11. Migrated cells are fixed and stained with Haematoxylin and independently counted in ten high-power fields.

Adapted from (33, 38, 39).

3.4. Cell Migration Expected Results

Upon completion of the assay, a comparison can be made between the cells remaining in the upper chamber or lower chamber through genetic analysis or by direct cell counting. Increasing the concentration of angiogenic stimuli in the lower chamber should increase the number of migratory endothelial cells that reach the lower chamber and so angiogenic factors can be equated by looking at cell number.

3.5. Adapting the Cell Migration Assay to Cancer Biology

Similar to the tubule formation assay, components from the tumour microenvironment can be introduced to the lower compartment to assess their ability to stimulate migration either in the form of extracts from the primary tumour or through media conditioned by tumour cells in culture.

It has been shown that cells derived from tumours such as GBM are able to increase the degree of migration of ECs in culture (34, 35). As before, the assay can be used to demonstrate the degree of migratory signals expressed by tumours to which endothelial cells respond, indicating the potential of tumour tissue to recruit and generate new vessels.

4. In Vivo Experimental Systems

Although studying vascular biology through *in vitro* cell culture methods has provided a good basic understanding of their structure, minimal functional and mechanistic information can be gained until they are studied in their natural environment. Due to limitations in obtaining primary human tumour tissue, it is essential to have experimental systems and animal models for use in studying these diseases. This also gives the opportunity to study the vasculature as a dynamic process and not a static event, elucidating further the mechanisms involved in the progression of tumour development and vessel formation. Currently, there are different types of animal models in use for the study of tumour biology and each has its own advantages and limitations.

4.1. Xenografts

Xenotransplantation is the transplantation of living cells, tissues or organs from one species to another. Such cells, tissues or organs are called xenografts. In the absence of large amounts of primary human tumour samples, well-established models which utilize minimal biopsy tissue or cell suspensions have become the gold standard in assessing tumour development in situ (35, 40–42). It should be noted that these xenograft models require immunocompromised models, which is an important component of tumour–stromal interaction important in regulating neoangiogenesis.

Xenografts can be implanted orthotopically (in the same anatomical location as the source of tissue) or heterotopically (at a different location), in order to look at the effects on the tumour in both environments. In most cases, heterotopic sites are located subcutaneously in the hind flank of the mouse where it is possible to propagate primary tumour tissue samples to expand the limited amount of tumour tissue available. Injections into either orthotopic (intracranial) or subcutaneous sites in studies of GBMs include established human GBM cell lines or freshly disassociated GBM explants, the latter reportedly retaining many of the molecular and growth characteristics of the human GBMs (43, 44). In addition, these GBM lines can be transfected with a luciferase reporter to facilitate repeated bioluminescence imaging (BLI) to monitor tumour growth and response to therapeutic intervention (Fig. 6).

Xenograft Subcutaneous Protocol

1. Immunocompromised mice 8–10 weeks old are anaesthetized with isofluorane at 4% for induction followed by 1.5–2% throughout the procedure with 0.5–1 μl O_2 per min. The mouse should be placed on a sterile towel and TearGel applied to the eyes to prevent corneal dehydration and abrasion.

For Tissue:

1. Once anaesthetized, the mouse is laid on its side.
2. Trim the hind limb of fur and wash with iodine and alcohol.
3. Make a small incision <0.5 cm in the right flank.
4. Using the scissors, create a subcutaneous pocket.

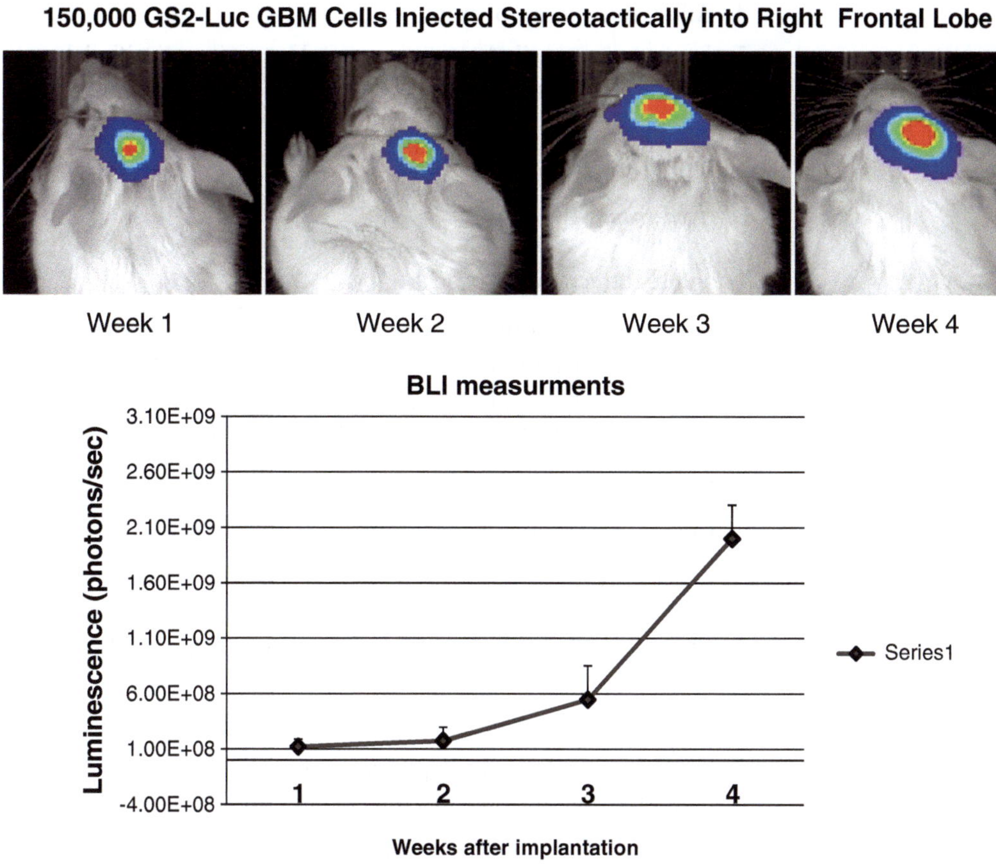

Fig. 6. Bioluminesence imaging (BLI) of intracranial GBM xenografts. Through the integration of a Luciferase constructs to the cells of interest, it is possible to visualize the tumour volume and area following injection of the substrate Luciferin to the mouse. Through a time course study, the level of luminescence is seen to increase with time following implantation, and through these kinetics studies it is possible to determine the equivalent tumour volume and allowing evaluation of various therapeutic interventions.

5. Insert a small 1 mm^3 section of tissue which appears to have a good blood supply of its own into the pocket.
6. Close the incision by suturing and return the mice to the cage to recover.
7. Leave mice until the tumour reaches a size specified under the institutional guidelines.
8. Sacrifice the mice and extract the tumour for propagation and analysis.

For Cells:

1. Once anaesthetized, the mouse is laid on its side.
2. Wash hind limb with iodine and alcohol.
3. Inject 1×10^7 cells in a volume of 300 μl of sterile PBS into the subcutaneous space on the right-hand flank.

4. Allow mice to recover in a cage.
5. Leave mice until the tumour reaches a size specified by the institutional guidelines.
6. Sacrifice the mice and extract the tumour for propagation or end point analysis.

Adapted from (40, 45).

Xenograft Intracranial Protocol

1. Immunocompromised mice 8–10 weeks old are anaesthetized intra-peritoneally with 0.5 mg/g Avertin and 5 mg/kg of the antipain drug Carprofen. TearGel should be applied to the eyes to prevent corneal dehydration and abrasion.

For Cells:

1. Once anaesthetized, position the mouse on a stereotactic frame, as previously described (21).
2. Trim the scalp and wash with alcohol.
3. Make a small incision <0.5 cm down the midline of the scalp to reveal the skull.
4. Position the needle tip over the frontal lobe coordinates.
5. With the needle moved upwards out of the way, a small 0.7-mm diameter dental drill is used to drill gently through the skull.
6. Load a 30-G Hamilton syringe with 2×10^5 cells in 10 μl PBS.
7. Once the needle is loaded into the frame, it is lowered to the skull and the frame is zeroed.
8. Injections are made 3-mm deep to the cortical surface and expelled slowly over the span of 1 min.
9. After injection, the needle should remain in place for 30 s before slowly withdrawing from the brain tissue.
10. Mice are left to recover in a cage.
11. Mice are sacrificed, according to the institutional guidelines, upon the observation of neurological symptoms mice are sacrificed with standard 4% PFA perfusion.
12. Brains are extracted and frozen in OCT compound prior to sectioning at the tumour site for end point analysis.

Adapted from (40–42, 45).

4.2. Limitations of Immortalized Cell Lines and Tissue Xenografts

Subcutaneous xenografts are faster and more readily established than those generated intracranially; however, due to the limitations created by the pseudo-environment, intracranial xenografts are more commonly used with injection of immortalized tumour cell lines.

Characteristically, the tumours derived from immortalized GBM cell lines like U87 or U251 cells are somewhat similar to those seen in primary human GBM tumours; however, certain critical aspects are not fully recapitulated, limiting their usefulness. These GBM cell line derived tumours tend to show a decrease in the pathological hallmarks which are used to classify Grade IV GBMs including invasion, necrosis and microvascular proliferation. In particular, the tumours derived from U87 cell lines do not induce endothelial proliferation and as such have a lower degree of vascular involvement, a problem that can be rectified with a different cell line such as U251; a GBM cell line that demonstrates a higher degree of vascular involvement. Using whole tissue derived directly from primary human samples negates the problems seen with cell lines and can sometimes give a more accurate recapitulation of solid tumours. However, these explants GBM orthotopic models do not grow always in a predictable manner, although certain laboratories have a panel of these GBM explants which are well characterized with molecular and growth characteristics (46) and other model systems do exists as reviewed recently (47).

4.3. Transgenic Mouse Models of Spontaneous Brain Tumours

Although xenograft tumour models offer a reliable alternative to primary tumour tissue, their limitations and variability in recapitulating all characteristics of primary tumours means that they do not accurately represent the cellular composition of the original tumour. Xenograft generation relies on the use of immune-deficient mice, which in itself is a false environment for tumour development and, as such, negates any responses mediated by the bone marrow/immune system. In order to support data derived from GBM xenografts, it is therefore important to use a secondary spontaneous model. A mouse model that develops spontaneous brain tumours that closely resembles the pathological features of human low- and high-grade gliomas (such as cell necrosis and increased vascularity) has been generated and extensively characterized by our laboratory (48) (Fig. 7). These mice express oncogenic Ras (^{12V}HRas) specifically within astrocytes and develop and die from low- and high-grade gliomas over 3–4 months, many with more than one tumour foci (48). These tumours are responsive to known GBM therapeutics and provide valuable information regarding the angiogenesis mechanisms and development of the tumour (49).

Animal models provide an important source of easily accessible material required to study the tumour vasculature and the mechanisms in the absence of primary tissue being available. All tumour specimens from these sources can be analyzed *ex vivo* to look at characteristics of the vasculature.

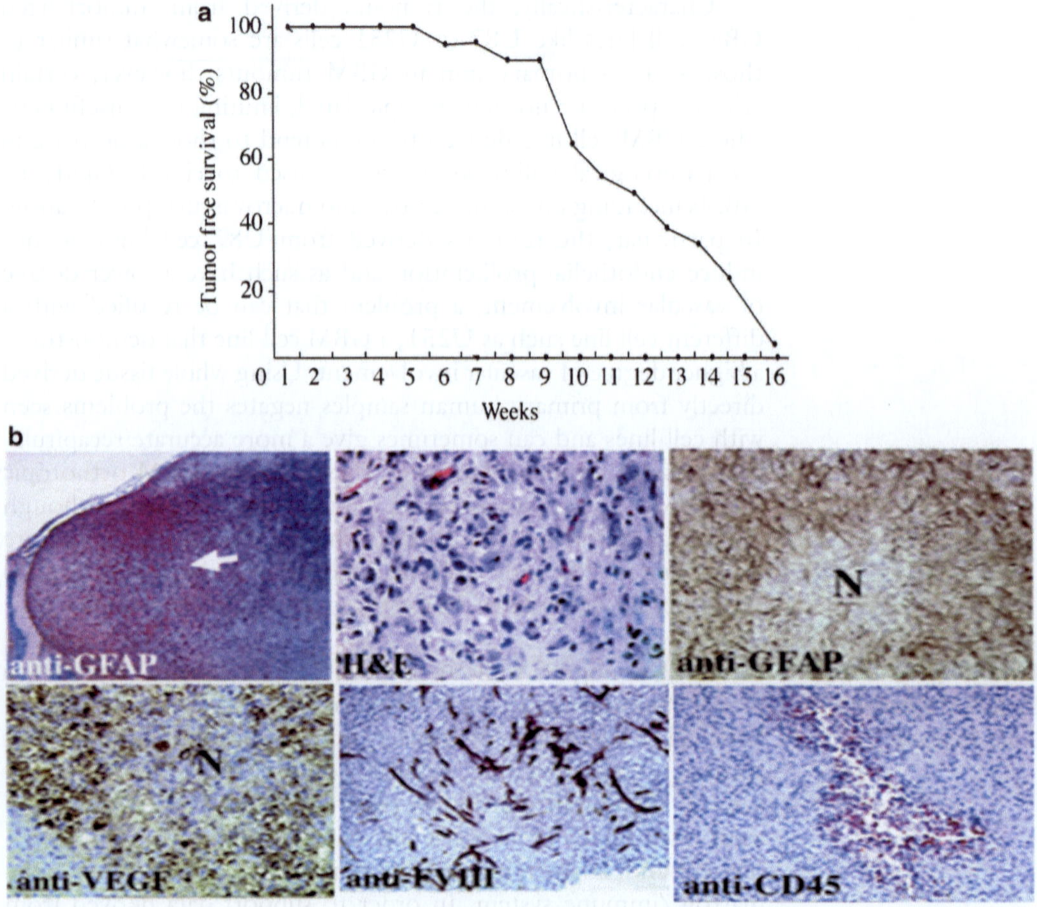

Fig. 7. GFAP: activated ^{12V}HRas model. (**a**) Mice engineered to overexpress activated ^{12V}HRas under the control of human 2.2 kB GFAP promoter, leads to development of both low- and high-grade astrocytomas. Mice are born normally with astrocytic hyperplasia and then go on to succumb to low- and high-grade astrocytomas and ultimate death by ~16 weeks of birth, as demonstrated by the survival curve. (**b**) The transgenic high-grade gliomas have similar pathological features to human GBMs. The tumours are invasive, have necrotic regions with surrounding GFAP+'ve pseudopalisading astrocytes and increased Factor VIII staining vascularity with VEGF production by tumour cells. The immune system is intact in these transgenic models, which is not in xenograft models, as evidenced by CD45 positive cells in the tumour vessels. *Taken from* (48).

5. Ex Vivo Angiogenesis Assays

Many assays have been developed to look at biopsied tissue from both experimental in vivo models and human samples. Whilst these systems and assays provide a more in situ analysis of vasculature formation during disease, they are still limited by being an end point analysis, not a real-time dynamic study limiting their potential to elucidate the mechanisms of vasculature formation with tumour progression. Ex vivo assays can demonstrate the presence of the vasculature and its interaction with host tissue, overcoming the limitations of *in vitro* culture based assays which look purely at cell behaviour. Of particular interest in the ex vivo assays is the ability to specifically examine the BBB which is significantly disrupted in the presence of disease and so can demonstrate the severity of the disease in question.

5.1. Evans Blue Assay

Evan's blue is an azo dye that can be administered intravenously to mice in order to assess the permeability of the BBB. The Evans blue dye binds to the protein albumin in the blood. Under normal physiological conditions, albumin is too large to cross the BBB and so is retained within the vessels along with the bound Evans blue dye so tissue remains white. Under pathological conditions, the BBB is disrupted and becomes leaky allowing the blood and Evans blue dye to diffuse through the vessels and into the tissue creating a blue area where the disruption has occurred. As well as highlighting the level of BBB damage, this assay may also be used to delineate the tumour borders (Fig. 8a).

Evans Blue Protocol

1. Mice are weighed and placed in a tail vein restrainer 30 min prior to assay.
2. The tail is cleaned with 70% alcohol and marked to signify the top surface of the tail, continually held by the handler to prevent movement.
3. Evans blue dye (2%) is drawn into a 27-G TB syringe, 2 ml/kg roughly 50 μl per 25 g mouse.
4. Evans blue solution should be injected into the lateral veins of the tail starting dorsal to the mouse and working towards the body in case of misses.
5. Maximum of three tries per vein can occur.
6. Mice are left for 30 min for the Evans blue to fully circulate.
7. Mice are sacrificed according to the institutional guidelines with standard 4% PFA perfusion.
8. Brains are extracted and frozen in OCT compound prior to sectioning at the tumour site for end point analysis.

Adapted from (40).

5.2. Evans Blue Expected Results

Evans blue can only diffuse where the BBB has been disrupted within the tumour area.

5.3. Microvessel Density Determination

Microvessel density (MVD) relies on staining the vessels of a tumour tissue sample extracted by using immunohistochemistry (IHC) techniques. Of particular interest are staining for endothelial cell markers CD31 and the von Willebrand factor (vWF). By looking at the number of CD31 or vWF positive vessels in high-powered fields,

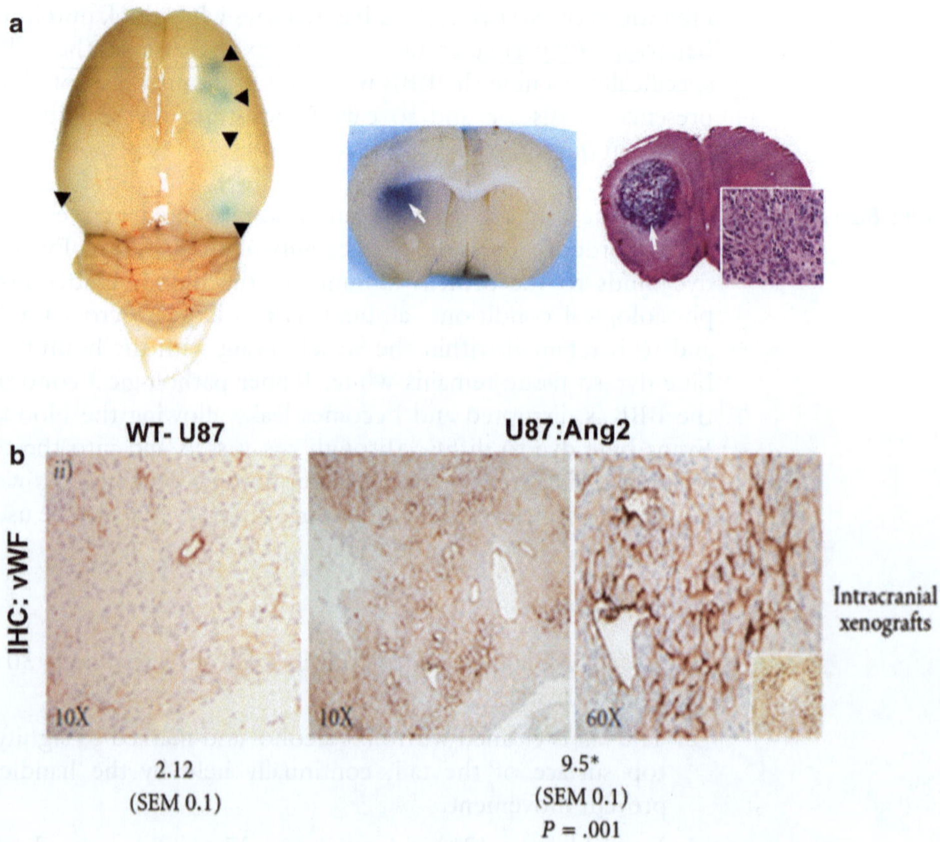

Fig. 8. Evans blue assay and microvascular density. (**a**) Evans blue dye is too large to penetrate the intact BBB; however, in areas where the BBB is compromised such as around a tumour it is able to leak through to the interstitial tissue. Visible on the surface of the brain, macro image, Evans blue clearly delineates specifically the tumour site as visible from the corresponding Haematoxylin stain. (**b**) Microvascular density (MVD) is a measure of the level of vasculature in a tissue section. Immunohistochemical staining of a vessel marker, such as CD31, CD34, Factor VIII (Von Willebrand factor), as in this image, is carried out and the number of vessels is counted, usually in ten high-powered fields by two blinded counters. In addition to MVD, the size, branching pattern and other aspects of the neoangiogenic process can be evaluated. *Taken from* (40).

the overall degree of vascularization of the sample can be quantified. It is also important to look at the difference in CD31 and vWF distribution in the sections to assess the type of vessels that are present in the tissue. By observing the size of vessels present in tumours and classifying them more discretely may highlight or elucidate mechanisms underlying the angiogenic switch indicating a critical tumour mass that is required to change mechanisms of angiogenesis.

MVD Protocol

1. Mount paraffin-embedded tissue sections (5 μm thick) on coated slides and dry at room temperature.
2. Deparaffinize tissue sections in three changes of Xylene and rehydrate through three 1-min changes of 100% ethanol, followed by two changes each of 95% and 75% ethanol.

3. Block tissue for up to 60 min in PBS supplemented with 0.2% Tween-20 and 3% bovine serum albumin (BSA).
4. Stain slides with primary antibodies directed at either CD-31 or vWF for 1 h at room temperature. Primary antibodies should be diluted in PBS supplemented with 3% BSA at a concentration previously optimized by user.
5. Wash slides three times for 5 min each in PBS.
6. Incubate the slides with secondary antibodies conjugated to biotin diluted in PBS with 3% BSA for 1 h.
7. Wash slides three times 5 min in PBS.
8. After washing, an avidin–biotin–peroxidase complex is applied to the sections (Vector Labs).
9. The slides are then developed with 0.5 mg/ml 3,3′-diaminobenzidine (DAB, Sigma, St. Louis, MO).
10. Slides are then incubated for 5 min at 37°C, rinsed, lightly counterstained with 2% methyl green (Sigma), dehydrated, and mounted for imaging.

MVD is calculated by counting the number of stained hollow vessels in ten high-power fields (×40) and in five high-power field at vascular "hot spots" predefined as having the highest proportion of vessels at ×4 magnification, for each slide. Counting should be done by two "blinded" independent observers on the same system to ensure no bias in the results (28–30) (Fig. 8b). Additional vasculature analysis such as average vessel diameter can also be compared in the same way.

6. Notes

1. Undiluted Matrigel solution is very viscous and quickly solidify at room temperature, so it MUST be kept on ice all the time. When pipetting, use a pre-chilled pipette tip to prevent any Matrigel formation during transfer.
2. Individual cell lines have different responses with immortalized endothelial cells developing tubules within 3 h of plating whilst primary purified endothelial cells such as human umbilical vein endothelial cells (HUVECs) can take up to 20 h to develop to the same point (32). Another important observation is that the source of tissue can affect the result, with mammary epithelial cells developing spheres with a central lumen in the culture, whilst metastatic tumour cells will invade and grow into the matrigel instead of forming tubules.

References

1. Louis DN, Ohgaki H, Wiestler OD, Cavenee WK, Burger PC, Jouvet A, Scheithauer BW, Kleihues P (2007) The 2007 WHO classification of tumours of the central nervous system. Acta Neuropathol 114:97–109
2. Plate KH, Mennel HD (1995) Vascular morphology and angiogenesis in glial tumors. Exp Toxicol Pathol 47:89–94
3. Onishi M, Ichikawa T, Kurozumi K, Date I (2011) Angiogenesis and invasion in glioma. Brain Tumor Pathol 28:13–24
4. Palmer AM (2010) The blood-brain barrier. Neurobiol Dis 37:1–2
5. Persidsky Y, Servio H, Ramirez J, Haorah J, Kanmogne GD (2006) Blood-brain barrier: structural components and function under physiologic and pathologic conditions. J Neuroimmune Pharmacol 1:223–236
6. Cardoso FL, Brites D, Brito MA (2010) Looking at the blood-brain barrier: molecular anatomy and possible investigation approaches. Brain Res Rev 64:328–363
7. Plate KH, Risau W (1995) Angiogenesis in malignant gliomas. Glia 15:339–347
8. Lopes S, Beatriz M (2003) Angiogenesis in brain tumors. Microsc Res Tech 60:225–230
9. Hanahan D, Weinberg RA (2011) Hallmarks of cancer: the next generation. Cell 144:646–647
10. Gimbrone MA Jr, Leapman SB, Cotran RS, Folkman J (1972) Tumor dormancy in vivo by prevention of neovascularisation. J Exp Med 136:261–276
11. Brem S (1976) The role of vascular proliferation in the growth of brain tumors. Clin Neurosurg 23:440–453
12. Coomber BL, Stewart PA, Hayakawa K, Farrell CL, Del Maestro RF (1987) Quantitative morphology of human glioblastoma mutliforme microvessels: structural basis of blood-brain barrier defect. J Neurooncol 5:299–307
13. Hillen F, Griffioen AW (2007) Tumour vascularization: sprouting angiogenesis and beyond. Cancer Metastasis Rev 26:489–502
14. Plate KH, Breier G, Weich HA, Risau W (1992) Vascular endothelial growth factor is a potential tumor angiogenesis factor in human gliomas in vivo. Nature 359:845–848
15. Shweiki D, Itin A, Soffer D, Keshet E (1992) Vascular endothelial growth factor induced by hypoxia may mediate hypoxia-initiated angiogenesis. Nature 359:843–845
16. Shweiki D, Neeman M, Itin A, Keshet E (1995) Induction of vascular endothelial growth factor expression by hypoxia and by glucose deficiency in multi-cell spheroids: implications for tumor angiogenesis. Proc Natl Acad Sci U S A 92:768–772
17. Koochekpour S, Merzak A, Pilkington G (1995) Vascular endothelial growth factor production is stimulated in response to growth-factors in human glioma-cells. Oncol Rep 2:1059–1061
18. Koga K, Todaka T, Morioka M, Hamada JI, Kai Y, Yano S, Okamura A, Takakura N, Suda T, Ushio Y (2001) Expression of angiopoietin-2 in human glioma cells and its role for angiogenesis. Cancer Res 61:6248–6254
19. Ribatti D (2004) The involvement of endothelial progenitor cells in tumor angiogenesis. J Cell Mol Med 8:294–300
20. Ahn GO, Brown JM (2009) Role of endothelial progenitors and other bone marrow-derived cells in the development of the tumor vasculature. Angiogenesis 12:159–164
21. Zhang HR, Chen FL, Xu CP, Ping XF, Wang QL, Liang ZQ, Wang JM, Bian XW (2009) Incorporation of endothelial progenitor cells into the neovasculature of malignant glioma xenograft. J Neurooncol 93:165–174
22. Du R, Lu K, Petritsch C, Liu P, Ganss R, Passegue E, Song H, Vandenberg S, Johnson R, Werb Z, Bergers G (2008) HIF1alpha induces the recruitment of bone marrow-derived vascular modulatory cells to regulate tumor angiogenesis and invasion. Cancer Cell 13(3):206–220
23. Ribatti D, Vacca A, Dammacco F (2003) New non-angiogenesis dependent pathways for tumour growth. Eur J Cancer 39:1835–1841
24. El Hallani S, Boisselier B, Peglion F, Rousseau CC, Idbaith A, Marie Y, Mokhtari K, Thomas JL, Eichmann A, Delattre JY, Maniotis AJ, Sanson M (2010) A new alternative mechanism in glioblastoma vascularization: tubular vasculogenic mimicry. Brain 133:973–982
25. Jain RK, Fukumura D (2007) Tumor microvasculature and microenvironment: targets for anti-angiogenesis and normalization. Microvasc Res 74:72–84
26. Chiao MT, Yang YC, Cheng WY, Shen CC, Ko JL (2011) CD133(+) glioblastoma stem-like cells induce vascular mimicry in vivo. Curr Neurovasc Res 8(3):210–219
27. Poon RT, Ng IO, Lau C, Yu WC, Yang ZF, Fan ST, Wong J (2002) Tumor microvessel density as a predictor of recurrence after resection of

hepatocellular carcinoma: a prospective study. J Clin Oncol 20:1775–1785

28. Choi WW, Lewis MM, Lawson D, Yin-Goen Q, Birdsong GG, Cotsonis GA, Cohen C, Young AN (2005) Angiogenic and lymphangiogenic microvessel density in breast carcinoma: correlation with clinicopathologic parameters and VEGF-family gene expression. Mod Pathol 18:143–152
29. Svagzdys S, Lesauskaite V, Pavalkis D, Nedzelskienė I, Pranys D, Tamelis A (2009) Microvessel density as new prognostic marker after radiotherapy in rectal cancer. BMC Cancer 9:95
30. Karademir S, Sökmen S, Terzi C, Sağol O, Ozer E, Astarcioğlu H, Coker A, Astarcioğlu I (2000) Tumor angiogenesis as a prognostic predictor in pancreatic cancer. J Hepatobiliary Pancreat Surg 7:489–495
31. Kubota Y, Kleinman H, Martin G, Lawley TJ (1988) Role of laminin and basement membrane in the morphological differentiation of human endothelial cells into capillary-like structures. J Cell Biol 107:1589–1598
32. Kleinman HK, Martin GR (2005) Matrigel: basement membrane matrix with biological activity. Semin Cancer Biol 15:378–386
33. Bootle-Wilbraham CA, Tazzyman S, Marshall JM, Lewis CE (2000) Fibrinogen E-fragment inhibits the migration and tubule formation of human dermal microvascular endothelial cells in vitro. Cancer Res 60:4719–4724
34. Folkins C, Shaked Y, Man S, Tang T, Lee CR, Zhu Z, Hoffman RM, Kerbel RS (2009) Glioma tumor stem-like cells promote tumor angiogenesis and vasculogenesis via vascular endothelial growth factor and stromal-derived factor 1. Cancer Res 69:7243–7251
35. Bao S, Wu Q, Sathornsumetee S, Hao Y, Li Z, Hjelmeland AB, Shi Q, McLendon RE, Bigner DD, Rich JN (2006) Stem cell-like glioma cells promote tumor angiogenesis through vascular endothelial growth factor. Cancer Res 66:7843–7848
36. Wang R, Chadalavada K, Wilshire J, Kowalik U, Hovinga KE, Geber A, Figelman B, Levershа M, Brennan C, Tabar V (2010) Glioblastoma stem-like cells give rise to tumour endothelium. Nature 468:829–833
37. Auerbach R, Lewis R, Shinners B, Akhtar N (2003) Angiogenesis assays: a critical overview. Clin Chem 49:32–40
38. Barendsz-Janson AF, Griffioen AW, Muller AD, van Dam-Miercas MC, Hillen HF (2000) In vitro tumor angiogenesis assays: plasminogen lysine binding site 1 inhibits in vitro tumor-induced angiogenesis. J Vasc Res 35:109–114
39. Auerbach R, Akhta N, Lewis RL, Shinners BL (2000) Angiogenesis assays: problems and pitfalls. Cancer Metastasis Rev 19:167–172
40. Zadeh G, Koushan K, Baoping Q, Shannon P, Guha A (2010) Role of angiopoietin-2 in regulating growth and vascularity of astrocytomas. J Oncol 2010:659231
41. Dinca EB, Sarkaria JN, Schroeder MA, Carlson BL, Voicu R, Gupta N, Berger MS, James CD (2007) Bioluminescence monitoring of intracranial glioblastoma xenograft: response to primary and salvage temozolomide therapy. J Neurosurg 107:610–616
42. Blouw B, Song H, Tihan T, Bosze J, Ferrara N, Gerber HP, Johnson RS, Bergers G (2003) The hypoxic response of tumors is dependent on their microenvironment. Cancer Cell 4:133–146
43. Pandita A, Aldape K, Zadeh G, Guha A, James C (2004) Contrasting *in vivo* and *in vitro* fates of glioblastoma cell subpopulations with amplified EGFR. Genes Chromosomes Cancer 39 (1):29–36
44. Sarkaria JN, Carlson BL, Schroeder MA, Grogan P, Brown P, Giannini C, Ballman KV, Kitange GJ, Guha A, Pandita A, James CD (2006) Use of an orthotopic xenograft model for assessing the effect of EGFR amplification on glioblastoma radiation response. Clin Cancer Res 12(7):2264–2271
45. Zadeh G, Koushan K, Pillo L, Shannon P, Guha A (2004) Role of Ang1 and its interaction with VEGF-A in astrocytomas. J Neuropathol Exp Neurol 63:978–989
46. Carlson BL, Pokorny JL, Schroeder MA, Sarkaria JN (2011) Establishment, maintenance and in vitro and in vivo applications of primary human glioblastoma mutliforme (GBM) xenograft models for translational biology studies and drug delivery. Curr Protoc Pharmacol 52:1–14
47. Candolfi M, Curtin JF, Nichols WS, Muhammad AG, King GD, Pluhar GE, McNiel EA, Ohlfest JR, Freese AB, Moore PF, Lerner J, Lowenstein PR, Castro MG (2007) Intracranial glioblastoma models in preclinical neurooncology: neuropathological characterization and tumor progression. J Neurooncol 85:133–148
48. Ding H, Roncari L, Shannon P, Wu X, Lau N, Karaskova J, Gutmann DH, Squire JA, Nagy A, Guha A (2001) Astrocyte-specific expression of activated p21-ras results in malignant astrocytoma formation in a transgenic mouse model of human gliomas. Cancer Res 61:3826–3836

49. Munoz D, Agnihotri S, Guha A (2009) Transgenic mouse models of CNS tumors: using genetically engineered murine models to study the role of p21-Ras in glioblastoma multiforme. In: Meir EG (ed) CNS cancer. Cancer drug discovery and development. Part 1. Humana Press, New York, pp 67–76

50. Armulik A, Genové G, Mäe M, Nisancioglu MH, Wallgard E, Niaudet C, He L, Norlin J, Lindblom P, Strittmatter K, Johansson BR, Betsholtz C (2010) Pericytes regulate the blood-brain barrier. Nature 468 (7323):557–561

Part III

Cellular, Molecular, and Genetic Basis of Brain Tumors: Data from Animal Models

Neuromethods (2013) 77: 203–215
DOI 10.1007/7657_2011_29

Published online: 13 March 2012

Using Genetically Engineered Mouse Models to Understand Low-Grade Glioma Development and Growth in Children

Aparna Kaul, Ibrahim Hussain, and David H. Gutmann

Abstract

Pilocytic astrocytoma (PA) is the most common brain tumor observed in children. These tumors can form sporadically in children with no underlying genetic disease or in 15–20% of children with neurofibromatosis type 1 (NF1), an inherited cancer predisposition syndrome. Though similar histologically, the genetic basis for PA formation in these two populations are distinct. In the general population, PAs likely arise in response to aberrant BRAF activation, whereas in the context of NF1, they result from bi-allelic inactivation of the *NF1* tumor suppressor gene. Since accurate rodent models of sporadic PA are currently under development, *Nf1* genetically engineered mouse models have served as tractable systems to study the role of aberrant intracellular signaling, nonneoplastic cells in the tumor microenvironment, and genomic modifiers on gliomagenesis. These small-animal models have also been used as platforms to discover next-generation targeted therapies and to evaluate the efficacy of these potential anticancer treatments prior to clinical trials for NF1-associated astrocytomas.

Key words: Astrocytoma, Pilocytic astrocytoma, Optic glioma, Brain tumor, Genetically engineered mice

1. Introduction

Brain tumors are the most common solid pediatric malignancy, affecting over 28,000 children every year and are among the leading causes of cancer-related death in the pediatric population (4, 5). Medulloblastoma is the most common malignant brain neoplasm in children under 4 years of age, whereas low-grade gliomas account for the majority of all primary brain tumors (6, 7). This heterogeneous group of glial neoplasms is classified by the World Health Organization (WHO) as WHO Grade I astrocytomas (pilocytic astrocytomas) or WHO Grade II astrocytomas (diffuse fibrillary astrocytomas). Pilocytic astrocytoma (PA) is the more frequently encountered subtype, accounting for 30% of all intracranial tumors in individuals 19 years of age and younger (7). While the majority of PAs form sporadically in patients with no family history of brain tumors, 15–20% of children with the neurofibromatosis-1 (NF1) inherited cancer predisposition syndrome develop PA, predominantly along the optic pathway. This genetic condition serves as an illustrative model for clinicians and scientists to understand the molecular and cellular foundations that underlie low-grade astrocytoma formation and growth in children. In this regard, the development of genetically engineered mouse (GEM) models of NF1-associated optic glioma (8) has been critical to the discovery of targets for therapeutic drug design as well as served as informative preclinical platforms to validate candidate therapies. Even though NF1-associated and sporadic PAs have distinct genetic etiologies (9, 10), the insights gained from the study of *Nf1* GEM models may elucidate cellular interactions and molecular targets critical for treating this common pediatric brain tumor in the general population.

2. Sporadic Pilocytic Astrocytomas

2.1. Clinical and Pathologic Features

Sporadic PAs are different from NF1-associated PAs in both their clinical course and natural history (11), but manifest symptomatically (based on location) and histologically in similar fashions. Frequent signs and symptoms of PAs include intracranial hypertension, macrocephaly, headaches, lethargy, and seizures. In the general population, 67–72% of PAs form in the cerebellum, where the median age of presentation ranges from 7 to 9 years (12–15). Clinically, sporadic PAs have excellent prognosis due to the low risk of recurrence following gross total resection. Cerebellar tumors are usually completely excised, while PAs that develop in the diencephalon, optic pathway, or brainstem are partially removed to avoid complications including damage to the visual pathways, endocrine abnormalities, or death. For this reason, sporadic PAs that develop in these areas are more likely to recur and associated with a higher rate of morbidity and mortality (13, 16).

PA tumors are usually well-circumscribed and cystic neoplasms that histologically exhibit biphasic arrangements of astrocytes with elongated processes alternating with loose hypocellular microcystic zones (17). These slow-growing tumors are composed of glial fibrillary acidic protein (GFAP)-immunoreactive cells (Fig. 1a) and in contrast to higher-grade glial neoplasms, have less mitotic activity and lower proliferative indices (18). Features strongly

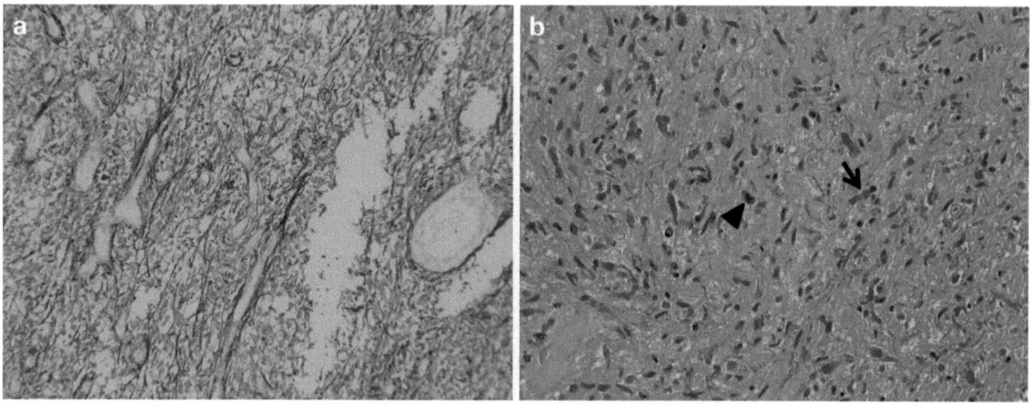

Fig. 1. Histopathology of a representative human pilocytic astrocytoma (WHO grade I). (**a**) Glial fibrillary acidic protein (GFAP) immunostaining demonstrates the presence of neoplastic glial cells. (**b**) Hematoxylin and eosin (H&E) stain reveals brightly staining Rosenthal fibers (*arrow*) and dark eosinophilic granular bodies (*arrowhead*).

associated with PAs include Rosenthal fibers and eosinophilic granular bodies (Fig.1b), while hyalinizied blood vessels, microcalcifications, oligodendroglioma-like cytology, and necrosis are associated with worse outcomes (14).

2.2. Molecular Genetics

There are few genetic alterations observed in sporadic PAs. Oncogenic *KRAS* (19, 20) and *TP53* (21–23) mutations have been identified, though these are rare events. Array-based comparative genomic hybridization studies have revealed copy number gains on chromosome 7q34, which includes the region spanning the *BRAF* locus. Further studies established that this gain represents a tandem duplication, resulting in fusions between an unknown gene (*KIAA1549*) and *BRAF* (1–3). The fusion transcripts contain the amino terminus of the *KIAA1549* gene and the BRAF kinase domain, creating a new molecule in which unregulated BRAF function leads to increased mitogen-activated protein kinase (MAPK) signaling.

Various studies have now confirmed that *BRAF* rearrangement is specific for sporadic PA, reinforced by the observation that two-thirds of sporadic PAs contain this fusion protein. This is in striking contrast to patients with NF1 who lack this rearrangement and instead harbor inactivation of the *NF1* tumor suppressor gene (1, 24). While this signature mutation represents a common genetic event in sporadic PA, the exact role of BRAF fusion protein expression in gliomagenesis is currently unknown. In this regard, the KIAA:BRAF fusion protein exhibits constitutive RAF kinase activity and can transform NIH3T3 fibroblasts; however, BRAF activation alone is not sufficient to drive glioma formation in mice (25). The fact that BRAF activation alone is not tumorigenic suggests that additional conditions must be required for glioma formation.

3. NF1-Associated Pilocytic Astrocytomas

3.1. Clinical Features

NF1 is an autosomal dominant condition occurring in one in 2,500–3,000 individuals worldwide (26). Children with NF1 often present with pigmentary features (café-au-lait spots, skinfold freckling, iris hamartomas), cardiovascular problems, orthopedic abnormalities (scoliosis, pseudarthrosis of the long bones), and learning disabilities. Importantly, individuals with NF1 have a predisposition to the development of peripheral and central nervous system tumors. Cutaneous and subcutaneous neurofibromas are benign peripheral nerve sheath tumors that develop in nearly all adults with NF1. One-third of individuals with NF1 will form a diffuse plexiform neurofibroma. These benign, but often extensive tumors, can impinge on neighboring nerves, blood vessels, or structures, and occasionally transform to malignant tumors (27). Upon transformation, their malignant counterparts, termed malignant peripheral nerve sheath tumors (MPNST), frequently result in metastasis and death (28).

Low-grade gliomas are the most common central nervous system tumor seen in children with NF1, and in contrast to their sporadic counterparts, most often develop along the optic pathway, including the optic nerves, chiasm, and hypothalamus (29, 30) (Fig. 2). Less commonly, gliomas may also arise in the brainstem or cerebellum (31). Optic pathway gliomas (OPG) can be asymptomatic (32, 33), but can result in decreased visual acuity, proptosis, and endocrine abnormalities (34–37). Compared to sporadic PAs, NF1-associated PAs are typically discovered earlier in childhood, with the majority being diagnosed within the first 6 years of life (32, 38). These NF1-associated optic gliomas usually have a more indolent clinical course and are less likely to progress or require treatment (39). There are rare reported cases of spontaneously regression (40, 41).

Treatment of NF1-associated optic gliomas rarely involves surgery, which is typically reserved for cosmetic purposes in a blind eye or for debulking large tumors that exert mass effect or cause hydrocephalus. Radiation therapy is avoided in individuals with NF1 due to the six- to eightfold increased risk of developing radiation-induced secondary malignant brain tumors as well as an elevated incidence of post-radiation neuropsychological problems and cerebral vasculopathies (moya-moya syndrome) (42, 43). Chemotherapy with carboplatin and vincristine is the first-line treatment for progressive symptomatic NF1-associated OPGs (44–46), and other agents, including vinblastine and temozolomide, represent second-line therapies for progressive tumors.

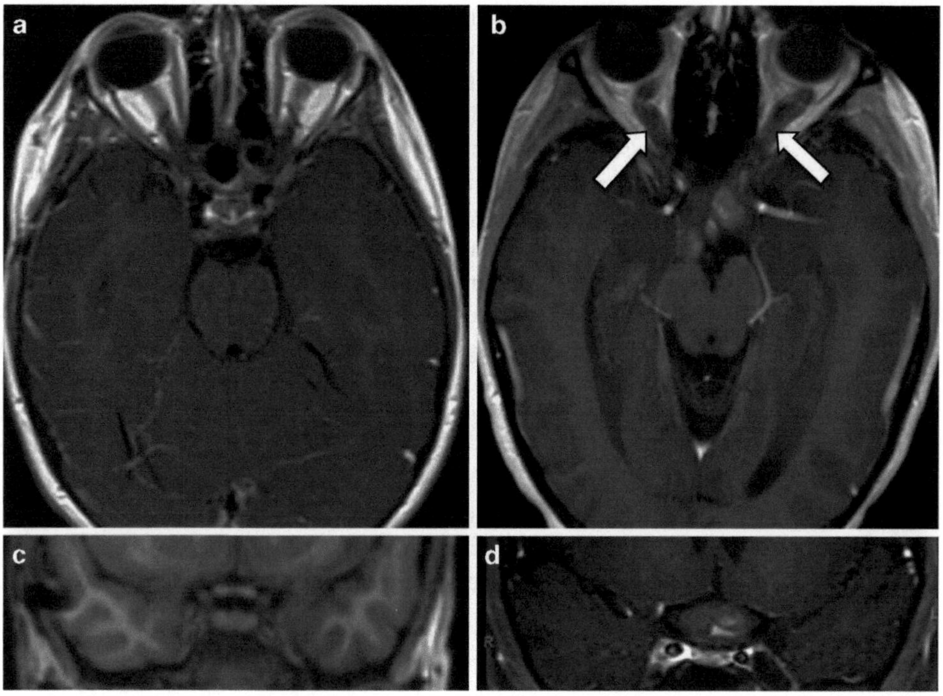

Fig. 2. MRI of a representative NF1-associated optic pathway glioma. Magnetic resonance images of the (**a**) optic nerves and (**c**) optic chiasm (magnified) from a child without an optic glioma are shown. By contrast, MRI reveals (**b**) an optic pathway glioma with expansion of the optic nerves (*arrows*) and (**d**) optic chiasm.

3.2. Neurofibromin Growth Control

The *NF1* gene on chromosome 17 encodes a large, 220–250 kDa, cytoplasmic protein, termed neurofibromin, which is expressed in many different cell types in the brain. Inspection of the predicted protein sequence of the *NF1* gene revealed that a small portion of the protein contained a 300 amino acid domain which functions as a GTPase activating protein (GAP) molecule (47), thereby establishing neurofibromin as a negative regulator of the Ras proto-oncogene. Similar to other GAP molecules, neurofibromin promotes the conversion of active GTP-bound Ras to inactive GDP-bound Ras, which reduces activation of downstream signaling molecules and attenuates cell growth (Fig. 3). In this regard, elevated Ras activity is mitogenic for many cell types, prompting early NF1-associated plexiform neurofibroma clinical trials using inhibitors of farnesyltransferase (FTPIs) to block Ras activity.

Further insights into the mechanism by which neurofibromin regulates cell growth in the brain have derived from studies using *Nf1* GEM models. Studies from our laboratory established mammalian target of rapamycin (mTOR) as a key Ras downstream signaling molecule critical for controlling *Nf1*-deficient astrocyte cell growth (48, 49). These studies demonstrated that the mTOR pathway may be a target for biologically based NF1-associated brain tumor therapies. Using *Nf1* optic glioma GEM strains, rapamycin, a macrolide antibiotic that inhibits mTOR activity, reduces tumor growth in vivo (50). Based on these promising results, clinical trials using rapamycin analogs are currently underway for patients with NF1-associated optic glioma.

Using high-throughput screening technologies, another Ras target downstream was identified. *Nf1*-deficient astrocytes have increased signal transducer and activator of transcription 3 (STAT3) activation, resulting from high levels of mTOR signaling (51). STAT3 regulates the transcription of numerous genes essential for cell cycle progression, including cyclin D1 (51, 52). While nonspecific STAT3 inhibitors have demonstrated a reduction in *Nf1*-deficient astrocytes in vitro and attenuated *NF1*-deficient glioma growth in vivo, more selective STAT3 inhibitors will need to be developed for use in children with NF1-associated OPG.

An additional mechanism by which neurofibromin deficiency leads to tumor development is through defective generation of intracellular cyclic-adenosine monophosphate (cAMP) (Fig. 3).

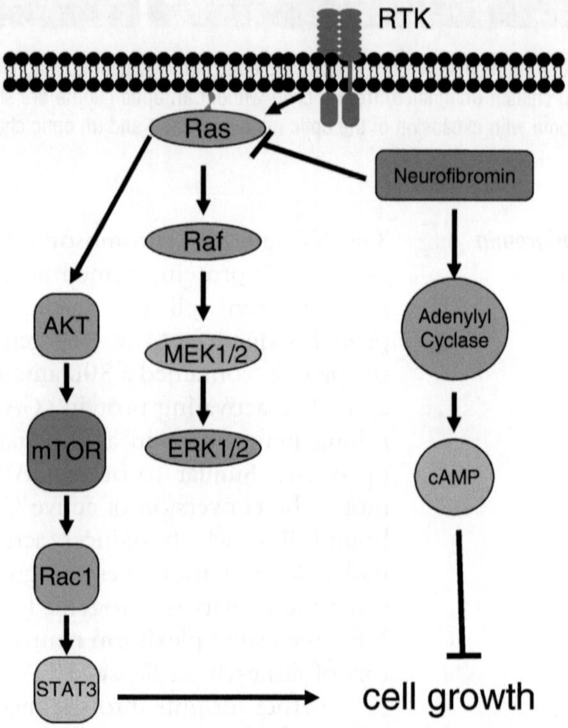

Fig. 3. Neurofibromin-regulated intracellular signaling pathways. The *NF1* gene product, neurofibromin, accelerates the conversion of the Ras proto-oncogene from its active GTP-bound form to its inactive GDP-bound form. GTP-bound Ras activates downstream signaling intermediates, including Raf/MEK/ERK and AKT/mTOR, leading to increased cell growth. Neurofibromin also increases cAMP activity through adenylyl cyclase to inhibit astrocyte cell growth.

Neurofibromin positively regulates cAMP production at the level of adenylyl cyclase (AC) (53), such that in the absence of neurofibromin, the lower intracellular cAMP levels enhance cell survival in response to chemokines (54). This enhanced survival of *Nf1−/−* astrocytes to chemokines, like CXCL12, is the opposite of what occurs in wild-type astrocytes, where CXCL12 treatment leads to increased apoptosis (54). The attenuated cAMP levels in *Nf1*-deficient astrocytes provide another therapeutic opportunity. Inhibition of cAMP degradation using Rolipram, a phosphodiesterase 4 (PDE4) inhibitor, restores cAMP levels in *Nf1*-deficient astrocytes in vitro and attenuates optic glioma growth in vivo (55, 56). Based on these encouraging findings, future strategies for treating NF1-associated gliomas may involve a combination of cAMP- and Ras/AKT/mTOR-targeted therapies.

3.3. *Nf1* GEM Models of Optic Glioma

Initial attempts at mouse modeling entailed the creation of mice lacking *Nf1* gene expression in all cells in their body. These *Nf1−/−* mice died in utero (57, 58), precluding their use for tumor modeling. However, *Nf1* conditional knockout mice developed by Dr. Luis Parada and his colleagues have enabled researchers to generate mice harboring *Nf1* inactivation in specific cell populations. Using this *Nf1* conditional knockout approach, previous studies revealed that loss of neurofibromin expression in central nervous system astrocytes results in increased cell proliferation, but is not sufficient for glioma formation (59). Since children with NF1 are born with one nonfunctional and one functional copy of the *NF1* gene in their body, but only develop tumors when both copies are nonfunctional in glial cells, *Nf1+/−* mice were generated that lack *Nf1* inactivation in astrocytes. The *Nf1* conditional knockout mice develop gliomas in the prechiasmatic optic nerves and chiasm (8). Optic gliomas form in these mice by 10–12 weeks of age (60). Gross inspection of the optic nerves in these mice reveals bilateral prechiasmal optic nerve and chiasm enlargement (Fig. 4b), similar to what is observed

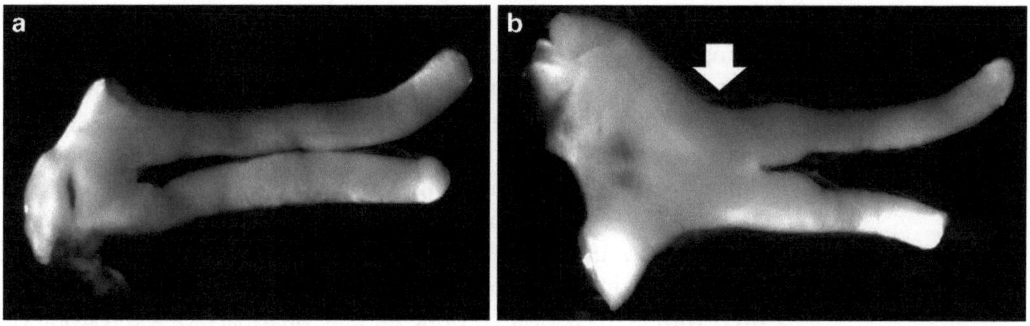

Fig. 4. Representative optic nerves and optic glioma from Nf1 genetically engineered mice. (**a**) The normal optic nerve and chiasm from a representative control mouse is shown. (**b**) A representative optic glioma in a genetically engineered mouse is demonstrated by the presence of swollen optic nerves and chiasm.

in children with NF1. Microscopically, the areas of optic nerve enlargement are accompanied by increased numbers of GFAP-positive astrocytes, elevated proliferative indices, and cells with atypical, hyperchromatic nuclei. In contrast to their human counterparts, however, the *Nf1* GEM gliomas lack microcystic spaces, Rosenthal fibers, and granular bodies (8).

The *Nf1* optic glioma GEM model has been employed in preclinical studies to demonstrate that conventional single-agent chemotherapy (temozolomide) reduces glioma size and proliferation (50). In addition, biologically based therapies, including rapamycin, have been tested for efficacy in this model, and the positive results were used to propel clinical trials in children with NF1-associated brain tumors.

3.4. The Role of the Tumor Microenvironment in Glioma Growth

NF1-associated PAs have a predilection for the optic pathway, which prompted a closer investigation of the nonneoplastic cell types and growth factors in this brain region. Endothelial cells and microglia, macrophage-like immune system cells, are among the noncancerous cells that produce growth factors that increase preneoplastic and neoplastic cell growth (61). Microglia with reduced *Nf1* gene expression (*Nf1+/−* microglia) are found in increased numbers prior to obvious optic glioma formation in *Nf1* optic glioma mice (60). Importantly, inhibition of microglial function attenuates optic glioma proliferation in vivo (62, 63). These findings underscore the importance of microglia and nonneoplastic cells in the tumor microenvironment to optic glioma growth, and present another target for future glioma therapies.

3.5. Visual Impairment Following Optic Glioma Formation

Nf1 optic glioma mice have been used to understand the cellular and molecular basis for visual loss in children with NF1-associated optic glioma. Visual impairment in children with NF1-associated OPG is frequently encountered, with only 20% of children exhibiting improvements in vision following chemotherapy (45). *Nf1* optic glioma mice exhibit reduced visual evoked potentials during early tumor formation, suggesting abnormal neuronal transmission (64). Moreover, there is evidence of axonal deterioration and increased retinal ganglion cell (RGC) apoptosis in response to glioma formation and progression (64–66). To gain insights into the impact of reduced *Nf1* gene expression on RGC function, *Nf1+/−* RGCs were grown in vitro. Surprisingly, *Nf1+/−* RGCs had a cell-autonomous decrease in neurite lengths, growth cones, and survival (56). These *Nf1+/−* RGC abnormalities were the result of reduced neurofibromin-regulated cAMP levels and could be rescued by restoring cAMP levels in vitro (56, 67). Relevant to visual impairment in children with NF1-associated optic glioma, pharmacological restoration of cAMP levels with Rolipram partly restored RGC survival in vivo (56), suggesting that future treatments for symptomatic NF1-associated OPG might involve neuroprotective strategies.

3.6. Genomic Influences on Nf1 GEM Tumorigenesis

Since the majority of children with NF1 do not develop brain tumors, recent studies have explored the influences of genomic variation on glioma susceptibility. *Nf1* GEM mice heterozygous for inactivating mutations involving both the *Nf1* and *Trp53* genes (NPCis mice) are predisposed to the development of high-grade astrocytomas (68, 69). NPCis mice maintained on the C57Bl/6 genetic background develop high-grade astrocytomas, whereas NPCis mice maintained on the 129 S4 genetic background are relatively resistant to astrocytoma formation (70). These findings suggest that genomic modifiers involving genes other than the *NF1* gene may influence glioma penetrance. The identification of these modifier loci in GEM models coupled with genome-wide association studies in children with NF1 could lead to the early identification of children most at risk for developing a brain tumor.

4. Conclusion

Nf1 GEM glioma models have greatly expanded our understanding of the cellular, molecular, and genomic basis for brain tumor development. These models have identified critical signaling pathways responsible for astrocyte proliferation and survival. Furthermore, these models have helped researchers understand how the nonneoplastic tumor microenvironment can dictate where and when brain tumors form, and how these nonneoplastic cells can be targeted to reduce glioma growth. The availability of robust small-animal models of NF1-associated glioma provides unique opportunities to develop treatments that target both neoplastic and nonneoplastic cells as well as protect neurons from damage. Importantly, these GEM models establish preclinical platforms for the rapid evaluation of promising therapies prior to their use in children with NF1. With continued progress in genomics and mouse modeling, we are nearing an era in which genetic information may identify at-risk individuals and facilitate the development of "personalized" treatment regimens.

Acknowledgments

We thank Dr. Sonika Dahiya and Mr. Scott Gianino for the images of human pilocytic astrocytoma specimens and mouse *Nf1* optic gliomas, respectively. Ibrahim Hussain is supported by funding from the Doris Duke Charitable Foundation to Washington University School of Medicine. David H. Gutmann is supported by funding from the National Cancer Institute, the National

Institutes of Health, the James S. McDonnell Foundation, and the Department of Defense.

References

1. Jones DT, Kocialkowski S, Liu L, Pearson DM, Backlund LM, Ichimura K, Collins VP (2008) Tandem duplication producing a novel oncogenic BRAF fusion gene defines the majority of pilocytic astrocytomas. Cancer Res 68:8673–8677
2. Bar EE, Lin A, Tihan T, Burger PC, Eberhart CG (2008) Frequent gains at chromosome 7q34 involving BRAF in pilocytic astrocytoma. J Neuropathol Exp Neurol 67:878–887
3. Pfister S, Janzarik WG, Remke M, Ernst A, Werft W, Becker N, Toedt G, Wittmann A, Kratz C, Olbrich H, Ahmadi R, Thieme B, Joos S, Radlwimmer B, Kulozik A, Pietsch T, Herold-Mende C, Gnekow A, Reifenberger G, Korshunov A, Scheurlen W, Omran H, Lichter P (2008) BRAF gene duplication constitutes a mechanism of MAPK pathway activation in low-grade astrocytomas. J Clin Invest 118:1739–1749
4. Porter KR, McCarthy BJ, Freels S, Kim Y, Davis FG (2010) Prevalence estimates for primary brain tumors in the United States by age, gender, behavior, and histology. Neuro Oncol 12:520–527
5. Dubuc AM, Northcott PA, Mack S, Witt H, Pfister S, Taylor MD (2010) The genetics of pediatric brain tumors. Curr Neurol Neurosci Rep 10:215–223
6. Giannini C, Scheithauer BW (1997) Classification and grading of low-grade astrocytic tumors in children. Brain Pathol 7:785–798
7. Butler SO, Btaiche IF, Alaniz C (2005) Relationship between hyperglycemia and infection in critically ill patients. Pharmacotherapy 25:963–976
8. Bajenaru ML, Hernandez MR, Perry A, Zhu Y, Parada LF, Garbow JR, Gutmann DH (2003) Optic nerve glioma in mice requires astrocyte Nf1 gene inactivation and Nf1 brain heterozygosity. Cancer Res 63:8573–8577
9. Rodriguez FJ, Giannini C, Asmann YW, Sharma MK, Perry A, Tibbetts KM, Jenkins RB, Scheithauer BW, Anant S, Jenkins S, Eberhart CG, Sarkaria JN, Gutmann DH (2008) Gene expression profiling of NF-1-associated and sporadic pilocytic astrocytoma identifies aldehyde dehydrogenase 1 family member L1 (ALDH1L1) as an underexpressed candidate biomarker in aggressive subtypes. J Neuropathol Exp Neurol 67:1194–1204
10. Kluwe L, Hagel C, Tatagiba M, Thomas S, Stavrou D, Ostertag H, von Deimling A, Mautner VF (2001) Loss of NF1 alleles distinguish sporadic from NF1-associated pilocytic astrocytomas. J Neuropathol Exp Neurol 60:917–920
11. Listernick R, Darling C, Greenwald M, Strauss L, Charrow J (1995) Optic pathway tumors in children: the effect of neurofibromatosis type 1 on clinical manifestations and natural history. J Pediatr 127:718–722
12. Fisher PG, Tihan T, Goldthwaite PT, Wharam MD, Carson BS, Weingart JD, Repka MX, Cohen KJ, Burger PC (2008) Outcome analysis of childhood low-grade astrocytomas. Pediatr Blood Cancer 51:245–250
13. Gajjar A, Sanford RA, Heideman R, Jenkins JJ, Walter A, Li Y, Langston JW, Muhlbauer M, Boyett JM, Kun LE (1997) Low-grade astrocytoma: a decade of experience at St. Jude Children's Research Hospital. J Clin Oncol 15:2792–2799
14. Tibbetts KM, Emnett RJ, Gao F, Perry A, Gutmann DH, Leonard JR (2009) Histopathologic predictors of pilocytic astrocytoma event-free survival. Acta Neuropathol 117:657–665
15. Burkhard C, Di Patre PL, Schuler D, Schuler G, Yasargil MG, Yonekawa Y, Lutolf UM, Kleihues P, Ohgaki H (2003) A population-based study of the incidence and survival rates in patients with pilocytic astrocytoma. J Neurosurg 98:1170–1174
16. Packer RJ, Ater J, Allen J, Phillips P, Geyer R, Nicholson HS, Jakacki R, Kurczynski E, Needle M, Finlay J, Reaman G, Boyett JM (1997) Carboplatin and vincristine chemotherapy for children with newly diagnosed progressive low-grade gliomas. J Neurosurg 86:747–754
17. Rath G, Schneider C, Langlois B, Sartelet H, Morjani H, Btaouri HE, Dedieu S, Martiny L (2009) De novo ceramide synthesis is responsible for the anti-tumor properties of camptothecin and doxorubicin in follicular thyroid carcinoma. Int J Biochem Cell Biol 41:1165–1172
18. Giannini C, Scheithauer BW, Burger PC, Christensen MR, Wollan PC, Sebo TJ, Forsyth PA, Hayostek CJ (1999) Cellular proliferation

in pilocytic and diffuse astrocytomas. J Neuropathol Exp Neurol 58:46–53
19. Sharma MK, Zehnbauer BA, Watson MA, Gutmann DH (2005) RAS pathway activation and an oncogenic RAS mutation in sporadic pilocytic astrocytoma. Neurology 65:1335–1336
20. Janzarik WG, Kratz CP, Loges NT, Olbrich H, Klein C, Schafer T, Scheurlen W, Roggendorf W, Weiller C, Niemeyer C, Korinthenberg R, Pfister S, Omran H (2007) Further evidence for a somatic KRAS mutation in a pilocytic astrocytoma. Neuropediatrics 38:61–63
21. Willert JR, Daneshvar L, Sheffield VC, Cogen PH (1995) Deletion of chromosome arm 17p DNA sequences in pediatric high-grade and juvenile pilocytic astrocytomas. Genes Chromosomes Cancer 12:165–172
22. Patt S, Gries H, Giraldo M, Cervos-Navarro J, Martin H, Janisch W, Brockmoller J (1996) p53 gene mutations in human astrocytic brain tumors including pilocytic astrocytomas. Hum Pathol 27:586–589
23. Hayes VM, Dirven CM, Dam A, Verlind E, Molenaar WM, Mooij JJ, Hofstra RM, Buys CH (1999) High frequency of TP53 mutations in juvenile pilocytic astrocytomas indicates role of TP53 in the development of these tumors. Brain Pathol 9:463–467
24. Yu J, Deshmukh H, Gutmann RJ, Emnett RJ, Rodriguez FJ, Watson MA, Nagarajan R, Gutmann DH (2009) Alterations of BRAF and HIPK2 loci predominate in sporadic pilocytic astrocytoma. Neurology 73:1526–1531
25. Robinson JP, VanBrocklin MW, Guilbeault AR, Signorelli DL, Brandner S, Holmen SL (2010) Activated BRAF induces gliomas in mice when combined with Ink4a/Arf loss or Akt activation. Oncogene 29:335–344
26. Friedman JM (1999) Epidemiology of neurofibromatosis type 1. Am J Med Genet 89:1–6
27. Williams VC, Lucas J, Babcock MA, Gutmann DH, Korf B, Maria BL (2009) Neurofibromatosis type 1 revisited. Pediatrics 123:124–133
28. Evans DG, Baser ME, McGaughran J, Sharif S, Howard E, Moran A (2002) Malignant peripheral nerve sheath tumours in neurofibromatosis 1. J Med Genet 39:311–314
29. Thiagalingam S, Flaherty M, Billson F, North K (2004) Neurofibromatosis type 1 and optic pathway gliomas: follow-up of 54 patients. Ophthalmology 111:568–577
30. King A, Listernick R, Charrow J, Piersall L, Gutmann DH (2003) Optic pathway gliomas in neurofibromatosis type 1: the effect of presenting symptoms on outcome. Am J Med Genet A 122A:95–99
31. Gutmann DH, Rasmussen SA, Wolkenstein P, MacCollin MM, Guha A, Inskip PD, North KN, Poyhonen M, Birch PH, Friedman JM (2002) Gliomas presenting after age 10 in individuals with neurofibromatosis type 1 (NF1). Neurology 59:759–761
32. Listernick R, Charrow J, Greenwald M, Mets M (1994) Natural history of optic pathway tumors in children with neurofibromatosis type 1: a longitudinal study. J Pediatr 125:63–66
33. Segal L, Darvish-Zargar M, Dilenge ME, Ortenberg J, Polomeno RC (2010) Optic pathway gliomas in patients with neurofibromatosis type 1: follow-up of 44 patients. J AAPOS 14:155–158
34. Steinbok P, Hentschel S, Almqvist P, Cochrane DD, Poskitt K (2002) Management of optic chiasmatic/hypothalamic astrocytomas in children. Can J Neurol Sci 29:132–138
35. Allen JC (2000) Initial management of children with hypothalamic and thalamic tumors and the modifying role of neurofibromatosis-1. Pediatr Neurosurg 32:154–162
36. Jahraus CD, Tarbell NJ (2006) Optic pathway gliomas. Pediatr Blood Cancer 46:586–596
37. Nicolin G, Parkin P, Mabbott D, Hargrave D, Bartels U, Tabori U, Rutka J, Buncic JR, Bouffet E (2009) Natural history and outcome of optic pathway gliomas in children. Pediatr Blood Cancer 53:1231–1237
38. Listernick R, Charrow J, Greenwald MJ, Esterly NB (1989) Optic gliomas in children with neurofibromatosis type 1. J Pediatr 114:788–792
39. Listernick R, Louis DN, Packer RJ, Gutmann DH (1997) Optic pathway gliomas in children with neurofibromatosis 1: consensus statement from the NF1 Optic Pathway Glioma Task Force. Ann Neurol 41:143–149
40. Astrup J (2003) Natural history and clinical management of optic pathway glioma. Br J Neurosurg 17:327–335
41. Piccirilli M, Lenzi J, Delfinis C, Trasimeni G, Salvati M, Raco A (2006) Spontaneous regression of optic pathways gliomas in three patients with neurofibromatosis type I and critical review of the literature. Childs Nerv Syst 22:1332–1337
42. Sharif S, Ferner R, Birch JM, Gillespie JE, Gattamaneni HR, Baser ME, Evans DG (2006) Second primary tumors in neurofibromatosis 1 patients treated for optic glioma: substantial risks after radiotherapy. J Clin Oncol 24:2570–2575
43. Grill J, Couanet D, Cappelli C, Habrand JL, Rodriguez D, Sainte-Rose C, Kalifa C (1999) Radiation-induced cerebral vasculopathy in

children with neurofibromatosis and optic pathway glioma. Ann Neurol 45:393–396

44. Packer RJ, Lange B, Ater J, Nicholson HS, Allen J, Walker R, Prados M, Jakacki R, Reaman G, Needles MN et al (1993) Carboplatin and vincristine for recurrent and newly diagnosed low-grade gliomas of childhood. J Clin Oncol 11:850–856
45. Dalla Via P, Opocher E, Pinello ML, Calderone M, Viscardi E, Clementi M, Battistella PA, Laverda AM, Da Dalt L, Perilongo G (2007) Visual outcome of a cohort of children with neurofibromatosis type 1 and optic pathway glioma followed by a pediatric neuro-oncology program. Neuro Oncol 9:430–437
46. Mahoney DH Jr, Cohen ME, Friedman HS, Kepner JL, Gemer L, Langston JW, James HE, Duffner PK, Kun LE (2000) Carboplatin is effective therapy for young children with progressive optic pathway tumors: a Pediatric Oncology Group phase II study. Neuro Oncol 2:213–220
47. Xu GF, O'Connell P, Viskochil D, Cawthon R, Robertson M, Culver M, Dunn D, Stevens J, Gesteland R, White R et al (1990) The neurofibromatosis type 1 gene encodes a protein related to GAP. Cell 62:599–608
48. Dasgupta B, Yi Y, Chen DY, Weber JD, Gutmann DH (2005) Proteomic analysis reveals hyperactivation of the mammalian target of rapamycin pathway in neurofibromatosis 1-associated human and mouse brain tumors. Cancer Res 65:2755–2760
49. Sandsmark DK, Zhang H, Hegedus B, Pelletier CL, Weber JD, Gutmann DH (2007) Nucleophosmin mediates mammalian target of rapamycin-dependent actin cytoskeleton dynamics and proliferation in neurofibromin-deficient astrocytes. Cancer Res 67:4790–4799
50. Hegedus B, Banerjee D, Yeh TH, Rothermich S, Perry A, Rubin JB, Garbow JR, Gutmann DH (2008) Preclinical cancer therapy in a mouse model of neurofibromatosis-1 optic glioma. Cancer Res 68:1520–1528
51. Banerjee S, Byrd JN, Gianino SM, Harpstrite SE, Rodriguez FJ, Tuskan RG, Reilly KM, Piwnica-Worms DR, Gutmann DH (2010) The neurofibromatosis type 1 tumor suppressor controls cell growth by regulating signal transducer and activator of transcription-3 activity in vitro and in vivo. Cancer Res 70:1356–1366
52. Yeh HH, Giri R, Chang TY, Chou CY, Su WC, Liu HS (2009) Ha-ras oncogene-induced Stat3 phosphorylation enhances oncogenicity of the cell. DNA Cell Biol 28:131–139
53. Dasgupta B, Dugan LL, Gutmann DH (2003) The neurofibromatosis 1 gene product neurofibromin regulates pituitary adenylate cyclase-activating polypeptide-mediated signaling in astrocytes. J Neurosci 23:8949–8954
54. Warrington NM, Woerner BM, Daginakatte GC, Dasgupta B, Perry A, Gutmann DH, Rubin JB (2007) Spatiotemporal differences in CXCL12 expression and cyclic AMP underlie the unique pattern of optic glioma growth in neurofibromatosis type 1. Cancer Res 67:8588–8595
55. Warrington NM, Gianino SM, Jackson E, Goldhoff P, Garbow JR, Piwnica-Worms D, Gutmann DH, Rubin JB (2010) Cyclic AMP suppression is sufficient to induce gliomagenesis in a mouse model of neurofibromatosis-1. Cancer Res 70:5717–5727
56. Brown JA, Gianino SM, Gutmann DH (2010) Defective cAMP generation underlies the sensitivity of CNS neurons to neurofibromatosis-1 heterozygosity. J Neurosci 30:5579–5589
57. Jacks T, Shih TS, Schmitt EM, Bronson RT, Bernards A, Weinberg RA (1994) Tumour predisposition in mice heterozygous for a targeted mutation in Nf1. Nat Genet 7:353–361
58. Brannan CI, Perkins AS, Vogel KS, Ratner N, Nordlund ML, Reid SW, Buchberg AM, Jenkins NA, Parada LF, Copeland NG (1994) Targeted disruption of the neurofibromatosis type-1 gene leads to developmental abnormalities in heart and various neural crest-derived tissues. Genes Dev 8:1019–1029
59. Bajenaru ML, Zhu Y, Hedrick NM, Donahoe J, Parada LF, Gutmann DH (2002) Astrocyte-specific inactivation of the neurofibromatosis 1 gene (NF1) is insufficient for astrocytoma formation. Mol Cell Biol 22:5100–5113
60. Bajenaru ML, Garbow JR, Perry A, Hernandez MR, Gutmann DH (2005) Natural history of neurofibromatosis 1-associated optic nerve glioma in mice. Ann Neurol 57:119–127
61. Hanisch UK (2002) Microglia as a source and target of cytokines. Glia 40:140–155
62. Daginakatte GC, Gutmann DH (2007) Neurofibromatosis-1 (Nf1) heterozygous brain microglia elaborate paracrine factors that promote Nf1-deficient astrocyte and glioma growth. Hum Mol Genet 16:1098–1112
63. Daginakatte GC, Gianino SM, Zhao NW, Parsadanian AS, Gutmann DH (2008) Increased c-Jun-NH2-kinase signaling in neurofibromatosis-1 heterozygous microglia drives microglia activation and promotes optic glioma proliferation. Cancer Res 68:10358–10366
64. Hegedus B, Hughes FW, Garbow JR, Gianino S, Banerjee D, Kim K, Ellisman MH, Brantley MA Jr, Gutmann DH (2009) Optic nerve dysfunction in a mouse model of neurofibromatosis-1

optic glioma. J Neuropathol Exp Neurol 68:542–551

65. Quigley HA, Nickells RW, Kerrigan LA, Pease ME, Thibault DJ, Zack DJ (1995) Retinal ganglion cell death in experimental glaucoma and after axotomy occurs by apoptosis. Invest Ophthalmol Vis Sci 36:774–786
66. Kim KY, Ju WK, Hegedus B, Gutmann DH, Ellisman MH (2010) Ultrastructural characterization of the optic pathway in a mouse model of neurofibromatosis-1 optic glioma. Neuroscience 170:178–188
67. Hegedus B, Dasgupta B, Shin JE, Emnett RJ, Hart-Mahon EK, Elghazi L, Bernal-Mizrachi E, Gutmann DH (2007) Neurofibromatosis-1 regulates neuronal and glial cell differentiation from neuroglial progenitors in vivo by both cAMP- and Ras-dependent mechanisms. Cell Stem Cell 1:443–457
68. Cichowski K, Shih TS, Schmitt E, Santiago S, Reilly K, McLaughlin ME, Bronson RT, Jacks T (1999) Mouse models of tumor development in neurofibromatosis type 1. Science 286:2172–2176
69. Vogel KS, Klesse LJ, Velasco-Miguel S, Meyers K, Rushing EJ, Parada LF (1999) Mouse tumor model for neurofibromatosis type 1. Science 286:2176–2179
70. Reilly KM, Loisel DA, Bronson RT, McLaughlin ME, Jacks T (2000) Nf1;Trp53 mutant mice develop glioblastoma with evidence of strain-specific effects. Nat Genet 26:109–113

Neuromethods (2013) 77: 217–225
DOI 10.1007/7657_2011_30

Published online: 24 March 2012

EGFR and Tumor Suppressor Function in Brain Cancer Development

Julie Lessard and Al Charest

Abstract

Genetically engineered mouse models (GEMM) of glioblastoma multiforme (GBM) provide a system in which de novo tumors arise and develop within the context of a proper tumor microenvironment. These models are considered superior to orthotopic xenograft systems because they allow for the development of tumors in situ and in an immune competent environment. More importantly, they allow for the assessment of specific genetic mutations on the etiology of GBM and their responses to therapeutic interventions. Using cutting edge Cre/Lox technologies, we created compound conditional transgenic strains based on a Cre-induced expression of mutant EGFRvIII with the concomitant loss of the Cdkn2a and PTEN tumor suppressor genes. The introduction of a conditional firefly *luciferase* transgene provides a means to monitor tumor initiation and growth noninvasively by bioluminescence imaging. The tumors that are generated in this model share numerous histopathological features with human GBMs. We provide here a detailed description of the methodologies used to initiate and monitor GBM tumors in this model system.

Key words: Glioblastoma multiforme, EGFR, PTEN, Genetically engineered mouse models

1. Introduction

Malignant glioma is a severe disease resulting in a median survival of 12–15 months for the ~14,000 new cases affected by the disease every year in the USA (1). These heterogeneous brain tumors are thought to arise from different cell types (astrocytes, oligodendrocytes, or both) and have been classified by the World Health Organization in four different grades according to their degree of malignancy (2, 3). Grades I and II tumors are relatively benign and grades III and IV represent very aggressive tumors with poor prognosis. Grade IV tumors are also referred to as glioblastoma multiforme (GBM). GBMs are divided into two subclasses based on clinical presentation and specific genetic events: primary GBMs are defined as tumors whose growth remain undetected until diagnosis, whereas secondary GBMs result from the evolution of low-grade to a high-grade glioma over time with a concomitant accumulation of genetic mutations (1, 3).

GBM results from acquisition of multiple genetic aberrations. Recently, genomic characterization of GBMs has lead to a subclassification based on their genetic profile. This classification has been put forward by The Cancer Genome Atlas (TCGA) consortium and others to improve our understanding of GBM heterogeneity (4–6). It is composed of four categories: neural, proneural, mesenchymal, and classical. Each category is predominantly characterized by a major genetic aberration: IDH1/IDH2 or PDGFRα (proneural), NF1 (mesenchymal), and EGFR (classical). The classical subtype of GBM is characterized by the amplification/overexpression of the EGFR gene on chromosome 7 and the loss of the Cdkn2a locus and loss of chromosome 10 in most tumors (4, 5, 7). EGFR overexpression or amplification is the most common event in GBM occurring in over 65% of all GBM. Many deletions/mutations in the EGF receptor have been reported like the well-characterized EGFRvIII, an EGFR mutant caused by an in-frame deletion of the exons 2–7 (8). This truncated form of EGFR lacks the domains I and II of the ligand-binding portion of the receptor and results in a constitutively active, ligand-independent receptor. Interestingly, around 50% of GBMs that have amplification of the EGFR locus also express the vIII mutant form and they cooperate together to promote tumor growth (9, 10). When over expressed in a genetically engineered mouse model (GEMM), both EGFR WT and vIII, alone or in combination, failed to induce tumor formation (11). This incapacity to initiate tumor is consistent with the fact that other events are required for transformation such as the loss of tumor suppressor gene function. Loss of the Cdkn2 locus, coding for the tumor suppressor genes $p16^{INK4a}$ and $p19^{ARF}$, occurs in 94% of GBM tumors with EGFR amplification/overexpression (5). $p16^{INK4a}$ negatively control the cyclin D1 protein, which feeds into the pRb pathway and essentially represses Rb, and $p19^{ARF}$ is a mediator of MDM2, which is a negative regulator of p53 (12, 13). PTEN, the negative regulator of the PI3K-Akt-mTOR pathway, is another tumor suppressor often loss in a wide variety of tumors including GBM (14).

Numerous mouse models, from xenografts to more complex GEMMs, have been generated to study GBM formation. Xenografts have value in that they consist in injecting tumor cells derived from human GBM directly in the brain of a mouse. However, they lack several considerations in the tumor initiation process including the role of the immune system and the fact that tumors derived from this technology do not arise from a single cell (15, 16). GEMMs give us the opportunity to control specifically, in time and tissue, the combination of genetic modifications and to validate in vivo the relevant data provided by the TCGA consortium about specific GBM genetic mutations.

Interested by the tumor initiation and progression induced by EGFR and the loss of tumor suppressor genes, we have generated a GEMM consisting of four genetic modifications that induce gliomagenesis. First, the mouse expresses a Cre/loxP-conditional system where the human EGFRvIII gene has been introduced in the mouse Collagen 1α1 gene locus (11). Then, firefly luciferase transgene, which is essential to follow tumor formation by BLI, is introduced in the genome under the control of the CRE/lox conditional expression system (17). Mice conditionally expressing EGFRvIII and luciferase have been mated with two different strains of mouse harboring loss of tumor suppressor genes: the previously described INK4a/ARF-null strain (18) and the conditional $PTEN^{2lox}$ strain (19). Following stereotactic injection of an adenovirus expressing CRE recombinase directly in the brain of these animals, this GEM model has demonstrated that overexpression of EGFRvIII cooperates with the loss of tumor suppressor genes to contribute to gliomagenesis and that tumors resulting from this system are highly aggressive and histopathologically similar to human GBM.

2. Materials

2.1. Stereotactic Intracranial Injection of Viruses

1. Mice of the indicated genotype
2. Ketamine-xylazine in 0.9% sodium chloride
3. 1-ml Syringes and 26 3/8 needles
4. Ophthalmic ointment
5. Commercial depilating cream
6. Gauzes
7. Betadine surgical scrub
8. 70% Ethanol
9. Stereotactic frame with nonpuncture ear (Harvard Apparatus Inc.)
10. Scalpel
11. Ideal Microdrill (Cell Point Scientific No. 67-1000 1249)
12. Microdrill Steel Burrs 0.5 mm (Fine Science Tools Cat. No. 19007-05)
13. 1-μl Hamilton syringe
14. Stereotaxic injector QSI Quint essential (Stoelting Co. No. 53311)
15. Adenovirus: Ad5CMVCre (Gene Transfer Vector Core, University of Iowa)
16. Sterile bone wax
17. 4-0 Gut suture
18. Surgical heat pad

2.2. BLI System Imaging

1. Isoflurane
2. Anesthesia chamber
3. Xenogen IVIS 100 imaging system
4. 15 mg/ml Firefly D-luciferin
5. Phosphate-buffered saline (PBS)
6. 1-ml Syringe
7. 263/8 Needles
8. Commercial depilating cream
9. Gauzes
10. Living Image software

2.3. Analysis of GBM

2.3.1. Samples Collection

1. PBS
2. Liquid nitrogen
3. Histology cassette
4. 4% Paraformaldehyde

2.3.2. Immunohistological Staining for GBM Markers EGFR, GFAP, NeuN, and S100

1. Paraffin-fixed tissue
2. Microtome (Micro HM310).
3. Xylenes
4. Alcohol 100, 95, 90, 80, and 70%
5. Microwave
6. Antigen target retrieval solution (Dao, S1699)
7. PBS
8. 0.3% H_2O_2 in methanol
9. Blocking solution (5% goat serum (SIGMA), 0.3% Triton X-100 in PBS)
10. Mouse-on-mouse blocking reagent (Vector Labs Inc., MKB-2213)
11. Primary antibodies: EGFR antibody (mouse mAb anti-EGFR, 31G7, Zymed Lab Inc.), anti-GFAP (Dao, Z0334), anti-S100 (Dao, Z0311), and anti-NeuN (Chemicon, MAB377)
12. Biotinylated secondary antibody (anti-rabbit or anti-mouse from Vector Labs Inc.)
13. Vectastain ABC kit from Vector Labs Inc.
14. Hematoxylin (SIGMA)
15. Coverslip
16. Microscope
17. Camera

3. Methods

3.1. Stereotactic Intracranial Injection of Viruses

1. Anesthetize >3-month-old adult mice EGFR$^{vIII/vIII}$; InkΔ2/3$^{-/-}$; PTEN$^{lox/lox}$ with intraperitoneal injections of 120 mg/kg ketamine and 10 mg/kg xylazine in 0.9% sodium chloride (see Notes 1 and 2).
2. Apply ophthalmic ointment on eyes and shave the scalp using commercial depilating cream.
3. Disinfect the shaved scalp with betadine surgical scrub and then ethanol 70% three times.
4. Place mouse in stereotactic frame.
5. Do a single incision from the anterior pole of the skull to the posterior ridge, skin can be retained using small clamps.
6. Drill a 0.5-mm hole in the skull using a microdrill at the coordinate of choice. For striatum injection, we typically inject 2.1-mm rostral to the bregma, 1.0 lateral to the midline and at 2.5-mm depth to the pia surface (see Notes 3 and 4).
7. Position the 1-μl Hamilton syringe in the hole and inject virus. Hamilton syringe is hooked onto Quint essential Stereotaxic injector and 1 μl of virus is injected at a rate of 0.1 μl/min (see Note 5).
8. After the virus injection, fill the hole with sterile bone wax and suture the skin (see Note 6).
9. Place the mouse on a surgical heated pad and monitor animal multiple times until completely ambulatory (see Note 7).

3.2. BLI System Imaging to Monitor Tumor Growth

1. Assess luciferase activity biweekly, ~2 weeks after virus injection.
2. Inject intraperitoneal 100 μl/10 g of body weight of D-luciferin (15 mg/ml) 5–15 min before imaging.
3. Place the mouse in an anesthesia chamber.
4. When mouse is anesthetized, shave the scalp using commercial depilatory cream.
5. Place the mouse in the Xenogen IVIS system, the nose in a nose cone delivering isoflurane.
6. Acquire image with 60 s of exposure time.
7. Analyze data using Living Image Software: region of interest is selected and BLI is quantified.

3.3. Analysis of GBM

3.3.1. Samples Collection

1. When tumor acquire signal strength of ~10^7 p/s/cm^2/sr, the size is sufficient for treatment and/or to sacrifice animal (Fig. 1) (see Note 8).
2. After cardiac perfusion with PBS of anesthetized animals, use one-half of the brain to collect tumor and to snap freeze tumors

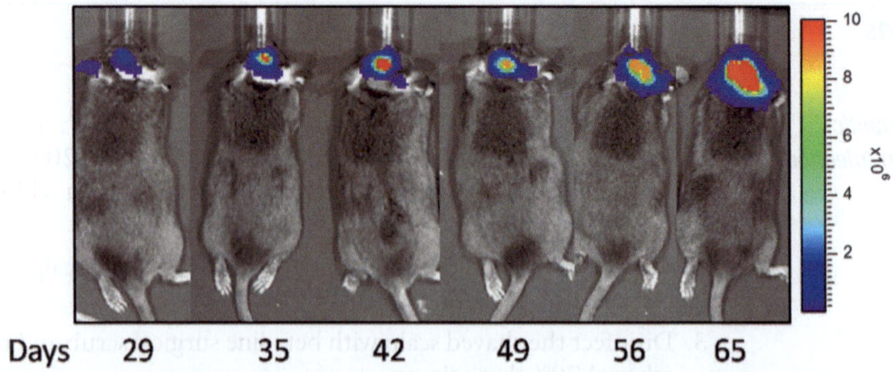

Fig. 1. Monitoring tumor growth using BLI system. Following stereotactic injection of Ad5CMVCre, the tumor growth is monitored by BLI. 10 min after injection of D-luciferin, the mouse is installed in the Xenogen IVIS system, and image is acquired with a 60-s-exposure time. The days post-surgery are indicated under the corresponding animal, and BLI corresponding to a region-of-interest is quantified to obtain a numeric value expressed in p/s/cm^2/sr.

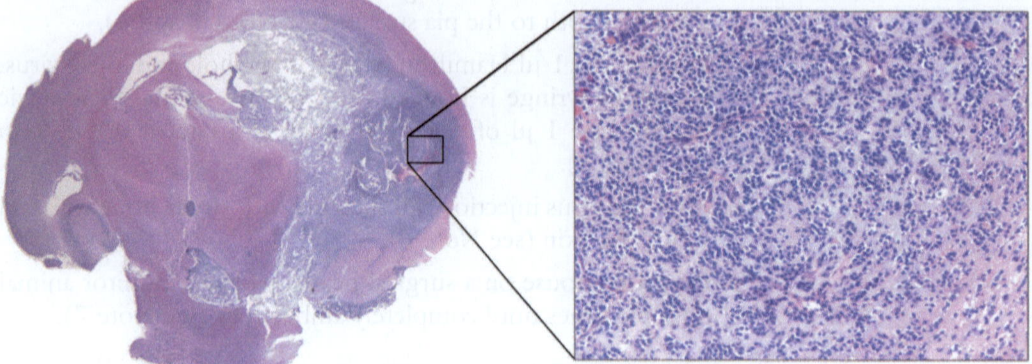

Fig. 2. Hematoxylin and eosin staining of mouse brain and GBM cells. Brain section has been captured with a dissection microscope (*left*). GBM cellular histopathology is show in an enlarged picture (20×) selected from the tumor (*right*).

in liquid nitrogen and then place at −80°C to further analysis of DNA, RNA, and proteins.

3. Fix the other half of the brain in 4% paraformaldehyde overnight at 4°C.
4. Cut the fixed brain in 2-mm sections and place sections in a histology cassette. Keep the cassette with sample in PBS at 4°C till paraffin fixation.
5. Paraffin block and hematoxylin and eosin (H&E) slides staining can be done by a Histology Core. H&E is necessary to visualize histopathological features of GBM (Fig. 2) (see Note 9).

3.3.2. Immunohistochemical Staining for GBM Markers EGFR, GFAP, NeuN, and S100

1. Cut paraffin-fixed tissue slides at 5–10 μM using a microtome (Micro HM310).
2. Melt slides at 55°C for 30 min.
3. Deparaffinize in xylene for 5 min.
4. Rehydrate in alcohol series. Wash twice in alcohol 100 and 95%, then once in 90, 80, 70% for 1 min.
5. Rinse 5 min in tap water.
6. To unmask the antigen, microwave sections in Antigen target retrieval solution (Dao, S1699) for 10 min at low power.
7. Allow cooling for 30 min.
8. Wash in PBS, three times for 5 min each.
9. Incubate sections in 0.3% H_2O_2 in methanol for 30 min.
10. Wash three times with PBS.
11. Incubate slides in blocking solution (5% goat serum (SIGMA), 0.3% Triton X-100 in PBS) for 1 h at room temperature.
12. If the primary antibody is a mouse monoclonal antibody, incubate slides in mouse-on-mouse blocking reagent (Vector Labs Inc., MKB-2213) for 1 h at room temperature.
13. Incubate with primary antibody. Antibodies are diluted in blocking solution and diluted and incubated as follows: EGFR antibody (mouse mAb anti-EGFR, 31 G7, Zymed Lab Inc.) 1:100 for 24 h, anti-GFAP (Dao) diluted 1:500 for 30 min, anti-S100 (Dao Z0311) for 30 min, and anti-NeuN (Chemicon, MAB377) for 24 h.
14. Incubate with biotinylated secondary antibody (anti-rabbit or anti-mouse from Vector Labs Inc.) diluted 1:500 in blocking solution for 1 h.
15. Determine staining with DAB (Vectastain ABC kit from Vector Labs Inc.).
16. Counterstain with hematoxylin for 5 min.
17. Dehydrate slides with alcohol grade (70–100%) and then twice in xylene, 5 min each time.
18. Cover with a coverslip.
19. Take pictures of each slide to analyze staining.

4. Notes

1. All procedures are done according to the recommendations of the Tufts University Institutional Animal Care and Use Committee (IACUC).

2. During procedure, verify frequently the animal depth of anesthesia. If animal react to stimulus, additional dose of ketamine (30 mg/kg of weight) may be injected intraperitoneal.
3. *The mouse brain in stereotaxic coordinates*, Paxinos G. and Franklin K.B.J., 2nd edition, Academic Press is the reference used to set the coordinates of injection.
4. Drilling hole can cause brain damages, what we want to reduce these damages as possible, so less pressure should be put on the drill. Blood and fluid can be absorbed using a paper point before injection.
5. The Cre adenovirus is bought from the Gene Transfer Vector Core, University of Iowa, 500 Newton Road, 221 Eckstein Medical Research Building Iowa City, IA 52242, USA. Tel: (319) 335-6726. The concentration of the virus preparation is 10^{12} viral particles per milliliter.
6. If skin incision is relatively short, tissue adhesive can be used instead of suture.
7. Following surgery, animal can receive analgesic for pain relief by intraperitoneal injection of buprenex at 0.05–0.1 mg/kg of body weight.
8. A signal strength of ~10^7 p/s/cm^2/sr indicates that tumor size is sufficient for treatment and/or to sacrifice animal, but monitoring animal for sickness is essential. Sometimes the animal is still healthy at signal strength of ~10^8 p/s/cm^2/sr. At this point, body condition can decrease rapidly requiring euthanasia and animal should be monitored frequently.
9. Our formaldehyde-fixed brains are processed by the Harvard Histopathology Core (Harvard University, Boston, MA, USA) for paraffin block and H&E slides staining.

Acknowledgment

We thank Hyun Jung Jun for technical advices.

References

1. Wen PY, Kesari S (2008) Malignant gliomas in adults. N Engl J Med 359:492–507
2. Louis DN, Ohgaki H, Wiestler OD, Cavenee WK, Burger PC, Jouvet A, Scheithauer BW, Kleihues P (2007) The 2007 WHO classification of tumours of the central nervous system. Acta Neuropathol 114:97–109
3. Huse JT, Holland EC (2010) Targeting brain cancer: advances in the molecular pathology of malignant glioma and medulloblastoma. Nat Rev Cancer 10:319–331
4. The Cancer Genome Atlas Network (2008) Comprehensive genomic characterization defines human glioblastoma genes and core pathways. Nature 455:1061–1068

5. Verhaak RG, Hoadley KA, Purdom E, Wang V, Qi Y, Wilkerson MD, Miller CR, Ding L, Golub T, Mesirov JP, Alexe G, Lawrence M, O'Kelly M, Tamayo P, Weir BA, Gabriel S, Winckler W, Gupta S, Jakkula L, Feiler HS, Hodgson JG, James CD, Sarkaria JN, Brennan C, Kahn A, Spellman PT, Wilson RK, Speed TP, Gray JW, Meyerson M, Getz G, Perou CM, Hayes DN (2010) Integrated genomic analysis identifies clinically relevant subtypes of glioblastoma characterized by abnormalities in PDGFRA, IDH1, EGFR, and NF1. Cancer Cell 17:98–110
6. Phillips HS, Kharbanda S, Chen R, Forrest WF, Soriano RH, Wu TD, Misra A, Nigro JM, Colman H, Soroceanu L, Williams PM, Modrusan Z, Feuerstein BG, Aldape K (2006) Molecular subclasses of high-grade glioma predict prognosis, delineate a pattern of disease progression, and resemble stages in neurogenesis. Cancer Cell 9:157–173
7. Frederick L, Wang XY, Eley G, James CD (2000) Diversity and frequency of epidermal growth factor receptor mutations in human glioblastomas. Cancer Res 60:1383–1387
8. Wong AJ, Ruppert JM, Bigner SH, Grzeschik CH, Humphrey PA, Bigner DS, Vogelstein B (1992) Structural alterations of the epidermal growth factor receptor gene in human gliomas. Proc Natl Acad Sci U S A 89:2965–2969
9. Inda MM, Bonavia R, Mukasa A, Narita Y, Sah DW, Vandenberg S, Brennan C, Johns TG, Bachoo R, Hadwiger P, Tan P, Depinho RA, Cavenee W, Furnari F (2010) Tumor heterogeneity is an active process maintained by a mutant EGFR-induced cytokine circuit in glioblastoma. Genes Dev 24:1731–1745
10. Huang PH, Xu AM, White FM (2009) Oncogenic EGFR signaling networks in glioma. Sci Signal 2:re6
11. Zhu H, Acquaviva J, Ramachandran P, Boskovitz A, Woolfenden S, Pfannl R, Bronson RT, Chen JW, Weissleder R, Housman DE, Charest A (2009) Oncogenic EGFR signaling cooperates with loss of tumor suppressor gene functions in gliomagenesis. Proc Natl Acad Sci USA 106:2712–2716
12. Serrano M, Hannon GJ, Beach D (1993) A new regulatory motif in cell-cycle control causing specific inhibition of cyclin D/CDK4. Nature 366:704–707
13. Zhang Y, Xiong Y, Yarbrough WG (1998) ARF promotes MDM2 degradation and stabilizes p53: ARF-INK4a locus deletion impairs both the Rb and p53 tumor suppression pathways. Cell 92:725–734
14. Lino MM, Merlo A (2010) PI3Kinase signaling in glioblastoma. J Neurooncol 103(3): 417–427
15. Hambardzumyan D, Parada LF, Holland EC, Charest A (2011) Genetic modeling of gliomas in mice: new tools to tackle old problems. Glia 59(8):1155–1168
16. Gopinathan A, Tuveson DA (2008) The use of GEM models for experimental cancer therapeutics. Dis Model Mech 1:83–86
17. Woolfenden S, Zhu H, Charest A (2009) A Cre/LoxP conditional luciferase reporter transgenic mouse for bioluminescence monitoring of tumorigenesis. Genesis 47:659–666
18. Serrano M, Lee H, Chin L, Cordon-Cardo C, Beach D, DePinho RA (1996) Role of the INK4a locus in tumor suppression and cell mortality. Cell 85:27–37
19. Lesche R, Groszer M, Gao J, Wang Y, Messing A, Sun H, Liu X, Wu H (2002) Cre/loxP-mediated inactivation of the murine Pten tumor suppressor gene. Genesis 32:148–149

Part IV

Therapies in the Treatment of Malignant Glioma: Data from Animal Models

Neuromethods (2013) 77: 229–243
DOI 10.1007/7657_2011_33

Published online: 13 March 2012

Rodent Glioma Models: Intracranial Stereotactic Allografts and Xenografts

Hikmat Assi, Marianela Candolfi, Pedro R. Lowenstein, and Maria G. Castro

Abstract

Modeling human disease in small animals has been fundamental in advancing our scientific knowledge and for the development of novel therapeutic strategies. In the case of brain cancer, implantable tumor models, both intracranial and also in the periphery, have been widely used and extensively characterized. These models can be used to better understand certain aspects of tumor biology such as growth, neovascularization, response to potential therapies, and interaction with the immune system. Brain tumors from patients as well as rodents have been cultured in vitro, in an attempt to establish permanent cell lines. Human glioma tumors have also been maintained by serial passage in the flanks of immune-deficient animals, as it has been shown that it is not feasible to continuously passage them in culture. In this chapter, we describe various gliomas that have been isolated from mice, rats, and humans and subsequently used as syngeneic or xenograft tumor models in vivo. The majority of the models that we present in this chapter arose either spontaneously or by administration of chemical carcinogens. We compare and contrast the histopathological, genetic, and invasive features of the tumor lines as well as identify novel treatment modalities that have been developed. Finally, we present the procedures for intracranial implantation of tumor cells in rodents using stereotactic surgical techniques. The use of this technique enables the generation of large numbers of animals harboring intracranial tumors with relative ease and the survival of tumor-bearing animals is highly reproducible. These characteristics make the use of these in vivo models very attractive when aiming to develop and test the effectiveness of novel anticancer therapies.

Key words: Xenograft, Allograft, Stereotactic, Glioma models, Brain cancer, Neurosurgery in rodents, Tumor implantation

1. Introduction

1.1. Brain Tumor Models

Human gliomas arise from glial cells and although rare, they are extremely malignant. Patients diagnosed with glioblastoma multiforme (GBM), the most common and aggressive form of brain tumors, face a high mortality rate (1). The highly invasive nature of CNS tumors as well as their anatomical location makes complete resection difficult if not impossible. Therefore, the preclinical development of novel therapies requires stringent brain cancer models that replicate the salient features of the human disease. Ideally, small animal models should retain the important features of human gliomas such as growth patterns, neuropathological features, vascularization, and genetic alterations to be considered useful models of brain cancer. These models must also be relatively easy to generate and highly reproducible. While various tumor models have been developed, there are none that perfectly emulate the human disease. Traditionally, tumors of the CNS have been generated by administration of carcinogenic chemicals such as *N*-ethyl-*N*-nitrosourea (ENU) and *N*-methyl-*N*-nitrosourea (MNU) (2–4). Currently, there are two main methods to develop brain tumor models: endogenous and transplantable tumor models. Endogenous brain tumors can be induced by intracranial administration of oncogenic viruses (5–7). Astrocytic brain tumors have been developed after the administration of vectors encoding Ras and AKT (8), EGFR or PDGF-B (9) into the brain of mice that are deficient in PTEN, Ink4a/Arf knockouts, or p53 (10, 11). Administration of retroviral vectors expressing PDGF-B also generates gliomas in rats (5). In all of these models, tumors appear within 3–12 months posttumor induction. These endogenous models exhibit the histopathological features and contain the genetic alterations of human GBM, constituting unique tools to study gliomagenesis. However, the variable latency and the fact that not all the animals develop tumors remain obstacles to assess the efficacy of novel therapeutic approaches using these endogenous models. Recently, a novel approach to induce endogenous GBM that uses the *Sleeping Beauty* (SB) transposable element has been described (12). This technique allows the integration of several oncogenes in cells of the mouse brain (13). Spontaneous brain tumors displaying the histological characteristics of human GBM were induced by injecting a DNA plasmid encoding SB-transposase in combination with plasmids containing the oncogenes AKT, NRAS, or EGRFvIII flanked by transposase recognition elements into the brain of neonatal mice (13). As a result, the mice developed tumors in roughly 3 weeks that mimic the genetic heterogeneity found in patients and structurally resembled human astrocytomas or glioblastomas.

Since transplantable rodent GBM models are technically easy to develop and very reproducible, they constitute a unique tool to evaluate novel therapies in vivo. In this protocol, we present several mouse, rat, and human GBM tumors that scientists have cultured in vitro, in an attempt to establish cell lines which have then been used to generate transplantable orthotopic glioma models (Figs. 1–3). Human GBM xenografts implanted in the brain of nude mice allow studying the response of human GBM cells to antiglioma approaches, such as cytotoxic and antiangiogenic agents in vivo in the context of the brain (14–18) and retain many of the genetic alterations present in the original specimen (19–21).

a

Glioma Model	Cell Line	Strain	Origin	Type	Histology	Genetic Alternations	Expression of CNS Markers	Invasive	Immuno-therapeutics
Mouse Allograft	Gl26	C57BL/6	MCA	GBM	Necrosis, atypical nuclei, hemorrhage, microvasculature proliferation [22]	*Undefined*	GFAP$^-$, Vimentin$^+$ [22]	+++ [22]	Ad.TK/Flt3L [28] MSCs expressing IL-12 [29]
	SMA 560	VM Dk (H-2b)	Sponta-neous [30]	Astro-cytoma	Nuclear atypia, foci of necrosis & vascularization [31]	↑TGFβ	GFAP$^+$, S100$^+$, Glutamine Synthase$^+$ [30]	++	Ad.TK/Flt3L [23] Anti TGFβ [32], sCD70 [33]
	B16-F10	C57BL/6 (H-2b)	Sponta-neous [34]	Melano-ma	Metastatic to lung, kidney & brain. Atypical nuclei & pigmented glandular ducts [35]	*Undefined*	*Undefined*	+++ [36]	Ad.TK/Flt3L [28] rIL-2, rINFalpha2b [37] Anti-CTLA-4 Ab [38]
	Gl261	C57BL/6	MCA [39]	GBM	Pleomorphic cells w/ atypical nuclei including necrosis and psuedopalisades [40]	↑c-Myc, ↑kRas [41]	GFAP$^+$, S100$^+$ [40]	++ [42]	Ad.TK/Flt3L [28] GM-CSF [43], HSV.TK [44] rIL-12 [45]

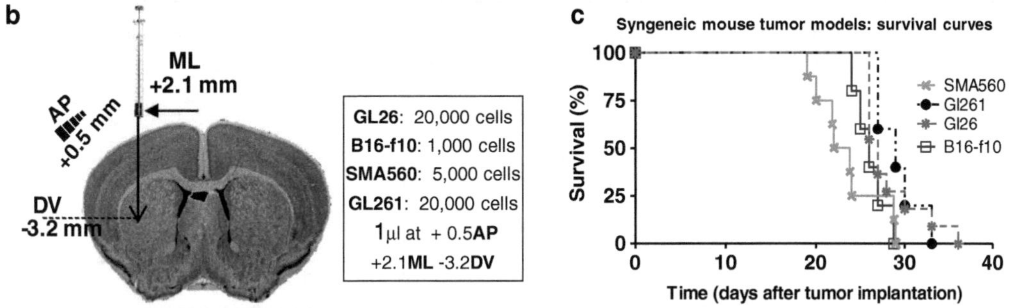

Fig. 1. (**a**) Table with information relating to the biological features of syngeneic mouse GBM models including effective therapies. (**b**) Diagram depicting the stereotactic coordinates for injection of syngeneic mouse glioma cell lines as well as the cell number. (**c**) Kaplan–Meyer survival curve of mice bearing syngeneic intracranial brain tumors. C57BL/6 or VM/Dk mice implanted in the striatum with Gl26, SMA560, B16-f10, or GL261 cells.

However, due to the lack of an intact immune system, this model is not suitable to assess antitumor immune responses and tumor–host interactions. On the other hand, although syngeneic tumor models are useful tools to test therapeutic approaches that stimulate the host immune response against the tumor, their anatomical pattern of invasion and genetic profile differs from human GBM. Thus, it is advisable to use several models to corroborate preclinical findings (22–24).

Stereotactic surgery is a well-established technique for targeting precise anatomical structures within the brain. This technique can be used to implant glioma cells in the brain and quickly generate tumor-bearing animals to study various aspects of neuro-oncology, such as angiogenesis, invasiveness, tumor-induced immune suppression, and efficacy of novel therapeutics. This protocol will describe how to perform intracranial implantation of cultured tumor cells in mice and rats using a stereotactic apparatus as well as the steps necessary for maintaining primary human tumor cultures.

a

Glioma Model	Cell Line	Strain	Origin	Type	Histology	Genetic Alternations	Expression of CNS Markers	Invasive	Immuno-therapeutics
Mouse *Xenograft*	U87	Human	N/A	GBM-Astro-ccytomas	Demarcated tumor with occasional foci of necrosis & vascularization (no pseudo-plasides) dissimilar to human GBM [22,46,47]	*PTEN, Ink4a, kRas (WT P53) [46, 48]*	GFAP$^-$, S100$^-$ Vimentin$^+$ [22,46]	- [46]	anti-RPTPβ [49]
	U251	Human	N/A	GBM	Necrosis, atypical nuclei, hemorrhage, microvasculature proliferation [22,46]	*P53, PTEN, p14ARF, p16, kRas,* ↑EGFR, ↑IL-13Rα2 *[48, 50]*	GFAP$^+$, S100$^+$ Vimentin$^+$ [22,46]	+++ [46]	Ad.hIL-13-PE toxin [27], γδT cells [51]
	GBM 12	Human	N/A	GBM	Necrosis, atypical nuclei, hemorrhage [19]	*P53,* ↑EGFR, ↑IL-13Rα2 *[19,52]*	*Undefined*	+++ [19]	Ad.hIL-13-PE toxin [27], IL-13 targeting Oncolytic virus [52]

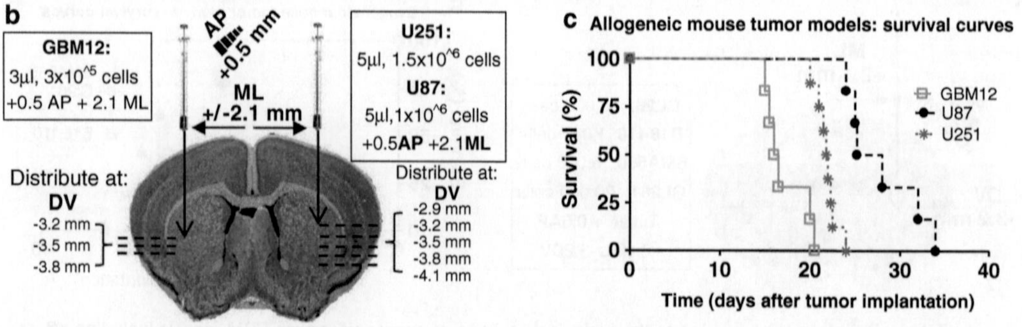

Fig. 2. (**a**) Table listing the biological features of human xenograft GBM models including effective therapies. (**b**) Diagram depicting the stereotactic coordinates for injection of human glioma cell lines as well as the cell number. (**c**) Kaplan–Meyer survival curve of mice bearing syngeneic intracranial brain tumors. Rag1-deficient mice were injected in the striatum with U87, U251, and GBM12 cells.

2. Materials

1. Glioma cell line of interest resuspended PBS or serum-free media (injectable volume ~1–5 µl). Cells can be kept for 2–3 h on ice during surgery.
2. Adult rat (250 g body weight. Fisher or Lewis Strain. Charles River Inc., http://www.criver.com).
3. Adult female mouse (20–25 g weight. 6–8 weeks old C57Bl/6, VM/Dk, Rag1$^{-/-}$ strain, Athymic Balb/c. Jackson Labs Inc., http://www.jax.org).
4. Drugs:
   (a) Ketamine (Bioniche Teoranta, http://www.bionichepharmausa.com)
   (b) Dexmedetomidine (Orion Corp., http://www.orion.fi)

a

Glioma Model	Cell Line	Strain	Origin	Type	Histology	Genetic Alternations	Expression of CNS Markers	Invasive	Immuno - therapeutics
Rat Allograft	CNS-1	Lewis	MNU [53]	Astrocytoma	Microvasculature proliferation, pseudopalisades, immune infiltrates [54]	*Undefined*	$GFAP^+$, $S100^{+/-}$, $Vimentin^+$, $NCAM^+$, RAR-α^+ [24,53,54]	++ [53]	Ad.TK/Flt3L [55, 56]
	F98	Fisher	ENU [57]	Anaplastic Glioma	Pleomorphic nuclei, areas of necrosis and vascular proliferation [54]	*Ink4a,* ↑RbRas, ↑PDGFβ, ↑EGFR, ↑Cyclin D1 & D2 [58]	$GFAP^+$, $Vimentin^+$ [58] [59]	+++ [60]	Ad.TK/Flt3L [24], LAK cells [61]
	9L	Fisher	MNU [62]	Gliosarcoma	Areas of hypoxia and necrosis, nuclear pleomorphism [60]	*P53,* ↑EGFR, ↑TGF [58]	$GFAP^-$, $S100^+$, $Vimentin^-$, $NCAM^+$, RAR-α^+ [58]	- [53, 63]	Ad.TK/Flt3L [24]
	RG2	Fisher	ENU [64]	GBM	Necrosis, atypical nuclei, hemorrhage [60]	*Ink4a,* ↑PDGF, ↑IGF-1, ↑Ras, ↑Her3 [58]	$GFAP^{+/-}$, $Vimentin^+$ [58]	+++ [65]	Refractory, INFγ [66]

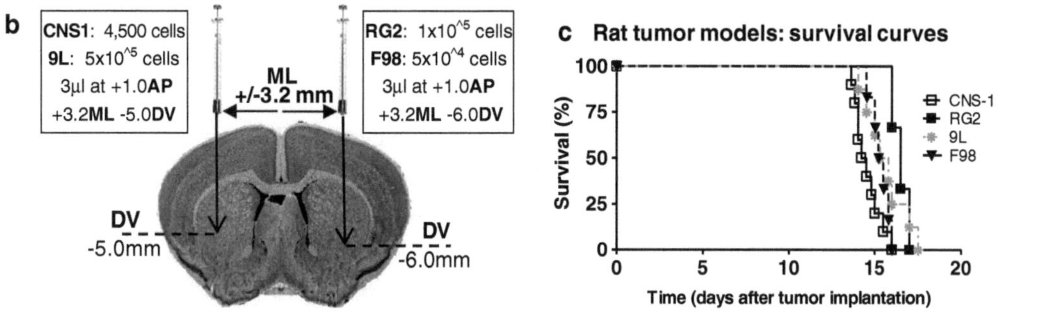

Fig. 3. (**a**) Table with information about the biological features of syngeneic rat GBM models including effective therapies. (**b**) Diagram depicting the stereotactic coordinates for injection of human glioma cell lines as well as the cell number. (**c**) Kaplan–Meyer survival curve of rats bearing intracranial tumors. Lewis or Fisher rats were injected in the striatum with CNS-1, F98, 9L, and RG2 cells.

(c) Carprofen (Hospira Inc., http://www.hospira.com)

(d) Xylazine (Animal Health International, http://www.pvpl.com)

(e) Atipamezole (Orion Corp., http://www.orion.fi)

(f) Buprenorphine (Animal Health International, http://www.pvpl.com)

5. Betadine (Bruce Medical, cat# FR-2100-88).
6. Puralube: Ophthalmic ointment (Amazon Inc., http://www.amazon.com).
7. Warm Lactated Ringer's solution (Hospira Inc., cat# 0409-7953-30).
8. Tissue culture:

(a) Accutase™ Enzyme Cell Detachment Medium (eBiosciences Inc., cat# 00-4555-56)

(b) Dulbecco's PBS, pH 7.4, without Ca or Mg (MediaTech Inc., cat# 21-040-CM)
(c) Dulbecco's modified Eagles media (DMEM), supplemented with 10% fetal bovine serum (MediaTech Inc., cat# 10-017-CM and Omega Scientific, cat# FB-01)
(d) T75 Flasks (VWR, cat# 82050-856)
(e) Matrigel™ (BD Pharmingen, cat# 356231)
(f) Hemocytometer (VWR, cat# 15170-263)

9. Stereotactic frame with rat and mouse adapters and blunt ear bars (Stoelting Inc.).
10. Stereomicroscope (e.g., Zeiss Stemi 1,000 zoom) equipped with 16× eyepieces and 0.4× auxiliary objective lens, and mounted on hinged coupling arm on a heavy foot stand (or equivalent).
11. Electric drill with 1.75-mm and 0.6-mm drill bit (Stoelting Inc., cat# 58610).
12. Digital scale for animal weight (Harvard Apparatus, cat# 724586).
13. Fiber-optic illuminator with twin goose-neck pipes (Leica Inc.).
14. 10-μl, 26-G Hamilton syringe with needle (Fisher, cat# 701RN).
15. 5–10 μl, 33-G Hamilton syringe with needle (Fisher, cat# 75RN).
16. 3-0 Nylon sutures (Ethicon Inc., cat# 1663H).
17. Petri dish, plastic (BD Pharmingen, 353002).
18. Surgical equipment:
    (a) Surgical shavers (Stoelting Inc., cat# 58610)
    (b) Scalpel and blades (Cardinal Health, cat# D2862-15)
    (c) Skin retractors (Fine Science Tools Inc., mouse cat# 17000-03, rat cat# 17000-04)
    (d) Cotton swabs (VWR, cat# 89031-270)
    (e) Curved and straight forceps
    (f) Holding scissors
    (g) Sharp scissors
    (h) Sterile gauze (VWR, cat# 95038-720)
    (i) Bead sterilizer (VWR, cat# IS-350)
    (j) Red lamp (Sylvania, cat# 14663)
    (k) Optional: infant incubator for recovery of nude mice

3. Methods

3.1. Glioma Cell Lines: Maintenance and Passage

Most of the tumor cell lines presented in the chapter (Figs. 1–3) can be passaged indefinitely in tissue culture and cryopreserved for long-term storage with the exception of GBM12. While the genetic features of the original human GBM specimen are maintained with serial tumor passage in vivo (19, 25), GBM12 cell passage in vitro for long periods of time results in loss of EGFR amplification and gain of MGMT promoter methylation (26). Thus, this human tumor must be serially transplanted in the flanks of immune-deficient mice, such as $Rag1^{-/-}$ mice or athymic nude mice. Short-term GBM12 cultures can be derived from flank tumor in order to perform in vitro studies or to establish intracranial tumors (26). To achieve highly reproducible intracranial GBM12 xenografts, it is necessary to implant the same number of cells in the brain of all the animals. Short-term GBM12 cells growing in a monolayers can be harvested and counted for intracranial implantation (see below) (19, 26, 27).

1. Swab the skin with Betadine to minimize the risk of infection and excise the tumor from the flank of the mouse using a sterile scalpel and place in a Petri dish with 4 ml of DMEM.
2. Mince the tumor into small 1–2 mm pieces using razor blade until tissue can pass through a 1 cc syringe. Draw up and expel the tissue through the syringe to break up the tumor into even smaller pieces.
3. At this point, the cells can be cultured in vitro or injected into the flank of new RAG1-deficient mice for in vivo propagation. Short-term cultures can be used for in vitro experiments or in preparation for intracranial implantation. Minced tissue preparations are difficult to count and inject accurately, therefore, they must be cultured in vitro first. GBM12 cultures can be harvested using trypsin and cells can be counted and resuspended in the concentration required to yield accurate survival curves when implanted in the brain (19, 26, 27).
4. To culture the cells, collect and seed in Matrigel™-coated T75 flasks with DMEM containing 10% fetal calf serum. Cells grow in monolayer and can be passaged for up to 1 month before injecting them into the brain. Their ability to grow in vivo decays and the genetic alterations present in the original human GBM specimen are lost when maintained in culture for longer periods of time.
5. Following short-term culture (2–3 weeks), the cells are harvested with Accutase™ Detachment Medium. To prepare cells for

injection, obtain a count using a hemocytometer and resuspended in DMEM (serum free) at 1×10^5 cells/μl (3 μl/mouse).

6. To propagate the tumor, transfer the minced tissue into a conical tube and allow the material to settle. Remove the excess media and add Matrigel™ to the wet tissue in a 1:1 ratio.
7. 200 μl of this mixture is injected into the flank of 2–3 immune-deficient mice using 1 cc syringes with 21 G needle. Once the tumors reach 1–2 cm in diameter, usually 1–2 months after tumor implantation, the animal is sacrificed and the flank tumor is excised to be re-implanted in the flank or cultured in vitro.

3.2. Glioma Models: Intracranial Grafts in Rodents

3.2.1. Animals and Surgical Preparation

1. Prior to performing any procedures involving an animal, the operator should ensure that the surgical area is clean, well organized, and contains all required instruments. Tools can be sterilized by placing them in a flash bead sterilizer for a short time. This sterilization technique is sufficient for rodent surgery but not for larger animals.
2. Lay out all the surgical instruments on an absorbent under pad in the order in which they will be used. Inspect the stereotactic frame to ensure that it is in proper operating condition. Ensure that all the manipulator arms move freely and smoothly with little to no sideways movement in the syringe holder. Direct the light beam at the ear bars and position the microscope such that the surgical area is visible through the eyepiece.
3. Place the mouse (female 6–8 weeks) under anesthesia by intraperitoneal (IP) injection of ketamine (75 mg/kg) and Dexmedetomidine (0.5 mg/kg). Before commencing surgery, administer Carprofen subcutaneously (5 mg/kg) as an analgesic and ensure that the animal is completely sedated by pinching the footpad or tail.
   - Rat Anesthesia dosage is as follows: Ketamine (75 mg/kg IP), Dexmedetomidine (0.25 mg/kg IP), and Carprofen (5 mg/kg SQ)
4. Using surgical shavers remove enough fur from the head of the animal to allow for aseptic procedures.
5. Carefully mount the animal onto the stereotactic frame and use the incisor bar to loosely immobilize the skull. Raise the animal to the level of the ears bars. Proceed by supporting the head of the animal and slide one of the ear bars into the ear canal and tighten in place. While keeping the head supported, slide the other ear bar into the proper position. The ears bars are meant to prevent any mediolateral head movement while maintaining the dorsoventral axis rotation free (see Note 2).

6. Use blunt forceps to open the mouth of the animal and insert the incisor bar. Press down gently to ensure the incisors are well seated. Swing the nose clamp over the nose of the animal and tighten gently. Ensure that there is no disruption in respiration. (see Notes 3 and 4).

3.2.2. Stereotactic Injection of Tumor Cells

7. Disinfect the incision area using alcohol wipes followed by Betadine. Apply ophthalmic lubricant to each of the eyes so they remain moist during surgery (see Note 4).
8. Using size 15 scalpels make a midline incision along the cranium roughly 1–2 cm in length beginning slightly posterior to the eyes and ending at the ears. Use the skin retractor to hold back the skin on both sides of the incision. The incision length for a rat is longer than a mouse as is the skin retractor.
9. Direct the light beams onto the exposed skull and focus the microscope on bregma; the junction of the sagittal and transverse sutures which should be visible under direct light. Use the blunt end of a scalpel and gently depress the parietal or frontal skull bone to assist in finding the junction (bregma) of the two bones (Fig. 4).
10. Position the syringe using the manipulator arms so that the needle tip is directly over bregma. Take the anteroposterior (AP) and mediolateral (ML) coordinates of bregma as these will serve as our starting coordinates.
11. Determine the coordinates of the injection site by adding or subtracting the appropriate lateral and anterior/posterior values. To inject into the striatum of a mouse move +0.5 mm AP then +2.2 mm ML from bregma. Coordinates for the rat striatum are +1.0 mm AP, +3.2 mm ML (see Note 5).

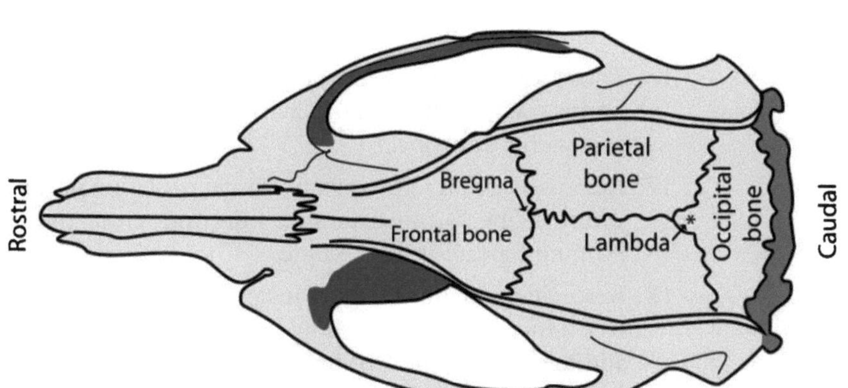

Fig. 4. Schematic of a rat skull depicting the position of bregma relative to the frontal and parietal skull bones and the position for placing of the ear bars. Adapted from: The Rat Brain in Stereotaxic Coordinates, Academic Press, New York, 1998.

12. Visualize the injection site through eyepiece and use the drill to itch a small mark in the skull where the needle will penetrate. This mark is used as a guide for the operator to drill in the correct position. The needle can be lifted so that it does not obstruct the area.

13. With caution, drill a small burr hole with a 0.6-mm bit for mice and 1.75-mm bit for rats. This must be performed with extreme precaution as to avoid severing any blood vessels. Use a bent 28 G needle to ensure penetration of the dura mater. The rat skull is thicker and therefore more difficult to drill. For this purpose, use a 1.75-mm drill bit and drill in short bursts to avoid generating heat. Also, wet the area frequently with fresh saline solution, to avoid excessive heat (see Notes 8 and 9).

14. Load the Hamilton syringe (33 G needle for mice, 26 G for rats) with 1–5 μl of cells of interest and expel a small amount onto a kimwipe to ensure that the syringe was properly loaded. Lower the needle such that it is level with the bur hole. Again read the dorsoventral coordinates and calculate the desired coordinate of the injection site. For injecting syngeneic mouse cells in the striatum slowly, move the needle down 3.5 mm and leave the needle in place for 1 min, then back up 0.3 mm to allow for a small pocket in which the cells can rest (−3.2 mm DV final). These coordinates will vary depending on the animal and cell line used. We have included the injection coordinates used by our laboratory to generate these glioma models in Figs. 1–3 and Notes 1 and 10.

15. Administer the injection slowly over the course of 3–5 min (~0.5–1 μl/min). Keep the needle in place for an additional minute to allow tumor cells to settle before slowly withdrawing from the brain. Rat brains can accommodate a larger volume; therefore, we typically deliver a 3 μl injection over the course of 3 min (~1 μl/min) (see Notes 6 and 7).

16. To remove debris or tumor cells that might have escaped the tumor tract, flush the skull cavity with sterile saline and dry the area with a kimwipe.

17. Remove the skin retractor and close the incision using 3-0 nylon suture. If performing another injection, clean the syringe by flushing repeatedly with saline.

18. Resuscitate the mouse by IP injection of atipamazole (1 mg/kg IP). Once the mouse shows signs of voluntary movement, administer buprinex (0.05–0.1 mg/kg SQ) as an analgesic. Monitor the mouse until it fully recovers from anesthesia and return them to their original cage. A red light can be used as a heating source to aid in recovery as well as IP injection of warm ringer's solution. When working with nude mice, recovery is improved when performed in an infant incubator.

19. Provide the animals with water-soaked chow in a Petri dish and continue to monitor the animal for the following 7 days for any surgical complications. Remove any remaining sutures by day 14.

4. Notes

1. Maintaining the cell preparation on ice is very important for viability. The viability of the cell preparations decreases quickly over time; therefore, we recommend not using cell preps that have been on ice for more than 3 h. Reproducibility can be assessed by examining the time to mortality on Kaplan–Meyer survival curve (Figs. 1–3).
2. To generate reproducible tumor models, it is critical that the animal is immobilized properly in the stereotactic frame. Correct placement of the ear bars is challenging but is crucial for proper alignment of the skull. The ear bars must be inserted slowly and carefully into the ear canals of the animal without causing injury.
3. The operator should ensure that the dorsoventral plane of the cranium is as flat as possible. This can be adjusted by changing the vertical position of the incisor bar.
4. Mounting of the animal onto the frame should be performed with extreme care so that the position of the animal does cause any respiratory issues. The breathing rate of the animal should be monitored during the entire procedure.
5. The stereotactic frame used is adequate for adult rats of 250 g body weight. Fitting larger animals into the frame may be difficult and the coordinates will also be inaccurate. The operator is advised to refer to a mouse or rat brain atlas for finding the coordinates of the injection site of interest (67).
6. The Hamilton syringes used to inject the cells must be clean and in good condition. The syringe must be clear of debris or blockage. The plunger should move smoothly up and down the barrel with little resistance. Hamilton syringe cleaner can be used to remove clogs. The syringe must be cleaned thoroughly with saline solution between each injection to avoid obstruction.
7. It is imperative that the injection bolus be administered over the course of several minutes to allow for the cells settle properly and not extravasate from the injection tract into the corpus callosum and external capsule.
8. The light source generates a significant amount of heat and should only be kept on when needed. The drill also generates a

lot of heat. This is more of an issue for the rats since their skull is thicker and difficult to drill. It is recommended to drill in short bursts and not in a single step. Saline can be used to cool the area down before continuing. Any blood or CSF can be removed with cotton applicators or kimwipes. During the drilling step, the operator will have to refocus the microscope to clearly visualize the surgical area.

9. The brain is well vascularized and the operator is likely to encounter blood vessels during the drill step. We recommended drilling slowly and in the vicinity around the blood vessel to avoid rupturing the vessel directly. Doing so will slightly alter the coordinates of the injection.

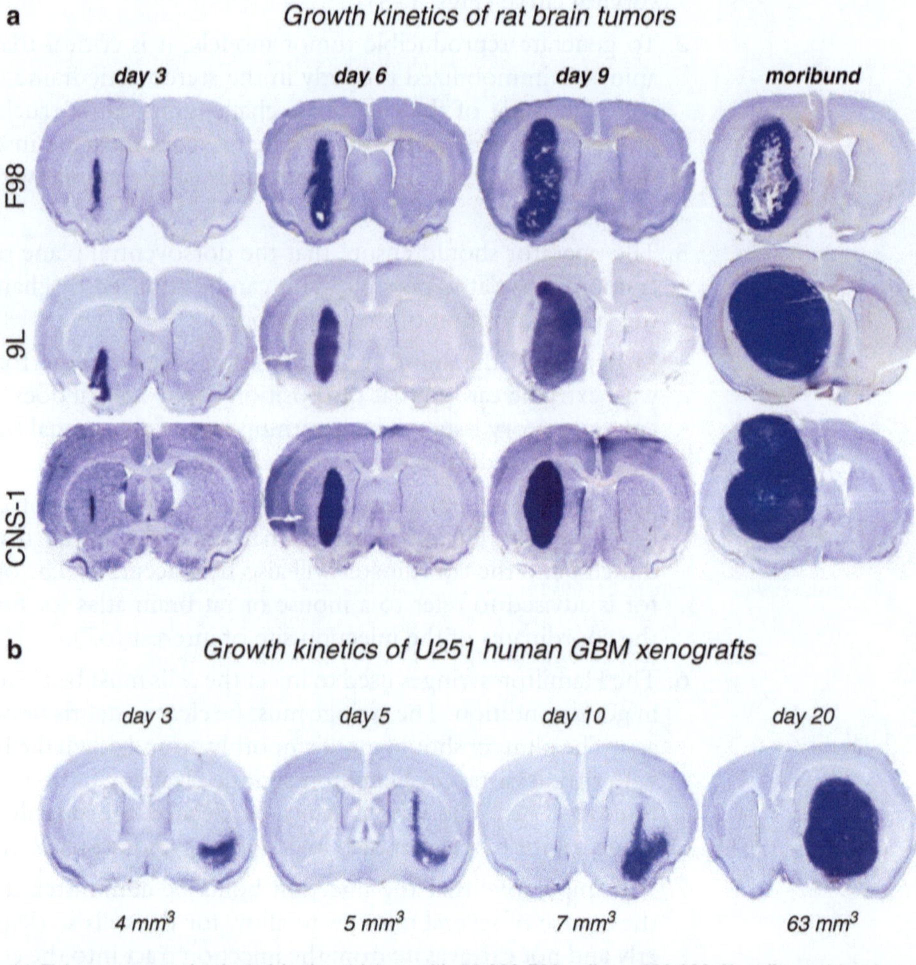

Fig. 5. (**a**) Fisher rats were implanted in the striatum with 50,000 F98 cells or 500,000 9L cells, and Lewis rats were injected with 4,500 CNS-1 cells. Brains were harvested at the indicated time points. Microphotographs show the appearance of representative brain tumor sections stained with Nissl. (**b**) Nude mice were implanted with 1×10^6 U251 human glioma cells. Animals were sacrificed at the specified time points. Microphotographs show the appearance of representative brain tumor sections stained with Nissl.

10. Each of the cell lines grow in a distinctive pattern in vivo. This explains why some cell lines are injected at different coordinates. Typically, we create a pocket for the tumor cells to settle during the injection. To achieve the aforementioned pocket, we lower the needle 0.3 mm deeper than where we wish to inject. Then we lift the needle (0.3 mm) and administer the injection. For example, if injecting CNS-1 cells, the needle is lowered to −5.3 mm DV and then lifted to −5.0 mm where the cells will be administered. This step is typically done for all the tumor models and yields more consistent tumor profiles.
11. To prevent the growth of extracranial tumors, it is imperative to flush the skull with saline to remove any cells that might have traveled up the needle tract. Tumors should grow and remain encapsulated in the striatum as seen in Fig. 5.

References

1. Burger PC, Green SB (1987) Patient age, histologic features, and length of survival in patients with glioblastoma multiforme. Cancer 59:1617–1625
2. Bilzer T, Reifenberger G, Wechsler W (1989) Chemical induction of brain tumors in rats by nitrosoureas: molecular biology and neuropathology. Neurotoxicol Teratol 11:551–556
3. Bulnes-Sesma S, Ullibarri-Ortiz de Zarate N, Lafuente-Sanchez JV (2006) Tumour induction by ethylnitrosourea in the central nervous system. Rev Neurol 43:733–738
4. Bosch DA (1977) Short and long term effects of methyl- and ethylnitrosourea (MNU & ENU) on the developing nervous system of the rat. I. Long term effects: the induction of (multiple) gliomas. Acta Neurol Scand 55:85–105
5. Assanah M, Lochhead R, Ogden A et al (2006) Glial progenitors in adult white matter are driven to form malignant gliomas by platelet-derived growth factor-expressing retroviruses. J Neurosci 26:6781–6790
6. Copeland DD, Talley FA, Bigner DD (1976) The fine structure of intracranial neoplasms induced by the inoculation of avian sarcoma virus in neonatal and adult rats. Am J Pathol 83:149–176
7. Steinbok P, Mahaley MS, U R et al (1979) Synergism between BCNU and irradiation in the treatment of anaplastic gliomas. An in vivo study using the avian sarcoma virus-induced glioma model. J Neurosurg 51:581–586
8. Holland EC, Varmus HE (1998) Basic fibroblast growth factor induces cell migration and proliferation after glia-specific gene transfer in mice. Proc Natl Acad Sci U S A 95:1218–1223
9. Dai C, Celestino JC, Okada Y et al (2001) PDGF autocrine stimulation dedifferentiates cultured astrocytes and induces oligodendrogliomas and oligoastrocytomas from neural progenitors and astrocytes in vivo. Genes Dev 15:1913–1925
10. Hu X, Pandolfi PP, Li Y et al (2005) mTOR promotes survival and astrocytic characteristics induced by Pten/AKT signaling in glioblastoma. Neoplasia 7:356–368
11. Uhrbom L, Hesselager G, Nister M et al (1998) Induction of brain tumors in mice using a recombinant platelet-derived growth factor B-chain retrovirus. Cancer Res 58:5275–5279
12. Ivics Z, Hackett PB, Plasterk RH et al (1997) Molecular reconstruction of Sleeping Beauty, a Tc1-like transposon from fish, and its transposition in human cells. Cell 91:501–510
13. Wiesner SM, Decker SA, Larson JD et al (2009) De novo induction of genetically engineered brain tumors in mice using plasmid DNA. Cancer Res 69:431–439
14. de Bouard S, Herlin P, Christensen JG et al (2007) Antiangiogenic and anti-invasive effects of sunitinib on experimental human glioblastoma. Neuro Oncol 9:412–423
15. Dinca EB, Lu KV, Sarkaria JN et al (2008) p53 Small-molecule inhibitor enhances temozolomide cytotoxic activity against intracranial glioblastoma xenografts. Cancer Res 68:10034–10039
16. Harding TC, Lalani AS, Roberts BN et al (2006) AAV serotype 8-mediated gene delivery

of a soluble VEGF receptor to the CNS for the treatment of glioblastoma. Mol Ther 13:956–966
17. Kitange GJ, Carlson BL, Schroeder MA et al (2009) Induction of MGMT expression is associated with temozolomide resistance in glioblastoma xenografts. Neuro Oncol 11:281–291
18. Kitange GJ, Carlson BL, Mladek AC et al (2009) Evaluation of MGMT promoter methylation status and correlation with temozolomide response in orthotopic glioblastoma xenograft model. J Neurooncol 92:23–31
19. Giannini C, Sarkaria JN, Saito A et al (2005) Patient tumor EGFR and PDGFRA gene amplifications retained in an invasive intracranial xenograft model of glioblastoma multiforme. Neuro Oncol 7:164–176
20. Sarkaria JN, Yang L, Grogan PT et al (2007) Identification of molecular characteristics correlated with glioblastoma sensitivity to EGFR kinase inhibition through use of an intracranial xenograft test panel. Mol Cancer Ther 6:1167–1174
21. Xie Q, Thompson R, Hardy K et al (2008) A highly invasive human glioblastoma pre-clinical model for testing therapeutics. J Transl Med 6:77
22. Candolfi M, Curtin JF, Nichols WS et al (2007) Intracranial glioblastoma models in preclinical neuro-oncology: neuropathological characterization and tumor progression. J Neurooncol 85:133–148
23. Candolfi M, Curtin JF, Yagiz K et al (2011) B cells are critical to T-cell mediated anti-tumor immunity induced by a combined immune-stimulatory/conditionally cytotoxic therapy for glioblastoma. Neoplasia 13(10):947–960
24. Ghulam Muhammad AK, Candolfi M, King GD et al (2009) Antiglioma immunological memory in response to conditional cytotoxic/immune-stimulatory gene therapy: humoral and cellular immunity lead to tumor regression. Clin Cancer Res 15:6113–6127
25. Carlson BL, Grogan PT, Mladek AC et al (2009) Radiosensitizing effects of temozolomide observed in vivo only in a subset of O6-methylguanine-DNA methyltransferase methylated glioblastoma multiforme xenografts. Int J Radiat Oncol Biol Phys 75:212–219
26. Carlson BL, Pokorny JL, Schroeder MA, et al (2011) Establishment, maintenance and in vitro and in vivo applications of primary human glioblastoma multiforme (GBM) xenograft models for translational biology studies and drug discovery. Current protocols in pharmacology/editorial board, S.J. Enna Chapter 14, Unit 14 16-23
27. Candolfi M, Xiong W, Yagiz K et al (2010) Gene therapy-mediated delivery of targeted cytotoxins for glioma therapeutics. Proc Natl Acad Sci U S A 107:20021–20026
28. Curtin JF, Liu N, Candolfi M et al (2009) HMGB1 mediates endogenous TLR2 activation and brain tumor regression. PLoS Med 6:e10
29. Ryu CH, Park SH, Park SA et al (2011) Gene therapy of intracranial glioma using interleukin 12-secreting human umbilical cord blood-derived mesenchymal stem cells. Hum Gene Ther 22:733–743
30. Serano RD, Pegram CN, Bigner DD (1980) Tumorigenic cell culture lines from a spontaneous VM/Dk murine astrocytoma (SMA). Acta Neuropathol 51:53–64
31. Sampson JH, Ashley DM, Archer GE et al (1997) Characterization of a spontaneous murine astrocytoma and abrogation of its tumorigenicity by cytokine secretion. Neurosurgery 41:1365–1372, discussion 1372–1373
32. Tran T-T, Uhl M, Ma JY et al (2007) Inhibiting TGF-beta signaling restores immune surveillance in the SMA-560 glioma model. Neuro Oncol 9:259–270
33. Miller J, Eisele G, Tabatabai G et al (2010) Soluble CD70: a novel immunotherapeutic agent for experimental glioblastoma. J Neurosurg 113:280–285
34. Briles EB, Kornfeld S (1978) Isolation and metastatic properties of detachment variants of B16 melanoma cells. J Natl Cancer Inst 60:1217–1222
35. Wosko TJ, Ferrara DT, Sartori LS (1984) Histological comparison of the B16 melanoma and its F1 variant. Cancer Lett 24:57–63
36. Adatia R, Albini A, Carlone S et al (1997) Suppression of invasive behavior of melanoma cells by stable expression of anti-sense perlecan cDNA. Ann Oncol 8:1257–1261
37. Kirkwood JM, Tarhini AA, Panelli MC et al (2008) Next generation of immunotherapy for melanoma. J Clin Oncol 26:3445–3455
38. Sarnaik AA, Weber JS (2009) Recent advances using anti-CTLA-4 for the treatment of melanoma. Cancer J 15:169–173
39. Seligman A, Shear MJ (1939) Experimental production of brain tumors in mice with methylcholanthrene. Am J Cancer 37:364–395
40. Newcomb EW, Zagzag D (2009) The murine GL261 glioma experimental model to assess novel brain tumor treatments. CNS Cancer 1:227–241
41. Szatmari T, Lumniczky K, Desaknai S et al (2006) Detailed characterization of the mouse glioma 261 tumor model for experimental glioblastoma therapy. Cancer Sci 97:546–553
42. Zagzag D, Miller DC, Chiriboga L et al (2003) Green fluorescent protein immunohistochemistry as a novel experimental tool for the

detection of glioma cell invasion in vivo. Brain Pathol 13:34–37

43. Yu JS, Burwick JA, Dranoff G et al (1997) Gene therapy for metastatic brain tumors by vaccination with granulocyte-macrophage colony-stimulating factor-transduced tumor cells. Hum Gene Ther 8:1065–1072
44. Miyatake S, Martuza RL, Rabkin SD (1997) Defective herpes simplex virus vectors expressing thymidine kinase for the treatment of malignant glioma. Cancer Gene Ther 4:222–228
45. Vetter M, Hofer MJ, Roth E et al (2009) Intracerebral interleukin 12 induces glioma rejection in the brain predominantly by CD8+ T cells and independently of interferon-gamma. J Neuropathol Exp Neurol 68:525–534
46. Radaelli E, Ceruti R, Patton V et al (2009) Immunohistopathological and neuroimaging characterization of murine orthotopic xenograft models of glioblastoma multiforme recapitulating the most salient features of human disease. Histol Histopathol 24:879–891
47. Rong Y, Durden DL, Van Meir EG et al (2006) 'Pseudopalisading' necrosis in glioblastoma: a familiar morphologic feature that links vascular pathology, hypoxia, and angiogenesis. J Neuropathol Exp Neurol 65:529–539
48. Ishii N, Maier D, Merlo A et al (1999) Frequent co-alterations of TP53, p16/CDKN2A, p14ARF, PTEN tumor suppressor genes in human glioma cell lines. Brain Pathol 9:469–479
49. Foehr ED, Lorente G, Kuo J et al (2006) Targeting of the receptor protein tyrosine phosphatase beta with a monoclonal antibody delays tumor growth in a glioblastoma model. Cancer Res 66:2271–2278
50. Furnari FB, Fenton T, Bachoo RM et al (2007) Malignant astrocytic glioma: genetics, biology, and paths to treatment. Genes Dev 21:2683–2710
51. Bryant NL, Gillespie GY, Lopez RD et al (2011) Preclinical evaluation of ex vivo expanded/activated gammadelta T cells for immunotherapy of glioblastoma multiforme. J Neurooncol 101:179–188
52. Allen C, Paraskevakou G, Iankov I et al (2008) Interleukin-13 displaying retargeted oncolytic measles virus strains have significant activity against gliomas with improved specificity. Mol Ther 16:1556–1564
53. Kruse CA, Molleston MC, Parks EP et al (1994) A rat glioma model, CNS-1, with invasive characteristics similar to those of human gliomas: a comparison to 9L gliosarcoma. J Neurooncol 22:191–200
54. Jacobs VL, Valdes PA, Hickey WF et al (2011) Current review of in vivo GBM rodent models: emphasis on the CNS-1 tumour model. ASN Neuro 3:e00063
55. King GD, Muhammad AKMG, Curtin JF et al (2008) Flt3L and TK gene therapy eradicate multifocal glioma in a syngeneic glioblastoma model. Neuro Oncol 10:19–31
56. Ali S, King GD, Curtin JF et al (2005) Combined immunostimulation and conditional cytotoxic gene therapy provide long-term survival in a large glioma model. Cancer Res 65:7194–7204
57. Ko L, Koestner A, Wechsler W (1980) Morphological characterization of nitrosourea-induced glioma cell lines and clones. Acta Neuropathol 51:23–31
58. Sibenaller ZA, Etame AB, Ali MM et al (2005) Genetic characterization of commonly used glioma cell lines in the rat animal model system. Neurosurg Focus 19:E1
59. Mathieu D, Lecomte R, Tsanaclis AM et al (2007) Standardization and detailed characterization of the syngeneic Fischer/F98 glioma model. Can J Neurol Sci 34:296–306
60. Barth RF, Kaur B (2009) Rat brain tumor models in experimental neuro-oncology: the C6, 9L, T9, RG2, F98, BT4C, RT-2 and CNS-1 gliomas. J Neurooncol 94:299–312
61. Tzeng JJ, Barth RF, Clendenon NR et al (1990) Adoptive immunotherapy of a rat glioma using lymphokine-activated killer cells and interleukin 2. Cancer Res 50:4338–4343
62. Benda P, Lightbody J, Sato G et al (1968) Differentiated rat glial cell strain in tissue culture. Science 161:370–371
63. Benda P, Someda K, Messer J et al (1971) Morphological and immunochemical studies of rat glial tumors and clonal strains propagated in culture. J Neurosurg 34:310–323
64. Koestner A, Swenberg JA, Wechsler W (1971) Transplacental production with ethylnitrosourea of neoplasms of the nervous system in Sprague-Dawley rats. Am J Pathol 63:37–56
65. Weizsacker M, Nagamune A, Winkelstroter R et al (1982) Radiation and drug response of the rat glioma RG2. Eur J Cancer Clin Oncol 18:891–895
66. Oshiro S, Liu Y, Fukushima T et al (2001) Modified immunoregulation associated with interferon-gamma treatment of rat glioma. Neurol Res 23:359–366
67. Paxinos G, Watson C (1998) The rat brain in stereotactic coordinates. Academic, New York

Neuromethods (2013) 77: 245–273
DOI 10.1007/7657_2012_38

Published online: 4 April 2012

Immunotherapy of High-Grade Gliomas: Preclinical In Vivo Experiments in Animal Models

Kristen A. Batich, Bryan D. Choi, and Duane A. Mitchell

Abstract

We have demonstrated that dendritic cells (DCs) loaded with total tumor RNA (ttRNA) represent a potent platform for the induction of antitumor immunity in preclinical murine glioma models. DCs are potent immunostimulatory cells that represent a promising contemporary biologic entity for realizing the potentials of immunotherapy. The use of RNA to encode tumor antigens for DCs was pioneered at our institution by Drs. Eli Gilboa and Smita Nair, but the ability of RNA-loaded DCs to stimulate potent antitumor immunity has been independently confirmed in several murine and human systems. In fact, there is accumulating evidence that RNA transfection represents a superior method for loading antigens onto DCs. In this chapter, we describe the methods involved in using DCs electroporated with ttRNA to potentiate the host immune response against infiltrating MGs. Furthermore, this chapter includes an in-depth review of the history of immunotherapy for high-grade gliomas using animal models as well as an overview of some of the more compelling immunotherapeutic strategies known to date.

Key words: Gliomas, Brain tumor, Immunotherapy, DC vaccines, Dendritic cells

1. Introduction

1.1. Immunotherapy of Malignant Gliomas

The prognosis for patients with primary malignant gliomas (MGs) remains very poor despite aggressive multimodal therapy, including surgical resection and radiotherapy plus temozolomide (TMZ) (5). Moreover, the nonspecific nature of conventional therapy for MGs often results in incapacitating damage to normal brain (6–8). This nonselectivity calls for novel therapeutic strategies that can precisely target invasive tumor cells while minimizing collateral damage to nearby brain parenchyma. The rationale for employing the immune system to target MGs is based on the premise that the inherent biologic specificity of the immune system can meet the clear and urgent need for a more precise medical therapy. Furthermore, the bioavailability of the immune system's cellular effector mechanisms and capacity to mount immunologic memory responses have been shown to be quite vast, even within the once-believed immunoprivileged CNS (9–12).

Cellular-based immunotherapy for MGs involves the ex vivo manipulation of cells in some fashion with the hopes of inducing an effective, specific, antitumor response upon transfer into the host. Cellular immunotherapy can be bifurcated into two modes of cell priming: (1) active immunotherapy, which involves vaccination of the host-modified tumor cells or with tumor antigen-loaded antigen presenting cells (APCs), or (2) adoptive immunotherapy, which utilizes the transfer of effector cells (such as natural killer cells or T cells) that are primed and expanded ex vivo and transferred back into the host to target the tumor (7). The generation of effective immunity against tumor-specific antigens (TSAs), has been shown, at least in preclinical models, to establish an immunologic repertoire of memory cells that can survey the host and prevent tumor recurrence, an additional benefit that no other targeted therapy possesses (13–15).

1.2. The Immunoprivileged Central Nervous System

The central nervous system (CNS) has previously been regarded as a site of limited immune surveillance. This concept is based on the primary studies of Medawar (16), Barker, and Billingham who demonstrated that allografts are tolerated better when implanted into the CNS compared to other tissues (17, 18). The "immunologic privilege" of the CNS has long been attributed to the blood–brain barrier (BBB) and the lack of a classic lymphatic system draining into the periphery (19).

However, over the past several years, the longstanding model that the CNS did not have the capability of interacting with activated lymphocytes, was void of specialized resident APCs, and lacked a lymphatic drainage system has been challenged. It is now accepted that lymphocytes traverse the BBB endothelial cells to ultimately reside in perivascular spaces or localize to the neuropil (20, 21). Moreover, it is also established that antigens efficiently drain into cervical lymph nodes via the cribroid plate and perineural sheath of cranial nerves (22, 23), and recent data suggest similar routes for APCs (24–26). Several studies in mice have supported the notion of a permissive barrier for immune cell trafficking. Adoptive transfer studies have revealed that T cell activation is sufficient to induce the migration of T cells into the CNS in the absence of an exogenous inflammatory stimulation of the endothelial barrier. One model suggested that inflammatory conditioning could induce endothelial cells to become competent in supporting cell trafficking by the induction of adhesion or chemoattractant molecules (27). In another study, transgenic GFAP-CCL21 mice with constitutively expressed CCL21 (a ligand that mediates lymphocyte homing to secondary lymphoid organs) on BBB endothelial cells facilitated the homeostatic CD4+ T cell proliferation of T cells isolated from lymph nodes draining the CNS. In another murine model, it was observed that CNS expression of CCL21 augmented the migration of CD4+ T cells from perivascular spaces into the CNS parenchyma of *T. gondii* infected mice (28). These findings contribute to the growing literature demonstrating that the immune privileged CNS can regulate T cell-mediated immunity.

Given the evidence for T cell trafficking into the CNS, the vast array of aberrantly expressed and mutated proteins present in tumor cells should permit them to be identified as foreign and ultimately rejected by the immune system. This concept of tumor surveillance as a normal function of the immune system has supporting evidence in that physiologic antitumor immune responses can often be detected in patients with cancer, although such responses are insufficient to prevent the progression of malignancy. Reasons for this insufficient response may be due to the debilitation of the immune system by locally and systemically-induced immunosuppression (29, 30). In a number of studies, MG cells have been shown to secrete a number of substances that suppress antitumor immunity, including TGF-β_2, prostaglandin (PGE_2), and IL-10 (31–33). Other mechanisms of immune evasion include production of extracellular matrix proteins, such as tenascin C and hyaluronic acid, disruption of the intracellular adhesion molecule (ICAM)-1, and downregulation of MHC class I antigens (34, 35).

1.3. Antigen Recognition and T Cell Trafficking Within the CNS

Microglia, which comprise about 10% of the total glial population in the CNS parenchyma (36), are a unique CNS cell population due to their mesodermic origin (as opposed to the neuroectodermally derived resident CNS cells). Having developed from a common precursor cell, microglia share many properties with macrophages (37, 38). Amidst virtually all inflammatory conditions in the CNS, parenchymal microglial cells rapidly upregulate MHC class II expression, suggesting that they may participate in antigen presentation to CD4+ T cells during the inflammatory process (38–42). Activated microglia also upregulate the co-stimulatory molecules CD40, CD80, and CD86, which render them capable for antigen presentation to T cells (21, 43–45). Furthermore, microglia have been shown to phagocytose myelin (46–48), indicating that these cells can activate myelin-specific CD4+ T cells. In preclinical models, microglia infected with Theiler's murine encephalomyelitis virus are capable of efficiently processing and presenting not only endogenous viral epitopes but also exogenous myelin epitopes to CD4+ T cells (49).

Naive T cells are not inherently located within the CNS due to their inability to pass through the BBB in the context of homeostatic conditions (i.e. the lack of an acute inflammatory response). When T cells have been activated against neurotrophic pathogens or CNS autoantigens, however, they cross the BBB and are restimulated upon encounter with their respective antigen on target cells and local APCs (50, 51). Recent studies of murine brain tumor models have shown that the tumor microenvironment promotes the proliferation of tumor-specific T cells and differentiation into cells with enhanced effector function (52). While activated T cells are thought to patrol the CNS in an antigen-independent manner, cells that encounter their cognate antigen are retained for longer periods within the CNS than those that do not encounter target antigen in the CNS (52). Chemokine receptors and αβ-integrin receptors together provide a coordinated homing signal for mediating the destination of trafficking lymphocytes. One such chemokine receptor, CXCR4, has been demonstrated to mediate neural stem cell localization to MGs (53, 54), and we have demonstrated that CXCR2 expression in T cells provides enhanced trafficking in response to glioma-secreted chemokines (IL-8 and GRO-1α) in vitro and in vivo (55).

1.4. A History of Animal Models: Proof-of-Concept for Immunotherapy

Beginning with preliminary studies by Medawar (16) and Burnet (56), the concept of immunological surveillance of tumor cells was based on immunogenicity studies utilizing implanted (57), carcinogen-induced (58, 59), or virus-induced (59) tumors. Using these models, it was believed that a tumor cell can express the genes encoding proteins or "factors" that the immune system can recognize to discriminate between normal and tumor cells. In 1970, Stutman, who was one of the contenders of this theory (60), investigated tumor development in immunodeficient (athymic "nude") mice that lacked adaptive immunity. Contrary to what was expected, the mice failed to show any notable acceleration in either induced carcinogenesis or growth of the implanted tumors. Thus, because of the absence of T and B cells in these mice, the conclusion was made that the immune system was not involved in tumor surveillance. It was later discovered that some immune surveillance components in nude mice, such as γδ T cells and natural killer (NK) cells, are fully functional.

Genetic recombination technologies employed in the 1990s allowed the generation of mouse lines in which the adaptive immune system was partially or entirely disrupted. These knockout lines allowed the role of adaptive immunity in protection against tumors to be more precisely analyzed. One model of particular utility was the generation of Recombination Activating Gene (RAG)-deficient murine model. These mice lack the ability to undergo somatic recombination, a process that is required for rearrangements of both immunoglobulin and T cell receptor genes to generate functional T and B cells (61). Another commonly used murine model with a disrupted immunity is the severe combined immunodeficient (SCID) model. Due to a mutation that inactivates the gene for the catalytic subunit of DNA-dependent protein kinase, this defect impairs DNA double-strand break repair and prohibits V(D)J recombination (62). Schreiber et al. (63) conducted experiments on spontaneous carcinogenesis with SCID and RAG−/− mice and with mice defective in signaling pathways involving IFN-γ, an important immunoregulatory cytokine for the formation of functional cytotoxic T lymphocytes (CTLs). An accelerated emergence of spontaneous and induced tumors was observed in these experiments as well as a shorter survival time. Within the field of tumor immunology, these findings from RAG−/− and SCID mice contributed to the revival of the immunological surveillance concept. It was also observed that when carcinogen-induced tumor cells were implanted into recipient mice, the tumor could be cleared in only a few recipients, indicating that some mice were incapable of effective immunological surveillance. From these findings, it was evident that the adaptive immune system can indeed recognize tumor cells and destroy some of them. Secondly, the pressure imposed by the immune system can select tumor cell variants that successfully escape immune recognition. As a result, the immunological properties of tumor cells can undergo immunoediting (64).

This conclusion has both positive and negative implications. The immune system combats tumor cells; hence, the understanding of the mechanisms behind such targeting would allow an increase in its efficiency. On the other hand, in spite of the resistance of the perfectly healthy immune system, tumor cells can change their properties and escape immune surveillance by several mechanisms (MHC downregulation, decreased tumor antigen expression (65), or induction of anergic CTLs). As it became evident from these knockout models, the immune system is capable of targeting tumors in vivo, but it attempts to do so amidst a great deal of immunosuppressive and immune escape avenues that are available to genetically unstable and biologically heterogeneous tumor cells (66, 67).

1.5. Applications of Syngeneic Versus Human Xenograft Models

The development of clinical approaches to cancer immunotherapy has relied heavily on preclinical data gathered from experimental animal models. An ideal preclinical model for the evaluation of immunotherapy would need to exhibit predictable and reproducible intracranial growth patterns, mirror the histopathological and biochemical aspects of human gliomas, and exhibit minimal endogenous antitumor immunity in tumor-bearing hosts. The implantation of rodent glioma cells has proven to be an excellent intracranial glioma model due to efficient tumorigenesis and reproducibly accelerated growth rates (68). Syngeneic murine models (i.e. GL26 cells in C57BL6 mice (69, 70), SMA-560 cells in VMDK mice (71), CNS-1 cells in Lewis rats (69, 72), and F98 and RG-2 cells in Fisher rats (73–75)) are all minimally immunogenic (do not elicit a response capable of rejecting tumors spontaneously in implanted hosts) and constitute excellent tools for studying immunotherapeutic responses of gliomas (69, 75–77). Furthermore, because these animal models have an intact immune system, they are a valuable tool to test cross-reactivity of immunotherapeutic approaches with normal antigens and potential autoimmune or inflammatory toxicities associated with treatment (76–78). As an extension to syngeneic models for immunotherapeutic evaluation, spontaneous genetically engineered mouse (GEM) glioma models (79, 80), based on loss or increased expression of relevant genes implicated in human glioma pathogenesis, provide a unique opportunity to examine the pathologic and molecular alterations associated with tumor development and progression, while still preserving the function of the host's immune system.

Human glioma xenografts implanted into immunocompromised mice have also been extensively employed in preclinical glioma research. Human xenograft models exhibit histopathological features that resemble human gliomas and retain the relevant gene amplifications (81). Also, these models are very useful for initial evaluation of novel imaging techniques (82) as well as new therapies for gliomas, including anti-angiogenic therapy (83, 84), chemotherapy (85), radiotherapy (86), targeted toxins (87), or oncolytic viruses (88–90). Although their xenogeneic nature impairs the study of immune-mediated antitumor strategies (due to the necessity of using an immunodeficient mouse to successfully engraft human tumor tissues), they do permit the assessment of the efficacy of therapeutic approaches in human glioma cells in an orthotopic animal model. The transfer of ex vivo manipulated lymphocytes has been used in a limited fashion for evaluation of immunotherapy in xenograft models and more recently, "humanized" mouse models consisting of immunodeficient mice reconstituted with human hematopoietic stem cells have been used to explore immunotherapeutic treatment of human cancers (91, 92).

1.6. Transgenic Mouse Models

Transgenic mouse models have allowed one to investigate the effects of different immunotherapy approaches on cancer. Human leukocyte antigen (HLA) transgenic mice can potentially overcome some of the main limitations inherent to human studies, such as the need for peripheral blood samples, the lack of controls, and the inability to control the influence of both MHC and non-MHC restricted genes on the immune responses. The HLA-A2 allele is one of the most common alleles in the general population (20–50% in Asian and Caucasian populations) and has been employed most frequently in transgenic mouse models (93, 94). In 1991 Vitelo et al. generated HLA A2.1/Kb transgenic mice (95), which have since been used by a number of groups to study CTL-mediated antitumor immunotherapy strategies (96). Because effector cells can recognize tumor cells and initiate their elimination, many transgenic strains have been activated to express modified human immune effector molecules. To study whether circulating naive CD4+ T cells could home to a tumor region and respond to idiotope (Id) peptides, Lauritzsen et al. established BALB/c mice transgenic for the αβ TCR specific for the $\lambda 2^{315}$ light-chain Id-peptide of a myeloma protein (97, 98). The transgenic CD4+ T cells displayed the characteristics of naive class II-restricted T cells in normal mice, proving that these mice were appropriate to study the role of naive T cells in immunosurveillance. By challenging the mice with different tumors, it was demonstrated that the transgenic mice were specifically protected against $\lambda 2^{315}$-expressing tumors.

Although B7-1 transgenic mice are usually generated to study the pathology of autoimmune diseases, the information provided by this model also holds implications in effective tumor immunotherapy. To explore the need for co-stimulation in autoimmunity, Allison et al. generated mice transgenic for B7-1 (99). Islets with IL-2 and confluent B7-1 expression were destroyed, whereas islets with IL-2 and patchy B7-1 expression showed selective killing on the B7-1+β cells. Islet-reactive T cells rejected syngeneic islet grafts, but only if these expressed B7-1. These results suggest that the eradication of a tumor might be dependent on continuously stimulated tumor-associated antigen (TAA)-specific CD8+ T cells locally by B7 molecules.

Preclinical findings from transgenic models have already found their ways to the patient, as they are currently being further evaluated in clinical settings. Transgenic lines expressing TAAs or oncogenes are extremely useful for the study of immunotherapeutic efficacy versus toxic effects. Furthermore, HLA and effector molecule transgenic mice permit the investigation of mechanisms involved in successful immunotherapy without the hindrance of tolerance. Crossing mice with both of these features will ultimately lead to the generation of animal models even more relevant to the treatment of human disease.

1.7. Oncolytic Viral Therapy Against Malignant Gliomas

The idea of "oncolytic viruses," which are viruses used to target cancerous cells, stems from anecdotal reports of spontaneous cancer remissions that occurred coincidentally with natural infection or upon the use of live attenuated vaccines (100). Since then, an improved understanding of the molecular basis for viral host cell tropism, cytotoxicity, and cell type-specificity has opened up avenues for the targeted design of virally based anti-neoplastic strategies. Currently, the requirements for viruses in cancer therapy need to fall under at least one of the following domains: an inherently low human pathogenic potential (i.e. the orphan reovirus (101)), a veterinary pathogen with unknown human pathogenicity (i.e. vesicular stomatitis virus (102)), or a human pathogen genetically engineered to selectively kill cancerous cells without collateral cytotoxicity in normal cells (i.e. herpes simplex virus-1 (103)).

One experimental cancer therapy using the poliovirus receptor nectin-like molecule-5 (Necl-5) has been employed to determine viral tumor tropism and to initiate glioma cell killing. Treatment of glioma xenografts with intratumoral inoculation of PVS-RIPO (a prototype nonpathogenic recombinant poliovirus) in one study produced rampant tumor cell death, potent host-mediated inflammatory reactions against infected tumors, and rapid tumor decline

(89, 104). These experiments documented two major components of oncolytic viral activity: direct viral tumor cell killing and resulting host inflammatory responses against the infected tumor.

Another preclinical study illustrated the potential usefulness of a replication-competent virus (reovirus) that killed glioma cells directly through cell lysis. The efficacy of this virus was demonstrated in established lines, in orthotopic (localized to where the cancer normally presents) and non-orthotopic glioma models, and in malignant glioma cell lines derived from surgical specimens. In contrast to SCID mice, no important toxic effects were found in nude mice, whose survival was dramatically prolonged after a single intracerebral administration of reovirus (105). The most important features of this potential anti-glioma treatment were its efficacy in an intracerebral location, its targeting of signaling pathways that are commonly activated in gliomas, and its ability to kill remote and invasive tumor cells.

Achieving success with oncolytic viruses in the clinical setting has been more challenging than the encouraging results achieved in many in vitro or animal studies. This translational difficulty stems from the fact that viruses depend on intricate relationships with their hosts that determine tropism for the intended tumor targets and the efficiency of targeted cell killing. Moreover, issues such as delivery, pre-existing immunity, or innate immune responses complicate therapeutic uses of viruses in patients (104).

1.8. Cellular-Based Immunotherapy Strategies in Animal Models

1.8.1. Active Immunotherapy: RNA-Transfected Dendritic Cells as Cancer Vaccines

Preclinical studies have shown DCs to be the most potent activators of de novo and recall responses in B and T cells and are thus regarded as one of the most promising entities for the realization of successful tumor immunotherapy (1–3). Our group has utilized the method of tumor-associated RNA transfection into DCs to prepare potent DC vaccines against MGs. A major advantage for using RNA-transfected DCs is the ability to extend this approach to a large number of patients despite limitations in obtaining tumor material for vaccine preparation. Building on existing technologies in molecular biology, we have shown that tumor mRNA can be isolated from a very small number of tumor cells, amplified, and used without loss of biological function (106). For instance, RNA isolated from tumor cells that were microdissected from pathology slides can be amplified and used to generate antitumor immune responses when transfected into DCs (107).

The efficacy of RNA transfection in eliciting immunity in vivo has been seen in several murine models, in which potent CTL responses developed in animals immunized with mRNA/lipid complexes (108–112). These responses were comparable to those achieved by infection with a live virus, indicating a physiologically significant level of immunity achieved by vaccination with RNA. In many studies, DCs transfected with RNA have proved superior to peptide-pulsed DCs in eliciting CTL responses against either peptide-pulsed experimental targets or intact tumor cells (4, 106, 107, 113). Furthermore, in tumor challenge experiments and stringent animal models of experimental metastasis, DCs transfected with in vitro transcribed RNA or ttRNA elicit protective antitumor immune responses (107, 112). Zhang et al. demonstrated the efficacy of tumor RNA-pulsed DCs in the 3LL lung carcinoma model and demonstrated that the vaccine's antitumor effect could be enhanced by expressing the chemokine lymphotactin in the DCs (111). Despite the presence of known normal antigens within ttRNA preparations, autoimmune toxicity has not been observed in mice receiving tumor protective vaccinations of ttRNA-loaded DCs (114). RNA transfection offers a means to expand the scope of vaccination to other classes of tumors by providing an effective and readily measured method for generating tumor-specific T cell responses and concurrently reducing the amount of tumor tissue needed for antigen isolation by a significant magnitude. Because tumor mRNA-transfected DCs are potent stimulators of T-cell immunity in vitro and in vivo (106, 107, 112) and can function as targets in CTL cytotoxicity assays, they also offer a practical substitute for tumor cells in assays to test the immunological effects of specific antigens. Therefore, the use of DCs transfected with RNA for tumor immunotherapy is an effective vaccination platform in the preclinical and clinical setting, and we describe the methods to generate ttRNA-pulsed DCs and subsequently activate tumor-specific T cells in Sect. 2.

1.8.2. Adoptive Immunotherapy

Adoptive immunotherapy (AIT), which comprises the ex vivo expansion of effector cells and the return of tumor-specific T cells back into tumor-bearing hosts, has been shown in recent clinical trials to be a potent modality for achieving markedly high and sustained levels of tumor-specific T cells (115–117). The effectiveness of adoptive T cell therapy relies on the ability for adoptively transferred lymphocytes to localize to tumor sites and mediate the destruction of antigen expressing tumor cells (118–120).

Our preclinical data on adoptive immunotherapy against MGs with systemically administered lymphocytes support the notion that intracerebral trafficking and retention of T cells may be a limiting factor in the effectiveness of treating MGs by adoptive immunotherapy (27, 121, 122). To enhance intratumoral trafficking, we have explored the transfer of the chemokine receptor, CXCR2, into antigen-specific T cells to increase migration toward chemokines secreted in abundance by MGs (123–125) and CMV-infected cells (126) (e.g. IL-8, GRO-1-α, and the CMV viral chemokine, UL146). Tumor-specific T cell migration studies in SCID mice using fluorescently labeled human lymphocytes have revealed that limited entry into the CNS may be a potential barrier to AIT against MGs. However, we have shown that CXCR2 RNA-mediated gene transfer significantly enhanced T cell infiltration into the CNS in response to a variety of CXCR2-specific ligands in vitro and in vivo (55). These findings support novel methods for enhancing tumor-antigen-specific T cell trafficking into the CNS, which may be potent enough to counter the limitations of the CNS barrier as well as the tumor-induced state of immunosuppression.

1.9. Brain Tumor Stem Cells: A Future Immunotherapeutic Target

Malignant gliomas display incredible intratumoral heterogeneity, not only morphologically but also by grades of differentiation. Tumor heterogeneity may be derived from both genetic and non-genetic/epigenetic causes. Growing evidence from hematopoietic malignancies and solid tumors has supported the hypothesis that a subpopulation of cancer cells in each malignancy has greater potential for tumor initiation and repopulation (127–129). Although the nomenclature of these cells has been highly controversial due to the lack of definitive criteria, glioma stem cells (GSCs) have been described by several groups as the cancer stem cell population of MGs (129, 130). They are functionally defined by the capacity for self-renewal measured by serial neurosphere assays and tumor propagation by in vivo intracranial limiting dilution assays. GSCs have been implicated in several malignant behaviors. GSCs have been found to express elevated levels of vascular endothelial growth factor (VEGF) to promote tumor angiogenesis (131). This is of particular significance as bevacizumab (Avastin), a humanized neutralizing anti-VEGF antibody, has demonstrated activity against gliomas and was recently approved by the United States Food and Drug Administration for the treatment of recurrent or progressive gliomas (132–134). GSCs have also been characterized to be relatively resistant to radiation due to preferential activation of the DNA damage checkpoint and lesion repair (135), and other groups have described relative resistance of GSCs to chemotherapies (136, 137). These studies may explain in part how patients with a promising radiographic response universally suffer recurrence and/or progression of their cancers. Therefore, direct targeting of GSCs may improve the efficacy of conventional cytotoxic therapies as well as anti-angiogenic therapies.

Although GSCs appear resistant to radiation and chemotherapy, their susceptibility to immunotherapeutic attack has not been formally tested. A recent study in a murine glioma model revealed that the immunologic targeting of brain tumors using DCs pulsed with lysates from tumor cells grown in stem-cell culture media was more efficacious than DCs pulsed with tumor lysates from cells grown in standard culture, suggesting that the targeting of GSCs may be of therapeutic benefit (138). Thus, the identification of antigens expressed in GSCs and the development of immunotherapy that effectively targets these cells is a promising approach toward improving the outcome of immunotherapy for malignant gliomas.

1.10. Future Directions

Although in the past, the identification of tumor antigens has been difficult for many types of cancers, the data generated by techniques such as serial analysis of gene expression (i.e. from the Cancer Genome Anatomy Project) have revealed many new potential targets within public gene expression databases (139). These data require augmentation and verification, but, when additional tumor-specific antigens are identified, antigen-specific DC vaccine strategies may become applicable to many tumor types, with a multitude of specifically identified targets (14).

Immunologic targeting of tumor-specific gene mutations has the capacity to allow more precise eradication of neoplastic cells. Most well-characterized tumor antigens are overexpressed normal proteins that have triggered immunologic tolerance to some degree. This compromises their effectiveness as tumor rejection antigens and poses a risk of autoimmunity if these normal proteins are effectively targeted. Conversely, tumor-specific antigens derived from mutations in somatic genes are less influenced by central tolerance and less likely to be associated with autoimmunity. These mutations, however, usually arise randomly as a result of the genetic instability of tumors (140). As such, these mutations tend to be patient specific, and many are incidental to the oncogenic process. The epidermal growth factor receptor (EGFR) mutation EGFRvIII, however, is a consistent tumor-specific mutation frequently seen in glioma patients (141–143) and a broad array of other common cancers (144–146).

Although there is data to support the possible benefits of tumor-specific DC vaccination, there remain a number of issues that must be addressed to optimize this therapeutic modality. For example, in a study of 21 biopsy samples of human gliomas, only 47% of the tumors tested positive for EGFRvIII (147), and expression patterns even within these tumors are not homogeneous. This being true, vaccinations that target only one antigen may not target all tumors or all cells comprised in a tumor, and these vaccinations may subsequently select for the survival and proliferation of those cells that do not express the targeted antigen. Although the specificity of tumor-specific vaccination therapies may have the possible advantage of minimizing autoimmune complications, the heterogeneity of malignant brain tumors may limit the effectiveness of vaccinations targeting only one tumor-specific antigen. Furthermore, cross-presentation of nontargeted antigens could also lead to deleterious autoimmune responses. Conversely, multi-antigenic vaccines in this patient population have shown robust immunologic responses and encouraging clinical results without the induction of autoimmunity, and catastrophic autoimmune responses have not been reported to date (146).

The advantages of vaccinating with total tumor-derived material are that the identity of the tumor antigen(s) need not be known and that the presence of multiple tumor antigens reduces the risk of antigen-negative escape mutants. The potential benefit of using ttRNA is that it can be amplified from a small number of tumor cells. Furthermore, isolating bona fide tumor cells from patient specimens by ex vivo purification methods and combing this with the use of RNA subtractive hybridization techniques may reduce the concentration of self, potentially autoreactive antigens in the vaccine preparation (112).

Another outstanding limitation to the realization and efficacy of cell-based immunotherapies is the level of host immunosuppression imposed by the infiltrating tumor. Despite being restricted to the intracranial compartment, MGs elicit defects in host systemic cellular immune responses that are notably severe (148). These defects are characterized by dramatic reductions in both CD4+ T cell numbers and function. As a result, patients with MG suffer severe CD4+ lymphopenia (30, 149) with substantial T cell anergy characterizing the remaining CD4+ compartment (150). We have recently shown that a disproportionate presence of regulatory T cells (Tregs) amid CD4+ T cells provides an explanation for diminished T-cell function in these patients (30). Furthermore, our intracranial MG murine model recapitulates these findings. Specifically, tumor-bearing mice show dramatically reduced CD4+ counts in the peripheral blood, spleen, and cervical lymph nodes (CLNs), as well as increased Treg fractions and reduced T cell responsiveness (151).

Specific monoclonal antibodies (mAbs) that block immunosuppressive pathways in tumor-bearing hosts have demonstrated considerable benefit in preclinical and clinical immunotherapy settings. CTLA-4, a receptor constitutively expressed on Tregs and upregulated during activation of effector T cells, mediates downregulation of the immunologic response in activated lymphocytes (152, 153). We have shown CTLA-4 blockade to be a potent method for mediating the immunologic rejection of advanced gliomas in our murine models and reversing glioma-induced immunosuppression (154) and are currently pursuing evaluation of the clinical safety and efficacy of this modality in patients with MGs. In mice, CTLA-4 inhibition led to resistance to Treg-mediated suppression in effector cells without affecting the suppressive capacity of Tregs from CTLA-4 treated animals, indicating that despite constitutive expression of CTLA-4 on Tregs, the major impact of CTLA-4 blockade is on activated effector cells. Strategies to block the function of Tregs have shown promise in other clinical settings, and recent studies from our laboratory and others have demonstrated the potential benefits of Treg blockade using anti-CD25 mAbs in enhancing immunotherapeutic treatment of MGs (151, 155).

Given the presented strategies for immunotherapeutic intervention in preclinical models, a future focus for the eradication of MGs in patients would be to incorporate immunotherapeutic strategies in a combinatorial fashion with current standard of care modalities. In general, conventional therapies involve tumor resection, irradiation, and systemic chemotherapy (156). Unfortunately, most patients develop recurrence or progression after radiation treatment, and tumor radioresistance renders re-irradiation treatment less effective and potentially more toxic (157). Thus, cellular-based immunotherapies that can act in a synergistic fashion with radiation and chemotherapy sensitivity may improve outcomes in these patients.

The synergistic effects of vaccines with chemotherapy have been examined in clinical trials from major healthcare institutions (158, 159). In one study, clinical outcomes (survival and progression times) were analyzed retrospectively in vaccinated and non-vaccinated de novo glioma patients receiving chemotherapy. Patients receiving post-vaccine chemotherapy had longer survival times and exhibited significantly longer times to tumor recurrence after chemotherapy, relative to their own previous recurrence times, as well as relative to patients receiving a vaccine or chemotherapy alone (159). Based on the evidence that DC vaccination induces specific CTL targeting of drug-resistance-related TAAs and the aforementioned clinical observations, it is possible that therapeutic vaccination acts synergistically with subsequent chemotherapy to elicit tangible clinical benefits for glioma patients via the sensitization of tumor cells to therapeutic drugs after immune selection that has depleted former drug-resistant cells.

Immunization with cancer vaccines in particular offers advantages that other cancer therapy strategies do not. First, it is highly specific for cancer cells and, therefore, low toxicity should be expected. Second, it recognizes and eliminates cancer cells independent of their cell cycle phase. Third, tumors that developed drug resistance would still be a suitable target for immunotherapy. Fourth, vaccination offers the unique potential for a durable antitumor effect due to the phenomenon of immunological memory, potentially obviating the need for prolonged, repetitive cycles of therapy. Finally, immunotherapy offers the possibility of preventative immunization of high-risk patients (160). Given that synergistic effects of DC immunotherapy followed by chemotherapy have been observed in the clinical setting, further investigation of appropriate dosing schedules in preclinical models are necessary to examine the most potent, long-lasting, and safest regimen for targeting gliomas.

2. Methods Section: ttRNA-Pulsed DCs Activate Tumor-Specific T Cells in Animal Models

Dendritic cells are specialized APCs that play a pivotal role in the induction of T and B cell immunity (161). These cells have the exceptional ability to activate naive CD4+ and CD8+ T cells in vitro and in vivo. Because DCs require the presence of an antigen in order to initiate cytotoxic responses (162), when they are pulsed with tumor homogenate (161) or tumor-associated RNA (106), they are able to elicit a primary CTL response in vitro. Subsequent studies in vivo demonstrated that immunization of mice with DCs pulsed with an antigen can prime a CTL response that is tumor specific and provides protective immunity in treated mice. DC-based therapy is currently being evaluated in clinical trials for a variety of tumors (164).

Immunization using defined tumor antigens is, however, limited at present to a handful of human tumor types in which candidates for tumor rejection antigens have been identified. The advantages of vaccinating with total tumor-derived material are that the identity of the tumor antigen(s) need not be known and that the presence of multiple tumor antigens reduces the risk of antigen-negative escape mutants. Another benefit of using total tumor antigens in the form of mRNA is that it can be amplified from a small number of tumor cells (106, 107). Hence, DC vaccination may be extended to patients with MGs from which only a small (and possibly microscopic) biopsy can be taken for diagnosis. Furthermore, isolating tumor cells from patient specimens by ex vivo purification methods and combing this with the use of RNA subtractive hybridization techniques may reduce the concentration of self, potentially autoreactive antigens in the vaccine preparation. This would be of crucial importance for vaccinations with CNS tumor-derived antigens, as it may diminish the risk of severe autoimmune complications (112).

3. Materials

The materials used are listed below. Comparable products from other suppliers should also be effective.

3.1. Tumor Cell Lines and Animal Models

1. SMA 560 cell line (spontaneous astrocytoma) and VM/Dk mice (71); 6–8 weeks of age (or syngeneic model of interest)
2. Cell media: DMEM zinc option (ZO) medium (GIBCO BRL, Gaithersburg, MD) containing 10% (vol/vol) fetal calf serum (FCS), 25 mM HEPES, 2 mM L-glutamine, and 1 mM sodium pyruvate

3.2. DC Generation from Bone Marrow Cultures

1. Syngeneic mice of desired number
2. Sterile scissors, Mayo scissors, sterile chucks, sterile gauze
3. Two pairs of sterile forceps
4. 70% Ethanol
5. Halothane or Isoflurane and gas chamber
6. 50-ml Conical tubes (BD Falcon #352098)
7. RPMI 1640 (Gibco #11875-085)

8. Penicillin/Streptomycin (P/S) (Gibco) (store in 5.5-ml aliquots at −20°C)
9. Heat-inactivated fetal calf serum (FCS) (Gibco)
10. 55 mM β-Mercaptoethanol (ME) (Invitrogen #21985-023)
11. 100 mM Sodium pyruvate (Gibco # 11360-070)
12. 10 mM Nonessential amino acids (NEAA) (Gibco # 11140-050)
13. 1 M HEPES buffer (Gibco # 15630080)
14. 200 mM L-Glutamine (Gibco # 25030081) (store in 4 ml aliquots at −20°C) (may be refrozen)
15. GM-CSF (Immunex) (store at −20°C)
16. IL-4 (Peprotech #214-14X) (store at −20°C)
17. Cell lysis buffer (10×; Cell Signaling Technology #9803) diluted 1:10 in sterile H20
18. 100 mm Tissue culture dishes (BD Falcon #353003)
19. 70 μm Cell screens (BD Falcon #352350)
20. 25 G Needles (BD #305122) and 10 ml syringes (BD #309604)
21. Trypan blue and hemocytometer
22. 6-Well tissue culture plates (BD Falcon #353046)
23. Pipet-aid with 5-, 10-, and 25-ml pipets
24. 60-mm Tissue culture dishes (BD Falcon #353002)
25. Opti-MEM media (Gibco # 11058-021)
26. 1.5-ml Eppendorfs (Brinkmann #220-36-320-4)
27. 2-mm Gene pulser cuvettes (BioRad #165-2086)
28. Electroporator (BTX model ECM 830)
29. Sterile disposable fine-tipped transfer pipets (Corning Samco #231-1S)
30. FACS tubes with caps (BD Falcon #352054)
31. Complete DC media (CDCM): To a 500-ml bottle of RPMI 1640, add the following:

    27.5 (5%)mL FCS, 5.5 ml Pen/Strep (stored frozen), 2.0 ml L-glutamine (stored frozen), 550 μl β-ME, 5.5 ml sodium pyruvate, 5.5 ml NEAA, 5.5 ml HEPES. Thaw 1.1×10^6 U GM-CSF (should be stored in frozen aliquots) and 10 μg IL-4 (also stored frozen in aliquots). Pour entire bottle of media over a 0.2 μm, 500 ml sterile vacuum filtration unit (see Note 1).

3.3. Total Tumor RNA Extraction

1. RNeasy kit (Qiagen, Valencia, CA)
2. 1.5-ml Eppendorf Safe-Lock tubes (for freezer storage of ttRNA)

3.4. Vaccination Using RNA-Pulsed DCs

1. 1× PBS (Gibco #10010-023)
2. 0.5-ml insulin syringes with needles (BD #305932)

3.5. CTL Priming and In Vitro Cytotoxicity Assay

1. Target cells (tumor cells of interest) (5–10 × 10^6)
2. Europium time-resolved fluorescence cytotoxicity assay kit
3. Complete RPMI 1640 media (Gibco #11875)

3.6. Implantation of Brain Tumors

1. Trypsin EDTA (0.25%)
2. 1× PBS (Gibco #10010-023)
3. Methylcellulose
4. 250-μl Hamilton syringes (Hamilton, Reno, NV)
5. Stereotactic frame (Kopf Instruments, Tujunga, CA)
6. Bone wax and surgical staples

3.7. Statistical Analysis for Survival Data

1. GraphPad Prism 5 (or a comparable statistical analysis program that can be used to generate survival curves)

4. Methods

4.1. Tumor Cell Lines and Animal Models

Using an appropriate tumor line and animal model, the effects of DC vaccines pulsed with total tumor RNA can be examined.

1. A syngeneic spontaneous high-grade glioma model with the manipulation of murine DCs would be an ideal platform. For the purpose of these methods, we describe the immunotherapeutic effects of DC vaccines using the SMA 560 cell line, which was derived from an intracerebral transplant of a spontaneous astrocytoma from a VM/Dk mouse (H-2b) (71).
2. Culture cell lines in ZO medium containing 5% (vol/vol) FCS.
3. For murine DC extraction, use 6–8-week-old female mice, and maintain mice in a virus-free environment in accordance with the Laboratory Animal Resources Commission standards.

4.2. DC Generation from Bone Marrow Cultures

The procedure used in these studies was the same as previously described (110).

Day 0

1. Sacrifice mouse in gas chamber. Perform cervical displacement on mice. Wash mouse by submersion in 70% ethanol.
2. Place mouse on sterile gauze on sterile chuck. Cut the Achilles tendon on one hind leg. Then, beginning at this site, use

scissors to extend skin opening up the entire rear length of the leg. Peel the skin from the ankle up the entire length of the leg, exposing the musculature. Cut the patellar tendon, found on the anterior surface. Turning the leg over again, bend the leg backwards at the knee, exposing the proximal tibia. Continue to push, exposing the tibia to the ankle. Holding the foot and tibia firmly in opposing hands, loosen the joint by gentle twisting until the tibia can be pulled in one swift motion from the foot. Clean the bone with gauze and place on ice in conical containing cold RPMI-P/S. Returning to the same leg, cut the remaining foot and lower leg muscles off at the knee, leaving only the upper leg. Slide the scissors in between the muscle and anterior surface of the femur. Spread the scissors to bluntly dissect the muscle off the bone. Repeat for the posterior surface of the bone. This will leave the distal femur and attached patella exposed. Grasp the exposed end of the femur and pull outwards from the body at a 45° angle to dislocate the hip. At this point, the scissors can be used to feel between the dislocated head of the femur and pelvis and to cut the femur away. Once the bone is removed, the patella may be grasped with the finger and folded back over the femur to remove it. Much of the musculature will come with the patella, and the femur may then be easily cleaned with gauze and placed in RPMI-Penicillin/Streptomycin (P/S). Repeat this procedure for each leg of each mouse.

3. Once both legs have been removed, the sternum may be removed. Make a horizontal incision in the skin, over the area of the xiphoid process. The proximal flap may then be grasped and pulled rostrally, exposing the rib cage and shoulders. Pectoral muscles should be detached from the chest wall. Then, make a horizontal incision in the peritoneum, over the xiphoid process. Sliding the scissors rostrally under the sternum, bluntly dissect the organs away from the sternum. Scissors may then be used to cut along each side of the sternum, towards the head. Cut horizontally proximal to the manubrium to detach the sternum. Place it on ice in the same conical tube containing the long bones in RPMI-P/S. [Repeat steps 1 through 4 for each mouse].
4. Following bone harvest, transport bones to tissue culture hood. Decant RPMI-P/S off the two conicals containing long bones and sternums. Replace with 70% ethanol, recap, and shake vigorously for at least one minto wash. Decant off ethanol and replace with cold RPMI-P/S. Shake again for 1 min.

5. Decant off RPMI-P/S and pour long bones and sternums separately into 100-mm dishes. Add enough fresh RPMI-P/S to submerge bones.
6. For long bones, use scissors or Mayos to cut off the distal (white) ends of the tibias and proximal (head) ends of the femurs. Place bones in fresh dish containing RPMI-P/S.
7. Fill a 10-ml syringe with RPMI-P/S from dish containing bones, and place a 25 G needle on the end. The needle can be inserted into the proximal (uncut) end of tibia or distal (uncut) end of femur (patella must have been removed) and used to squirt marrow out the cut end. Bones should be white when emptied. Subsequently insert the needle into the opposite (cut) end of bone to ensure efficient marrow harvest. The marrow will appear as long red strings in dish. [If all RPMI is expelled from the syringe, remove the needle, refill from the same dish, and attach a fresh needle].
8. Once all long bones are emptied and discarded, use just the 10-cc syringe to draw the marrow and RPMI up and down in the dish until marrow is dispersed and dissociated.
9. Transfer marrow through a 70-μm cell screen fitted on a 45° angle into a 50-ml conical tube. Let sit until sternum marrow is harvested.
10. For sternums, cut away xiphoid and loose muscle still attached. Transfer to fresh dish containing RMPI-P/S.
11. Holding sternums at one end with forceps (keep submerged), crush the sternum from one end to the other with Mayos, squeezing out the marrow (as if from a tube of toothpaste). Repeat until sternum is white. Discard sternums.
12. Repeat steps 12 and 13 for sternums. If space allows, marrow may be added to same conical and cell screen as long bones.
13. Spin the cell suspension at 500 × g × 5 min × 4°C.
14. Resuspend pellet (combine pellets if there are more than one) in 5 ml lysis buffer per five mice harvested. Incubate in 37°C bath for 3–4 min.
15. Stop reaction by adding 25–30 ml cold RPMI. Spin as in step 13.
16. Resuspend pellet in 10 ml CDCM. Run cells over another 70-μ m cell screen into a new 50-ml conical on ice. This will remove any lysed cell aggregates.
17. Count cell numbers via trypan blue exclusion (using a 1:20 dilution of cell suspension: trypan blue).
18. Adjust cell concentration to 1×10^6/ml and plate in 6-well plates at 3 ml per well. Incubate at 37°C, 5% CO_2, humidified (see Note 2).

Day 3 Wash: Removal of Nonadherents

1. Harvest media from wells, three at a time, using a Pipet-aid with 10-ml pipet. (Place tip down the side of each well to draw up media—do not scrape the bottom of the well with the pipet).
2. With the Pipet-Aid on "slow," expel media in circle down inside wall of well and draw back up. Do this once per well for all three wells.
3. Discard media and replace with 3 ml fresh CDCM per well. Repeat these steps for all wells and reincubate in Day 0 Step 18 conditions.

Day 7: Re-plating of Nonadherents

1. With Pipet-Aid on "fast" and a 10-ml pipet, harvest media from wells three at a time.
2. For each well, expel media in circular fashion down sides of well and draw back up. Perform this wash three times for each well in the three-well set, one after the other.
3. *Save the collected media* (see Note 3) and cells by transferring to a 50-ml conical on ice. Discard 6-well plates. Repeat for all wells, and pool cells and media in conicals.
4. Spin cells 500 × *g* × 5 min × 4°C, resuspend pellet in 10 ml CDCM, and count via trypan blue.
5. Adjust cell concentration to 1×10^6/ml and replate into 60-mm tissue culture dishes (not Petri dishes) at 5 ml/dish. Reincubate in Day 0 Step 18 conditions.

Day 8: Electroporation of DCs

1. With Pipet-Aid on "fast" and a 10-ml pipet, wash each culture dish with its own media three times, and transfer cells and media to a 50-ml conical on ice. Repeat for each dish and pool cells and media.
2. Spin cells 500 × *g* × 5 min × 4°C. Following spin, *save the decanted media (will be used later)*.
3. Resuspend in 10 ml OptiMEM and count via trypan blue.
4. Respin cells and resuspend pellet in OptiMEM at 2.5×10^7/ml.
5. Transfer cells in 200-μl aliquots (5×10^6 cells each) into Eppendorfs and keep cold.
6. Obtain total tumor mRNA to be electroporated. *Keep on ice when not using*.
7. Mix RNA (25 μg ttRNA) with a single aliquot of cells by pipetting up and down briefly (do not allow bubbles to form).

8. Transfer individual cell/RNA mix *IMMEDIATELY* into gene pulser cuvette. Place cuvette in electroporator and pulse at 300 V for 500 μs (ECM 830 electroporator; BTX, San Diego, CA) (see Note 4).
9. Using sterile disposable dropper, transfer electroporated cells and RNA *IMMEDIATELY* into 5 ml 50% conditioned media (2.5 ml fresh CDCM + 2.5 ml *saved* media from step 2) in a 60-mm tissue culture dish. Repeat steps 5–9 for each aliquot. Reincubate cells for 24 h.

Day 9: Prepare DCs for Vaccination

1. Harvest cells as on Day 8.
2. Pool cells that were electroporated in 50-ml conicals on ice. Spin 500 × *g* × 5 min × 4°C.
3. Resuspend pellet in 5 ml 1× PBS and count. Respin cells and resuspend in PBS at desired concentration for vaccination. Transfer to capped FACS tubes and place on ice. Be sure to agitate cells prior to drawing into syringe for vaccination. At this point, 2–3 × 10^6 cells from each group may be set aside on ice for phenotyping (using murine antibodies for CD11c, CD80, and CD86). Alternatively, cells leftover after vaccination may be brought back for phenotyping.

4.3. Total Tumor RNA Extraction

1. Extract ttRNA from your glioma cell line using RNeasy Mini Kit (Qiagen, Valencia, CA).
2. Store ttRNA in a −80°C freezer until use for electroporation.

4.4. Vaccination Using RNA-Pulsed DCs

1. Using the electroporated DCs from Sect. 4.2, Day 9, resuspend cells in 1× PBS (10^5 RNA-pulsed DCs in 500 μl PBS/mouse) for intraperitoneal immunizations.
2. Inject a final volume of 500 μl PBS/RNA-pulsed DCs intraperitoneally into each C57BL/6 mouse.

4.5. CTL Priming and In Vitro Cytotoxicity Assay

The in vitro cell-mediated cytotoxicity assay can be performed using standard procedures as previously described (110).

1. Obtain splenocytes 12–14 days after the final immunization.
2. Deplete RBCs with lysis buffer and culture.
3. Culture 10^7 splenocytes with 5 × 10^5 stimulator tumor cells pretreated with IFN-γ and irradiated at 7,500 rads in 5 ml of Iscove's modified Dulbecco's medium with 10% FCS, 1 mM sodium pyruvate, 100 IU/ml penicillin, 100 mg/ml streptomycin, and 5 × 10^{-5} M β-mercaptoethanol/well in a 6-well tissue culture plate.
4. Culture cells for 5 days at 37°C and 5% CO_2.

5. Harvest effector cells on day 5 on Histopaque 1,083 gradient prior to use in the CTL assay.
6. Label target cells (5–10 × 10^6 tumor cells) with europium for 20 min at 4°C.
7. Incubate europium-labeled targets (103) and serial dilutions of effector cells at varying effector-to-target (E:T) ratios in 200 μl of complete RPMI 1640 in U-bottom 96-well plates.
8. Centrifuge plates at 500 × g for 3 min and incubate at 37°C for 4 h.
9. Harvest 50 μl of the supernatant, and measure europium release by time-resolved fluorescence. Determine specific cytotoxic activity by the following formula:

$$\%\ \text{Specific release} = \frac{[\text{Experimental release} - \text{Spontaneous release}]}{[\text{Total release} - \text{Spontaneous release}]} \times 100$$

(Spontaneous release of the target cells is designated as <25% of total release by detergent in all assays).

4.6. Implantation of Brain Tumors

1. Harvest tumor cells (depending on line used for ttRNA extraction) by trypsinization; wash twice in Dulbecco's 1× PBS, and mix suspension with an equal volume of 10% methylcellulose in zinc option medium.
2. Bring cells to a final concentration of 500 cells/5 μl.
3. Load tumor cells into loaded into 250-μl Hamilton syringes (Hamilton, Reno, NV) using a stereotactic frame (Kopf Instruments, Tujunga, CA).
4. Position the injection needle 2 mm to the right of bregma and 4 mm below the surface of the skull. After injection, close the skull hole with bone wax, and close the wound with surgical staples (Stoelting Co., Wood Dale, IL) (see Notes 5 and 6).

4.7. Statistical Analysis for Survival Data

1. Determine survival using the Kaplan and Meier method.
2. Compare survival data between groups using Wilcoxon's log-rank test (with a designated significance level at $p < 0.05$).

5. Notes

1. Complete DC media (CDCM) should be made on Day 0 and used only for that DC prep (i.e. on Days 0–9). It should not be stored for multiple uses. Occasionally, bone marrow culture yields will be high enough to warrant more than one 500-ml bottle per prep.

2. Each DC bone marrow prep is outlined to use five mice for each harvest and DC vaccination experiment. Therefore, one can expect a generation of 350×10^6 total marrow cells to be extracted from a total of five mice on Day 0. Upon Day 7, the usual number of DCs cultivated should reach 50×10^6 cells total. These numbers are the average numbers generated for Days 0 and 7 for a 5-mouse bone marrow procedure.
3. The difference in the technique for Day 3 and Day 7 washes in the DC bone marrow procedure reflects the morphologic changes in cultured DCs with time. On Day 3, it is absolutely critical that cells are washed with the Pipet-Aid on "slow." However, on Day 7, the DCs alter their adherent properties and can be found in the supernatant of the wells. This is why on Day 7 the nonadherent cell mixture is saved and replated. For Day 7 washes, the Pipet-Aid should be operated on the "fast" setting.
4. For the RNA electroporation step, the highest number of electroporated DCs will be seen if the operator moves as quickly as possible. The quickest and most effective procedure for electroporation is to first remove the 200-μl aliquot of DCs into the electroporator cuvette with a P-200 pipet. Then, take a P-10 pipet and mix thoroughly (3–5 times) 10 μl of the RNA in the Eppendorf tube. Inject 10 μl of the RNA into the deep corner of the floor of the cuvette with the DCs. Quickly transfer to the P-200 and mix this RNA-DC suspension (8 times is sufficient) without generating bubbles in the cuvette. This entire process (from the insertion of the cells to the implantation of the cuvette into the electroporator) should not exceed 15 s.
5. For the tumor implantation procedure, the entire process should be completed in less than 3 h to minimize variation in tumor cell viability.
6. When performing the tumor cell intracranial injections, be sure to randomize the mice to the different injection doses so as to account for time differences in cell viability (i.e. half of the mice in a group receiving a 50-cell dose in the first hour and the remaining half receiving that same 50-cell dose during the second hour).

References

1. Steinman RM (1991) The dendritic cell system and its role in immunogenicity. Annu Rev Immunol 9:271–296
2. Gilboa E, Nair SK, Lyerly HK (1998) Immunotherapy of cancer with dendritic-cell-based vaccines. Cancer Immunol Immunother 46:82–87
3. Nair SK, Boczkowski D, Snyder D, Gilboa E (1997) Antigen-presenting cells pulsed with unfractionated tumor-derived peptides are potent tumor vaccines. Eur J Immunol 27:589–597
4. Paglia P, Chiodoni C, Rodolfo M, Colombo MP (1996) Murine dendritic cells loaded in vitro with soluble protein prime cytotoxic T lymphocytes against tumor antigen in vivo. J Exp Med 183:317–322
5. Mitchell DA, Fecci PE, Sampson JH (2008) Immunotherapy of malignant brain tumors. Immunol Rev 222:70–100

6. Imperato JP, Paleologos NA, Vick NA (1990) Effects of treatment on long-term survivors with malignant astrocytomas. Ann Neurol 28:818–822
7. Mitchell DA, Fecci PE, Sampson JH (2003) Adoptive immunotherapy for malignant glioma. Cancer J 9:157–166
8. Stupp R, Dietrich PY, Ostermann Kraljevic S, Pica A, Maillard I, Maeder P et al (2002) Promising survival for patients with newly diagnosed glioblastoma multiforme treated with concomitant radiation plus temozolomide followed by adjuvant temozolomide. J Clin Oncol 20:1375–1382
9. Sehgal A, Berger MS (2000) Basic concepts of immunology and neuroimmunology. Neurosurg Focus 9:1–6
10. Stevens A, Kloter I, Roggendorf W (1988) Inflammatory infiltrates and natural killer cell presence in human brain tumors. Cancer 61:738–743
11. Bullard DE, Gillespie GY, Mahaley MS, Bigner DD (1986) Immunobiology of human gliomas. Semin Oncol 13:94–109
12. Owens T, Renno T, Taupin V, Krakowski M (1994) Inflammatory cytokines in the brain: does the CNS shape immune responses? Immunol Today 15:566–571
13. Heimberger AB, Crotty LE, Archer GE, McLendon RE, Friedman A, Dranoff G et al (2000) Bone marrow-derived dendritic cells pulsed with tumor homogenate induce immunity against syngeneic intracerebral glioma. J Neuroimmunol 103:16–25
14. Heimberger AB, Archer GE, Crotty LE, McLendon RE, Friedman AH, Friedman HS et al (2002) Dendritic cells pulsed with a tumor-specific peptide induce long-lasting immunity and are effective against murine intracerebral melanoma. Neurosurgery 50:158–164, discussion 164–156
15. Mitchell DA, Sampson JH (2009) Toward effective immunotherapy for the treatment of malignant brain tumors. Neurotherapeutics 6:527–538
16. Medawar PB (1948) Immunity to homologous grafted skin. III. The fate of skin homografts transplanted to the brain, to subcutaneous tissue, and to the anterior chamber of the eye. Br J Exp Pathol 29:58–69
17. Barker CF, Billingham RE (1977) Immunologically privileged sites [Review]. Adv Immunol 25:1–54
18. Billingham RE, Boswell T (1953) Studies on the problem of corneal homografts. Proc R Soc Lond B Biol Sci 141:392–406
19. Becher B, Bechmann I, Greter M (2006) Antigen presentation in autoimmunity and CNS inflammation: how T lymphocytes recognize the brain. J Mol Med 84:532–543
20. Hickey WF (2001) Basic principles of immunological surveillance of the normal central nervous system. Glia 36:118–124
21. Bechmann I (2005) Failed central nervous system regeneration: a downside of immune privilege? Neuromolecular Med 7:217–228
22. Cserr HF, Knopf PM (1992) Cervical lymphatics, the blood-brain barrier and the immunoreactivity of the brain: a new view. Immunol Today 13:507–512
23. Goldmann J, Kwidzinski E, Brandt C, Mahlo J, Richter D, Bechmann I (2006) T cells traffic from brain to cervical lymph nodes via the cribroid plate and the nasal mucosa. J Leukoc Biol 80:797–801
24. Carson MJ, Sutcliffe JG, Campbell IL (1999) Microglia stimulate naive T-cell differentiation without stimulating T-cell proliferation. J Neurosci Res 55:127–134
25. Hatterer E, Davoust N, Didier-Bazes M, Vuaillat C, Malcus C, Belin MF et al (2006) How to drain without lymphatics? Dendritic cells migrate from the cerebrospinal fluid to the B-cell follicles of cervical lymph nodes. Blood 107:806–812
26. Karman J, Ling C, Sandor M, Fabry Z (2004) Initiation of immune responses in brain is promoted by local dendritic cells. J Immunol 173:2353–2361
27. Lees JR, Archambault AS, Russell JH (2006) T-cell trafficking competence is required for CNS invasion. J Neuroimmunol 177:1–10
28. Ploix CC, Noor S, Crane J, Masek K, Carter W, Lo DD et al (2011) CNS-derived CCL21 is both sufficient to drive homeostatic CD4+ T cell proliferation and necessary for efficient CD4+ T cell migration into the CNS parenchyma following Toxoplasma gondii infection. Brain Behav Immun 25:883–896
29. Fine HA (1995) Novel biologic therapies for malignant gliomas. Antiangiogenesis, immunotherapy, and gene therapy. Neurol Clin 13:827–846
30. Fecci PE, Mitchell DA, Whitesides JF, Xie W, Friedman AH, Archer GE et al (2006) Increased regulatory T-cell fraction amidst a diminished CD4 compartment explains cellular immune defects in patients with malignant glioma. Cancer Res 66:3294–3302
31. McVicar DW, Davis DF, Merchant RE (1992) In vitro analysis of the proliferative potential of T cells from patients with brain tumor: glioma-associated immunosuppression

unrelated to intrinsic cellular defect. J Neurosurg 76:251–260

32. Peluso I, Fantini MC, Fina D, Caruso R, Boirivant M, MacDonald TT et al (2007) IL-21 counteracts the regulatory T cell-mediated suppression of human CD4+ T lymphocytes. J Immunol 178:732–739
33. Yamagiwa S, Gray JD, Hashimoto S, Horwitz DA (2001) A role for TGF-beta in the generation and expansion of CD4 + CD25+ regulatory T cells from human peripheral blood. J Immunol 166:7282–7289
34. Gomez GG, Kruse CA (2006) Mechanisms of malignant glioma immune resistance and sources of immunosuppression. Gene Ther Mol Biol 10:133–146
35. Skog J (2006) Glioma-specific antigens for immune tumor therapy. Expert Rev Vaccines 5:793–802
36. Perry VH (1998) A revised view of the central nervous system microenvironment and major histocompatibility complex class II antigen presentation. J Neuroimmunol 90:113–121
37. Schmidtmayer J, Jacobsen C, Miksch G, Sievers J (1994) Blood monocytes and spleen macrophages differentiate into microglia-like cells on monolayers of astrocytes: membrane currents. Glia 12:259–267
38. Streit WJ, Graeber MB (1993) Heterogeneity of microglial and perivascular cell populations: insights gained from the facial nucleus paradigm. Glia 7:68–74
39. McGeer PL, Kawamata T, Walker DG, Akiyama H, Tooyama I, McGeer EG (1993) Microglia in degenerative neurological disease. Glia 7:84–92
40. Becher B, Antel JP (1996) Comparison of phenotypic and functional properties of immediately ex vivo and cultured human adult microglia. Glia 18:1–10
41. Sedgwick JD, Ford AL, Foulcher E, Airriess R (1998) Central nervous system microglial cell activation and proliferation follows direct interaction with tissue-infiltrating T cell blasts. J Immunol 160:5320–5330
42. Ulvestad E, Williams K, Bo L, Trapp B, Antel J, Mork S (1994) HLA class II molecules (HLA-DR, -DP, -DQ) on cells in the human CNS studied in situ and in vitro. Immunology 82:535–541
43. McMahon EJ, Bailey SL, Castenada CV, Waldner H, Miller SD (2005) Epitope spreading initiates in the CNS in two mouse models of multiple sclerosis. Nat Med 11:335–339
44. Becher B, Prat A, Antel JP (2000) Brain–immune connection: immuno-regulatory properties of CNS-resident cells. Glia 29:293–304
45. Aloisi F (2001) Immune function of microglia. Glia 36:165–179
46. Li H, Newcombe J, Groome NP, Cuzner ML (1993) Characterization and distribution of phagocytic macrophages in multiple sclerosis plaques. Neuropathol Appl Neurobiol 19:214–223
47. Slobodov U, Reichert F, Mirski R, Rotshenker S (2001) Distinct inflammatory stimuli induce different patterns of myelin phagocytosis and degradation in recruited macrophages. Exp Neurol 167:401–409
48. Williams K, Ulvestad E, Waage A, Antel JP, McLaurin J (1994) Activation of adult human derived microglia by myelin phagocytosis in vitro. J Neurosci Res 38:433–443
49. Olson JK, Eagar TN, Miller SD (2002) Functional activation of myelin-specific T cells by virus-induced molecular mimicry. J Immunol 169:2719–2726
50. Serafini B, Columba-Cabezas S, Di Rosa F, Aloisi F (2000) Intracerebral recruitment and maturation of dendritic cells in the onset and progression of experimental autoimmune encephalomyelitis. Am J Pathol 157:1991–2002
51. Irani DN (1998) The susceptibility of mice to immune-mediated neurologic disease correlates with the degree to which their lymphocytes resist the effects of brain-derived gangliosides. J Immunol 161:2746–2752
52. Masson F, Calzascia T, Di Berardino-Besson W, de Tribolet N, Dietrich PY, Walker PR (2007) Brain microenvironment promotes the final functional maturation of tumor-specific effector CD8+ T cells. J Immunol 179:845–853
53. Ehtesham M, Winston JA, Kabos P, Thompson RC (2006) CXCR4 expression mediates glioma cell invasiveness. Oncogene 25:2801–2806
54. Ehtesham M, Yuan X, Kabos P, Chung NH, Liu G, Akasaki Y, Black KL, Yu JS (2004) Glioma tropic neural stem cells consist of astrocytic precursors and their migratory capacity is mediated by CXCR4. Neoplasia 6:287–293
55. Mitchell DA, Karikari I, Cui X, Xie W, Schmittling R, Sampson JH (2008) Selective modification of antigen-specific T cells by RNA electroporation. Hum Gene Ther 19:511–521
56. Burnet FM (1970) The concept of immunological surveillance. Prog Exp Tumor Res 13:1–27
57. Foley EJ (1952) Immunity of C3H mice to lymphosarcoma 6-C3H-Ed following regression of the implanted tumor. Proc Soc Exp Biol Med 80:675–677

58. Foley EJ (1953) Antigenic properties of methylcholanthrene-induced tumors in mice of the strain of origin. Cancer Res 13:835–837
59. Sjogren HO (1961) Further studies on the induced resistance against isotransplantation of polyoma tumors. Virology 15:214–219
60. Stutman O (1974) Tumor development after 3-methylcholanthrene in immunologically deficient athymic nude mice. Science 183:534–536
61. Mombaerts P, Iacomini J, Johnson RS, Herrup K, Tonegawa S, Papaioannou VE (1992) RAG-1-deficient mice have no mature B and T lymphocytes. Cell 68:869–877
62. Kirchgessner CU, Patil CK, Evans JW, Cuomo CA, Fried LM, Carter T, Oettinger MA, Brown JM (1995) DNA-dependent kinase (p350) as a candidate gene for the murine SCID defect. Science 267:1178–1183
63. Shankaran V, Ikeda H, Bruce AT, White JM, Swanson PE, Old LJ et al (2001) IFNγ and lymphocytes prevent primary tumor development and shape tumor immunogenicity. Nature 410:1107–1111
64. Dunn GP, Old LJ, Schreiber RD (2004) The immunobiology of cancer immunosurveillance and immunoediting. Immunity 21:137–148
65. Seliger B, Ritz U, Ferrone S (2006) Molecular mechanisms of HLA class I antigen abnormalities following viral infection and transformation. Int J Cancer 118:129–138
66. O'Garra A, Vieira P (2004) Regulatory T cells and mechanisms of immune system control. Nat Med 10:801–805
67. Bierie B, Moses HL (2006) Tumor microenvironment: TGFβ: The molecular Jekyll and Hyde of cancer. Nat Rev Cancer 6:506–520
68. King GD, Curtin JF, Candolfi M, Kroeger K, Lowenstein PR, Castro MG (2005) Gene therapy and targeted toxins for glioma. Curr Gene Ther 5:535–557
69. Candolfi M, Curtin JF, Nichols WS, Muhammad AKMG, King GD, Pluhar GE et al (2007) Intracranial glioblastoma models in preclinical neurooncology: neuropathological characterization and tumor progression. J Neurooncol 85:133–148
70. Kim C, Hong M, Park S, Kim C, Park M, Sohn H et al (2006) Enhancement of anti-tumor immunity specific to murine glioma by vaccination with tumor cell lysate-pulsed dendritic cells engineered to produce interleukin-12. Cancer Immunol Immunother 55:1309–1319
71. Sampson JH, Ashley DM, Archer GE, Fuchs HE, Dranoff G, Hale LP et al (1997) Characterization of a spontaneous murine astrocytoma and abrogation of its tumorigenicity by cytokine secretion. Neurosurgery 41:1365–1373
72. Kruse CA, Molleston MC, Parks EP, Schiltz PM, Kleinschmidt-DeMasters BK, Hickey WF (1994) A rat glioma model, CNS-1, with invasive characteristics similar to those of human gliomas: a comparison to 9 L gliosarcoma. J Neurooncol 22:191–200
73. Tzeng JJ, Barth RF, Orosz CG, James SM (1991) Phenotype and functional activity of tumor-infiltrating lymphocytes isolated from immunogenic and nonimmunogenic rat brain tumors. Cancer Res 51:2373–2378
74. Barth RF (1998) Rat brain tumor models in experimental neuro-oncology: the 9 L, C6, T9, F98, RG2 (D74), RT-2 and CNS-1 gliomas. J Neurooncol 36:91–102
75. Muhammad AKMG, Candolfi M, King GD, Yagiz K, Foulad D, Mineharu Y et al (2009) Anti-glioma immunological memory in response to conditional cytotoxic/immune-stimulatory gene therapy: humoral and cellular immunity lead to tumor regression. Clin Cancer Res 15:6113–6127
76. Ali S, Curtin JF, Zirger JM, Xiong W, King GD, Barciaet C et al (2004) Inflammatory and anti-glioma effects of an adenovirus expressing human soluble Fms-like tyrosine kinase 3 ligand (hsFlt3L): treatment with hsFlt3L inhibits intracranial glioma progression. Mol Ther 10:1071–1084
77. Ali S, King GD, Curtin JF, Candolfi M, Xiong W, Liu C (2005) Combined immunostimulation and conditional cytotoxic gene therapy provide long-term survival in a large glioma model. Cancer Res 65:7194–7204
78. Candolfi M, Yagiz K, Foulad D, Alzadeh GE, Tesarfreund M, Muhammad AK et al (2009) Release of HMGB1 in response to proapoptotic glioma killing strategies: efficacy and neurotoxicity. Clin Cancer Res 15: 4401–4414
79. Weiss W, Israel M, Cobbs C, Holland E, James D, Louis D et al (2002) Neuropathology of genetically engineered mice: consensus report and recommendations from an international forum. Oncogene 21:7453–7463
80. Ding H, Nagy A, Gutmann D, Guha A (2000) Astrocytoma models. Neurosurg Focus 8:1–7
81. Giannini C, Sarkaria JN, Saito A, Uhm JH, Galanis E, Carlson BL et al (2005) Patient tumor EGFR and PDGFRA gene amplifications retained in an invasive intracranial

xenograft model of glioblastoma multiforme. Neuro Oncol 7:164–176

82. Cho SY, Ravasi L, Szajek LP, Seidel J, Green MV, Fine HA et al (2005) Evaluation of (76) Br-FBAU as a PET reporter probe for HSV1-tk gene expression imaging using mouse models of human glioma. J Nucl Med 46:1923–1930
83. Kirsch M, Strasser J, Allende R, Bello L, Zhang J, Black PM (1998) Angiostatin suppresses malignant glioma growth in vivo. Cancer Res 58:4654–4659
84. Lund EL, Bastholm L, Kristjansen PE (2000) Therapeutic synergy of TNP-470 and ionizing radiation: effects on tumor growth, vessel morphology, and angiogenesis in human glioblastoma multiforme xenografts. Clin Cancer Res 6:971–978
85. Tentori L, Leonetti C, Scarsella M, D'Amati G, Vergati M, Portarena I et al (2003) Systemic administration of GPI 15427, a novel poly(ADP-ribose) polymerase-1 inhibitor, increases the antitumor activity of temozolomide against intracranial melanoma, glioma, lymphoma. Clin Cancer Res 9:5370–5379
86. Shen S, Khazaeli MB, Gillespie GY, Alvarez VL (2005) Radiation dosimetry of 131I-chlorotoxin for targeted radiotherapy in glioma-bearing mice. J Neurooncol 71:113–119
87. Kawakami K, Kawakami M, Kioi M, Husain SR, Puri RK (2004) Distribution kinetics of targeted cytotoxin in glioma by bolus or convection-enhanced delivery in a murine model. J Neurosurg 101:1004–1011
88. Cirielli C, Inyaku K, Capogrossi MC, Yuan X, Williams JA (1999) Adenovirus-mediated wild-type p53 expression induces apoptosis and suppresses tumorigenesis of experimental intracranial human malignant glioma. J Neurooncol 43:99–108
89. Gromeier M, Lachmann S, Rosenfeld MR, Gutin PH, Wimmer E (2000) Intergeneric poliovirus recombinants for the treatment of malignant glioma. Proc Natl Acad Sci USA 97:6803–6808
90. Gromeier M, Solecki D, Patel DD, Wimmer E (2000) Expression of the human poliovirus receptor/CD155 gene during development of the central nervous system: implications for the pathogenesis of poliomyelitis. Virology 273:248–257
91. Ahmed N, Ratnayake M, Savoldo B, Perlaky L, Dotti G, Wels WS et al (2007) Regression of experimental medulloblastoma following transfer of HER2-specific T cells. Cancer Res 67:5957–5964
92. Vatakis DN, Koya RC, Nixon CC, Wei L, Kim SG, Avancena P et al (2011) Antitumor activity from antigen-specific CD8 T cells generated in vivo from genetically engineered human hematopoietic stem cells. Proc Natl Acad Sci USA 108:E1408–E1416
93. Sette A, Sidney J (1999) Nine major HLA class I supertypes account for the vast preponderance of HLA-A and -B polymorphism. Immunogenetics 50:201–212
94. Harvald B (1989) Genetic epidemiology of Greenland. Clin Genet 36:364–367
95. Vitiello A, Marchesini D, Furze J, Sherman LA, Chesnut RW (1991) Analysis of the HLA-restricted influenza-specific cytotoxic T lymphocyte response in transgenic mice carrying a chimeric human-mouse class I major histocompatibility complex. J Exp Med 173:1007–1015
96. McLaughlin PM, Kroesen BJ, Harmsen MC, de Leij L (2001) Cancer immunotherapy: insights from transgenic animal models. Crit Rev Oncol Hematol 40:53–76
97. Lauritzsen GF, Hofgaard PO, Schenck K, Bogen B (1998) Clonal deletion of thymocytes as a tumor escape mechanism. Int J Cancer 78:216–222
98. Lauritzsen GF, Weiss S, Dembic Z, Bogen B (1994) Naive idiotypespecific CD4+ T cells and immunosurveillance of B-cell tumors. Proc Natl Acad Sci USA 91:5700–5704
99. Allison J, Stephens LA, Kay TW, Kurts C, Heath WR, Miller JF et al (1998) The threshold for autoimmune T cell killing is influenced by B7-1. Eur J Immunol 28:949–960
100. Bluming AZ, Ziegler JL (1971) Regression of Burkitt's lymphoma in association with measles infection. Lancet 2:105–106
101. Coffey MC, Strong JE, Forsyth PA, Lee PW (1998) Reovirus therapy of tumors with activated Ras pathway. Science 282:1332–1334
102. Stojdl DF, Lichty B, Knowles S, Marius R, Atkins H, Sonenberg N et al (2000) Exploiting tumor-specific defects in the interferon pathway with a previously unknown oncolytic virus. Nat Med 6:821–825
103. Martuza RL, Malick A, Markert JM, Ruffner KL, Coen DM (1991) Experimental therapy of human glioma by means of a genetically engineered virus mutant. Science 252:854–856
104. Dobrikova EY, Broadt T, Poiley-Nelson J, Yang X, Soman G, Giardina S et al (2008) Recombinant oncolytic poliovirus eliminates glioma in vivo without genetic adaptation to a pathogenic phenotype. Mol Ther 16:1865–1872
105. Wilcox ME, Yang W, Senger D, Rewcastle NB, Morris DG, Penny MA et al (2001)

Reovirus as an oncolytic agent against experimental human malignant gliomas. J Natl Cancer Inst 93:903–912
106. Boczkowski D, Nair SK, Snyder D, Gilboa E (1996) Dendritic cells pulsed with RNA are potent antigen-presenting cells in vitro and in vivo. J Exp Med 184:465–472
107. Boczkowski D, Nair SK, Nam JH, Lyerly HK, Gilboa E (2000) Induction of tumor immunity and cytotoxic T lymphocyte responses using dendritic cells transfected with messenger RNA amplified from tumor cells. Cancer Res 60:1028–1034
108. Hoerr I, Obst R, Rammensee HG, Jung G (2000) In vivo application of RNA leads to induction of specific cytotoxic T lymphocytes and antibodies. Eur J Immunol 30:1–7
109. Martinon F, Krishnan S, Lenzen G, Magné R, Gomard E, Guillet JG et al (1993) Induction of virus-specific cytotoxic T lymphocytes in vivo by liposome-entrapped mRNA. Eur J Immunol 23:1719–1722
110. Zhou WZ, Hoon D, Huang SKS, Fujii S, Hashimoto K, Morishita R et al (1999) RNA melanoma vaccine: induction of antitumor immunity by human glycoprotein 100 mRNA immunization. Hum Gene Ther 10:2719–2724
111. Zhang W, He L, Yuan Z, Xie Z, Wang J, Hamada H, Cao X (1999) Enhanced therapeutic efficacy of tumor RNA-pulsed dendritic cells after genetic modification with lymphotactin. Hum Gene Ther 10:1151–1161
112. Ashley DM, Faiola B, Nair S, Hale LP, Bigner DD, Gilboa E (1997) Bone marrow-generated dendritic cells pulsed with tumor extracts or tumor RNA induce antitumor immunity against central nervous system tumors. J Exp Med 186:1177–1182
113. Nair SK, Boczkowski D, Morse M, Cumming RI, Lyerly HK, Gilboa E (1998) Induction of primary carcinoembryonic antigen (CEA)-specific cytotoxic T lymphocytes in vitro using human dendritic cells transfected with RNA. Nat Biotechnol 16:364–369
114. Mitchell DA, Nair SK (2000) RNA-transfected dendritic cells in cancer immunotherapy. J Clin Invest 106:1065–1069
115. Dudley ME, Wunderlich JR, Robbins PF, Yang JC, Hwu P, Schwartzentruber DJ et al (2002) Cancer regression and autoimmunity in patients after clonal repopulation with antitumor lymphocytes. Science 298:850–854
116. Huang J, Khong HT, Dudley ME, El-Gamil M, Li YF, Rosenberg SA et al (2005) Survival, persistence, and progressive differentiation of adoptively transferred tumor-reactive T cells associated with tumor regression. J Immunother 28:258–267
117. Rosenberg SA, Dudley ME (2004) Cancer regression in patients with metastatic melanoma after the transfer of autologous antitumor lymphocytes. Proc Natl Acad Sci USA 101(Suppl 2):14639–14645
118. Cole DJ, Taubenberger JK, Pockaj BA, Yannelli JR, Carter C, Carrasquillo J et al (1994) Histopathological analysis of metastatic melanoma deposits in patients receiving adoptive immunotherapy with tumor-infiltrating lymphocytes. Cancer Immunol Immunother 38:299–303
119. Pockaj BA, Sherry RM, Wei JP, Yannelli JR, Carter CS, Leitman SF et al (1994) Localization of 111indium-labeled tumor infiltrating lymphocytes to tumor in patients receiving adoptive immunotherapy. Augmentation with cyclophosphamide and correlation with response. Cancer 73:1731–1737
120. Zhou J, Dudley ME, Rosenberg SA, Robbins PF (2005) Persistence of multiple tumor-specific T-cell clones is associated with complete tumor regression in a melanoma patient receiving adoptive cell transfer therapy. J Immunother 28:53–62
121. Engelhardt B, Ransohoff RM (2005) The ins and outs of T-lymphocyte trafficking to the CNS: anatomical sites and molecular mechanisms. Trends Immunol 26:485–495
122. Plautz GE, Shu S (2001) Adoptive immunotherapy of CNS malignancies. Cancer Chemother Biol Response Modif 19:327–338
123. Charalambous C, Pen LB, Su YS, Milan J, Chen TC, Hofman FM (2005) Interleukin-8 differentially regulates migration of tumor-associated and normal human brain endothelial cells. Cancer Res 65:10347–10354
124. Morita M, Kasahara T, Mukaida N, Matsushima K, Nagashima T, Nishizawa M et al (1993) Induction and regulation of IL-8 and MCAF production in human brain tumor cell lines and brain tumor tissues. Eur Cytokine Netw 4:351–358
125. Nitta T, Allegretta M, Okumura K, Sato K, Steinman L (1992) Neoplastic and reactive human astrocytes express interleukin-8 gene. Neurosurg Rev 15:203–207
126. van Den Pol AN, Mocarski E, Saederup N, Vieira J, Meier TJ (1999) Cytomegalovirus cell tropism, replication, and gene transfer in brain. J Neurosci 19:10948–10965
127. Al-Hajj M, Wicha MS, Benito-Hernandez A, Morrison SJ, Clarke MF (2003) Prospective identification of tumorigenic breast cancer cells. Proc Natl Acad Sci U S A 100:3983–3988

128. Bonnet D, Dick JE (1997) Human acute myeloid leukemia is organized as a hierarchy that originates from a primitive hematopoietic cell. Nat Med 3:730–737
129. Galli R, Binda E, Orfanelli U, Cipelletti B, Gritti A, De Vitis S et al (2004) Isolation and characterization of tumorigenic, stem-like neural precursors from human glioblastoma. Cancer Res 64:7011–7021
130. Hemmati HD, Nakano I, Lazareff JA, Masterman-Smith M, Geschwind DH, Bronner-Fraser M et al (2003) Cancerous stem cells can arise from pediatric brain tumors. Proc Natl Acad Sci U S A 100:15178–15183
131. Bao S, Wu Q, Sathornsumetee S, Hao Y, Li Z, Hjelmeland AB et al (2006) Stem cell-like glioma cells promote tumor angiogenesis through vascular endothelial growth factor. Cancer Res 66:7843–7848
132. Friedman HS, Prados MD, Wen PY, Mikkelsen T, Schiff D, Abrey LE et al (2009) Bevacizumab alone and in combination with irinotecan in recurrent glioblastoma. J Clin Oncol 27:4733–4740
133. Vredenburgh JJ, Desjardins A, Herndon JE 2nd, Dowell JM, Reardon DA, Quinn JA et al (2007) Phase II trial of bevacizumab and irinotecan in recurrent malignant glioma. Clin Cancer Res 13:1253–1259
134. Vredenburgh JJ, Desjardins A, Herndon JE 2nd, Marcello J, Reardon DA, Quinn JA et al (2007) Bevacizumab plus irinotecan in recurrent glioblastoma multiforme. J Clin Oncol 25:4722–4729
135. Bao S, Wu Q, McLendon RE, Hao Y, Shi Q, Hjelmeland AB et al (2006) Glioma stem cells promote radioresistance by preferential activation of the DNA damage response. Nature 444:756–760
136. Johannessen TC, Wang J, Skaftnesmo KO, Sakariassen PO, Enger PO, Petersen K et al (2009) Highly infiltrative brain tumours show reduced chemosensitivity associated with a stem cell-like phenotype. Neuropathol Appl Neurobiol 35:380–393
137. Liu G, Yuan X, Zeng Z, Tunici P, Ng H, Abdulkadir IR et al (2006) Analysis of gene expression and chemoresistance of CD133+ cancer stem cells in glioblastoma. Mol Cancer 5:67
138. Pellegatta S, Poliani PL, Corno D, Menghi F, Ghielmetti F, Suarez-Merino B et al (2006) Neurospheres enriched in cancer stem-like cells are highly effective in eliciting a dendritic cell-mediated immune response against malignant gliomas. Cancer Res 66:10247–10252
139. Lal A, Lash AE, Altschul SF, Velculescu V, Zhang L, McLendon RE et al (1999) A public database for gene expression in human cancers. Cancer Res 59:5403–5407
140. Lennerz V, Fatho M, Gentilini C, Frye RA, Lifke A, Ferel D et al (2005) The response of autologous T cells to a human melanoma is dominated by mutated neoantigens. Proc Natl Acad Sci U S A 102:16013–16018
141. Wikstrand CJ, Hale LP, Batra SK, Hill ML, Humphrey PA, Kurpad SN et al (1995) Monoclonal antibodies against EGFRvIII are tumor specific and react with breast and lung carcinomas and malignant gliomas. Cancer Res 55:3140–3148
142. Frederick L, Wang XY, Eley G, James CD (2000) Diversity and frequency of epidermal growth factor receptor mutations in human glioblastomas. Cancer Res 60:1383–1387
143. Heimberger AB, Hlatky R, Suki D, Yang D, Weinberg J, Gilbert M et al (2005) Prognostic effect of epidermal growth factor receptor and EGFRvIII in glioblastoma multiforme patients. Clin Cancer Res 11:1462–1466
144. Aldape KD, Ballman K, Furth A, Buckner JC, Giannini C, Burger PC et al (2004) Immunohistochemical detection of EGFRvIII in high malignancy grade astrocytomas and evaluation of prognostic significance. J Neuropathol Exp Neurol 63:700–707
145. Cunningham MP, Essapen S, Thomas H, Green M, Lovell DP, Topham C et al (2005) Coexpression, prognostic significance and predictive value of EGFR. EGFRvIII and phosphorylated EGFR in colorectal cancer. Int J Oncol 27:317–325
146. Sampson JH, Archer GE, Mitchell DA, Heimberger AB, Herndon JE 2nd, Lally-Goss D et al (2009) An epidermal growth factor receptor variant III-targeted vaccine is safe and immunogenic in patients with glioblastoma multiforme. Mol Cancer Ther 8:2773–2779
147. Wikstrand CJ, McLendon RE, Friedman AH, Bigner DD (1997) Cell surface localization and density of the tumor-associated variant of the epidermal growth factor receptor, EGFRvIII. Cancer Res 57:4130–4140
148. Dix AR, Brooks WH, Roszman TL, Morford LA (1999) Immune defects observed in patients with primary malignant brain tumors. J Neuroimmunol 100:216–232
149. Hughes MA, Parisi M, Grossman S, Kleinberg L (2005) Primary brain tumors treated with steroids and radiotherapy: low CD4 counts and risk of infection. Int J Radiat Oncol Biol Phys 62:1423–1426

150. Morford LA, Elliott LH, Carlson SL, Brooks WH, Roszman TL (1997) T cell receptor-mediated signaling is defective in T cells obtained from patients with primary intracranial tumors. J Immunol 159:4415–4425
151. Fecci PE, Sweeney AE, Grossi PM, Nair SK, Learn CA, Mitchell DA et al (2006) Systemic anti-CD25 monoclonal antibody administration safely enhances immunity in murine glioma without eliminating regulatory T cells. Clin Cancer Res 12:4294–4305
152. Chambers CA, Krummel MF, Boitel B, Hurwitz A, Sullivan TJ, Fournier S et al (1996) The role of CTLA-4 in the regulation and initiation of T-cell responses. Immunol Rev 153:27–46
153. Brunet JF, Denizot F, Luciani MF, Roux-Dosseto M, Suzan M, Mattei MG et al (1987) A new member of the immunoglobulin superfamily–CTLA-4. Nature 328:267–270
154. Fecci PE, Ochiai H, Mitchell DA, Grossi PM, Sweeney AE, Archer GE et al (2007) Systemic CTLA-4 blockade ameliorates glioma-induced changes to the CD4+ T cell compartment without affecting regulatory T-cell function. Clin Cancer Res 13:2158–2167
155. Velicu S, Han Y, Ulasov I, Brown IE, El Andaloussi A, Gajewski TF et al (2006) Cross-priming of T cells to intracranial tumor antigens elicits an immune response that fails in the effector phase but can be augmented with local immunotherapy. J Neuroimmunol 174:74–81
156. Sawaya R (1999) Extent of resection in malignant gliomas: a critical summary. J Neurooncol 42:303–305
157. Rich JN, Bigner DD (2004) Development of novel targeted therapies in the treatment of malignant glioma. Nat Rev Drug Discov 3:430–446
158. Gribben JG, Ryan DP, Boyajian R, Urban RG, Hedley ML, Beach B et al (2005) Unexpected association between induction of immunity to the universal tumor antigen CYP1B1 and response to next therapy. Clin Cancer Res 11:4430–4436
159. Wheeler CJ, Das A, Liu G, Yu JS, Black KL (2004) Clinical responsiveness of glioblastoma multiforme to chemotherapy after vaccination. Clin Cancer Res 10:5316–5326
160. Emens LA, Jaffee EM (2003) Toward a breast cancer vaccine: work in progress. Oncology 17:1200–1211
161. Liau LM, Prins RM, Kiertscher SM, Odesa SK, Kremen TJ, Giovannone AJ et al (2005) Dendritic cell vaccination in glioblastoma patients induces systemic and intracranial T-cell responses modulated by the local central nervous system tumor microenvironment. Clin Cancer Res 11:5515–5525
162. Kobayashi T, Yamanaka R, Homma J, Tsuchiya N, Yajima N, Yoshida S et al (2003) Tumor mRNA-loaded dendritic cells elicit tumor-specific CD8(+) cytotoxic T cells in patients with malignant glioma. Cancer Immunol Immunother 52:632–637
163. Mitchell DA, Nair SK (2000) RNA transfected dendritic cells as cancer vaccines. Curr Opin Mol Ther 2:176–181
164. Boon T, De Plaen E, Lurquin C, Van den Eynde B, van der Bruggen P, Traversari C et al (1992) Identification of tumour rejection antigens recognized by T lymphocytes. Cancer Surv 13:23–37

Neuromethods (2013) 77: 275–298
DOI 10.1007/7657_2012_39

Published online: 4 April 2012

Assessing Mechanisms of Glioblastoma Invasion

Stephen M. Robbins and Donna L. Senger

Abstract

Glioblastomas are one of the deadliest and most invasive cancers in humans. This heterogenous population of tumours interact with the surrounding brain parenchyma, disrupting physical barriers such as basement membranes, extracellular matrices and cell–cell contact while activating cellular processes that enable infiltration of cells long distances away from the original tumour mass. The protocols describe in this chapter aim to establish experimental models, both in vitro and in vivo, as tools for discovery and evaluation of the underlying molecular mechanisms of cellular invasion by providing experimental methodology that recapitulate, as close as possible, the tumour microenvironment in humans. Although established for the assessment of glioma migration and invasion, these methods provide experimental platforms that can be easily adapted for different types of cancers and for the screening of candidate cancer therapeutics.

Key words: Human glioma, Brain tumours, Invasion, Migration, Orthotopic brain tumour models, Murine model of human glioma

1. Introduction

1.1. Disease Reservoirs of Malignant Glioma

Human malignant glioma (MG) is one of the most common primary central nervous system tumours in adults with a median survival of 12–14 months and long-term survivors (i.e. surviving $\geq$ 3 years) are rare (1–3). A major barrier to effective malignant glioma treatment is the highly invasive nature of these tumours, where tendrils of tumour extend into the normal surrounding brain parenchyma rendering these tumours incurable by local therapies such as surgery or radiotherapy (4–6). Specifically, the tumour cells activate a number of coordinate cellular programs including those necessary for migration and invasion (7). These cells, along with a second highly invasive cell population, the recently described "brain tumour initiating cells" (BTICs), act as "disease reservoirs" which elude our current treatments and ultimately lead to disease relapse (8–14). Although the exact characterization of the BTIC cell population is still debated (15), these cells, when implanted into the brains of immunocompromised mice, result in a highly invasive and tumourigenic tumours that closely resemble the human disease (10, 16, 17). The invasive traits of these two cell populations together with the ability of invading glioma cells to reduce proliferation and upregulate survival pathways (5, 18–20) makes them highly resistant to conventional treatments (11, 12, 20). Identifying the biological properties of these distinct populations of cells within malignant glioma and specifically the underlying invasive mechanisms will aid in the development of new effective treatments (21).

1.2. Molecular Aspects of Glioma Progression and Invasion

Glioma invasion is not random but rather occurs along white matter tracts (optic radiations, corpus callosum or anterior commissure), basement membranes of blood vessels, along the subependyma and adjacent to neurons. In contrast to other organs, there are few "classical" extracellular matrix (ECM) proteins (collagen, laminin, vitronectin, fibronectin) in the brain except around blood vessels and the glia limitans externa. Instead the ECM of the brain is mostly composed of glycosoaminoglycans (GAGs) and a few structural glycoproteins, with the main GAGs being hyaluronic acid (HA), chondroitin, keratin and heparan sulphates (22). In addition, the glioma cells themselves synthesize many ECM components including tenascin, vitronectin, collagen types I, III, IV and VI, fibronectin, laminin, hyaluronan, chondroitin sulphate and heparan sulphate proteoglycans.

Over the years many individual genes have been implicated in glioma invasion and recently gene expression profiling has identified a subclass of gliomas expressing several genes with specific roles in cell migration and invasion that strongly correlate with poor patient survival (23–26). At present, no comprehensive model of glioma invasion exists; however, invasion can be conceptualized into discrete steps that include adhesion to the ECM, degradation of the ECM to allow passage, and glioma cell movement. Several cell surface receptors have been implicated in glioma adhesion and invasion, including the integrins (27–30), immunoglobulin superfamily members (e.g. NCAM) (4), cadherins (4, 31), CD44 (32, 33) , BEHAB (34, 35), elastin binding protein (36), SPARC (37), the ephrin (Eph) receptor–ligand system (38–41), semaphorins (42, 43), plexins (43) and the neurotrophin receptor $p75^{NTR}$ (44, 45). In addition to activating downstream signal cascades important for cell movement, engagement of many of these receptors results in the activation of several classes of proteases including matrix metalloproteinases (MMPs), ADAMs (a disintegrin and metalloproteinase), plasmin and cathepsins reviewed in (46–48) necessary for the degradation of the ECM.

1.3. Development of Assays to Evaluate Glioma Invasion

As a result of intense investigation in the area of cell invasion, numerous techniques have been establishment to identify molecular events required for the invasive ability. In the area of glioma invasion, strategies to directly identify genes and proteins important for invasion have included isolation of invasive cells from human cell lines in vitro (18, 19), the use of organotypic brain slice cultures (36, 49–53) and the collection of tumour and invasive cells from the main tumour mass in frozen glioblastoma patient specimens using laser capture microdissection (54–57). In addition, in vitro assays have been generated to assess the role of individual proteins or processes in glioma invasion, many of which will be discussed in this chapter. Although these techniques have been used successfully, no single technique is sufficiently comprehensive and the selection of assay(s) performed in a given study should be chosen with care. Here we will outline several systems that we have previously used to either verify or identify the underlying mechanisms involved in glioma invasion both in vitro and in vivo.

2. Materials and Methods

2.1. In Vivo Selection of Highly Invasive Glioma Variants

Background

One of the problems in xenotransplanting human glioma cells into the brains of immunocompromised mice is that the resulting tumours are circumscribed with very little cell infiltration into the brain parenchyma (58) (see Fig. 1, for example). To generate an orthotopic model that more closely mimics the human disease and allows for the identification of molecular determinants of glioma invasion in a global and unbiased manner, we hypothesized that using an in vivo selection strategy similar to the one developed by Richard Hynes and colleagues for metastatic melanoma (59), we could develop an in vivo glioma model that more closely recapitulates the human condition thus providing a strategy to isolate genetically and phenotypically distinct population of glioma cells that are highly invasive. As proof of concept, using the in vivo selection strategy outlined below, we previously isolated highly invasive glioma variants from two established human glioma cell lines U87MG and U251N (44). Stable transfectants of U87MG and U251N expressing green fluorescent proteins (GFP) and a neomycin resistant gene were serially transplanted into the brain of immunocompromised mice. Cells from the main tumour mass, and those that migrated to the contralateral hemisphere were isolated, grown and re-injected into the brains of a second mouse. Using this approach, both highly tumourigenic and highly invasive populations of cells were isolated and characterized for their phenotypic and genotypic changes (44).

Materials

All cell culture protocols outlined in this chapter use DMEM Complete Media that contains Dulbecco's modified Eagle's

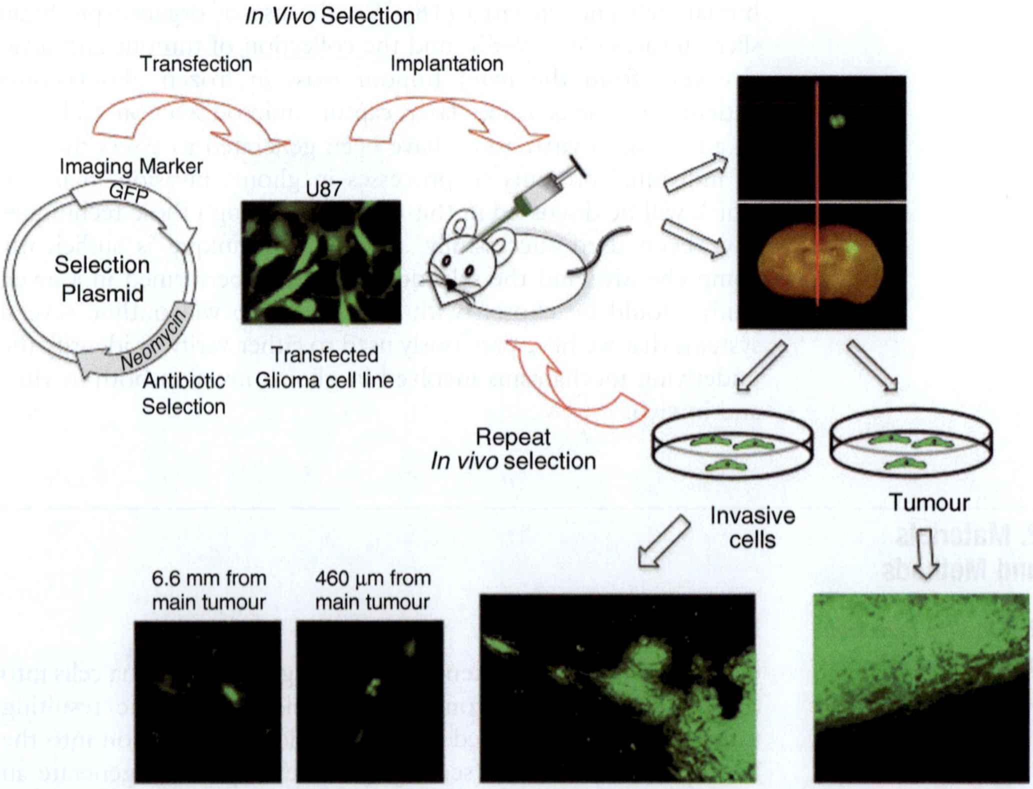

Fig. 1. Schematic of an in vivo selection strategy for the isolation of highly invasive glioma variants from an established glioma cell line. (*Top panel*) Transfect a non-invasive human glioma cell line with a plasmid containing an imaging marker (e.g. GFP) to visual the tumour cells in vivo, and a drug resistance or selection gene (e.g. Neomycin) to selectively isolate and grow the tumour cells in culture. Implant cells into the putamen of the brain of an immunocompromised animal using a stereotactic apparatus. Allow the tumours to establish for 4–6 weeks at which time the animals are sacrificed and the ipsilateral side of the brain containing a grossly visible tumour is separated from the contralateral side containing only isolated invasive glioma cells. Dissociate the brain tissue to single cells. Grow the cell suspension in media containing the appropriate antibiotic for the plasmid drug selection, then re-implant the isolated tumour cells into a second immunocompromised animal repeating the process to select for increasingly non-invasive or invasive glioma cells. The ipsilateral side of the brain contains a grossly visible tumour with a well demarcated border (*right lower panel*) while the tumour cells isolated from the contralateral side form tumours with highly infiltrative borders, with individual cells migrating substantial distances from the main tumour mass (*lower left 3 panels*).

medium F12 supplemented with 10% heat-inactivated foetal bovine serum, 1% antibiotic/antimycotic, 0.1 mM non-essential amino acids, 2 mM L-glutamine and 1 mM sodium pyruvate (Gibco BRL, Gaithersburg, MD, USA). The choice of culture media may vary depending on the cell line being assessed; thus, the DMEM Complete Media should be substituted with the appropriate media for each individual cell line used. All other products used are included in the methods outlined below. Comparable products from other suppliers should also be effective.

2.1.1. Establishment of Invasive Variants from a Human Glioma Cell Line by In Vivo Selection

1. Select a human glioma cell line that has the ability to establish intracranial tumours in immunocompromised mice (e.g. U87MG or U251 obtained from ATCC). Grow and maintain cells using standard tissue culture protocols in DMEM Complete Media or the appropriate media for each individual cell line depending on the growth requirements, at 37°C in a humidified 5% CO_2 incubator.
2. Establish stable cell lines expressing a selection plasmid that allows for the re-isolation of the glioma cells once they have been passaged through the animal. Choose a selection plasmid that expresses both GFP (to visualize the implanted glioma cells) and an antibiotic resistance gene (e.g. neomycin for recovery of the tumour cells post-intracranial implantation via antibiotic selection) such as the pEGFP-N1 vector from Clontech (Palo Alto, CA, USA). Seed cells at 2×10^5 cells/well in a 6-well tissue culture plate in DMEM Complete Media and incubate overnight at 37°C in a humidified 5% CO_2 incubator. Introduce the pEGFP-N1 vector using FuGENE™ 6 transfection reagent (Roche Diagnostic Co., Indianapolis, IN, USA) according to the manufacturers instructions. The plasmid and FuGENE™ reagent can be combined in various ratios in serum-free media and incubated at room temperature for 30 min to allow for the formation of transfection complexes. Add the plasmid/FuGENE™ mixture drop wise to the cells and incubate the cells in a humidified 5% CO_2 incubator at 37°C overnight.

   Note: Using a linearized plasmid usually helps increase the efficiency of establishing a stable cell line.
3. The following day change the media to fresh DMEM Complete Media containing G418, or the appropriate drug selection antibiotic for the transfection plasmid used, to select for cells that have taken up the vector. Determination of the appropriate concentration for the selection drug should be determined prior to expression vector transfection by performing a drug sensitivity curve for each cell line, selecting the minimum concentration of drug that gives 100% cell death (see below, Sect. 2.1.2); usually in the range of 100–400 μg for established glioma lines. Grow the cells under antibiotic selection until the wells are confluent. GFP-transfected cells can be identified by fluorescent microscopy and GFP expression of >95% can be obtained by fluorescence-activated cell sorting (FACScan flow cytometer; Becton, Dickinson and Company, Oakville, ON, Canada).
4. Stable transfectants of glioma cells can be maintained in the same media, with the addition of G418 at the concentration determined above (Invitrogen, Carlsbad, CA, USA). It is important to have stable clones prior to implantation in the

animals so that the in vivo selected glioma cells maintain expression of the plasmid.

5. For in vivo implantation, harvest actively growing GFP/neomycin glioma cells using trypsin–EDTA (0.25%), wash twice in sterile PBS [137 mM NaCl, 8.1 mM Na_2HPO_4, 2.68 mM KCl and 1.47 mM KH_2PO_4 (pH 7.5)] and resuspend 10^7 cells in 100 μl PBS.
6. For the stereotactic injection, use a 10-ml syringe (801RN Hamilton Co, Lot#436284) with a 30-gauge needle (RN NDL Lot#440603) mounted on a Kopf stereotactic apparatus (Model 900 Small Animal Stereotaxic Instrument, Kopf Instruments, Tujanga, CA; http://www.kopfinstruments.com) outfitted with a mouse adaptor (Model 926 Mouse Adaptor includes a nose and tooth (palate) bar assembly with zygoma ear cups).
7. Anesthetize the mouse using an intraperitoneal administration of ketamine 200 mg/kg and xylazine 10 mg/kg (MTC pharmaceuticals, Cambridge ON, Canada) in sterile water. In addition, give a subcutaneous pre-operative dose of buprenorphine 0.05 mg/kg for pain. Implant 1×10^5 cells/mouse in 2–3 μl sterile PBS, intracerebrally into the right putamen of 6- to 8-week-old female SCID mice at a depth of 3 mm through a scalp incision and a 0.5 mm burr hole made 1.5–2 mm right of the midline and 0.5–1 mm posterior to the coronal suture. To ensure the cells are injected into the putamen, the head position should be flat or parallel to horizontal, if available, use ear bar stabilizers (Kopf Instruments, Model 921 Zygoma Ear Cups) to maintain correct head position. The implantation depth of 3 mm is the depth after the needle has entered the brain tissue. Once a depth of 3 mm has been reached, go a further 1 mm forward, then withdraw the needle back 1 mm to the original 3 mm depth; this provides additional space for the cells to be deposited. To prevent the withdrawal of tumour cells along the needle tract, inject cells slowly (30 s to 1 min), wait 3–5 min to allow the cells to adhere to the brain tissue and then slowly withdraw the needle allowing the tissue to close behind the needle. It is important to observe any deviation of the implantation site caused by the needle that is not perpendicular to the horizontal, this could result in an implantation site too close to the lateral ventricle.
8. Once the needle has been withdrawn, the incision is closed using a 9-mm Clay Adams wound clip (BD #427631).
9. House the animals according to your institutions animal regulations. In our case, animals are housed in groups of three to five and maintained on a 12-h light/dark schedule with a

temperature of 22 ± 1°C and a relative humidity of 50 ± 5%. Food and water should be available ad libitum.

10. Following surgery, weigh and record the body weight of each animal daily and observe the animal for any signs of discomfort.
11. Allow at least 21 days for the formation of the tumour. The actual duration will depend on the health of the mouse and the type of cells injected.
12. To isolate the tumour and invasive glioma cell populations, sacrifice the mouse, divide the brain coronally close to the needle tract and examine using fluorescence to visualize the GFP-expressing glioma cells (Fig. 1). Ensure that the main tumour mass is completely embedded within the normal tissue of the brain. This is to make certain that the tumour cells present in the contralateral hemisphere arrived there by invading through the normal brain parenchyma.
13. For tissue culture, separate the hemisphere containing the main tumour mass from the contralateral hemisphere. Independently isolate cells from each hemisphere as described below (Sect. 2.1.3).
14. Resuspend the tissue homogenate and gently force through a 100-μm mesh cell strainer (BD Bioscience; BD Falcon™ Cell Strainer, Catalogue # 352360) into a 50-ml conical centrifuge tube. The resulting cell suspension is then centrifuged at low speed (400–500 × *g*) for 10 min.
15. Remove the supernatant and resuspend the cell pellet in DMEM Complete Media containing G418 (in our case 400 μg/ml) to select for the GFP-transfected tumour cells. Cells obtained from the hemisphere containing the main tumour mass are the "tumour" cells and cells isolated from the contralateral hemisphere are the "invasive" cells. Maintain the culture at 37°C with 5% CO_2. Once the cells have reached 80% confluence, passage the cells to expand them prior to implantation into the brain of a second immunocompromised mouse. Depending on the number of invasive cells recovered from the contralateral hemisphere, the invasive cultures may need to be maintained and expanded for 2–3 months to obtain enough cells to be re-inoculated into a SCID mouse. Repeat the procedure until two distinct in vivo tumour populations are observed.

After 2–3 rounds of in vivo selection, the brains of the immunocompromised mice implanted with the non-invasive "tumour" cells should form large tumours with sharply demarcated edges (Fig. 1; lower right) while the brains of mice implanted with the "invasive" cells should form highly infiltrative tumours with poorly defined edges (Fig. 1; lower left). Using this experimental strategy, we have previously isolated two populations of cells from a single parental cell line that are phenotypically and genotypically distinct with gene expression differences between the non-invasive and highly invasive in vivo selected glioma cells (44).

2.1.2. Determination of G418 Sensitivity

1. Plate 2×10^4 cells/well into a 6-well plate (10 wells total).
2. Grow cells for 24 h at 37°C in a humidified 5% CO_2 incubator.
3. Remove media and then add media containing varying concentrations of G418 (0, 50, 100, 200, 300, 400, 500, 600, 800, 1,000 μg/ml).
4. Replenish the selection medium every 3–4 days and observe the percentage of surviving cells over time.
5. Select the minimum concentration that kills 100% of the cells in 2 weeks (For U87MG and U251N, we have determined that 400 μg/ml of G418 is optimal).

Similar strategy can also be used for other selection drugs.

2.1.3. Preparation of Tumour and Invasive Cells from Mouse Brain Injected with GFP/Neomycin Transfected Human Glioma Cells

The following experimental procedure should be performed under sterile tissue culture conditions.

1. Once the brain has been removed, wash the brain tissue three times in sterile PBS in 60-mm petri dish by transferring from one dish to another using a sterile 18-gauge needle.
2. Transfer the tissue into a clean 60-mm dish containing 2 ml of sterile PBS.
3. Holding the tissue with a needle, use a sterile scalpel to mince the tissue into small pieces.
4. Transfer all of the tissue into a 15-ml conical tube using a 10-ml pipette.
5. Add 500 μl of 10× trypsin–EDTA (2.5% stock; Invitrogen; Catalogue # 15400-054) and 500 μl of 1 mg/ml DNAse I (Sigma; Catalogue # D5025) to the tissue suspension.
6. Incubate the tissue suspension in 37°C water bath for 30 min, mixing occasionally.
7. Place 100-μm mesh cell strainer (BD Bioscience; BD Falcon™ Cell Strainer, Catalogue # 352360) on top of a 50-ml conical centrifuge tube. Resuspend the tissue homogenate and gently force the homogenate through the mesh using the plunger of a 3-cc syringe into the centrifuge tube.
8. Wash the mesh strainer four times with 4 ml of DMEM Complete Media, continuing to gently force the tissue through the mesh with the syringe plunger.
9. Centrifuge the cell suspension at low speed (400–500 × *g*) for 10 min to pellet cells.

10. Remove the supernatant and resuspend cells in 12 ml of DMEM Complete Media containing G418, or the appropriate selection drug for the plasmid used, at the determined concentration (e.g. 400 μg/ml for U87MG).
11. Plate tumour cells in a 100-mm tissue culture dish in DMEM Complete Media containing G418.
12. Place the cultures in a tissue culture incubator at 37°C for 2 days.
13. Replace 2/3 of the volume of media with fresh media containing G418 (e.g. 400 μg/ml for U87MG) and return to the tissue culture incubator.
14. Replace all media with fresh media containing G418 after 1 or 2 days.
15. Replace media every 2–3 days to keep G418 active.
16. Confirm the presence of GFP expressing cells under fluorescent light.
17. Expand the cells until there are enough cells to inject a second set of immunocompromised mice [recommend injecting 4–5 mice (1×10^5 cells/mouse) for each round of in vivo selection].

2.2. Assessing Glioma Migration In Vitro

2.2.1. In Vitro Circular Migration Assay

Migration assays described here are performed using a microliter-scale radial monolayer assay adapted from Berens et al. (60).

Equipment

10-Well Teflon-masked microscope slides (Creative Scientific Methods Phoenix, ZA, USA); autoclaved

Cell Sedimentation Manifold (Creative Scientific Methods, http://www.creative-sci.com, Phoenix, AZ, USA)

Hemocytometer

Image analysis software

2.3. Method

1. Coat 10-well Teflon-masked microscope slides with 50 μl of 20 μg/ml laminin (Sigma-Aldrich) or ECM of choice, for 1 h at 37°C. ECM chosen will dependent on the cell line used.
2. Remove laminin and wash the microscope slide once with 100 μl of sterile phosphate buffered saline.
3. Remove cells from the tissue culture dish by rinsing once with sterile PBS followed by Puck's EDTA (1 mM EDTA, 10 mM HEPES, 5 mM KCl, 140 mM NaCl, 4 mM $NaHCO_3$, and 6 mM Dextrose, pH 7.3).

   Note: Cells are released from the culture dish using Puck's EDTA to prevent cleavage and activation of cell surface receptors (e.g. the family of proteinase-activated receptors) that may occur with Trypsin–EDTA.

Fig. 2. Shown is the Model 900 Small Animal Stereotaxic Instrument from Kopf Instruments (http://www.kopfinstruments.com) outfitted with a mouse adaptor [Model 926 Mouse Adaptor with a nose and tooth (palate) bar assembly with zygoma ear cups]. The stereotactic apparatus is outfitted with a 10-ml Hamilton syringe (*blue*; 801RN Hamilton Co) with a 30-gauge needle.

4. Incubate cells at 37°C for ~5 min.
5. Collect cells by washing the tissue culture dish containing the cells with DMEM Complete Media and placing the cell suspension in a 15-ml conical centrifuge tube.
6. Centrifuge cells at 400–500 × *g* for 5 min to pellet the cells, then remove the supernatant. Resuspend the cell pellet in a minimal amount of DMEM Complete Media and count cells using a hemocytometer.
7. Prepare a cell suspension of 1,000 cells/μl media (2 μl of cell suspension is required per well).
8. Place the Cell Sedimentation Manifold (CSM Inc., Phoenix, AZ, USA) over the laminin-coated slides to establish a circular 1-mm diameter confluent monolayer (Fig. 2).
9. Add 50 μl of media to each well of the sedimentation manifold.
10. Seed 2 μl (2,000 cells) into the central lumen of the sedimentation cylinder (5 wells per cell type/culture condition or treatment).
11. Place slides on ice for 60 min and then place the sedimentation manifolds at 37°C in a humidified 5% CO_2 incubator for ~6 h.
12. Confirm that the cells have completely attached to the slide and remove the sedimentation manifold.

13. Remove the media and incubate in fresh media plus/minus any additional experimental treatments.
14. Using a phase contrast or fluorescent (if labelled with a fluorescent marker) microscope, take a digital image of each well as the starting point of the assay (before migration).
15. Acquire digital images at desired time points.
16. Measure the radius of the best-fit circles using image analysis software (e.g. Axiovision 4.2 imaging software; Carl Zeiss Canada Ltd.) around the area covered by cells (illustrated in Fig. 2).
17. Calculate the quantitative migration scores as the increase of the radius beyond the initial radius of the object.

2.4. Monitoring Glioma Invasion In Vitro

2.4.1. In Vitro Invasion Assay

Equipment

The assays described in this section use a traditional transwell insert for a 24-well plate (6.5-mm diameter) with an 8.0-μm pore; however, several membrane diameters and pore sizes are available, but for mammalian cells that are not neutrophils, a pore size of 8.0 μm is needed to allow the nucleus to move through the membrane. The self-centering hanging design keeps the membrane about a millimetre off the bottom of the well.

Method

1. The day prior to setting up the in vitro invasion assay, remove culture media from cells to be used, wash with sterile PBS and replace with serum-free media. Incubate cultures overnight at 37°C in a humidified 5% CO_2 incubator.
2. Prepare Matrigel^{TM} to be used by diluting one part Matrigel^{TM} (BD Biosciences, Catalogue # 354234) with two parts cold serum-free media using cold pipette tips (−20°C).
3. Insert an 8-μm pore size transwell invasion chamber (BD Bioscience, Mississauga, ON, Canada) (Fig. 3) into a 24-well tissue culture dish containing 500 μL of matrix to wet the outside of the membrane, removing any excess liquid. Layer 30 μl of Matrigel^{TM} into the cup of the transwell and let dry in a tissue culture hood. To avoid the formation of a meniscus and to get a more even layer of Matrigel^{TM}, a layer of sterile ddH_2O can be placed on top of the Matrigel^{TM} before letting it dry. The transwells can be stored 4°C for up to 1 week. Alternatively, transwells can be coated with ECM of choice; e.g. 20 μg/ml laminin, Purecol^{TM} collagen mix (#5005-B) used at nine parts collagen to one part 10× DMEM Complete Media, or brain-tumour-like-matrix (see Sect. 2.4.4).

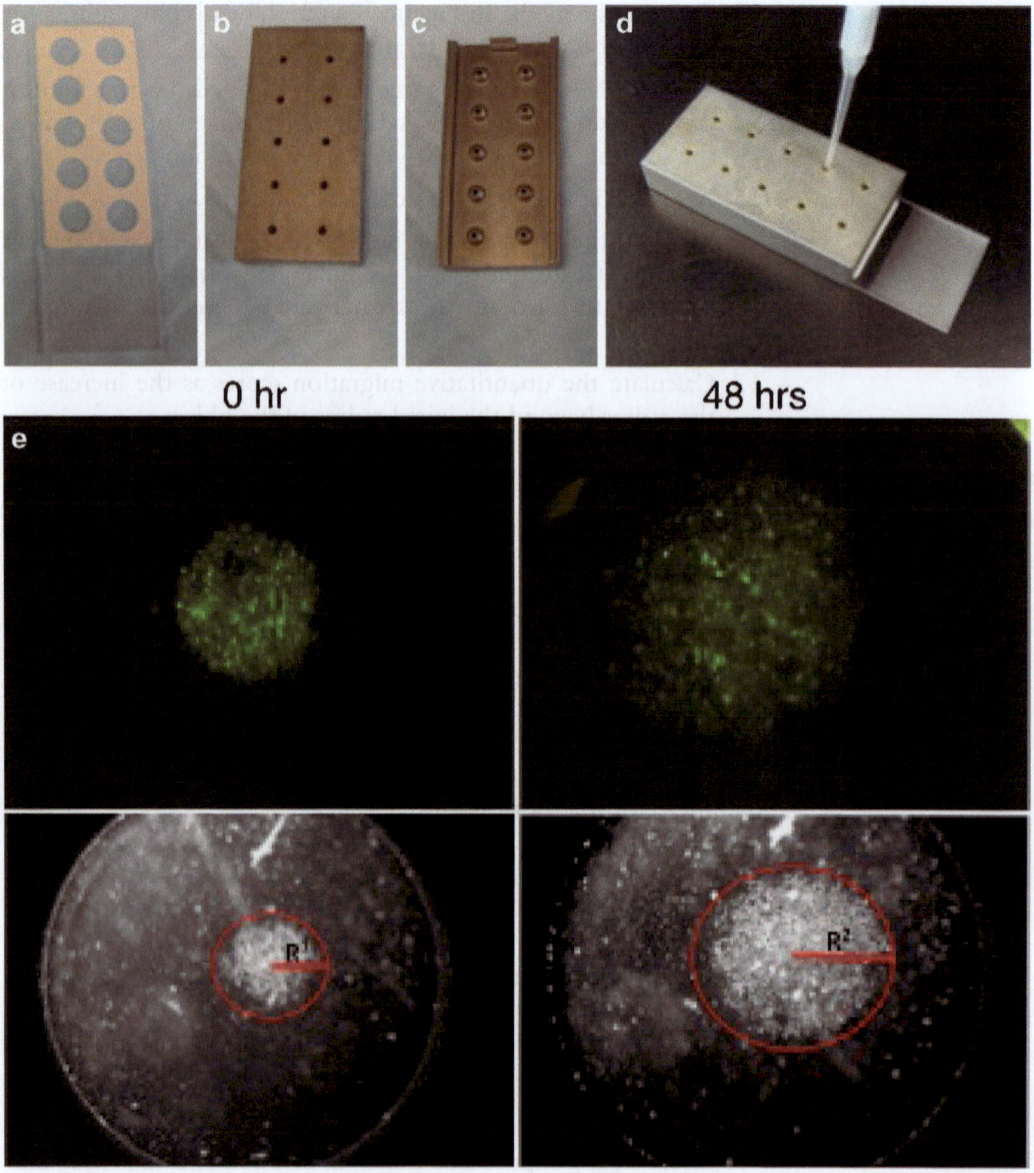

Fig. 3. Images in (**a**)–(**d**) show the components of the circular migration assay that contains a 10-well Teflon-masked microscope slide (**a**), and the Cell Sedimentation Manifold, upper, lower and side view (**b**–**d**). Images shown in (**e**) are fluorescent micrographs (*top panel*) and photomicrographs (*lower panel*) of U87MG glioma cells expressing GFP plated in 10% FBS and grown on 20 mg/ml Laminin. Images on the *left* were taken immediately after the removal of the Cell Sedimentation Manifold while images on the *right* were taken 48 h later. Distance travelled by the glioma cells is measured as the difference in the radius between the two time points (R2 − R1; *red*). The time of the assay can be modified based on the migratory rates of the cells to be assessed.

Note: As different cell lines and different treatments can result in a wide range of invasive potential, it is important when choosing a matrix to be aware of the matrices properties. For example, Matrigel^{TM} is a basement membrane (contains collagen IV and laminin) and will cause polarization/differentiation of the test cells. Cells in this state will have decreased migration providing a more rigorous measure of invasion. Collagen I/III can be used as a more general representation of stromal ECM allowing one to measure the movement of cells without having to overcome the integrin polarization signals, or the more specific "brain-tumour-like-matrix" described for the 3D invasion assay (Sect. 2.4.4) can be used to more closely mimic the in vivo tumour environment.

4. Allow the Matrigel™ to solidify at room temperature for 1 h, once solidified, rinse with serum-free media.
5. Remove cells to be assayed from the tissue culture dishes using Puck's EDTA (1 mM EDTA, 10 mM HEPES, 5 mM KCl, 140 mM NaCl, 4 mM $NaHCO_3$ and 6 mM Dextrose, pH 7.3) at 37°C. Cells should be released from the culture plate within ~5 min. Transfer cells to a 15-ml conical centrifuge tube and centrifuge on low speed (400–500 × *g*) for 5 min to remove the EDTA. Rinse the cells once with media containing 1% FBS, repeating centrifugation.

   Note: Cells are released from the culture dish using Puck's EDTA to prevent cleavage and activation of cell surface receptors (e.g. the family of proteinase-activated receptors) that may occur with Trypsin–EDTA.
6. Resuspend the cells in media containing 1% FBS and count cells using a hemocytometer.
7. Dilute cells to a concentration of 1.5×10^5 cells/ml in 1% FBS media with appropriate treatments that may include inhibitors, antibodies, peptides, etc. (500 μl/insert).
8. Rinse insert with 500 μl of serum-free medium.
9. Add 500 μl of the treatments to empty wells of a 24-well plate.
10. Lower insert into the well at an angle to avoid trapping bubbles.
11. Add 100–200 μl of cells to the insert (2.5×10^4 cells/insert).
12. Incubate cells at 37°C in a humidified 5% CO_2 incubator for 3–6 h depending on the cell line.

Staining

1. Remove media from the chamber inserts and wells.
2. Wipe out the transwell a second time using a clean, moistened swab.
3. Place the insert in methanol for 2 min to fix the migrated cells (3.5% formaldehyde in PBS for 15 min can also be used).
4. Remove excess methanol and place in hematoxylin for 4 min (this stains primarily the nuclei and is useful if a second

molecule is to be stained using conventional immunohistochemical techniques).

5. Remove the insert from the hematoxylin stain and rinse well with water.
6. Place the insert into PBS to blue the hematoxylin stain.
7. Alternatively, remove media and stain the cells with Crystal Violet by adding 180 μl of 1% crystal violet (stains the entire cell; Sigma catalogue # c-3886) in 95% EtOH into the lower chamber ensuring that the membrane is touching the stain.
8. Aspirate the crystal violet and wash the transwells three times in PBS.
9. Remove Matrigel^{TM} and cells that have not migrated through using a PBS-soaked cotton-tipped swab.
10. Migrated cells can be visualized on an inverted microscope by placing the transwell insert on top of a glass slide, or using a new scalpel blade, cut around the membrane.
11. Clean a glass slide with 70% ethanol and add a drop of mounting media and place the membrane insert onto mounting media (Vector Laboratories) with cells facing upward.
12. Add another drop of mounting media and cover with a round glass coverslip.
13. Determine the number of migrated cells by counting the stained cells adherent to the lower side of the membranes in five random microscopic fields of the chamber with a 20× objective (Fig. 3).

2.4.2. Modified Boyden Chamber Assay

An in vitro invasion assay can also be performed using a modified Boyden chamber from Neuroprobe Inc. (Neuroprobe AP48 Micro-chemotaxis Chamber; http://neuroprobe.com; Fig. 4). Briefly, a 25 × 80 mm polyvinylpyrrolidone (PVP)-free poretics polycarbonate membrane with an 8-μm pore (Neuroprobe Inc.) can be coated with the ECM chosen as described above. Prepare the modified Boyden chamber by filling the bottom wells with 25 μl of DMEM complete media and cover with the ECM-coated membrane. A slight positive meniscus in the well helps prevent bubble formation when the membrane is applied. Place the silicone gasket on top of the membrane and then the upper chamber plate tightening down with the thumb-nuts. Prepare your test cells and adjust the volume to plate 2.5×10^4 cells/50 μl of serum-free DMEM into each well of the upper chamber plate. Incubate the assembled apparatus at 37°C in a humidified incubator with 5% CO_2 for 6–8 h depending on the invasive potential of each individual cell line. At the end of the incubation period, aspirate fluid from the top wells, loosen thumb-nuts and remove the upper chamber plate and gasket. Using forceps carefully remove the membrane insuring not to

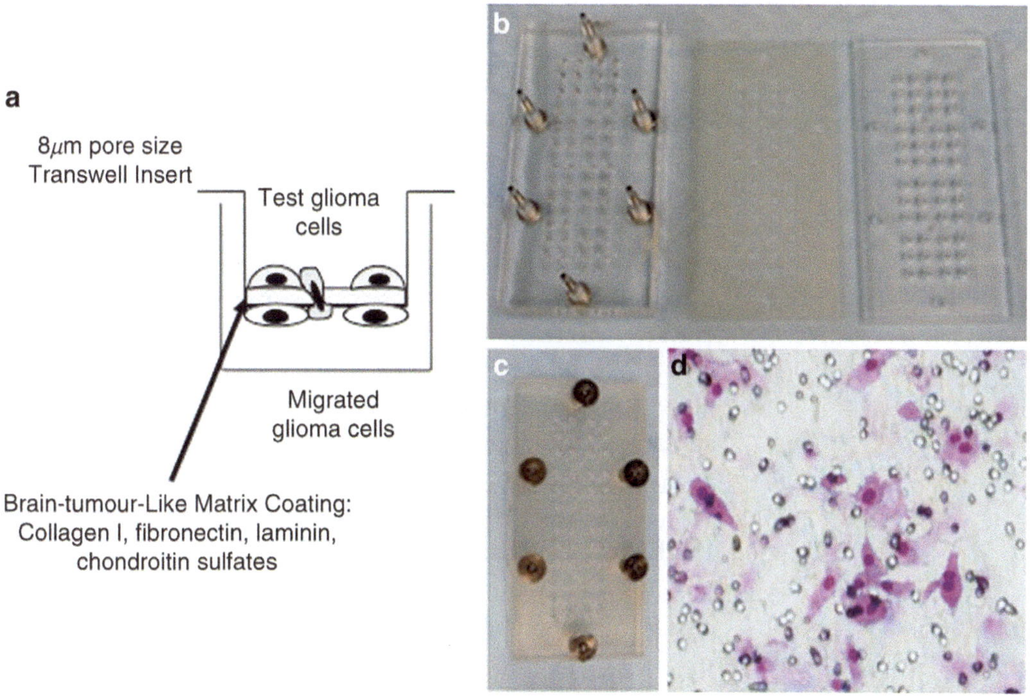

Fig. 4. (**a**) Schematic diagram of the Transwells Chamber. Images in (**b**) and (**c**) show the components of the Neuroprobe AP48 Micro-Chemotaxis Chamber (http://neuroprobe.com) that contains a silicon gasket, acrylic top and bottom plates, and assembly hardware. (**d**) Photomicrograph of Hemacolor® stained U87MG glioma cells that have migrated through a laminin-coated Micro-Chemotaxis Chamber. Timing of the assay normally ranges from 2 to 6 h depending on the migratory/invasive properties of the cell type to be assessed.

touch the cells attached to the membrane. Carefully rinse the membrane with PBS and remove the non-migrated cells from the upper side of the membrane using a wiper blade. Fix in methanol and stain cells with Hemacolor® kit, 5 min in each solution (Merck Chemicals International; http://www.merck-chemicals.com). Wash with PBS, pH 7.4. Place the membrane, cell side up, on a clean microscope slide and quantify the migrated cells by counting the stained cells adherent to the lower side of the membrane in five fields (under ×100 magnification); 3–4 wells should be used for each experimental group (see Fig. 4d, for example).

2.4.3. In Vitro Invasion Assay for Primary Human Glioma Cells

Short-Term Culture of Primary Human Glioma Cells

1. Immediately following surgical removal of glioma tumour tissue of a patient, place specimen into DMEM-F12 culture medium in sterile 35-mm tissue culture dish.
2. Using sterile forceps and scalpel, remove any visible blood clots, necrotic regions or connective tissue.

3. Rinse tissue well using sterile PBS, place the rinsed tissue in a clean 35-mm tissue culture dish and cut the tumour into pieces of approximately 0.5–1 mm^2.
4. Place the dissected tissue into a 15-ml conical centrifuge tube (or 50-ml tube depending on the volume of tumour tissue).
5. Add sterile PBS containing trypsin (0.25%) and 50 μg/ml DNase I (Roche Diagnostics, Laval, Quebec, Canada), and place in a 37°C water bath for 30 min, gently inverting every 5–10 min to ensure sufficient exposure of all tissue to the digestive enzymes.
6. Strain the digested cell suspension through a 100-μm-mesh filter (BD Biosciences, Mississauga, ON, Canada) to remove any debris, washing the filter with sterile PBS.
7. Centrifuge the cell suspension at low speed (400–500 × *g*) for 10 min to pellet the cells and wash the cell pellet twice with DMEM-F12 media.
8. Remove any contaminating red blood cells using RBC lysis buffer (described below).
9. Remove lysis solution by washing the remaining cells with sterile PBS, centrifuge to pellet the cells (400–500 × *g*; 5 min) and resuspend in DMEM Complete Media containing 20% FBS.
10. Plate tumour cells in a 100-mm tissue culture dish and place in a tissue culture incubator for 1–3 weeks changing media as required.

RBC Lysis Buffer (10× Stock)

16.58 g NH_4Cl

2 g $KHCO_3$

0.074 g Na_2EDTA

1. Dissolve in a beaker containing 180 ml ddH_2O.
2. Adjust pH to 7.4.
3. Transfer to a cylinder and add ddH_2O to 200 ml.
4. Filter through a 0.45-μm filter.
5. Stock solution can be aliquoted into sterile tubes and stored at −20°C.

Transwell Motility Assay of Primary Human Glioma Cells

1. Serum "starve" primary human glioma cells that have been cultured for less than 3 weeks by removing culture media and replacing it with media containing 1% FBS for 2 h at 37°C and 5% CO_2.
2. Release cells from the culture dish using Puck's EDTA (1 mM EDTA, 10 mM HEPES, 5 mM KCl, 140 mM NaCl, 4 mM $NaHCO_3$, and 6 mM Dextrose, pH 7.3) at 37°C.

3. Resuspend cells in 1% FBS media and plate 2.5 × 10^4 cells/well in 8-well chamber slides (Nalge Nunc International, Rochester, NY, USA) for immunocytochemistry or in transwell chambers (Costar™ #3422) coated with 20 μg/ml laminin or the ECM chosen above (see Sect. 2.2 above).
4. For transwell chambers, place 1% FBS containing media in the bottom chamber and incubate the transwells at 37°C and 5% CO_2 for 6 h.
5. Fix the remaining migrated cells for 15 min with 4% paraformaldehyde.
6. The migrated cells can now be stained for the invasive marker of interest (e.g. $p75^{NTR}$; described in Johnston et al. (44)) and counterstained with hematoxylin.
7. Remove the media, and the cells that did not migrate by scraping the top of the membrane using a cotton-tipped swab as described above (Sect. 2.2).
8. Migration of the tumour cells can be quantified for (1) the percentage of invading cells or (2) the percentage of invading cells expressing a specific invasive marker by counting the number of invasive-marker-positive cells (e.g. $p75^{NTR}$; IHC brown) and invasive-marker-negative cells (hematoxylin blue) in the original population (tumour cells plated in the chamber slide; Step 3) as compared to the migratory population (cells that migrated through the transwell; Step 7). For illustration, see Fig. 4.

2.4.4. 3D-ECM Invasion Assay

To provide an in vitro model that more closely recapitulates the in vivo environment of the brain, we have established a three-dimensional invasion assay where glioma cells are plated within a brain-tumour-like-matrix.

Note: The brain-tumour-like-matrix described here can also be used as an ECM for the 2D invasion assay (Sect. 2.4.1).

1. Prepare brain-tumour-like-matrix by combining a collagen gel solution (Chemicon International 3D Collagen cell culture system, Cat# ECM675, http://www.chemicon.com) with 5× DMEM F12 media on ice.
2. Add the neutralization solution (40:1) and additional ECM proteins at a concentration of 10 μg/ml (Laminin, Fibronectin, Chondroitin Sulphate Proteoglycan, Chemicon).
3. Suspend 1 × 10^5 cells in 350 μl of brain-tumour-like-matrix and pipette 70 μl of the brain-tumour-like-matrix and cell mixture into transwell chambers with 8.0-μm pore size polycarbonate membrane (Costar, http://www.costar.com) recommending five chambers for each cell line or treatment. Immediately transfer to a 37°C incubator for 60 min to allow the matrix to polymerize.

4. Once polymerized, add 100 μl of serum-free DMEM to the upper chamber and 1.0 ml of 10% FBS in DMEM Complete Media to the lower chamber.
5. Incubate the transwell chambers at 37°C in a humidified 5% CO_2 incubator for 6 h.
6. Remove media and wash chambers with PBS.
7. Fix cells with acid–alcohol (3:1 methanol: glacial acetic acid) for 15 min at room temperature.
8. Remove excess acid–alcohol and place in hematoxylin for 4 min.
9. Remove contents in the top chamber and mount membranes on a glass slides.
10. Count the number of invading cells in five random microscopic fields of the membrane with a 20× objective.

2.5. Assessing Glioma Growth and Invasion In Vivo

2.5.1. Assessing the Invasive Ability of Glioma Cells In Vivo

1. Implant glioma cells to be assessed into the putamen of immunocompromised mice as described in Sect. 2.1.1 (items 1–10).
2. Let the tumours grow to the desired size (generally from 2 to 6 weeks post-injection). Optional: 24 h prior to sacrificing the animals Bromodeoxyuridine (BrdU; 50 mg/kg) (61) can be given by intraperitoneal injection and incorporation of BrdU can be used to assess the proliferation rate of the tumour cells in vivo.
3. Remove the brain of the mouse, rinse in sterile PBS and freeze the brains in Tissue-Tek® OCT compound (Electron Microscopy Sciences, http://www.emsdiasum.com).
4. Cryosection the brain into 7–9 μm sections onto a glass microscope slide for examination by immunohistochemistry.
5. Air-dry frozen sections at room temperature for 30 min.
6. Fix the whole brain sections with ice-cold acetone for 15 min at –20°C.
7. Rinse sections four times with Wash Buffer [Tris-buffered saline (TBS) + 0.05% Tween 20] for 2 min.
8. Remove endogenous peroxidases in the sections by inactivating with 0.075% H_2O_2/methanol (made fresh) for 10 min.
9. Block non-specific binding by incubating the sections with 10% normal goat serum (or serum from the host in which the secondary was made) in PBS for 1 h.
10. Incubate sections with the primary antibody of choice (e.g. mouse monoclonal anti-human nuclei to detect all human tumour cells (Chemicon) or antibody against BrdU to determine

the percentage of cells that have divided during the 24 h prior to sacrifice) in the presence of blocking buffer (10% normal goat serum in PBS) overnight at 4°C. Incubate in a closed container to ensure that the slides do not dry out.

11. Wash sections four times in Wash Buffer for 2 min.
12. Incubate the sections with the appropriate biotinylated secondary antibody (Vector Laboratories, http://www.vectorlabs.com) for 1 h at room temperature.
13. Wash sections four times in Wash Buffer for 2 min.
14. Detect the secondary antibody using avidin-biotin peroxidase complexes that are formed using the VECTASTAIN Elite ABC kit (Vector Laboratories) and detect by the addition of SIGMAFAST™ DAB (3,3′-diaminobenzidine tetrahydrochloride; Sigma-Aldrich), which converts to a brown reaction product by the peroxidase (usually occurs within 2 min).
15. Counterstain sections using Hematoxylin for 15 min. Blue the slides in PBS for 3 min.
16. Wash slides in running lukewarm water for 3 min (do not let water run directly onto slides).
17. Dehydrate sections in a series of ethanol/xylene: 70% ethanol; 95% ethanol; 100% ethanol; 100% xylene.
18. Mount and coverslip the sections using Entellan mounting media (Electron Microscopy Sciences).

3. General Discussion for the In Vitro and In Vivo Assessment of Malignant Glioma

In this chapter, we have described a number of techniques that we have used to define invasive features of cells including malignant glioma (44, 45, 62, 63). While many of the past experiments have focused on the use of established cell lines, we have been able to utilize all techniques described herein for the phenotypic comparison of primary patient specimens including the use of the newly described population of glioma stem cells or as referred to here, brain tumour initiating cells (17, 44).

All of the experiments described in this chapter come with their inherent advantages and disadvantages that warrant further discussion. While the in vitro assays for migration and invasion (through a matrix) are surrogates for assessing the migratory properties of tumour cells, these matrices do not contain the exact complex or unique composition of the ECM within the brain or the in vivo tumour microenvironment. It is interesting to note that clinically, glioma cells tend to follow white matter tracts within the brain such as the myelin-rich nerve bundles of the corpus callosum that connects the two hemispheres of the brain. While the in vitro assays certainly do not recapitulate these features, in vivo orthotropic xenograft models have the advantage of providing a more "natural" environment to assess the invasive potential of any given glioma cell line. The in vivo models are not, however, without their downside as many of the established glioma cell lines form highly demarcated tumours and thus do not accurately mimic the patient situation. As a consequence, many investigators are now using orthotopic xenograft models established using patient BTIC lines, these cells have been shown to be highly invasive in vivo with many having a propensity to migrate along the corpus callosum (44, 45). Whether the established lines have lost key migratory abilities remains to be determined. Another feature within the tumour microenvironment that is not mimicked in the described in vitro assays is the effect of hypoxia. Hypoxia is a well-recognized phenomenon in glioma patients and is believed to attribute to the poor clinical outcome associated with this disease. One could consider modifying the in vitro assays to include the use of hypoxic chambers to test whether reduced oxygen levels alter the invasive behaviour of glioma cells (64–68). The constant development of new assays and cell attachment substrates is being driven by the need to produce an in vitro environment that resembles the in vivo situation. Assays presented here have been developed to assess migration and invasion of glioma cells. They can easily be adapted to other cell types or substratum.

All of the in vitro experiments described in this chapter that utilize the transwell or the modified Boyden chamber are performed with a pore size of 8 μM, a size that is optimal for the brain tumour cells. These assays can also be modified to purify pseudopodial protrusions of invasive cancer cells by using a membrane with a 1-μM pore size, a size that selectively permits the passage of these protrusions but not the cell body (69, 70). This modified strategy maybe useful in defining key molecular components of the glioma cells that explain their highly migratory properties.

In collaboration with our colleagues, we have successfully used the in vivo orthotopic models described here for the preclinical assessment of a number of novel therapeutics as therapeutic modalities for the treatment of patients with malignant glioma including gamma-secretase inhibitors (45) and several oncolytic viruses (71–76). All of the in vivo assays described have been established for the implantation of human glioma cells directly into the brains of immunocompromised mice. It should be noted that we have also used syngenic models including both racine and murine models that have the intrinsic advantage of having a host immune system (71, 72, 75). The role of the immune system for the invasive behaviour of the tumour is not clear and thereby the use of these models may provide additional information that may otherwise be missed. This becomes germane when assessing certain therapeutic modalities including immune-based therapy.

Acknowledgments

We thank Dr. Roberta Gamma, Dr. Xiaoguang Hao, Wale Oyelusi, Dr. Jennifer Rahn and Hayley Robbins for helping to optimize all of the experimental protocols described herein and for providing some of the figures. Work performed in the authors' laboratories has been funded by the Canadian Institutes of Health Research, the Alberta Cancer Foundation and Alberta Innovates Health Solutions through funds from the Alberta Cancer Prevention Legacy Act. S.M.R. is a scientist of the Alberta Heritage Foundation for Medical Research and a Canada Research Chair in Cancer Biology. D.L.S. is funded by the Alberta Cancer Foundation.

References

1. Stupp R, Mason WP, van den Bent MJ, Weller M, Fisher B, Taphoorn MJ, Belanger K, Brandes AA, Marosi C, Bogdahn U, Curschmann J, Janzer RC, Ludwin SK, Gorlia T, Allgeier A, Lacombe D, Cairncross JG, Eisenhauer E, Mirimanoff RO (2005) Radiotherapy plus concomitant and adjuvant temozolomide for glioblastoma. N Engl J Med 352:987–996
2. Wen PY, Kesari S (2008) Malignant gliomas in adults. N Engl J Med 359:492–507
3. Senger D, Cairncross JG, Forsyth PA (2003) Long-term survivors of glioblastoma: statistical aberration or important unrecognized molecular subtype? Cancer J 9:214–221
4. Berens ME, Giese A (1999) "...those left behind". Biology and oncology of invasive glioma cells. Neoplasia 1:208–219
5. Giese A, Bjerkvig R, Berens ME, Westphal M (2003) Cost of migration: invasion of malignant gliomas and implications for treatment. J Clin Oncol 21:1624–1636
6. Giese A, Loo MA, Tran N, Haskett D, Coons SW, Berens ME (1996) Dichotomy of astrocytoma migration and proliferation. Int J Cancer 67:275–282
7. Demuth T, Berens ME (2004) Molecular mechanisms of glioma cell migration and invasion. J Neurooncol 70:217–228
8. Singh SK, Clarke ID, Hide T, Dirks PB (2004) Cancer stem cells in nervous system tumors. Oncogene 23:7267–7273
9. Singh SK, Clarke ID, Terasaki M, Bonn VE, Hawkins C, Squire J, Dirks PB (2003) Identification of a cancer stem cell in human brain tumors. Cancer Res 63:5821–5828
10. Singh SK, Hawkins C, Clarke ID, Squire JA, Bayani J, Hide T, Henkelman RM, Cusimano MD, Dirks PB (2004) Identification of human brain tumour initiating cells. Nature 432:396–401
11. Bao S, Wu Q, McLendon RE, Hao Y, Shi Q, Hjelmeland AB, Dewhirst MW, Bigner DD, Rich JN (2006) Glioma stem cells promote radioresistance by preferential activation of the DNA damage response. Nature 444:756–760
12. Bleau AM, Holland EC (2009) Chemotherapeutic treatment of gliomas increases the amount of cancer stem-like cells. Med Sci (Paris) 25:775–777
13. Bleau AM, Huse JT, Holland EC (2009) The ABCG2 resistance network of glioblastoma. Cell Cycle 8:2936–2944
14. Lefranc F, Brotchi J, Kiss R (2005) Possible future issues in the treatment of glioblastomas: special emphasis on cell migration and the resistance of migrating glioblastoma cells to apoptosis. J Clin Oncol 23:2411–2422
15. Rowan K (2009) Are cancer stem cells real? After four decades, debate still simmers. J Natl Cancer Inst 101:546–547
16. Lee J, Kotliarova S, Kotliarov Y, Li A, Su Q, Donin NM, Pastorino S, Purow BW, Christopher N, Zhang W, Park JK, Fine HA (2006) Tumor stem cells derived from glioblastomas cultured in bFGF and EGF more closely mirror the phenotype and genotype of primary tumors than do serum-cultured cell lines. Cancer Cell 9:391–403
17. Kelly JJ, Stechishin O, Chojnacki A, Lun X, Sun B, Senger DL, Forsyth P, Auer RN, Dunn JF, Cairncross JG, Parney IF, Weiss S (2009) Proliferation of human glioblastoma stem cells occurs independently of exogenous mitogens. Stem Cells 27:1722–1733

18. Mariani L, Beaudry C, McDonough WS, Hoelzinger DB, Demuth T, Ross KR, Berens T, Coons SW, Watts G, Trent JM, Wei JS, Giese A, Berens ME (2001) Glioma cell motility is associated with reduced transcription of proapoptotic and proliferation genes: a cDNA microarray analysis. J Neurooncol 53:161–176
19. Joy AM, Beaudry CE, Tran NL, Ponce FA, Holz DR, Demuth T, Berens ME (2003) Migrating glioma cells activate the PI3-K pathway and display decreased susceptibility to apoptosis. J Cell Sci 116:4409–4417
20. Bleau AM, Hambardzumyan D, Ozawa T, Fomchenko EI, Huse JT, Brennan CW, Holland EC (2009) PTEN/PI3K/Akt pathway regulates the side population phenotype and ABCG2 activity in glioma tumor stem-like cells. Cell Stem Cell 4:226–235
21. Nakada M, Nakada S, Demuth T, Tran NL, Hoelzinger DB, Berens ME (2007) Molecular targets of glioma invasion. Cell Mol Life Sci 64:458–478
22. Giese A, Westphal M (1996) Glioma invasion in the central nervous system. Neurosurgery 39:235–250, discussion 250–252
23. Freije WA, Castro-Vargas FE, Fang Z, Horvath S, Cloughesy T, Liau LM, Mischel PS, Nelson SF (2004) Gene expression profiling of gliomas strongly predicts survival. Cancer Res 64:6503–6510
24. Liang Y, Diehn M, Watson N, Bollen AW, Aldape KD, Nicholas MK, Lamborn KR, Berger MS, Botstein D, Brown PO, Israel MA (2005) Gene expression profiling reveals molecularly and clinically distinct subtypes of glioblastoma multiforme. Proc Natl Acad Sci U S A 102:5814–5819
25. Li A, Walling J, Ahn S, Kotliarov Y, Su Q, Quezado M, Oberholtzer JC, Park J, Zenklusen JC, Fine HA (2009) Unsupervised analysis of transcriptomic profiles reveals six glioma subtypes. Cancer Res 69:2091–2099
26. Verhaak RG, Hoadley KA, Purdom E, Wang V, Qi Y, Wilkerson MD, Miller CR, Ding L, Golub T, Mesirov JP, Alexe G, Lawrence M, O'Kelly M, Tamayo P, Weir BA, Gabriel S, Winckler W, Gupta S, Jakkula L, Feiler HS, Hodgson JG, James CD, Sarkaria JN, Brennan C, Kahn A, Spellman PT, Wilson RK, Speed TP, Gray JW, Meyerson M, Getz G, Perou CM, Hayes DN (2010) Integrated genomic analysis identifies clinically relevant subtypes of glioblastoma characterized by abnormalities in PDGFRA, IDH1, EGFR, and NF1. Cancer Cell 17:98–110
27. Rutka JT, Muller M, Hubbard SL, Forsdike J, Dirks PB, Jung S, Tsugu A, Ivanchuk S, Costello P, Mondal S, Ackerley C, Becker LE (1999) Astrocytoma adhesion to extracellular matrix: functional significance of integrin and focal adhesion kinase expression. J Neuropathol Exp Neurol 58:198–209
28. Hynes RO (1992) Integrins: versatility, modulation, and signaling in cell adhesion. Cell 69:11–25
29. Hynes RO (2002) A reevaluation of integrins as regulators of angiogenesis. Nat Med 8:918–921
30. D'Abaco GM, Kaye AH (2007) Integrins: molecular determinants of glioma invasion. J Clin Neurosci 14:1041–1048
31. Maret D, Gruzglin E, Sadr MS, Siu V, Shan W, Koch AW, Seidah NG, Del Maestro RF, Colman DR (2010) Surface expression of precursor N-cadherin promotes tumor cell invasion. Neoplasia 12:1066–1080
32. Merzak A, Koocheckpour S, Pilkington GJ (1994) CD44 mediates human glioma cell adhesion and invasion in vitro. Cancer Res 54:3988–3992
33. Rutka JT, Ackerley C, Hubbard SL, Tilup A, Dirks PB, Jung S, Ivanchuk S, Kurimoto M, Tsugu A, Becker LE (1998) Characterization of glial filament-cytoskeletal interactions in human astrocytomas: an immuno-ultrastructural analysis. Eur J Cell Biol 76:279–287
34. Nutt CL, Zerillo CA, Kelly GM, Hockfield S (2001) Brain enriched hyaluronan binding (BEHAB)/brevican increases aggressiveness of CNS-1 gliomas in Lewis rats. Cancer Res 61:7056–7059
35. Zhang H, Kelly G, Zerillo C, Jaworski DM, Hockfield S (1998) Expression of a cleaved brain-specific extracellular matrix protein mediates glioma cell invasion In vivo. J Neurosci 18:2370–2376
36. Jung S, Hinek A, Tsugu A, Hubbard SL, Ackerley C, Becker LE, Rutka JT (1999) Astrocytoma cell interaction with elastin substrates: implications for astrocytoma invasive potential. Glia 25:179–189
37. Rempel SA, Golembieski WA, Fisher JL, Maile M, Nakeff A (2001) SPARC modulates cell growth, attachment and migration of U87 glioma cells on brain extracellular matrix proteins. J Neurooncol 53:149–160
38. Nakada M, Drake KL, Nakada S, Niska JA, Berens ME (2006) Ephrin-B3 ligand promotes glioma invasion through activation of Rac1. Cancer Res 66:8492–8500
39. Nakada M, Niska JA, Miyamori H, McDonough WS, Wu J, Sato H, Berens ME (2004) The phosphorylation of EphB2 receptor regulates migration and invasion of human glioma cells. Cancer Res 64:3179–3185

40. Nakada M, Anderson EM, Demuth T, Nakada S, Reavie LB, Drake KL, Hoelzinger DB, Berens ME (2010) The phosphorylation of ephrin-B2 ligand promotes glioma cell migration and invasion. Int J Cancer 126:1155–1165
41. Miao H, Li DQ, Mukherjee A, Guo H, Petty A, Cutter J, Basilion JP, Sedor J, Wu J, Danielpour D, Sloan AE, Cohen ML, Wang B (2009) EphA2 mediates ligand-dependent inhibition and ligand-independent promotion of cell migration and invasion via a reciprocal regulatory loop with Akt. Cancer Cell 16:9–20
42. Neufeld G, Shraga-Heled N, Lange T, Guttmann-Raviv N, Herzog Y, Kessler O (2005) Semaphorins in cancer. Front Biosci 10:751–760
43. Li X, Law JW, Lee AY (2012) Semaphorin 5A and plexin-B3 regulate human glioma cell motility and morphology through Rac1 and the actin cytoskeleton. Oncogene 31 (5):595–610
44. Johnston AL, Lun X, Rahn JJ, Liacini A, Wang L, Hamilton MG, Parney IF, Hempstead BL, Robbins SM, Forsyth PA, Senger DL (2007) The p75 neurotrophin receptor is a central regulator of glioma invasion. PLoS Biol 5:e212
45. Wang L, Rahn JJ, Lun X, Sun B, Kelly JJ, Weiss S, Robbins SM, Forsyth PA, Senger DL (2008) Gamma-secretase represents a therapeutic target for the treatment of invasive glioma mediated by the p75 neurotrophin receptor. PLoS Biol 6:e289
46. Yong VW, Power C, Forsyth P, Edwards DR (2001) Metalloproteinases in biology and pathology of the nervous system. Nat Rev Neurosci 2:502–511
47. Yong VW, Krekoski CA, Forsyth PA, Bell R, Edwards DR (1998) Matrix metalloproteinases and diseases of the CNS. Trends Neurosci 21:75–80
48. Rao JS (2003) Molecular mechanisms of glioma invasiveness: the role of proteases. Nat Rev Cancer 3:489–501
49. Holtkamp N, Afanasieva A, Elstner A, van Landeghem FK, Konneker M, Kuhn SA, Kettenmann H, von Deimling A (2005) Brain slice invasion model reveals genes differentially regulated in glioma invasion. Biochem Biophys Res Commun 336:1227–1233
50. Jung S, Ackerley C, Ivanchuk S, Mondal S, Becker LE, Rutka JT (2001) Tracking the invasiveness of human astrocytoma cells by using green fluorescent protein in an organotypical brain slice model. J Neurosurg 94:80–89
51. Beadle C, Assanah MC, Monzo P, Vallee R, Rosenfeld SS, Canoll P (2008) The role of myosin II in glioma invasion of the brain. Mol Biol Cell 19:3357–3368
52. Caspani EM, Echevarria D, Rottner K, Small JV (2006) Live imaging of glioblastoma cells in brain tissue shows requirement of actin bundles for migration. Neuron Glia Biol 2:105–114
53. Ohnishi T, Matsumura H, Izumoto S, Hiraga S, Hayakawa T (1998) A novel model of glioma cell invasion using organotypic brain slice culture. Cancer Res 58:2935–2940
54. Hoelzinger DB, Mariani L, Weis J, Woyke T, Berens TJ, McDonough WS, Sloan A, Coons SW, Berens ME (2005) Gene expression profile of glioblastoma multiforme invasive phenotype points to new therapeutic targets. Neoplasia 7:7–16
55. Mariani L, McDonough WS, Hoelzinger DB, Beaudry C, Kaczmarek E, Coons SW, Giese A, Moghaddam M, Seiler RW, Berens ME (2001) Identification and validation of P311 as a glioblastoma invasion gene using laser capture microdissection. Cancer Res 61:4190–4196
56. Mariani L, Beaudry C, McDonough WS, Hoelzinger DB, Kaczmarek E, Ponce F, Coons SW, Giese A, Seiler RW, Berens ME (2001) Death-associated protein 3 (Dap-3) is overexpressed in invasive glioblastoma cells in vivo and in glioma cell lines with induced motility phenotype in vitro. Clin Cancer Res 7:2480–2489
57. Kislin KL, McDonough WS, Eschbacher JM, Armstrong BA, Berens ME (2009) NHERF-1: modulator of glioblastoma cell migration and invasion. Neoplasia 11:377–387
58. Tysnes BB, Mahesparan R (2001) Biological mechanisms of glioma invasion and potential therapeutic targets. J Neurooncol 53:129–147
59. Clark EA, Golub TR, Lander ES, Hynes RO (2000) Genomic analysis of metastasis reveals an essential role for RhoC. Nature 406:532–535
60. Berens ME, Rief MD, Loo MA, Giese A (1994) The role of extracellular matrix in human astrocytoma migration and proliferation studied in a microliter scale assay. Clin Exp Metastasis 12:405–415
61. Miller MW, Nowakowski RS (1988) Use of bromodeoxyuridine-immunohistochemistry to examine the proliferation, migration and time of origin of cells in the central nervous system. Brain Res 457:44–52
62. Campbell TN, Attwell S, Arcellana-Panlilio M, Robbins SM (2006) Ephrin A5 expression promotes invasion and transformation of murine fibroblasts. Biochem Biophys Res Commun 350:623–628
63. Zaman SN, Resek ME, Robbins SM (2008) Dual acylation and lipid raft association of Src-family protein tyrosine kinases are required for SDF-1/CXCL12-mediated chemotaxis in

the Jurkat human T cell lymphoma cell line. J Leukoc Biol 84:1082–1091

64. Szalad A, Katakowski M, Zheng X, Jiang F, Chopp M (2009) Transcription factor Sp1 induces ADAM17 and contributes to tumor cell invasiveness under hypoxia. J Exp Clin Cancer Res 28:129

65. Mendez O, Zavadil J, Esencay M, Lukyanov Y, Santovasi D, Wang SC, Newcomb EW, Zagzag D (2010) Knock down of HIF-1alpha in glioma cells reduces migration in vitro and invasion in vivo and impairs their ability to form tumor spheres. Mol Cancer 9:133

66. Ali MA, Reis A, Ding LH, Story MD, Habib AA, Chattopadhyay A, Saha D (2009) SNS-032 prevents hypoxia-mediated glioblastoma cell invasion by inhibiting hypoxia inducible factor-1alpha expression. Int J Oncol 34:1051–1060

67. Zheng X, Jiang F, Katakowski M, Kalkanis SN, Hong X, Zhang X, Zhang ZG, Yang H, Chopp M (2007) Inhibition of ADAM17 reduces hypoxia-induced brain tumor cell invasiveness. Cancer Sci 98:674–684

68. Fujiwara S, Nakagawa K, Harada H, Nagato S, Furukawa K, Teraoka M, Seno T, Oka K, Iwata S, Ohnishi T (2007) Silencing hypoxia-inducible factor-1alpha inhibits cell migration and invasion under hypoxic environment in malignant gliomas. Int J Oncol 30:793–802

69. Jia Z, Barbier L, Stuart H, Amraei M, Pelech S, Dennis JW, Metalnikov P, O'Donnell P, Nabi IR (2005) Tumor cell pseudopodial protrusions. Localized signaling domains coordinating cytoskeleton remodeling, cell adhesion, glycolysis, RNA translocation, and protein translation. J Biol Chem 280:30564–30573

70. Stuart HC, Jia Z, Messenberg A, Joshi B, Underhill TM, Moukhles H, Nabi IR (2008) Localized Rho GTPase activation regulates RNA dynamics and compartmentalization in tumor cell protrusions. J Biol Chem 283:34785–34795

71. Lun X, Alain T, Zemp FJ, Zhou H, Rahman MM, Hamilton MG, McFadden G, Bell J, Senger DL, Forsyth PA (2010) Myxoma virus virotherapy for glioma in immunocompetent animal models: optimizing administration routes and synergy with rapamycin. Cancer Res 70:598–608

72. Lun X, Chan J, Zhou H, Sun B, Kelly JJ, Stechishin OO, Bell JC, Parato K, Hu K, Vaillant D, Wang J, Liu TC, Breitbach C, Kirn D, Senger DL, Forsyth PA (2010) Efficacy and safety/toxicity study of recombinant vaccinia virus JX-594 in two immunocompetent animal models of glioma. Mol Ther 18:1927–1936

73. Lun X, Senger DL, Alain T, Oprea A, Parato K, Stojdl D, Lichty B, Power A, Johnston RN, Hamilton M, Parney I, Bell JC, Forsyth PA (2006) Effects of intravenously administered recombinant vesicular stomatitis virus (VSV (deltaM51)) on multifocal and invasive gliomas. J Natl Cancer Inst 98:1546–1557

74. Lun X, Yang W, Alain T, Shi ZQ, Muzik H, Barrett JW, McFadden G, Bell J, Hamilton MG, Senger DL, Forsyth PA (2005) Myxoma virus is a novel oncolytic virus with significant antitumor activity against experimental human gliomas. Cancer Res 65:9982–9990

75. Lun XQ, Jang JH, Tang N, Deng H, Head R, Bell JC, Stojdl DF, Nutt CL, Senger DL, Forsyth PA, McCart JA (2009) Efficacy of systemically administered oncolytic vaccinia virotherapy for malignant gliomas is enhanced by combination therapy with rapamycin or cyclophosphamide. Clin Cancer Res 15:2777–2788

76. Wilcox ME, Yang W, Senger D, Rewcastle NB, Morris DG, Brasher PM, Shi ZQ, Johnston RN, Nishikawa S, Lee PW, Forsyth PA (2001) Reovirus as an oncolytic agent against experimental human malignant gliomas. J Natl Cancer Inst 93:903–912

Part V

Imaging Technologies in Animal Tumor Models: Measuring Brain Tumor Growth and Metabolism

Neuromethods (2013) 77: 301–327
DOI 10.1007/7657_2012_36

Published online: 27 March 2012

Positron Emission Tomography-Based Molecular Imaging

Yannic Waerzeggers, Thomas Viel, Sonja Schäfers, Stefan Wagner, Andreas Faust, Burkhard Riemann, Klaus Kopka, Otmar Schober, Klaus Schäfers, and Andreas H. Jacobs

Abstract

Positron emission tomography (PET) is a nuclear imaging technique that can yield information on the temporal and spatial dynamics of various disease-specific or drug-induced biological alterations in intact tissues and living organisms. In this chapter we elucidate how this information can be generated and used in brain tumor animal models to give insight into tumor characteristics and guide anticancer treatment.

Key words: PET imaging, Neurooncology, Tumor metabolism, Gene expression, Bioluminescence imaging

1. Introduction

Positron emission tomography (PET) is a nuclear imaging technique that allows noninvasive in vivo measurements of various biochemical, physiological, pharmacological, and molecular processes in the brain. Processes that can be studied include blood flow (perfusion), metabolism (glucose, oxygen) and cell function, density and affinity of receptors (presynaptic and postsynaptic), neurotransmitter release, enzyme activity, drug delivery and uptake, and gene expression. Especially in neuro-oncology this imaging method opened up new avenues for studying brain tumor pathophysiology as it can provide relevant and site-specific measurements of molecular pathways critical to cancer cell function and thus has the potential to close the gap between in vitro and in vivo disease biology. Furthermore, compared to conventional diagnostic imaging such as computer tomography (CT) and magnetic resonance imaging (MRI), that mainly address morphologic tumor characteristics, PET imaging has a higher sensitivity regarding detection of tumor and early monitoring of response to treatment (1).

Human brain disorders and potential therapeutic strategies are often studied using small laboratory animals, primarily rodents. Because the most popular rodents to be used as model animals are the mouse and the rat, we focus on rat and mouse studies.

Recently, high-resolution small-animal PET scanners (so called μPET scanners) have been developed and commercially produced, making novel in vivo imaging studies of rodents possible. Because small-animal PET measurements of rodent brains have a higher resolution than PET measurements in humans (2–4 mm for clinical PET scanners, submillimeter resolution for preclinical animal scanners), small animal PET brain imaging is optimal for detailed regional analysis in rodents. In addition, small-animal PET imaging allows repetitive follow-up in the same animal, which reduces the number of animals needed and saves time and labor when investigating therapeutic interventions or when characterizing diseases in rodents (2).

Over the past decades many in vitro (3–5) and in vivo (6, 7) models of brain tumors have been developed and they increased our understanding of the underlying pathophysiological mechanisms. An ideal tumor model should be able to accurately recapitulate all aspects of human tumor physiology such as angiogenesis, tumor–stromal and tumor–host interaction and tumor environment and should also address the various factors that underlie the molecular and genetic lesions associated with tumor heterogeneity (8). In this regard especially animal models are being used in cancer research as they enable insights into the biological mechanisms leading to cancer development at the molecular, cellular and tissue level within a complex living organism. Such animal models can be classified as xenograft tumor models or models of spontaneous tumor formation in genetically engineered animals (8, 9). Especially the latter more closely resemble the genetic and cellular heterogeneity of primary human brain tumors, which are thought to be major determinants of tumor aggressiveness and treatment resistance. For a more in depth description of available animal models to offer insights into human brain neoplasms or to understand glioma biology see also the chapters on specific animal models: glioblastoma, oligodendroglioma, medulloblastoma, retinoblastoma, ependymoma, meningioma, pituitary and angiogenesis models.

Brain tumor animal models are valuable tools for in vivo tumor characterization and to develop and test safety and efficacy of anticancer treatment.

2. Principles of Positron Emission Tomography

PET imaging enables the assessment and quantification of the intensity and regional/spatial distribution, e.g. of the expression of endogenous or exogenous genes coding for enzymes, transporters, antigens or receptors by measuring the accumulation or binding of the respective enzyme, antigen or transporter substrates or receptor binding compounds (Fig. 1).

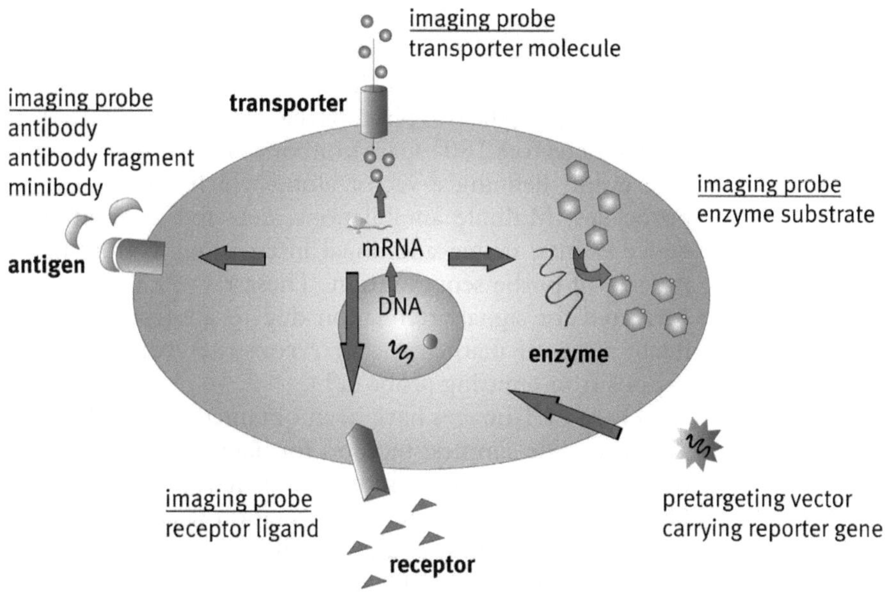

Fig. 1. Schematic drawing of PET image signal generation. An intravenously injected imaging or reporter probe, such as a radiolabelled PET tracer, can be an enzyme substrate, a receptor ligand, an antibody, or a transporter molecule, which is retained at the target organ through (1) enzymatic modification followed by mechanical or chemical entrapment; (2) binding to a receptor or antigen; (3) cell uptake by a membrane transporter; or a combination of the above. The enzymes, receptors, antigens, or transporters can be expressed by endogenous genes or can be the product of an exogenous gene that has been introduced into the target tissue by a pretargeting vector.

These substrates or binding compounds, called PET (radio) tracers, probes or ligands, are labeled with PET decaying nuclides/ positron emitters such as ^{11}C, ^{13}N, ^{15}O, ^{18}F, ^{64}Cu or ^{124}I. As such nuclei are commonly found in biological tissue, PET tracers can be chemically identical to or close analogues of naturally occurring molecules, making them especially suitable to interrogate in vivo biological processes.

A positron emitted from a PET probe travels a few millimeters in tissue before being annihilated by collision with a target electron, resulting in a pair of high-energy (511 keV) photons or gamma rays. The two gamma rays are emitted at nearly 180° from one another and travel in opposite directions throughout the subject before they are detected simultaneously by the PET detector(s) that are arranged in a ring-shaped array surrounding the subject. There exist two major principles for the technical design of PET detectors. The first one uses photon-sensitive scintillation crystals that convert the incoming gamma ray photons into light flashes and further into electrical signals by the photomultiplier tubes surrounding the scintillation crystals. The second principle is based on a multi-wire gas chamber in which the 511 keV photons are converted into electrons. These electrons are then multiplied within the gas by means of an electrical field (avalanche effect) and finally detected by anode wires. This detector principle is used in the quadHIDAC PET scanner (Oxford Positron Systems Ltd., Weston-on-the-Green, UK), which is used in our institution. A more detailed description of the technical aspects of this scanner can be found in (10). For both principles, the two photons that elicit signals at nearly the same time in detectors 180° apart from one another constitute a recordable event, defining a vector along which the positron-emitting probe lies. Multiple annihilations, detected as additional coincidence events, define additional intersecting vectors that identify the site of probe sequestration. These raw PET datasets are then corrected for signal attenuation due to absorption by tissue and finally used to mathematically reconstruct volumetric images of the positron-emitting probe (11).

Once PET images have been obtained, the PET data must be analyzed. The simplest method for data interpretation is qualitative imaging, which is mostly sufficient for tumor detection. However, qualitative data interpretation fails to give insight into in vivo tumor biology and to take full advantage of the inherently quantitative nature of PET data (12). The most practical method for PET image quantification is by calculating the standardized uptake value (SUV), defined as the ratio of tissue radioactivity uptake and the injected dose per unit subject weight. SUV describes the level of activity in a particular spot compared to activity elsewhere in the body. A SUV reading of one [1] is baseline or normal cellular activity. SUV is vulnerable to several major sources of variability, such as body composition and plasma glucose levels, length of uptake period and time of SUV evaluation, partial volume and spillover effect (13). Major reason for the continuous usage of SUV in preclinical research is that blood sampling is not necessary. The most detailed quantitative approach uses dynamic PET imaging and kinetic analysis. In this approach the blood clearance curve is obtained from dynamic images of a blood pool structure or from blood sampling and serves as the input into a compartmental model from which relevant kinetic parameters can be estimated (12, 14).

The most commonly used positron-emitting isotopes ^{11}C, ^{13}N, ^{15}O and ^{18}F have relatively short half-lives, in the order of minutes to hours, whereas the ^{64}Cu and ^{124}I have longer half-lives, from hours to days. Short-lived and long-lived isotopes have distinct advantages and disadvantages such as the necessity to have an on-site cyclotron and radiochemistry laboratory for tracer production in the case of short-lived isotopes.

Positron-emitting radionuclides are primarily used to tag small molecules that are recognized by enzymes or bind to cell receptors or membrane transporters. These enzymes, receptors or transporters can be endogenously expressed and this methodology is used in conventional PET imaging, e.g. imaging of cell metabolism with 2-deoxy-2-[^{18}F]fluoro-D-glucose ([^{18}F]FDG) targeted to the enzyme hexokinase or imaging of the activity of amino acid transporters with L-[S-methyl-^{11}C]Methionine ([^{11}C]MET) in subjects with brain tumors. These molecules can also be expressed by exogenously introduced genes (marker genes) and used for the evaluation of newly developed experimental therapies such as gene therapy protocols.

Advantages of PET imaging compared with other noninvasive imaging techniques are its (1) very high sensitivity (within the picomolar range), (2) excellent signal-to-background ratios, (3) quantitative and inherently tomographic nature with unlimited depth penetration, (4) availability of a broad range of applicable probes and (5) absence of toxicity due to the low absolute amount of substance used (15). However, nuclear imaging has (1) only moderate spatial resolution (currently in the submillimeter range for preclinical applications and of 2–4 mm in clinical practice) when compared to magnetic resonance (MR) imaging (1–10 μm for ultrahigh-field MR scanners), (2) demands sophisticated instrumentation with committed personnel, and usually can (3) only be performed at facilities that have an on-site or nearby cyclotron for delivery of radioisotopes and the required radiochemical laboratories for the preparation of radiotracers. Furthermore, the exposure of studied subjects to radiation leads to the need of dosimetry and may limit the number of examinations that can be performed (16). As a consequence, regardless of the very promising potential of PET imaging especially in the field of molecular-genetic imaging, the technique is still not widely available compared to other noninvasive imaging modalities such as CT, MRI or ultrasound.

3. PET Imaging of Tumor Biology

PET imaging has the unique ability to provide not only noninvasively and quantitatively but also longitudinally disease-specific information related to disease initiation, characterization, progression and response to therapy.

This information can be allocated to more general biochemical and physiological processes related with tumor growth (such as energy or membrane metabolism, protein or DNA synthesis, vascular function or hypoxia) or to processes related to disease-specific molecular-genetic alterations involved in apoptosis, angiogenesis and other signaling pathways.

As a result of oncogenic mutations tumor cells undergo "metabolic transformation" leading to enhanced bioenergetic and biosynthetic activities (17, 18). Due to their metabolic autonomy tumor cells are taking up nutrients (glucose and glutamine) and allocate them into metabolic pathways that convert them into biosynthetic precursors (amino acids, nucleotides and lipids) that support their growth and proliferation. Tumor cells can achieve this phenotype through changes in the expression of enzymes that determine metabolic flux rates, including nutrient transporters (e.g. GLUT1 glucose transporter, amino acid transporters) and enzymes (e.g. hexokinase, thymidine kinase 1, choline kinase).

[^{18}F]FDG PET is a standard PET imaging modality used to measure local or regional glucose metabolism. [^{18}F]FDG PET is based on the uptake principle of the [^{18}F]FDG tracer, a direct analogue of glucose. As tumor cells rely on aerobic glycolysis for a major part of their energy consumption, [^{18}F]FDG PET detects the increased energy demand associated with tumor growth. In neuro-oncology it has been shown that an increased intratumoral glucose consumption, as measured with [^{18}F]FDG PET imaging, correlates with tumor grade, cell density, biological aggressiveness and patient survival, both in primary and recurrent gliomas and that [^{18}F]FDG PET imaging can be used to detect the metabolic differences between normal brain tissue, low- and high-grade gliomas and radionecrosis (19). [^{18}F]FDG PET can also be used to predict the outcome to chemo- and radiotherapy (20). However, due to the high background uptake of [^{18}F]FDG in normal brain tissue as well as in inflammatory cells, the specificity of [^{18}F]FDG PET is relatively low.

Due to the enhanced biosynthetic activity in tumor cells these cells display an increased protein, DNA and membrane synthesis. As a consequence, the demand for amino acids, nucleosides and lipids, the respective building blocks for proteins, DNA and membranes, is increased and radio-labeled analogues of these molecules have been used in the management of malignancies (e.g. for tumor detection, staging and restaging and therapy follow-up). Nearly all amino acids have been radio-labeled but only a few are used systematically for cancer research purposes. Among these, labeled methionine and tyrosine are the most commonly used due to their ease of production and limited number of in vivo metabolites (21, 22). Uptake of [^{11}C]MET and [^{18}F]fluoroethyltyrosine ([^{18}F]FET) have been shown to correlate with in vitro cell proliferation, Ki-67 expression, nuclear antigen expression and microvessel density (23) and can be used for glioma grading (24), to differentiate recurrent tumor from radiation necrosis (25) and to evaluate therapy response and predict outcome (19, 26). Also a variety of nucleic acids have been radio-labeled as PET tracers to study DNA turn-over and thus reflect cellular proliferation (1, 27). Of these, ^{18}F-labeled 3′-fluoro-3′-deoxythymidine (FLT) is the most intensively studied. FLT accumulates in proliferating tissues following phosphorylation to FLT-monophosphate (FLT-MP) by the S-phase specific thymidine kinase-1. It has been demonstrated in many types of cancer, including malignant glioma, that [^{18}F]FLT uptake in vivo indeed is a measure of tumor proliferation activity as there exists a significant correlation between [^{18}F]FLT uptake and the in vitro proliferation marker Ki-67 (28–34).

Both [^{18}F]FLT and [^{11}C]MET are more selective in detecting brain tumors compared to [^{18}F]FDG, which is nonspecifically retained in brain tissue, due to its high glucose metabolism. [^{18}F]FLT and [^{11}C]MET, on the contrary, reveal only a low uptake in healthy brain tissue (19). Furthermore, amino acid and nucleoside analogue tracers have been shown to be superior to [^{18}F]FDG to differentiate between tumor and inflammation (35, 36), which is of high importance as the increased glucose metabolism of inflammatory tissue is the main source of false-positive [^{18}F]FDG PET findings in oncology.

High cellular uptake of choline, an important precursor in the synthesis of cell membranes, is a common feature in tumors and reflects the increased cell proliferation associated with malignant transformation and radio-labeled choline variants have been used for PET imaging of cancer (37). Although choline PET imaging can be considered as a general tumor imaging modality, to date studies mostly focused on the evaluation of prostate cancer and more recently also of brain tumors. In malignant glioma, areas of increased uptake of ^{18}F-labeled choline seem to correspond with areas of elevated choline metabolites as determined by MR spectroscopy (38) and are able to differentiate between recurrent tumor and radiation necrosis (39, 40).

Tumor angiogenesis and tumor hypoxia, both related with poor disease prognosis, are other key factors in tumorigenesis that are tightly correlated and that can be evaluated with PET imaging. Tumor-associated angiogenic processes result from the imbalance between pro- and antiangiogenic factors that are triggered by hypoxia or genetic anomalies such as expression of oncogenes or tumor suppressor gene mutations (41). Early PET imaging studies have used tumor perfusion and blood volume imaging with radio-labeled water or carbon monoxide as a measure of tumor vasculature and angiogenesis (12). More recently PET probes have been developed that can target directly the molecular events associated with angiogenesis, such as the vascular endothelial growth factor (VEGF)/VEGF receptor (VEGFR) signaling pathway, matrix metalloproteinases or integrins (42). Tumor hypoxia can be detected by the use of PET tracers sensitive to low tissue oxygen such as radio-labeled misonidazole derivatives (43) or by direct imaging of hypoxia-related molecular events such as the induction of hypoxia-inducible factor-1α (HIF-1α) (44).

Aberrant apoptosis is another integral component in cancer growth and is regulated by a complex proteolytic cascade triggered by cell surface death receptors (extrinsic pathway) or via mitochondria (intrinsic pathway). Both pathways eventually lead to activation of caspases (e.g. caspase-3) and to redistribution of phosphatidylserine to the outer cell membrane and represent processes that can be targeted by specific PET probes (45).

Recent advances in genomics, proteomics and metabolomics have revealed several other signaling pathways involved in tumorigenesis and much effort has been put in the development of imaging probes specifically directed to these pathways (46). One of the best studied is the epidermal growth factor (EGF)/epidermal growth factor receptor (EGFR) pathway and EGFR dysregulations have been identified in up to 60% of patients with glioblastoma and correlated with poor prognosis (47). Over the last decade several novel molecular therapies targeted to this signaling pathway have been developed and tested (e.g. monoclonal antibodies against EGFR) with contradictory results (48, 49). Moreover, a number of PET probes have been developed to directly detect the expression activity of EGFR and thus enable a specific imaging read-out for anti-EGFR therapy (50–52). For instance, Lee et al. radiolabeled a chimeric antibody (ch806) that selectively binds to the constitutively active EGFR mutant de2-7 EGFR (or EGFRvIII) and demonstrated that this radioligand (^{124}I-IMP-R4-ch806) allows high-resolution imaging and quantification of de2-7 EGFR expression in xenograft gliomas (50). Precise evaluation of tumor-related molecules that can be targeted by specific drugs will not only allow evaluation and quantification of treatment efficacy, but will also allow efficient patient selection on the basis of the presence/absence of the drug target and guide optimal drug dosage and time course for treatment.

4. PET Imaging of Marker Gene Expression

The treatment of cancer using gene therapy approaches continues to attract interest of many researchers and promising results have also been reported for brain tumors in selected cases (53).

As the success of gene therapy protocols mainly depends on the specific and adequate delivery of therapeutic genes to target tissue and the local temporal and spatial control and regulation of the level of gene expression, the development of methods capable of imaging vector distribution and gene expression noninvasively, sensitively and longitudinally have received much attention over the last years (54).

Molecular biologists and cell biologists have, for many years, been using marker genes such as beta-galactosidase or green fluorescent protein (GFP) to analyze gene expression and regulation in cell cultures. With the development of dedicated small animal PET scanners, marker (aka reporter) genes have been developed whose expression can be monitored noninvasively in living animals.

Several genes have been proposed as potential marker genes for radiotracer-based molecular imaging (11, 55, 56). These marker genes can be broadly classified as genes whose gene product is an intracellular protein (e.g. enzymes) or is associated to the cell membrane (membrane receptors or transporters). The most widely used PET reporter gene encoded enzyme is the herpes simplex virus type 1 thymidine kinase enzyme, HSV-1-TK (57). HSV-1-TK can, like mammalian thymidine kinase, convert thymidine to its phosphorylated form. However, HSV-1-TK can also convert acycloguanosines (such as acyclovir, penciclovir and ganciclovir) and uracil derivatives (such as 2′-fluoro-2′deoxy-1-β-D-arabinofuranosyluracil—FIAU) to their phosphorylated derivatives, unlike the endogenous thymidine kinase. Once phosphorylated these derivatives are unable to exit the cells. Thus, diffusible positron-emitting acycloguanosines or uracil derivatives (such as [^{18}F]-labeled 9-[4-fluoro-3-(hydroxymethyl)butyl]guanine, [^{18}F]FHBG, or [^{18}F]-labeled FIAU, resp.) are converted to trapped substrates by the HSV-1 thymidine kinase PET reporter gene product enabling the site-specific quantification of the HSV-1-TK enzyme activity and thus reflecting the level of reporter gene activity. The main advantage of the use of the herpes simplex virus type 1 thymidine kinase gene (HSV-1-*tk*) in gene therapy protocols is the fact that this gene can be simultaneously used as an imaging gene and as a therapeutic gene. The phosphorylated acycloguanosines, following conversion to their di- and tri-phosphate forms by HSV-1-TK, can kill cells either by blocking DNA synthesis or by causing chain termination. Most commonly ganciclovir is used as prodrug in this suicide gene therapy paradigm. However, at the trace levels utilized for PET reporter gene imaging, no physiological consequences of radiolabeled acycloguanosine administration are observable. Therefore, HSV-1-*tk* has been studied extensively for gene therapy purposes due to the fact that it can serve as a suicide gene and reporter gene simultaneously depending on the administered substrate (ganciclovir (GCV) and [^{18}F]FHBG or [^{18}F]FIAU) (58–60). The HSV-1-*tk* reporter gene has been used to track tumor cells, immune cells or stem cells, to assess tumor growth and to evaluate response to all kinds of therapies (cytotoxic, cytostatic, radiation therapy, chemotherapy, gene therapy) (54, 61) even in clinical application (58). Other PET reporter genes used for gene therapy purposes include the dopamine D2 receptor (D2R), the somatostatin type 2 receptor (SSTr2) and the sodium iodide transporter (NIS) genes (54). Especially the latter two have attracted much interest as these genes can also be used as therapeutic genes (radionuclide therapy) (62, 63).

However, in most imaging-directed gene therapy paradigms the marker gene does not have a therapeutic role itself but is coupled to an independent therapeutic gene, e.g. cytosine deaminase, interleukines or other cytokines. In such a vector system the expression of the marker gene is directly linked to the expression of the therapeutic gene. This means that direct localization and quantification of the marker gene reveals an indirect localization and quantification of the proportionally co-expressed therapeutic gene and allows the noninvasive determination of the tissue-dose of vector-mediated gene expression and, hence, the gene therapeutic capacity of the delivery system (60, 64, 65). The co-expression of 2 or more transgenes is generally achieved by using dual promoters, by inserting an internal ribosomal entry site or by fusing the transgenes into a single translational cassette within the gene construct or vector system.

One of the major disadvantages of reporter gene imaging is the fact that the reporter gene must be delivered ("pretargeted") to the target tissue by the use of transfection or transduction (Fig. 1) and such manipulations may hamper their clinical applicability. A reporter gene construct usually includes transcriptional control elements that can initiate and regulate reporter gene expression. Two general groups of promoter–enhancer control elements are recognized: one group comprises "constitutive" promoters that result in a continuous expression of reporter genes, the other group comprises "inducible" or "tissue-specific" promoters that respond to specific endogenous transcription factors and transcription-regulating complexes (56). The former have typically been used to monitor gene delivery and expression. The latter are generally used to monitor changes in cellular or physiological processes in tumors, which affect the activity of the inducible promoter and thus the concentration of an imaging-detectable marker. Gene reporter imaging can thus be made cell or tissue specific, providing a valuable insight into the specific molecular/genetic machinery of tumors (16). Moreover, the use of tissue-specific promoters within the vector system allows for a more refined tissue transduction, reducing side-effects to neighboring nontumor cells.

5. Some Applications of PET Imaging in Preclinical Brain Tumor Models

Noninvasive, repetitive, and quantitative imaging of biological properties has been employed in the clinic for cancer detection, staging, and management for many years using structural and functional imaging techniques such as CT, MRI, ultrasound, SPECT and PET. However, noninvasive imaging techniques have not been employed very extensively in the study of small animal models of cancer (11). The recent development of dedicated instrumentation (μPET tomographs) and improved reporter genes for noninvasive imaging have paved the way for new paradigms to study cancer in small animal models with the potential for extrapolation to the clinic (translational research). Moreover, dedicated small-animal instrumentation allows well-proven techniques in clinical routine to be refined in preclinical research (reverse translation). For instance, for accurate and specific estimation of cell proliferation, a parameter that is necessary to monitor and characterize brain malignancies and evaluate treatment responses, dynamic imaging and kinetic analysis methods are necessary. Only by applying compartmental analysis to estimate tracer rate constants can the exact contribution of tracer delivery (dependent on tissue perfusion and blood–brain barrier permeability) and tracer metabolic retention (real indicator of tumor proliferation status) to the total tracer uptake be measured (23, 34). Such dynamic PET imaging is widely applied in clinical studies and could recently also be achieved in a preclinical glioma model. Bradbury et al. demonstrated the feasibility of a 3-compartment, 4-parameter model to characterize [^{18}F]FLT kinetics in vivo in a genetically engineered mouse model of high-grade gliomas (66). The authors developed and implemented a compartmental modeling approach where tracer rate constants were estimated by use of image-derived input (left ventricular blood clearance) and tissue time-activity data (tumor uptake) and that was able to clarify the extent to which tracer transport and metabolism contribute to overall tumor uptake. Such exact and noninvasive quantification of tumor proliferation will be of great value to evaluate tumor response of novel targeted therapies.

Traditionally response to anticancer therapy is evaluated by analyzing only morphologic changes (Response Evaluation Criteria in Solid Tumors—RECIST criteria) induced at the target lesion (67). Although this approach is feasible for the assessment of cytotoxic therapies, the value of anatomic criteria for the evaluation of the efficacy of novel signaling pathway inhibitors is rather poor as these therapies often don't provoke changes in tumor volume. Especially here lays the value of PET imaging with its ability to evaluate therapy-induced changes in tumor metabolic activity, gene expression profile, cellularity and vascularity and to evaluate new targets to be exploited as imaging biomarkers in cancer (68). One such new target is the translocator protein (TSPO), also referred to as peripheral benzodiazepine receptor (PBR). The TSPO receptor is located at the outer mitochondrial membrane and is involved in steroidogenesis (cholesterol transport), oxidative processes, immunomodulation (inflammatory responses and microglial activation), cellular proliferation and apoptosis (TSPO ligands may be useful anticancer drugs by inducing cell-cycle arrest and apoptosis) (69). This protein has been extensively used as a biomarker of brain injury and inflammation (70). Recently an increased expression of TSPO in astrocytoma has been established and TSPO expression could be correlated to the grade of tumor malignancy and patient survival (71) opening up the feasibility of TSPO specific ligands to be used as imaging biomarkers for diagnosis and follow-up as well as for anticancer treatment due to their pro-apoptotic effect (72). Most TSPO-ligand PET studies have been performed with the gold standard ligand [^{11}C]-(R)-PK 11195, which however has some important drawbacks such as low signal-to-noise ratio ($\leq$2:1

tumor-to-normal tissue ratio), low blood–brain barrier permeability and relative high nondisplacable tracer binding both in tumor and in normal brain tissue (73). More selective TSPO tracers have been developed over the last years (74, 75) and Buck et al. characterized the high-affinity ligand [^{18}F]-*N*-fluoroacetyl-*N*-(2,5-dimethoxybenzyl)-2-phenoxyaniline ([^{18}F]PBR06), as a candidate probe for the quantitative assessment of TSPO expression in glioma (76). In C6 glioma-bearing rats, [^{18}F]PBR06 was found to preferentially accumulate in the tumors with only modest accumulation in contralateral brain tissue, could easily be displaced by cold ligand (75% displacement) and [^{18}F]PBR06 uptake correlated well with TSPO protein levels as assessed by Western blotting and quantitative immunohistochemistry in tumor and normal tissue. Another target that has gained increased attention over the last years, especially since the advent of effective antiangiogenic drugs are integrins. Integrin $\alpha_v\beta_3$ is highly expressed on activated endothelial cells and solid tumor cells, not expressed on mature vessels or on non-neoplastic epithelium and expression is highly related to tumor angiogenesis and metastasis (77). Furthermore, $\alpha_v\beta_3$ integrins are a well-known therapeutic target for blocking tumor angiogenesis (78, 79) and $\alpha_v\beta_3$-targeted radiolabeled ligands thus have the potential to become highly valuable imaging biomarkers for the evaluation of treatment response. Many radiolabeled arginine–glycine–aspartic acid sequence (RGD)-containing peptides have been developed, which show a good correlation between tracer uptake and in vivo $\alpha_v\beta_3$ integrin expression (15). Radio-labeled RGD-peptides have been used to longitudinally follow brain tumor growth and associated angiogenesis (80) and to evaluate the effect of antiangiogenic therapy (81). Battle et al. demonstrated that an [^{18}F]-labeled small peptide containing the $\alpha_v\beta_3$ integrin-binding RGD sequence can detect changes in human glioblastoma xenograft uptake after treatment with the antiangiogenic agent sunitinib as early as 2 weeks after treatment initiation and that this treatment effect can be detected significantly earlier than a concomitant reduction in tumor size (81).

Although imaging paradigms that directly monitor drug targets exploit the full potential of PET imaging, most imaging studies just monitor the metabolic or proliferative effect of molecular pathway inhibitory drugs or other experimental treatment therapies such as gene- or cell-based therapy. For instance, Veeravagu et al. evaluated the efficacy of a novel $\alpha_v\beta_3$ integrin-targeted radioimmunotherapeutic agent ([^{90}Y]Abegrin) in murine xenograft glioblastoma models by imaging the effect on tumor metabolic activity ([^{18}F]FDG) and cell proliferation ([^{18}F]FLT) (82) and Assadian et al. performed [^{18}F]FDG imaging to predict the therapeutic efficacy of three drugs with a distinct mode of action, alone or in combination (83). In the former study, [^{18}F]FDG uptake was significantly reduced in animals bearing orthotopic U87MG tumors and treated with [^{90}Y]Abegrin or unlabeled Abegrin, whereas [^{18}F]FLT uptake was only reduced in the [^{90}Y]Abegrin treated group. Animals receiving saline or [^{90}Y]IgG treatment or animals bearing tumors with low $\alpha_v\beta_3$ integrin expression (HT29 colorectal carcinoma cells) did not show any difference in [^{18}F]FLT or [^{18}F]FDG uptake, indicating the specificity of the treatment. In the latter study that tested an HIF-1α inhibitor (YC-1), a proapoptotic agent (an activator of calcium-activated potassium channels—NS1619) and an alkylating agent (temozolomide), the authors showed that the therapeutic efficacy of YC-1 can be determined as early as 3 days after therapy by means of [^{18}F] FDG PET imaging. Recently, the synergistic effect of paclitaxel, a nonspecific antimicrotuble agent, and tumor necrosis factor-related apoptosis inducing ligand (TRAIL) on glioblastoma tumor xenografts could be monitored with [^{18}F]FDG PET imaging (84). The PET scans performed 5–6 days post therapy revealed continued tumor metabolic activity above background in mock-treated control tumors, and tumors exposed to paclitaxel or TRAIL monotherapy. However, tumors treated with combined paclitaxel and TRAIL showed no apparent increase in metabolic activity above background. Only in mice treated with combined paclitaxel and TRAIL a decrease in tumor volume could be identified on microCT images performed 3 weeks post therapy, in all other treatment arms tumor volumes increased.

Several gene- and cell-based therapies have been developed and tested for the treatment of brain tumors, even in clinical application (85–87). Although very promising in preclinical models, these treatment paradigms showed moderate and contradictory results in clinical application. Inefficient spatial or temporal tumor targeting, partial tumor coverage and/or tumor heterogeneity could have been reasons for these sparse results. Noninvasive imaging techniques, and especially PET imaging, have the ability to increase the efficiency of gene- and cell-based therapies by guiding all steps of such experimental protocols, from the guidance of vehicle application in relevant tumor areas, over biodistribution and delineation of the ideal therapeutic time window to response assessment of these experimental therapies (54) (Fig. 2a). PET imaging using tumor metabolism directed tracers (such as [^{18}F]FDG, [^{11}C] MET or [^{18}F]FLT) can identify viable tumor tissue as the only meaningful target sites in case of direct application of viral vectors or cell vehicles (59, 60) and can monitor the therapy-induced changes in tumor metabolism. Furthermore, in treatment paradigms using reporter genes linked to suicide genes the exact localization, distribution and activity of the suicide genes can be evaluated by radiotracers specific for the used reporter gene, such as [^{18}F]FHBG/[^{18}F]FIAU in case of HSV-1-tk reporter genes or radiolabeled iodine or somatostatin analogues in case of NIS or SSTr2. More importantly, the level of reporter gene activity as measured by reporter probe uptake can be linked to the level of

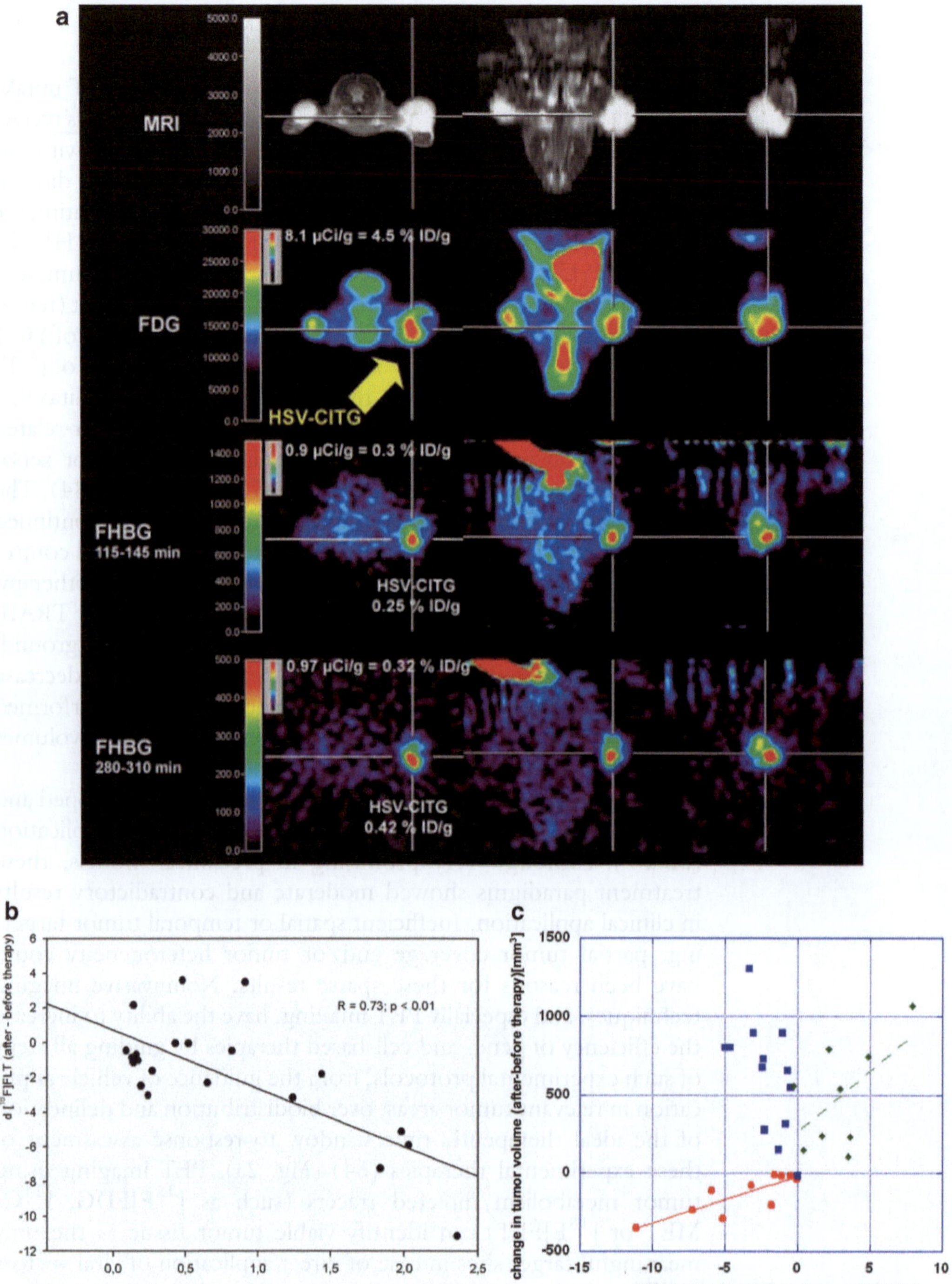

Fig. 2. Imaging-guided gene therapy paradigm. (**a**) Experimental protocol for identification of viable target tissue and assessment of vector-mediated gene expression in vivo in a mouse model with three subcutaneous gliomas. *Row 1*: localization of tumors is displayed by MRI. *Row 2*: the viable target tissue is displayed by [^{18}F]FDG-PET; note the signs of necrosis in the lateral portion of the left-sided tumor (*arrow*). *Rows 3 + 4*: following vector-application into the medial viable portion of the tumor (*arrow*) the tissue dose of vector-mediated gene expression is quantified by [^{18}F]FHBG-PET. *Row 3* shows an image acquired early after tracer injection, which is used for co-registration; *row 4* displays a late image

treatment response as measured by metabolic tracer uptake as early as 3 days after start of therapy (Fig. 2b) (60, 88). Thus PET imaging can monitor the expression level of therapeutic genes, which dependent on the efficacy of tumor transduction, and its correlation to the induced therapy response. The clinical applicability and usefulness of such imaging protocols could already been shown in glioma patients (58, 89). One of the major advantages of using combined reporter and suicide genes in cell-based therapies is the possibility to identify cell-induced tumorigenic changes, such as terratoma formation, in an early phase (90), leaving the opportunity to use the suicide gene as a backup safety measure by effectively eliminating the introduced cells by prodrug treatment as could be impressively shown by Cao and coworkers (91).

6. Suggested Protocol for Small-Animal μPET Imaging of Energy Metabolism ([^{18}F]-FDG), Proliferation ([^{18}F]-FLT), and Amino Acid Uptake ([^{11}C]MET)

For research institutions without an own cyclotron, most [^{18}F] labeled or other long-lived radiotracers can be either purchased from PET facilities with on-site production of radiotracers or be manufactured after purchase of the radio-labeled isotope, e.g. [^{18}F] fluoride, from commercial vendors.

6.1. Radiotracer

For radiotracers with short-lived isotopes, such as ^{11}C or ^{15}O, an on-site cyclotron and radiochemistry laboratory for ad hoc production of the required radiotracers are mandatory. At our institution no carrier added [^{18}F]fluoride is produced via proton bombardment of [^{18}O]H_2O (nuclear reaction $^{18}O(p, n)^{18}F$) in the cyclotron. After [^{18}F]fluoride production and separation from [^{18}O]H_2O a Kryptofix^{TM} 2.2.2 solution (K222), that activates the [^{18}F]fluoride anion for nucleophilic reactions in organic solvents, is added. For the radiosynthesis of [^{18}F]FDG the precursor 1,3,4,6-tetra-*O*-acetyl-2-*O*-trifluoromethane-sulfonyl-β-D-mannopyranose reacts by nucleophilic substitution (S_N2) with the activated [^{18}F] fluoride yielding the target compound after hydrolytic (basic (92) or acidic (93)) deprotection of the acetyl groups. After purification via several cartridges [^{18}F]FDG is diluted in a physiological buffer ready for injection. Similarly, one of the most effective [^{18}F]FLT radiosynthesis starts with a S_N2 reaction of an appropriate precursor

Fig. 2. (Continued) with specific tracer accumulation in the tumor that is used for quantification (adapted from Jacobs, Rueger et al., with permission (60)). (**b**) Response to gene therapy correlates to therapeutic gene expression. The intensity of cdIREStk39gfp expression, which is equivalent to transduction efficiency and tissue-dose of vector mediated therapeutic gene expression, is measured by [^{18}F]FHBG PET (in %ID/g), and the induced therapeutic effect is measured by [^{18}F]FLT PET ($R = 0.73$, $p < 0.01$). Therapeutic effect ([^{18}F]FLT) was calculated as the difference between [^{18}F]FLT accumulation after and before therapy (adapted from Jacobs, Rueger et al., with permission (60). (**c**) Relation between changes in volumetry and FLT uptake. There is a strong correlation between volumetry and change in FLT uptake ($R = 0.83$) for those tumors responding to therapy (complete responders) and a weak correlation ($R = 0.57$) for those tumors not responding to therapy (nonresponders) (adapted from Jacobs, Rueger et al., with permission (60).

(e.g. (5′-*O*-(4,4′-dimethoxytriphenylmethyl)-2′-deoxy-3′-*O*-nosyl-β-D-*threo*-pentofuranosyl)-3-*N*-Boc-thymine) and [^{18}F]fluoride. After radiofluorination the intermediate is deprotected under acidic conditions and the resulting solution containing [^{18}F]FLT is purified by semipreparative HPLC to yield [^{18}F]FLT for injection (94).

A common method for the preparation of [^{11}C]MET is based on the *S*-methylation of L-homocysteine with the reactive methylating agent [^{11}C]iodomethane. Here, cyclotron-derived [^{11}C]carbon dioxide (nuclear reaction $^{14}N(p, \alpha)^{11}C$) is transformed in a two step procedure to the radiosynthon [^{11}C]iodomethane that can transfer his methyl-group to L-homocysteine even under mild reaction conditions using potassium fluoride on aluminum oxide as base. Finally solid phase extraction based purification and formulation provides the [^{11}C]MET solution for injection (95).

6.2. PET Scanner

For small animal PET imaging a dedicated μPET scanner or a high-resolution clinical scanner is recommended. At our institution imaging is performed on the quadHIDAC PET scanner. This device acquires data within a large field-of-view of 16 cm (transaxial) and 28 cm (axial). Images are reconstructed into slices with voxel dimensions of $0.4 \times 0.4 \times 0.4$ mm^3. The spatial resolution of the quadHIDAC scanner is approximately 1 mm full width at half maximum (FWHM) in axial and transaxial direction using filtered backprojection image reconstruction, 0.7 mm FWHM (axial and transaxial) using one-pass listmode expectation maximization image reconstruction (96).

6.3. Animal Handling and PET Imaging

1. Fasten animal for at least 4 h, measure body weight and blood glucose
2. Anesthetize animal with 3.0% isoflurane vaporized in a mix of O_2 and N_2O (1:2) in a knock-out box
3. Position the animal on an automated, feedback-controlled heating blanket with temperature feedback in order to maintain a constant body temperature
4. Keep animal anesthetized with 1.5–2% isoflurane vaporized in a mix of O_2 and N_2O (1:2)
5. Place a catheter into the tail vain of the animal (rat: 24G, mouse: 26G)
6. Fill a syringe with the tracer solution and measure activity of the syringe ($A_{\text{syringe full}}$) with a beta-counter (register time: $t_{\text{syringe full}}$)
7. Adapt the volume of tracer solution in order to have the desired activity in the syringe, which is dependent on type of radiotracer and animal:

	Rat	Mouse
[^{18}F]FDG	12 MBq	7.5 MBq
[^{18}F]FLT	12 MBq	7.5 MBq
[^{11}C]MET	12 MBq	7.5 MBq

8. Inject the tracer into the rat/mouse (register time: $t_{\mathrm{injection}}$), for dynamic PET measurements tracer injection should be performed with the animal already laying on the scanner bed and centered within the PET scanner, for static measurements injection should be performed on the heating blanket.
9. Measure syringe activity ($A_{\mathrm{syringe\ empty}}$) after injection (register time: $t_{\mathrm{syringe\ empty}}$)
10. Inject physiological serum into the rat/mouse in order to reach a total constant volume of 300 μl
11. For dynamic PET imaging, image acquisition is started immediately after the tracer administration and continued for at least 1 h for [^{18}F]-labeled tracers and 30 min for [^{11}C]-labeled tracers. For static PET imaging the time point of image acquisition is dependent on the injected tracer:

	Time point of image acquisition after tracer injection (register time: $t_{\mathrm{start\ measurement}}$) (min)	Duration of image acquisition (min)
[^{18}F]FDG PET	60	15
[^{18}F]FLT PET	60	30
[^{11}C]MET PET	20	40

6.4. Data Acquisition and Reconstruction

In our institution all acquisitions are performed in list mode format and reconstructed into a 450 × 450 × 700 voxel image (voxel size of 0.4 × 0.4 × 0.4 mm^3) with one-pass listmode EM reconstruction method. The image volumes are commonly reduced to the central part of 150 × 150 × 300 voxels for a mouse-sized object and corrected for decay.

6.5. Data Analysis

1. PET images can be opened using a dedicated software like VINCI (97) (Fig. 3a).
2. Draw a volume of interest (VOI) in the area under investigation, e.g. tumor.
3. Draw a VOI in a reference region (e.g. healthy contra lateral brain hemisphere) (Fig. 3b).

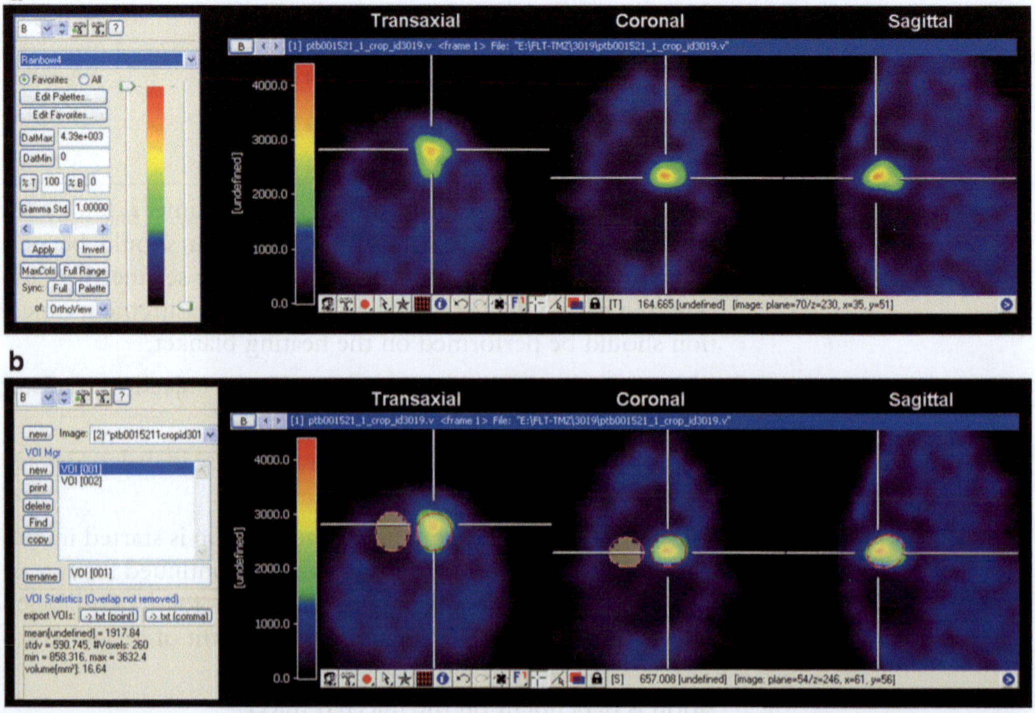

Fig. 3. (**a**) Illustration of an [^{18}F]FLT PET image displayed with the software program VINCI offering multiple options for image processing. (**b**) Illustration of a VOI drawn over a region of interest (here an intracranial malignant glioma) with the software program VINCI.

4. Calculate percent of injected dose in the tumor area (%ID/cc), Standard uptake value of the tumor area (SUV) and tumor to background ratio (T/B) as follows:
   - $\%\text{ID/cc} = 100 \times A_{\text{measured in VOI}}/A_{\text{injected}}\,(t_{\text{start measurement}})$
   - $\text{SUV} = 100 \times A_{\text{measured in VOI}}/A_{\text{injected}}\,(t_{\text{start measurement}})/$ body weight
   - $\text{T/B} = A_{\text{measured in tumor VOI}}/A_{\text{measured in reference region VOI}}$

In those equations:

(a) $A_{\text{measured in VOI}}$ is indicated by the software and expressed in Bq/cc being decay corrected to the start time of the scan ($t_{\text{start measurement}}$).

(b) A_{injected} ($t_{\text{start measurement}}$) can be calculated using the registered following values: $A_{\text{syringe full}}$ ($t_{\text{syringe full}}$), $A_{\text{syringe empty}}$ ($t_{\text{syringe empty}}$), $t_{\text{syringe full}}$, $t_{\text{syringe empty}}$, $t_{\text{start measurement}}$ and the following formulas:

$$A_{syringefull}(t_{startmeasurment}) = A_{syringefull}(t_{syringefull}) \times 2^{-(t_{syringefull} - t_{startmeasurement})/t_{1/2} radio-element}$$

$$A_{syringeempty}(t_{startmeasurement}) = A_{syringeempty}(t_{syringeempty}) \times 2^{-(t_{syringeempty} - t_{startmeasurement})/t_{1/2} radio-element}$$

$$A_{injected}(t_{startmeasurement}) = A_{syringefull}(t_{startmeasurement}) - A_{syringeempty}(t_{startmeasurement})$$

7. Combination with Other Imaging Technologies

PET is a valuable tool to monitor molecular events in living organisms. When being used in combination with other imaging modalities complementary information can be gathered.

CT and MRI provide morphological details with high spatial resolution. When using these modalities together with PET, radiotracer accumulation can be exactly localized. Furthermore, the use of ionizing or paramagnetic contrast agents and specific image sequences can add physiological information. In the case of gliomas for instance, MRI can provide information on the status of the blood brain barrier or the presence of edema (Viel et al., submitted). These data help in further characterizing these highly heterogeneous tumors.

Next to PET there are also other imaging modalities providing molecular details. Instead of labeling PET tracers with radioactivity, these molecules can be tagged with a fluorescence probe, allowing the follow up of tracer distribution and accumulation with fluorescence reflectance imaging (FRI). Also nonspecific or activatable probes can be used for FRI. Due to the scatter and absorption of light in tissue this optical imaging method is limited to preclinical research and invasive or superficial clinical applications. Furthermore, even though much research is done in this field, most acquired data are two-dimensional only, not allowing the exact localization of the tracer. But optical imaging does have some advantages over PET. It is inexpensive and quick and several animals can be imaged simultaneously. There is no radioactivity involved so that exposure to radioisotopes can be avoided and you do not depend on an on-site cyclotron.

Another optical imaging technique next to FRI is bioluminescence imaging (BLI). This method is based on the fact that luciferases, a specific class of enzymes, catalyze chemical reactions that result in the production of light. In the case of the luciferase from the firefly *Photinus pyralis* the substrate luciferin is oxidized in an ATP dependent manner. Injected luciferin is capable of crossing the blood brain barrier, allowing the imaging of brain tumors that have been genetically modified to express firefly luciferase (Fig. 4).

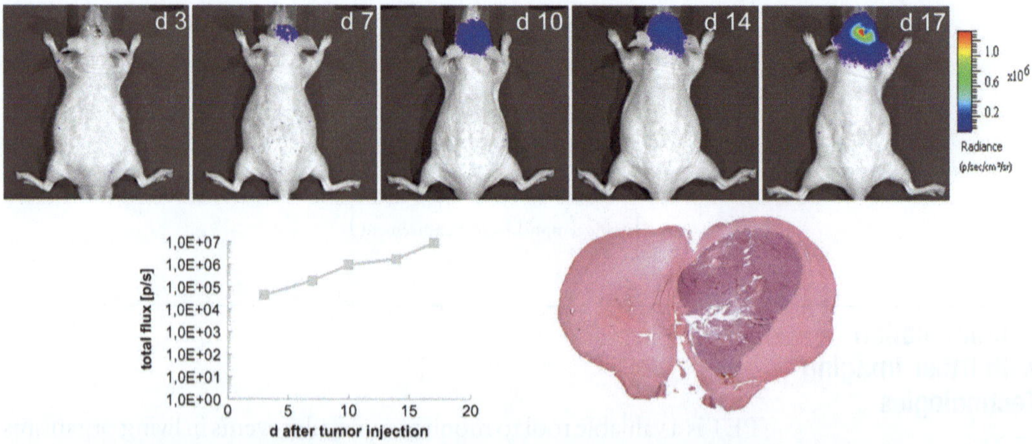

Fig. 4. BLI of orthotopic glioma growth. The human glioma cell line Gli36dEGFR-LITG is genetically modified to express firefly luciferase. After intracranial injection of these cells into NMRI nude mice exponential tumor growth can be monitored by BLI after administration of the substrate luciferin. Histology confirms the identity of the tumor.

In this case the luciferase activity is a measure for tumor viability, and hence can be used to assess efficiency of tumor therapy. Other applications of BLI include the follow up of stem cell migration (90) and the monitoring of gene expression. The latter is possible if the luciferase is controlled by a promoter, which depends on the presence and activity of a gene of interest. This allows e.g. the monitoring of gene expression in response to chemotherapy (98, 99) or specific promoter activating molecules, such as hormones or drugs (100). However, the same limitations hold through as described in section 4 on PET imaging of marker gene expression: the reporter gene (i.e. luciferase) must be delivered to the target tissue.

Even though the translation to clinical situations is limited optical imaging is a useful modality to rapidly and cost effectively screen therapeutic models or agents in preclinical settings.

7.1. Suggested Protocol for In Vivo Bioluminescence Imaging

In the following the measurement of in vivo bioluminescence using the IVIS Spectrum and the Living Image 4.0 Software (Caliper Life Sciences, Hopkinton, MA, USA) is described. This method can also be utilized with other optical imaging devices. However, minor changes might have to be applied.

In vivo light emission of the luciferase reaction peaks in a plateau phase several minutes after intraperitoneal injection of luciferin. This plateau might be different for each tumor and each animal model since kinetics may be dependent on tissue and vascularization. Therefore, to get the most quantitative result several subsequent images are acquired in the range between 6 min and 20 min after substrate injection to determine the peak emission for every experiment.

1. Anesthetize the mouse using isoflurane (2% isoflurane, 0.5 l/min oxygen).
2. Inject 200 μl of a 20 mg/ml stock solution of D-luciferin (e.g. Synchem) intraperitoneally.
3. Arrange the mouse in the imaging chamber so that the bioluminescence signal is as close to the lens as possible. When looking at signals arising from the inside of the body or from an unknown destination it might be advisable to image the mouse in several positions. Make sure that the mouse stays narcotized in the imaging chamber.
4. Acquire a first image about 6 min after substrate injection. Choose a small f/stop (this increases the sensitivity taking into account that the depth of focus is reduced), and a medium binning (small binning results in a better spatial resolution but a reduced sensitivity; hence, medium binning is advisable when dealing with rather weak signals like in BLI). Exposure time might be set to automatic. If doing so it is essential to correct for the exposure time during the data analysis. This can be done by choosing radiance (p/s/cm^2/sr) as readout.
5. Continue to take images every 2 min (create a sequence) up to 20 min after injection of the substrate. In an initial experiment you might want to extend this time to 60 min to make sure that the time window between 6 min and 20 min post injection is appropriate for your model.
6. For data analysis superimpose gray scale photographic images and bioluminescence color images and check the location of the signal. Draw a region of interest (ROI) around the luminescence signal. Since the camera automatically corrects for instrument background and no background signal arises from the investigated animal it is not necessary to subtract a background ROI. However, since the emitted light is scattered in the tissue it is advisable to draw a rather big ROI. Determine the total flux (in photons per second; p/s) arising from this ROI and use the highest value of each sequence for quantification purposes.

8. Conclusion

The use of PET in oncology is based on its unique ability to localize and evaluate tumor-related genetic and metabolic processes in vivo. The techniques that are used routinely in the clinical setting to evaluate patients with primary and metastatic brain tumors are also commonly employed in rodent brain tumor models. Moreover, over the last years several new PET imaging probes that are directed to newly identified tumor-specific targets involved in tumorigenesis have been developed and tested in these rodent models. Together with continuous refinements in available brain tumor models and improvements in small animal imaging scanners, these accomplishments allow the early identification of morphological, pathophysiological, functional and metabolic changes associated with tumor initiation and progression. Early tumor characterization allows optimized patient selection and treatment evaluation and animal brain tumor models are especially valuable to guide the design and refinement of effective therapeutic interventions. Eventually the increased knowledge gained by the combined use of noninvasive molecular imaging and adequate tumor models will lead to more effective and personalized patient care.

Acknowledgments

Our work is supported in part by the 6th Framework EU grants EMIL (LSHC-CT-2004-503569), DiMI (LSHB-CT-2005-512146) and CliniGene NoE (LSHB-CT-2006-018933), the Innovative Medicines Initiative (IMI) grant QuIC-ConCePT (Quantitative Imaging in Cancer: Connecting Cellular Processes with Therapy), the European Strategy Forum on Research Infrastructures (ESFRI) project Euro-BioImaging and by the Bundesministerium für Bildung und Forschung (MoBiMed 01EZ0811). We thank Nina Gerigk (EIMI, Münster) for the design of Fig. 1.

References

1. Buck AK, Herrmann K, Shen C, Dechow T, Schwaiger M, Wester HJ (2009) Molecular imaging of proliferation in vivo: positron emission tomography with [18F]fluorothymidine. Methods 48:205–215
2. Xi W, Tian M, Zhang H (2011) Molecular imaging in neuroscience research with small-animal PET in rodents. Neurosci Res 70: 133–143
3. Corcoran A, De Ridder LI, Del Duca D, Kalala OJ, Lah T, Pilkington GJ, Del Maestro RF (2003) Evolution of the brain tumour spheroid model: transcending current model limitations. Acta Neurochir (Wien) 145: 819–824
4. Holtkamp N, Afanasieva A, Elstner A, van Landeghem FK, Konneker M, Kuhn SA, Kettenmann H, von Deimling A (2005) Brain slice invasion model reveals genes differentially regulated in glioma invasion. Biochem Biophys Res Commun 336:1227–1233
5. Ohnishi T, Matsumura H, Izumoto S, Hiraga S, Hayakawa T (1998) A novel model of glioma cell invasion using organotypic brain slice culture. Cancer Res 58:2935–2940
6. Pilkington GJ, Bjerkvig R, De Ridder L, Kaaijk P (1997) In vitro and in vivo models for the study of brain tumour invasion. Anticancer Res 17:4107–4109
7. Barth RF, Kaur B (2009) Rat brain tumor models in experimental neuro-oncology: the C6, 9L, T9, RG2, F98, BT4C, RT-2 and CNS-1 gliomas. J Neurooncol 94:299–312
8. Fomchenko EI, Holland EC (2006) Mouse models of brain tumors and their applications in preclinical trials. Clin Cancer Res 12: 5288–5297
9. Huse JT, Holland EC (2009) Genetically engineered mouse models of brain cancer and the promise of preclinical testing. Brain Pathol 19:132–143

10. Schnockel U, Hermann S, Stegger L, Law M, Kuhlmann M, Schober O, Schafers K, Schafers M (2010) Small-animal PET: a promising, non-invasive tool in pre-clinical research. Eur J Pharm Biopharm 74:50–54
11. Herschman HR (2004) PET reporter genes for noninvasive imaging of gene therapy, cell tracking and transgenic analysis. Crit Rev Oncol Hematol 51:191–204
12. Yu EY, Mankoff DA (2007) Positron emission tomography imaging as a cancer biomarker. Expert Rev Mol Diagn 7:659–672
13. Huang SC (2000) Anatomy of SUV. Standardized uptake value. Nucl Med Biol 27: 643–646
14. Shoghi KI (2009) Quantitative small animal PET. Q J Nucl Med Mol Imaging 53:365–373
15. Beer AJ, Kessler H, Wester HJ, Schwaiger M (2011) PET Imaging of Integrin alphaVbeta3 Expression. Theranostics 1:48–57
16. Neves AA, Brindle KM (2006) Assessing responses to cancer therapy using molecular imaging. Biochim Biophys Acta 1766:242–261
17. Deberardinis RJ, Sayed N, Ditsworth D, Thompson CB (2008) Brick by brick: metabolism and tumor cell growth. Curr Opin Genet Dev 18:54–61
18. Vander Heiden MG, Cantley LC, Thompson CB (2009) Understanding the Warburg effect: the metabolic requirements of cell proliferation. Science 324:1029–1033
19. Dhermain FG, Hau P, Lanfermann H, Jacobs AH, van den Bent MJ (2010) Advanced MRI and PET imaging for assessment of treatment response in patients with gliomas. Lancet Neurol 9:906–920
20. Herholz K, Coope D, Jackson A (2007) Metabolic and molecular imaging in neuro-oncology. Lancet Neurol 6:711–724
21. Schober O, Meyer GJ, Stolke D, Hundeshagen H (1985) Brain tumor imaging using C-11-labeled L-methionine and D-methionine. J Nucl Med 26:98–99
22. Langen KJ, Hamacher K, Weckesser M, Floeth F, Stoffels G, Bauer D, Coenen HH, Pauleit D (2006) O-(2-[18F]fluoroethyl)-L-tyrosine: uptake mechanisms and clinical applications. Nucl Med Biol 33:287–294
23. Jacobs AH, Thomas A, Kracht LW, Li H, Dittmar C, Garlip G, Galldiks N, Klein JC, Sobesky J, Hilker R, Vollmar S, Herholz K, Wienhard K, Heiss WD (2005) 18F-fluoro-L-thymidine and 11C-methylmethionine as markers of increased transport and proliferation in brain tumors. J Nucl Med 46: 1948–1958
24. Popperl G, Kreth FW, Mehrkens JH, Herms J, Seelos K, Koch W, Gildehaus FJ, Kretzschmar HA, Tonn JC, Tatsch K (2007) FET PET for the evaluation of untreated gliomas: correlation of FET uptake and uptake kinetics with tumour grading. Eur J Nucl Med Mol Imaging 34:1933–1942
25. Terakawa Y, Tsuyuguchi N, Iwai Y, Yamanaka K, Higashiyama S, Takami T, Ohata K (2008) Diagnostic accuracy of 11C-methionine PET for differentiation of recurrent brain tumors from radiation necrosis after radiotherapy. J Nucl Med 49:694–699
26. Galldiks N, Kracht LW, Burghaus L, Thomas A, Jacobs AH, Heiss WD, Herholz K (2006) Use of 11C-methionine PET to monitor the effects of temozolomide chemotherapy in malignant gliomas. Eur J Nucl Med Mol Imaging 33:516–524
27. Alexiou GA, Tsiouris S, Kyritsis AP, Argyropoulou MI, Voulgaris S, Fotopoulos AD (2010) Assessment of glioma proliferation using imaging modalities. J Clin Neurosci 17:1233–1238
28. Vesselle H, Grierson J, Muzi M, Pugsley JM, Schmidt RA, Rabinowitz P, Peterson LM, Vallieres E, Wood DE (2002) In vivo validation of 3′deoxy-3′-[(18)F]fluorothymidine ([(18)F]FLT) as a proliferation imaging tracer in humans: correlation of [(18)F]FLT uptake by positron emission tomography with Ki-67 immunohistochemistry and flow cytometry in human lung tumors. Clin Cancer Res 8:3315–3323
29. Wells P, Gunn RN, Alison M, Steel C, Golding M, Ranicar AS, Brady F, Osman S, Jones T, Price P (2002) Assessment of proliferation in vivo using 2-[(11)C]thymidine positron emission tomography in advanced intra-abdominal malignancies. Cancer Res 62:5698–5702
30. Buck AK, Bommer M, Stilgenbauer S, Juweid M, Glatting G, Schirrmeister H, Mattfeldt T, Tepsic D, Bunjes D, Mottaghy FM, Krause BJ, Neumaier B, Dohner H, Moller P, Reske SN (2006) Molecular imaging of proliferation in malignant lymphoma. Cancer Res 66:11055–11061
31. Buck AK, Schirrmeister H, Hetzel M, Von Der Heide M, Halter G, Glatting G, Mattfeldt T, Liewald F, Reske SN, Neumaier B (2002) 3-deoxy-3-[(18)F]fluorothymidine-positron emission tomography for noninvasive assessment of proliferation in pulmonary nodules. Cancer Res 62:3331–3334
32. Wagner M, Seitz U, Buck A, Neumaier B, Schultheiss S, Bangerter M, Bommer M, Leithauser F, Wawra E, Munzert G, Reske SN (2003) 3′-[18F]fluoro-3′-deoxythymidine ([18F]-FLT) as positron emission tomography

tracer for imaging proliferation in a murine B-Cell lymphoma model and in the human disease. Cancer Res 63:2681–2687

33. Kenny LM, Vigushin DM, Al-Nahhas A, Osman S, Luthra SK, Shousha S, Coombes RC, Aboagye EO (2005) Quantification of cellular proliferation in tumor and normal tissues of patients with breast cancer by [18F] fluorothymidine-positron emission tomography imaging: evaluation of analytical methods. Cancer Res 65:10104–10112
34. Backes H, Ullrich R, Neumaier B, Kracht L, Wienhard K, Jacobs AH (2009) Noninvasive quantification of 18F-FLT human brain PET for the assessment of tumour proliferation in patients with high-grade glioma. Eur J Nucl Med Mol Imaging 36:1960–1967
35. van Waarde A, Cobben DC, Suurmeijer AJ, Maas B, Vaalburg W, de Vries EF, Jager PL, Hoekstra HJ, Elsinga PH (2004) Selectivity of 18F-FLT and 18F-FDG for differentiating tumor from inflammation in a rodent model. J Nucl Med 45:695–700
36. Lee TS, Ahn SH, Moon BS, Chun KS, Kang JH, Cheon GJ, Choi CW, Lim SM (2009) Comparison of 18F-FDG, 18F-FET and 18F-FLT for differentiation between tumor and inflammation in rats. Nucl Med Biol 36:681–686
37. Mertens K, Slaets D, Lambert B, Acou M, De Vos F, Goethals I (2010) PET with (18)F-labelled choline-based tracers for tumour imaging: a review of the literature. Eur J Nucl Med Mol Imaging 37:2188–2193
38. Kwee SA, Coel MN, Lim J, Ko JP (2004) Combined use of F-18 fluorocholine positron emission tomography and magnetic resonance spectroscopy for brain tumor evaluation. J Neuroimaging 14:285–289
39. Rottenburger C, Hentschel M, Kelly T, Trippel M, Brink I, Reithmeier T, Tobias Meyer P, Nikkhah G (2011) Comparison of C-11 Methionine and C-11 Choline for PET imaging of brain metastases: a prospective pilot study. Clin Nucl Med 36:639–642
40. Kwee SA, Ko JP, Jiang CS, Watters MR, Coel MN (2007) Solitary brain lesions enhancing at MR imaging: evaluation with fluorine 18 fluorocholine PET. Radiology 244:557–565
41. Jouanneau E (2008) Angiogenesis and gliomas: current issues and development of surrogate markers. Neurosurgery 62:31–50, discussion 50
42. Haubner R, Beer AJ, Wang H, Chen X (2010) Positron emission tomography tracers for imaging angiogenesis. Eur J Nucl Med Mol Imaging 37(Suppl 1):S86–103
43. Bruehlmeier M, Roelcke U, Schubiger PA, Ametamey SM (2004) Assessment of hypoxia and perfusion in human brain tumors using PET with 18F-fluoromisonidazole and 15O-H2O. J Nucl Med 45:1851–1859
44. Serganova I, Doubrovin M, Vider J, Ponomarev V, Soghomonyan S, Beresten T, Ageyeva L, Serganov A, Cai S, Balatoni J, Blasberg R, Gelovani J (2004) Molecular imaging of temporal dynamics and spatial heterogeneity of hypoxia-inducible factor-1 signal transduction activity in tumors in living mice. Cancer Res 64:6101–6108
45. Blankenberg FG (2008) In vivo detection of apoptosis. J Nucl Med 49(Suppl 2):81S–95S
46. Waerzeggers Y, Monfared P, Viel T, Faust A, Kopka K, Schäfers M, Tavitian B, Winkeler A, Jacobs A (in press) Specific biomarkers of receptors, pathways of inhibition and targeted therapies. Part II: pre-clinical developments. Br J Radiol
47. Shinojima N, Tada K, Shiraishi S, Kamiryo T, Kochi M, Nakamura H, Makino K, Saya H, Hirano H, Kuratsu J, Oka K, Ishimaru Y, Ushio Y (2003) Prognostic value of epidermal growth factor receptor in patients with glioblastoma multiforme. Cancer Res 63:6962–6970
48. Mellinghoff IK, Wang MY, Vivanco I, Haas-Kogan DA, Zhu S, Dia EQ, Lu KV, Yoshimoto K, Huang JH, Chute DJ, Riggs BL, Horvath S, Liau LM, Cavenee WK, Rao PN, Beroukhim R, Peck TC, Lee JC, Sellers WR, Stokoe D, Prados M, Cloughesy TF, Sawyers CL, Mischel PS (2005) Molecular determinants of the response of glioblastomas to EGFR kinase inhibitors. N Engl J Med 353:2012–2024
49. Stragliotto G, Vega F, Stasiecki P, Gropp P, Poisson M, Delattre JY (1996) Multiple infusions of anti-epidermal growth factor receptor (EGFR) monoclonal antibody (EMD 55,900) in patients with recurrent malignant gliomas. Eur J Cancer 32A:636–640
50. Lee FT, O'Keefe GJ, Gan HK, Mountain AJ, Jones GR, Saunder TH, Sagona J, Rigopoulos A, Smyth FE, Johns TG, Govindan SV, Goldenberg DM, Old LJ, Scott AM (2010) Immuno-PET quantitation of de2-7 epidermal growth factor receptor expression in glioma using 124I-IMP-R4-labeled antibody ch806. J Nucl Med 51:967–972
51. Gelovani JG (2008) Molecular imaging of epidermal growth factor receptor expression-activity at the kinase level in tumors with positron emission tomography. Cancer Metastasis Rev 27(4):645–53

52. Miyagawa T, Gogiberidze G, Serganova I, Cai S, Balatoni JA, Thaler HT, Ageyeva L, Pillarsetty N, Finn RD, Blasberg RG (2008) Imaging of HSV-tk Reporter gene expression: comparison between [18F]FEAU, [18F]FFEAU, and other imaging probes. J Nucl Med 49:637–648
53. Germano IM, Binello E (2009) Gene therapy as an adjuvant treatment for malignant gliomas: from bench to bedside. J Neurooncol 93:79–87
54. Waerzeggers Y, Monfared P, Viel T, Winkeler A, Voges J, Jacobs AH (2009) Methods to monitor gene therapy with molecular imaging. Methods 48:146–160
55. Serganova I, Blasberg R (2005) Reporter gene imaging: potential impact on therapy. Nucl Med Biol 32:763–780
56. Serganova I, Ponomarev V, Blasberg R (2007) Human reporter genes: potential use in clinical studies. Nucl Med Biol 34:791–807
57. Jacobs A, Tjuvajev JG, Dubrovin M, Akhurst T, Balatoni J, Beattie B, Joshi R, Finn R, Larson SM, Herrlinger U, Pechan PA, Chiocca EA, Breakefield XO, Blasberg RG (2001) Positron emission tomography-based imaging of transgene expression mediated by replication-conditional, oncolytic herpes simplex virus type 1 mutant vectors in vivo. Cancer Res 61:2983–2995
58. Jacobs A, Voges J, Reszka R, Lercher M, Gossmann A, Kracht L, Kaestle C, Wagner R, Wienhard K, Heiss WD (2001) Positron-emission tomography of vector-mediated gene expression in gene therapy for gliomas. Lancet 358:727–729
59. Miletic H, Fischer Y, Litwak S, Giroglou T, Waerzeggers Y, Winkeler A, Li H, Himmelreich U, Lange C, Stenzel W, Deckert M, Neumann H, Jacobs AH, von Laer D (2007) Bystander killing of malignant glioma by bone marrow-derived tumor-infiltrating progenitor cells expressing a suicide gene. Mol Ther 15:1373–1381
60. Jacobs AH, Rueger MA, Winkeler A, Li H, Vollmar S, Waerzeggers Y, Rueckriem B, Kummer C, Dittmar C, Klein M, Heneka MT, Herrlinger U, Fraefel C, Graf R, Wienhard K, Heiss WD (2007) Imaging-guided gene therapy of experimental gliomas. Cancer Res 67:1706–1715
61. Alauddin MM, Gelovani JG (2010) Radiolabeled nucleoside analogues for PET imaging of HSV1-tk gene expression. Curr Top Med Chem 10:1617–1632
62. Hingorani M, Spitzweg C, Vassaux G, Newbold K, Melcher A, Pandha H, Vile R, Harrington K (2010) The biology of the sodium iodide symporter and its potential for targeted gene delivery. Curr Cancer Drug Targets 10:242–267
63. De Jong M, Valkema R, Jamar F, Kvols LK, Kwekkeboom DJ, Breeman WA, Bakker WH, Smith C, Pauwels S, Krenning EP (2002) Somatostatin receptor-targeted radionuclide therapy of tumors: preclinical and clinical findings. Semin Nucl Med 32:133–140
64. Shah K, Jacobs A, Breakefield XO, Weissleder R (2004) Molecular imaging of gene therapy for cancer. Gene Ther 11:1175–1187
65. Jacobs AH, Winkeler A, Hartung M, Slack M, Dittmar C, Kummer C, Knoess C, Galldiks N, Vollmar S, Wienhard K, Heiss WD (2003) Improved herpes simplex virus type 1 amplicon vectors for proportional coexpression of positron emission tomography marker and therapeutic genes. Hum Gene Ther 14:277–297
66. Bradbury MS, Hambardzumyan D, Zanzonico PB, Schwartz J, Cai S, Burnazi EM, Longo V, Larson SM, Holland EC (2008) Dynamic small-animal PET imaging of tumor proliferation with 3′-deoxy-3′-18F-fluorothymidine in a genetically engineered mouse model of high-grade gliomas. J Nucl Med 49:422–429
67. Wahl RL, Jacene H, Kasamon Y, Lodge MA (2009) From RECIST to PERCIST: Evolving Considerations for PET response criteria in solid tumors. J Nucl Med 50(Suppl 1):122S–150S
68. Serkova NJ (2011) Translational imaging endpoints to predict treatment response to novel targeted anticancer agents. Drug Resist 14:224–235
69. Papadopoulos V, Baraldi M, Guilarte TR, Knudsen TB, Lacapere JJ, Lindemann P, Norenberg MD, Nutt D, Weizman A, Zhang MR, Gavish M (2006) Translocator protein (18 kDa): new nomenclature for the peripheral-type benzodiazepine receptor based on its structure and molecular function. Trends Pharmacol Sci 27:402–409
70. Chen MK, Guilarte TR (2008) Translocator protein 18 kDa (TSPO): molecular sensor of brain injury and repair. Pharmacol Ther 118:1–17
71. Vlodavsky E, Soustiel JF (2007) Immunohistochemical expression of peripheral benzodiazepine receptors in human astrocytomas and its correlation with grade of malignancy, proliferation, apoptosis and survival. J Neurooncol 81:1–7

72. Chelli B, Lena A, Vanacore R, Da Pozzo E, Costa B, Rossi L, Salvetti A, Scatena F, Ceruti S, Abbracchio MP, Gremigni V, Martini C (2004) Peripheral benzodiazepine receptor ligands: mitochondrial transmembrane potential depolarization and apoptosis induction in rat C6 glioma cells. Biochem Pharmacol 68:125–134
73. Pappata S, Cornu P, Samson Y, Prenant C, Benavides J, Scatton B, Crouzel C, Hauw JJ, Syrota A (1991) PET study of carbon-11-PK 11195 binding to peripheral type benzodiazepine sites in glioblastoma: a case report. J Nucl Med 32:1608–1610
74. Chauveau F, Van Camp N, Dolle F, Kuhnast B, Hinnen F, Damont A, Boutin H, James M, Kassiou M, Tavitian B (2009) Comparative evaluation of the translocator protein radioligands 11C-DPA-713, 18F-DPA-714, and 11C-PK11195 in a rat model of acute neuroinflammation. J Nucl Med 50:468–476
75. Van Camp N, Boisgard R, Kuhnast B, Theze B, Viel T, Gregoire MC, Chauveau F, Boutin H, Katsifis A, Dolle F, Tavitian B (2010) In vivo imaging of neuroinflammation: a comparative study between [(18)F]PBR111, [(11)C]CLINME and [(11)C]PK11195 in an acute rodent model. Eur J Nucl Med Mol Imaging 37:962–972
76. Buck JR, McKinley ET, Hight MR, Fu A, Tang D, Smith RA, Tantawy MN, Peterson TE, Colvin D, Ansari MS, Baldwin RM, Zhao P, Guleryuz S, Manning HC (2011) Quantitative, preclinical PET of translocator protein expression in glioma using 18F-N-fluoroacetyl-N-(2,5-dimethoxybenzyl)-2-phenoxyaniline. J Nucl Med 52:107–114
77. Hood JD, Cheresh DA (2002) Role of integrins in cell invasion and migration. Nat Rev Cancer 2:91–100
78. Cox D, Brennan M, Moran N (2010) Integrins as therapeutic targets: lessons and opportunities. Nat Rev Drug Discov 9:804–820
79. Cai W, Chen X (2006) Anti-angiogenic cancer therapy based on integrin alphavbeta3 antagonism. Anticancer Agents Med Chem 6:407–428
80. Chen X, Park R, Khankaldyyan V, Gonzales-Gomez I, Tohme M, Moats RA, Bading JR, Laug WE, Conti PS (2006) Longitudinal microPET imaging of brain tumor growth with F-18-labeled RGD peptide. Mol Imaging Biol 8:9–15
81. Battle MR, Goggi JL, Allen L, Barnett J, Morrison MS (2011) Monitoring tumor response to antiangiogenic sunitinib therapy with 18F-Fluciclatide, an 18F-labeled {alpha}V{beta}3-Integrin and {alpha}V{beta}5-Integrin imaging agent. J Nucl Med 52 (3):424–30
82. Veeravagu A, Liu Z, Niu G, Chen K, Jia B, Cai W, Jin C, Hsu AR, Connolly AJ, Tse V, Wang F, Chen X (2008) Integrin alphavbeta3-targeted radioimmunotherapy of glioblastoma multiforme. Clin Cancer Res 14:7330–7339
83. Assadian S, Aliaga A, Del Maestro RF, Evans AC, Bedell BJ (2008) FDG-PET imaging for the evaluation of antiglioma agents in a rat model. Neurooncology 10:292–299
84. Dorsey JF, Mintz A, Tian X, Dowling ML, Plastaras JP, Dicker DT, Kao GD, El-Deiry WS (2009) Tumor necrosis factor-related apoptosis-inducing ligand (TRAIL) and paclitaxel have cooperative in vivo effects against glioblastoma multiforme cells. Mol Cancer Ther 8:3285–3295
85. Kroeger KM, Muhammad AK, Baker GJ, Assi H, Wibowo MK, Xiong W, Yagiz K, Candolfi M, Lowenstein PR, Castro MG (2010) Gene therapy and virotherapy: novel therapeutic approaches for brain tumors. Discov Med 10:293–304
86. Van Gool S, Maes W, Ardon H, Verschuere T, Van Cauter S, De Vleeschouwer S (2009) Dendritic cell therapy of high-grade gliomas. Brain Pathol 19:694–712
87. Rath P, Shi H, Maruniak JA, Litofsky NS, Maria BL, Kirk MD (2009) Stem cells as vectors to deliver HSV/tk gene therapy for malignant gliomas. Curr Stem Cell Res Ther 4:44–49
88. Rueger MA, Ameli M, Li H, Winkeler A, Rueckriem B, Vollmar S, Galldiks N, Hesselmann V, Fraefel C, Wienhard K, Heiss WD, Jacobs AH (2011) [(1)F]FLT PET for noninvasive monitoring of early response to gene therapy in experimental gliomas. Mol Imaging Biol 13:547–557
89. Yaghoubi SS, Jensen MC, Satyamurthy N, Budhiraja S, Paik D, Czernin J, Gambhir SS (2009) Noninvasive detection of therapeutic cytolytic T cells with 18F-FHBG PET in a patient with glioma. Nat Clin Pract Oncol 6:53–58
90. Waerzeggers Y, Klein M, Miletic H, Himmelreich U, Li H, Monfared P, Herrlinger U, Hoehn M, Coenen HH, Weller M, Winkeler A, Jacobs AH (2008) Multimodal imaging of neural progenitor cell fate in rodents. Mol Imaging 7:77–91
91. Cao F, Drukker M, Lin S, Sheikh AY, Xie X, Li Z, Connolly AJ, Weissman IL, Wu JC (2007) Molecular imaging of embryonic stem cell misbehavior and suicide gene ablation. Cloning Stem Cells 9:107–117

92. Füchtner F, Steinbach J, Mäding P, Johannsen B (1996) Basic hydrolysis of 18F]fluoro-1,3,4,6-tetra-O-acetyl-d-glucose in the preparation of 2-[18F]fluoro-2-deoxy-d-glucose. Appl Radiat Isot 47:61–66
93. Hamacher K, Coenen HH, Stocklin G (1986) Efficient stereospecific synthesis of no-carrier-added 2-[18F]-fluoro-2-deoxy-D-glucose using aminopolyether supported nucleophilic substitution. J Nucl Med 27:235–238
94. Yun M, Oh SJ, Ha HJ, Ryu JS, Moon DH (2003) High radiochemical yield synthesis of 3′-deoxy-3′-[18F]fluorothymidine using (5′-O-dimethoxytrityl-2′-deoxy-3′-O-nosyl-beta-D-threo pentofuranosyl)thymine and its 3-N-BOC-protected analogue as a labeling precursor. Nucl Med Biol 30:151–157
95. Schmitz F, Plenevaux A, Del-Fiore G, Lemaire C, Comar D, Luxen A (1995) Fast routine production of l-[11C-methyl]methionine with Al2O3KF. Appl Radiat Isot 46:893–897
96. Schafers KP, Reader AJ, Kriens M, Knoess C, Schober O, Schafers M (2005) Performance evaluation of the 32-module quadHIDAC small-animal PET scanner. J Nucl Med 46:996–1004
97. Vollmar SCJ, Sue M, Klein J, Jacobs AH, Herholz K (2004) In: Kremer K, Macho V (eds) VINCI—Volume Imaging in Neurological Research, Co-Registration and ROIs Forschung und wissenschaftliches Rechnen 2003. Gesellschaft für wissenschaftliche Datenverarbeitung, Göttingen, pp 115–131
98. Monfared P, Winkeler A, Klein M, Li H, Klose A, Hoesel M, Waerzeggers Y, Korsching S, Jacobs AH (2008) Noninvasive assessment of E2F-1-mediated transcriptional regulation in vivo. Cancer Res 68:5932–5940
99. Klose A, Waerzeggers Y, Monfared P, Vukicevic S, Kaijzel EL, Winkeler A, Wickenhauser C, Lowik CW, Jacobs AH (2011) Imaging bone morphogenetic protein 7 induced cell cycle arrest in experimental gliomas. Neoplasia 13:276–285
100. Winkeler A, Sena-Esteves M, Paulis LE, Li H, Waerzeggers Y, Ruckriem B, Himmelreich U, Klein M, Monfared P, Rueger MA, Heneka M, Vollmar S, Hoehn M, Fraefel C, Graf R, Wienhard K, Heiss WD, Jacobs AH (2007) Switching on the lights for gene therapy. PLoS ONE 2:e528

Neuromethods (2013) 77: 329–365
DOI 10.1007/7657_2011_31

Published online: 6 April 2012

In Vivo Magnetic Resonance Spectroscopic Imaging and Ex Vivo Quantitative Neuropathology by High Resolution Magic Angle Spinning Proton Magnetic Resonance Spectroscopy

Rui V. Simões, Ana Paula Candiota, Margarida Julià-Sapé, and Carles Arús

Abstract

The applications of two magnetic resonance techniques to the study of brain tumours are discussed. Multivoxel MR spectroscopic imaging (MRSI) can be performed in vivo in animal models and HRMAS is performed ex vivo. The first one is able to provide "molecular images" of tumours and the second one gives rich metabolomic information from excised biopsies. The application of both techniques yields a high amount of multidimensional data, which can be analysed with complex statistical methods, such as those provided by pattern recognition techniques.

Key words: Magnetic resonance spectroscopy, Magnetic resonance spectroscopic imaging, High resolution magic angle spinning proton magnetic resonance spectroscopy, Brain tumours, Pattern recognition, Glioblastoma

1. Introduction to Preclinical Brain Tumour MR Work

1.1. In Vivo Multivoxel MR Spectroscopic Imaging of Preclinical Brain Tumours

Like in MRI, most of in vivo applications of multivoxel MR spectroscopic imaging (MRSI) are performed with proton (^{1}H), although other metabolically relevant nuclei can be studied. Most early work on ^{1}H-MRSI of animal models of brain tumours concentrated in murine models, mice and rats, and focused into producing maps of different metabolites or substances. The types of brain tumours investigated were mostly allograft stereotactic models and xenografts of human tumours in immunocompromised animals, although lately, the use of spontaneous tumours appearing in genetically engineered mice (GEM) is increasing. Pioneering work from research teams in Grenoble (1) and Würzburg

(2) produced usable single "metabolite" content images from C6 high-grade glioma in rats based on 2D spectroscopy correlation peak imaging. This demonstrated relative increases inside the tumour area with respect to non-infiltrated brain for lactate, alanine, hypotaurine and phosphoethanolamine. In addition, NAA and glucose were undetectable in the tumour while well seen in uninvolved brain. Further work also produced pH images using exogenous probes and compared them to lactate maps in C6 glioma tumours in rat brain (3, 4).

A different approach was used by a group in Kuopio—see, e.g. (5)—which acquired short TE (5 ms) MRSI grids for mobile lipids (ML) and total choline detection, quantification and image representation. ML increases were correlated with induced apoptosis in rat BT4C gliomas (cells positive for herpes simplex thymidine kinase, TK) by gancyclovir therapy.

Nevertheless, improvements in hardware (stronger gradient coils and efficient water cooling systems) and new shimming methods have enabled MRSI to be performed also in the mouse brain (6–11), with applications to brain tumour studies having also been reported (12, 13). Moreover, a recent methodological development has been applied to monitoring brain tumour response in preclinical models (14). Essentially, it uses ^{13}C MRSI maps obtained after hyperpolarised 1-C ^{13}C pyruvate injection and lactate or lactate/pyruvate ratio representation. The lactate/pyruvate ratio has been shown to decrease upon therapy, possibly sampling decreased lactate dehydrogenase activity upon therapy in the tumour mass.

MRSI grids of unsuppressed water intensities have also been transformed into temperature maps profiting from the earlier described phenomena of temperature-dependent chemical shift of water-exchangeable protons of metabolites in brain tissue (15) or directly of water itself (16–18). Methodological improvements in post-processing, e.g. (19), have allowed researchers to obtain temperature change maps in rat brains subjected to ischaemia (20).

We have also applied a similar approach to control for tissue temperature upon induced, mild hypothermia, in GL261 glioma tumours in C57 mice (21). Finally, MRSI maps can also be obtained from mice harbouring brain tumours after perturbing their basal metabolome pattern. This has been dubbed for perturbation-enhanced MRSI (PE-MRSI) (22) and shows promise for non-invasive tumour type prediction. Finally, other agents like DMSO (23, 24) have potential interest as MRSI-based contrast materials for phenotyping brain tumours.

In summary, it is expected that preclinical MRSI will generate information of interest for differentiating among tumour types or even their molecular sub-types, allowing tumour progression to be monitored, differentiating tumour from non-tumour abnormal brain mass and predicting and/or tracking response to therapy.

1.2. HRMAS Studies of Brain Tumour Biopsies

The HRMAS data acquisition methodology allows analysing brain tumour patterns at high resolution without resorting to tissue extraction and with a resolution comparable to liquid state NMR (25). The advantage with preclinical models is that animal sacrifice prior to biopsy sampling can be carried out with focused microwave (FMW) irradiation and this preserves the in vivo pattern (Fig. 1) without the ischaemia time which unavoidably afflicts human biopsies. Even that some HRMAS studies on biopsies from animal models of brain tumours have been described (21, 26), most development work has been carried out with human brain tumour biopsies, where the effect of post-surgery ischaemia time upon tissue pattern cannot be avoided. Still, useful information about

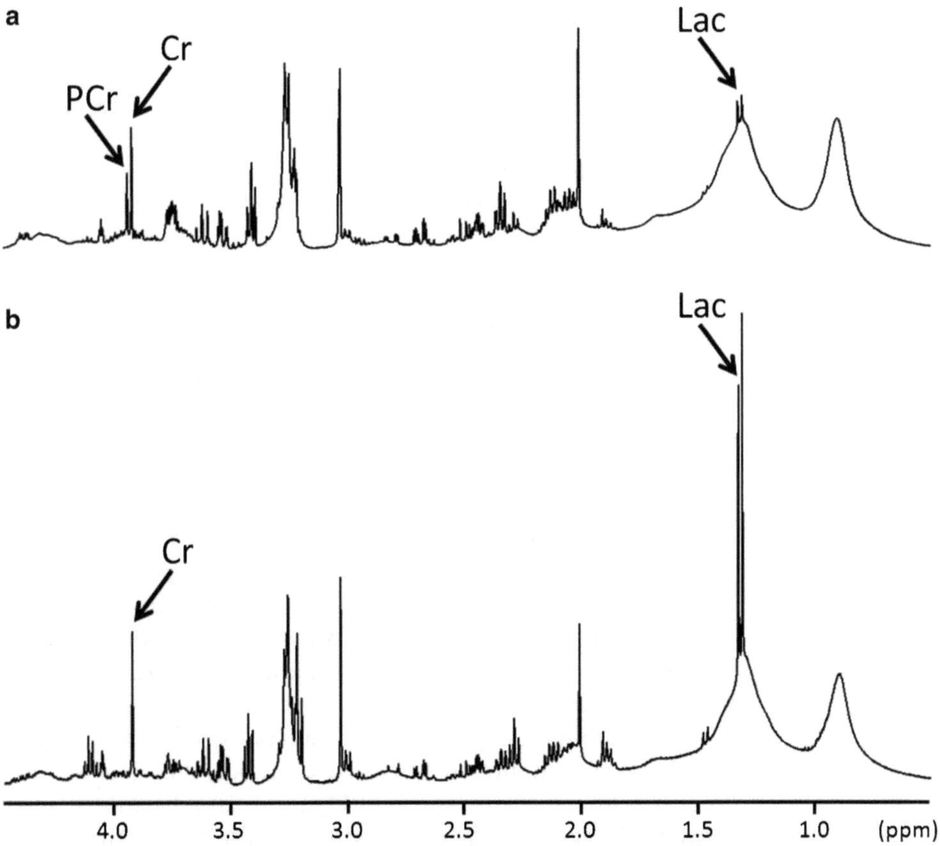

Fig. 1. HRMAS spectra from normal mouse brain (C57/BL6) acquired at 37°C and 9.4 T (pulse-and-acquire sequence) with 6,000 Hz spinning rate, in 17 min. (**a**) Mouse euthanised with focused microwave (FMW) irradiation and (**b**) mouse euthanised with anaesthesia overdose. Note differences (*arrows*) in creatine (Cr, 3.93 ppm), phosphocreatine (PCr, 3.95 ppm) and lactate (Lac, 1.32 ppm) depending on the sacrifice method used. See Sect. 4.1 for further details.

tumour typing, grading and heterogeneity has been obtained with HRMAS.

In this respect, ample literature exists about using HRMAS as an ex vivo typing tool for human brain tumours (27–30), but much less for animal models of these brain tumours. For example, total spinning time has been shown to have non-negligible effects over the metabolome pattern (31) and even tissue architecture (32). Apart from this, varying tissue temperature produces mostly reversible pattern changes between 0 and 37°C, which do not seem to affect pattern recognition-based discrimination of major tumour types (33).

2. Technical Requirements for Successful MRSI and HRMAS Acquisition

The physiological and MRSI parameters herewith discussed will refer to studies carried out with mice, unless otherwise indicated.

2.1. Anaesthesia and Other Basic Monitoring Requirements for In Vivo MRSI

In order to obtain proper MRI/MRS/MRSI data, any movements during the MR exploration should be avoided. To achieve this, animals are usually explored under the effect of anaesthesia, which also reduces stress for the animal. General anaesthesia is a state of general depression of the CNS involving analgesia, suppression of reflex activity and relaxation of voluntary muscle. Convenient anaesthesia may be achieved both by means of inhalational and injectable agents. The main advantages of inhalational agents are that the depth of anaesthesia can be adjusted fast, animals recover from it quickly and agents used are either exhaled unchanged or metabolised only in relatively small proportions by the liver, therefore being less likely to interfere with experimental results (34). On the other hand, environmental pollution with inhalational anaesthetics must be considered a hazard to lab personnel and due precautions to avoid fully open administration methods should be considered. It is important to remark that both strain and genetic modifications in mice could cause variations in their susceptibility to anaesthesia-associated morbidity and mortality (35).

In our work, anaesthesia is always performed using isofluorane at 2.5–4.0% (induction in closed chamber) and 1–2.5% (maintenance with an open system) in O_2 using an inhalational apparatus (Matrx VME2, Midkmark, Versailles, Ohio, USA). After induction, animals are moved to the Biospec 70/30 holder and MRI/ MRSI exploration is carried out as described in Sect. 3. During all experimental procedures, animals are always housed, handled and transported according to protocols previously approved by the institutional Ethics and Animal Welfare Committee and also according to regional and state legislation.

2.1.1. Monitoring Vital Signs

Minimal requirements for monitoring and physiologic support during anaesthesia depend on many factors including the health status of the animal, the anaesthetic employed and the objective of the imaging procedure. For example, more intensive monitoring and support would be indicated for an animal weakened by prior experimental manipulations, especially when undergoing lengthy functional imaging procedures (36).

Temperature Control

Differences between ambient and body temperature must be minimised, since the hypothalamic heat-regulating mechanism is depressed during anaesthesia and the animal is no longer able to shiver. As rodents have large surface area to body mass ratio, heat losses will be correspondingly greater than in bigger animals (34). Low body temperature has profound effects in modifying drug activity, although in some cases it might exert a protective effect, particularly in the CNS, that may be wise to consider (21, 34). This fact has been put to effective use in protecting organs whilst the blood supply is temporarily suspended. Body temperature can be monitored with rigid or flexible probes, which are available in multiple sizes allowing their use with small rodents including mice. Temperature probes most frequently are placed in the rectum. The usual rectal temperature in mice is 37.5°C (35.5–39°C) (34).

A self-regulating heating device generally consists of three units: a probe, a temperature controller and a warming source (such as a warm water recirculating blanket). Changes in the body temperature are then automatically adjusted for by altering the temperature of the warming source, thereby maintaining the animal within a very narrow temperature range (36). In our case, a heated water blanket incorporated into the MR system is used to avoid hypothermia and/or control the desired temperature, in case that a moderate hypothermia is desired. In studies carried out at normothermia, temperature is maintained between 36.5 and 37.5°C. On the other hand, in studies carried out with moderate hypothermia, the body temperature is adjusted to 28.5–29.5°C. In case finer local temperature monitoring for brain is required, this may be achieved by MRSI as described in Sect. 3.6.

Breathing Rhythm Control

Most agents which depress CNS activity are also respiratory depressants (36, 37). Essential organs, particularly the CNS and liver, may be severely damaged by relatively brief periods of O_2 deprivation and the respiratory depression during spontaneous breathing usually becomes irreversible when it falls to about one-third of the normal rate (34). The respiratory frequency is usually evaluated by counting breaths per minute. The physiological breath rhythm expected for an awake mouse is 160–180 breaths/min (range 80–100 to 230–250) (34, 37). Changes in breathing and, consequently, in oxygenation can affect results, especially during functional imaging, by altering drug metabolism or cerebral blood flow. Respiratory rate can be approximated through chest movement detected by a small compressible pillow integrated with a pressure transducer. The animal's respiratory movement compresses the pillow and affects the pressure transducer that is linked to a computer that provides a graphical display of movement and the calculated respiratory rate. The respiratory rhythm in our experiments with anesthesized mice is maintained at 60–80 breaths/min.

Both temperature and breathing rhythm are monitored by a control/gating module from SA Instruments Inc. (Stone Brook, NY, USA). Data gathered by the module are transferred to a personal PC (Dell Inspiron 510m) and monitored with the PC-SAM 32 software (version 6.26, Small Animal Instruments, Incorporated).

2.1.2. Post-anaesthetic Management

Too often, there is a swift decrease in the attention devoted to the animal as soon as the experimental procedure is completed, when there is still considerable risk of animal death. Careful attention must be paid until the animal is fully conscious. A warm recovery environment is essential and should be prepared before the animal is anaesthetised. This assumes that body temperature is normal at the time of anaesthetic procedures. If the animal is allowed to become hypothermic, the metabolic rate will be correspondingly depressed and recovery may be delayed by slow detoxification of the anaesthetic agent (34). Also, we should pay attention because in our case, preclinical models of brain tumours, the animals are not in their optimum health state. Biological systems which are already subject to pathological changes may be particularly affected by anaesthesia, and this is more pronounced during long anaesthesia periods as usually needed for combined MRI/MRSI experiments. In our experiments, animals are maintained in a warm environment and monitored until total recovery (usually between 5 and 7 min), and after that period they are returned to their cage.

3. Recording Strategies for MRSI Experiments

In Sects. 3.1 through 3.4, specific details are provided on how to acquire highly resolved ^{1}H-MRSI data from preclinical mouse models of human brain tumours, while Sect. 3.5 considers post-processing requirements. In Sect. 3.6, hyperpolarised ^{13}C-MRSI is briefly considered, dynamic versus basal MRSI discussed and MRSI thermometry described.

3.1. ^{1}H-MRSI: Shimming Quality, Water Suppression, Volume of Interest Selection, k-Space Sampling and Echo Time

The principles of ^{1}H-MRSI are very similar to those of MRI as far as phase encoding and basic pulse sequences. The main difference is an additional frequency axis—the *chemical shift* dispersion (Figs. 2 and 3). Hence MRSI is often called chemical shift imaging or CSI. In vivo ^{1}H-MRSI of the brain is challenging for four main reasons: large signals, like extracranial lipids, can overwhelm small metabolite signals; the water resonance is several orders of magnitude larger than signals produced by the low concentration of metabolites (often more than 10,000 times); its low sensitivity makes the detection of low concentration metabolites a compromise between time resolution and signal-to-noise ratio (SNR) and there is a large overlap between different metabolite resonances. To overcome the first two problems, at least partially, efficient spatial localisation and water suppression methods are required, respectively.

Restricting signal detection to a defined region of interest, usually named volume of interest (VOI) for ^{1}H-MRSI (voxel for ^{1}H-MRS), has several advantages. Not only it removes unwanted signals from the outside and minimises partial-volume effects (contamination of signal from one compartment by signal from another compartment) but also reduces B_0 and B_1 field variations within the region of interest, allowing better resolved spectra to be obtained. The standard technique for in vivo ^{1}H-MRSI localisation in preclinical and clinical settings is PRESS (point resolved spectroscopy)

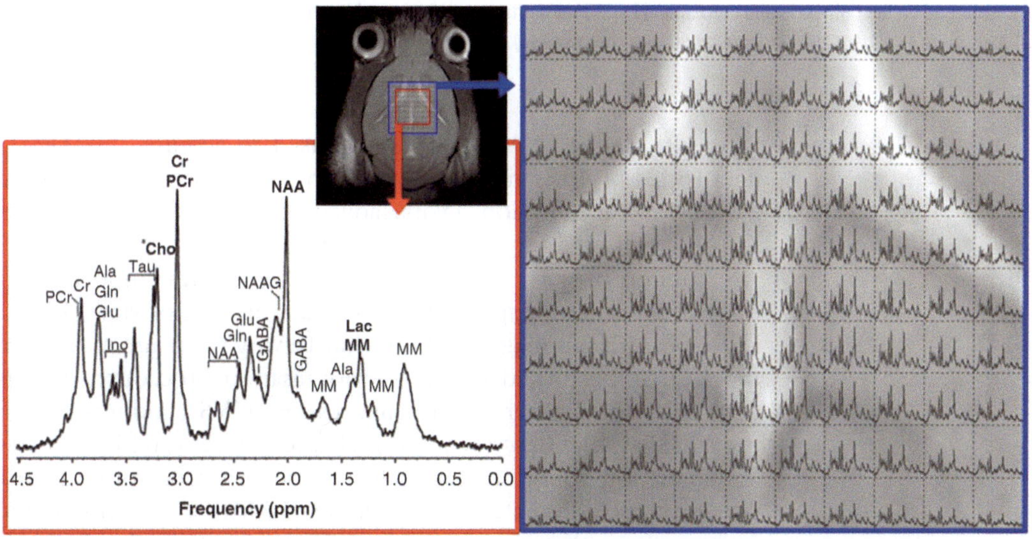

Fig. 2. Single voxel (SV) ^{1}H-MRS (on the left, *red contour line*) and ^{1}H-MRSI (on the right, *blue contour line*) obtained from a normal C57BL/6 mouse brain in vivo, with PRESS localisation, at 7.0 T. The *upper panel* shows (*small centre inset*) a reference T2-W image of the brain and the PRESS voxel positions, in the centre of the FOV: ^{1}H-MRS, 27 μL voxel volume; ^{1}H-MRSI, 4.8 μL nominal voxel size, Fourier interpolated to 0.3 μL as displayed. Data were acquired with TR/TE = 2,500/12 ms, VAPOR water suppression and outer volume suppression. Details about resonance assignment can mostly be obtained from (103). *Ala* alanine, *Cho* choline, *Cr* creatine, *GABA* γ-aminobutyric acid, *Gln* glutamine, *Glu* glutamate, *Ino* myo-inositol, *Lac* lactate, *MM* macromolecules, *NAA* *N*-acetyl aspartate, *NAAG* *N*-acetylaspartyl glutamate, *PCr* phosphocreatine, *Tau* taurine.

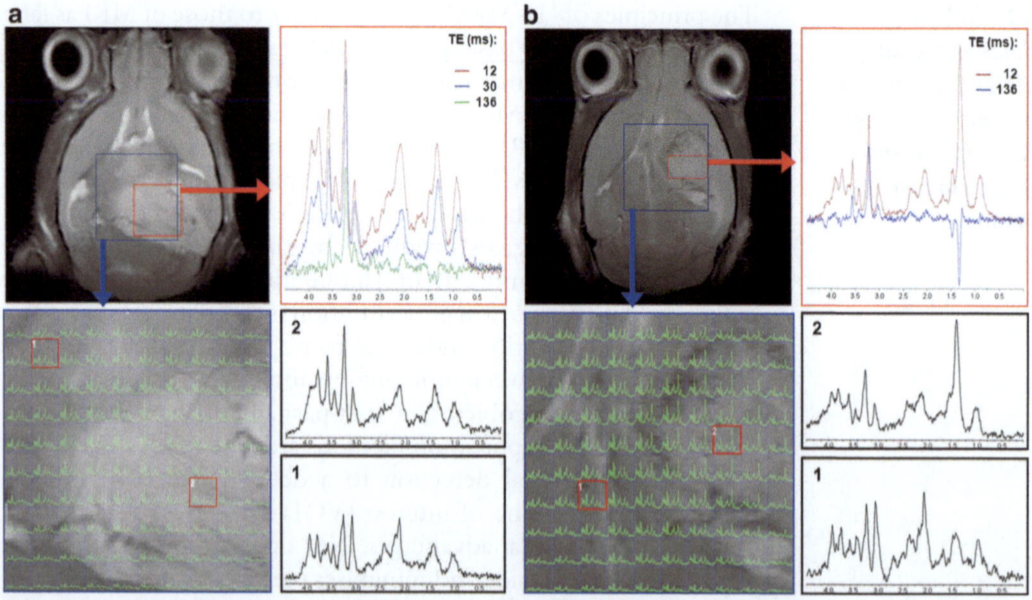

Fig. 3. ^{1}H-MRS and ^{1}H-MRSI obtained in vivo at 7.0 T from two C57BL/6 brain tumour-afflicted mice: (**a**) spontaneous grade 2 oligodendroglioma detected in a genetically engineered mouse model: s100ß-v-erbB;Ink4a/Arf(+/−) (22); (**b**) high grade IV astrocytoma detected in an allograft model, intracranial stereotactic injection of GL261 cells. For (**a**) and (**b**), SV ^{1}H-MRS at different TEs (12–136 ms) with T2-W reference image on upper-left showing the SV MRS voxel position (*red square*). ^{1}H-MRSI, at 12 ms TE, with 10 × 10 voxels within the VOI region is also highlighted on the T2-W reference image (*blue square*); spectra highlighted with *red squares* over the enlarged MRSI matrix, bottom-left in (**a**) and (**b**), are also enlarged on the right-side and correspond to normal/peritumoural brain and tumour (1 and 2, respectively). SV voxel in (**a**) and (**b**) have *square* and *rectangular* shapes, respectively.

(38), although STEAM (39) is also used to acquire ^{1}H-MRS(I) data (6, 21). Other methods have also been described to achieve MRSI voxel localisation, taking advantage of adiabatic RF pulses (40–43) and used for mouse brain MRSI (7).

To improve SV localisation, it is sometimes important to remove unwanted magnetisation outside the field of view (FOV), i.e. to perturb this magnetisation while leaving the magnetisation in the VOI unperturbed during the localisation procedure. This is normally named outer volume suppression (OVS), and it is used in the MRSI studies described in this chapter. Since water is the most abundant compound in mammalian tissue, it is no surprise that its two protons dominate the individual ^{1}H-MRSI patterns in the region where they resonate (ca. 4.75 ppm). This also leads to baseline distortions and artefacts, i.e. water sidebands due to vibration-induced signal modulation, which compromise the detection of certain metabolite resonances. Therefore, it is necessary to remove or suppress the water resonance in order to obtain reliable metabolite spectra. Although there are different techniques available to achieve this, two of the most commonly described in the literature are CHESS (chemical shift selective "water suppression"

(44, 45)) and, more recently, VAPOR (variable pulse powers and optimised relaxation delays (46)). VAPOR essentially combines T1-based water suppression, i.e. uses T1 relaxation to discriminate between water and other resonances, and optimised frequency-selective perturbations, to provide excellent water suppression with a large insensitivity towards T1 and B_1 inhomogeneity.

As far as acquisition, two basic parameters are important in any MR spectroscopy technique: the number of scans, i.e. number of times that the sample is excited and the signal is recorded (free induction decay or FID); and the repetition time (TR), i.e. the interval between consecutive scans during the experiment, when the nuclear spins generate the MR signal (FID), and are allowed to relax, which defines the total duration of each experiment. In the specific case of in vivo localised MR spectroscopy, the transversal relaxation of the nuclei due to intrinsic sample and instrumental causes is too fast to allow recording a usable FID, and another parameter comes into play—the echo time (TE). This is the same as in standard spin-echo MRI sequences, such as RARE, and basically refers to the time elapsed between excitation of the nuclei and their refocusing for recovering most of the initially lost signal. By choosing specific values for this parameter, one can select the brain molecules seen on the spectral profiles according to their intrinsic T2 values and possible J-couplings, e.g. filtering out most MR-visible lipids (frequently abundant in tumours) or allowing to observe the characteristic inversion of Lactate at 1.32 ppm due to J-coupling-induced modulation at long echo times of 135–144 ms.

All of the above applies both to MRSI and to conventional MRS. MRSI is, however, much more technically demanding than MRS, essentially due to: significant magnetic field inhomogeneities across the entire sample, particularly in the mouse brain as compared, e.g. to the rat brain; spectral degradation due to intervoxel contamination ("voxel bleeding"); long data acquisition times and processing of large, multidimensional datasets (2D, 3D or even 4D). Concerning magnetic field homogeneity adjustments (shimming, a process that consists in adjusting the current at a series of small coils placed around the sample area), fully automated procedures based on B_0 mapping have been described (47, 48) which considerably reduce the time and effort during this procedure while producing excellent results. With respect to intervoxel contamination, this is a typical problem in Fourier imaging modalities and results from the Cartesian sampling of k-space. This contamination in MRSI spectra from adjacent voxels is explained by the shape of the spatial response function (SRF, displays the spatial origin of the signal of a pixel) which is not square but, instead, a sinc-like function (49). It has been described, however, that acquisition-weighted CSI, a non-Cartesian method of sampling k-space, consisting in applying a k-space filter, e.g. Hanning window, which gives more weight to the phase-encode steps at the centre of k-space than to those in the outer regions, reduces this contamination substantially with no penalty in sensitivity or spatial resolution (2, 50).

The next sections describe how to acquire highly resolved PRESS ^{1}H-MRSI data from the mouse brain, or mouse brain tumours, using acquisition-weighted sampling of k-space. An outcome example is provided in Fig. 3.

3.2. ^{1}H-MRSI, Hardware

The following hardware and software configuration is advised in order to acquire routinely good quality ^{1}H-MRSI data from rodent brains, in particular, from the mouse brain and mouse brain tumours:

1. A high-field magnet, at least 7 T, preferably horizontal.
2. Robust gradient and shim systems, with at least 300 mT/m and nine channels, respectively, capable of handling high duty cycles (strong shim currents).
3. A proton mouse head RF coil with good SNR, e.g. receiving quadrature surface coil decoupled from a transmitting resonator.
4. An anaesthesia induction chamber, with anaesthetic gas circulation (e.g. isoflurane) and proper exhaustion system.
5. A robust animal holder, allowing efficient head restraining (stereotactic type, with fixation points at both ear cavities and a biting tooth bar), optimal circulation of anaesthetic gas (e.g. isofluorane) and body temperature control (e.g. recirculating heated water system).
6. Medical tape, eye lubricant and vaseline.
7. Real-time monitoring of basic physiologic parameters of the animal, specifically the respiratory rate with, e.g. chest/abdominal sensor, and the body temperature with, e.g. rectal probe; see Sect. 3.4. for additional comments.
8. Automated protocol(s) for localised adjustment of first- and second-order shims.
9. The MRSI protocol, including,
   - Spin-echo acquisition mode.
   - Hanning filter for phase encoding steps.
   - PRESS localisation.
   - VAPOR water suppression.

Protocols for standard anatomical MRI sequences are also required but normally available in any commercial MR spectrometer.

3.3. ^{1}H-MRSI, Protocol

The following protocol describes all the steps required to generate highly resolved ^{1}H-MRSI data from the mouse brain and mouse brain tumours, as reported (21), i.e. using a Bruker 70/30 BioSpec magnet running with Paravision 4.0 software.

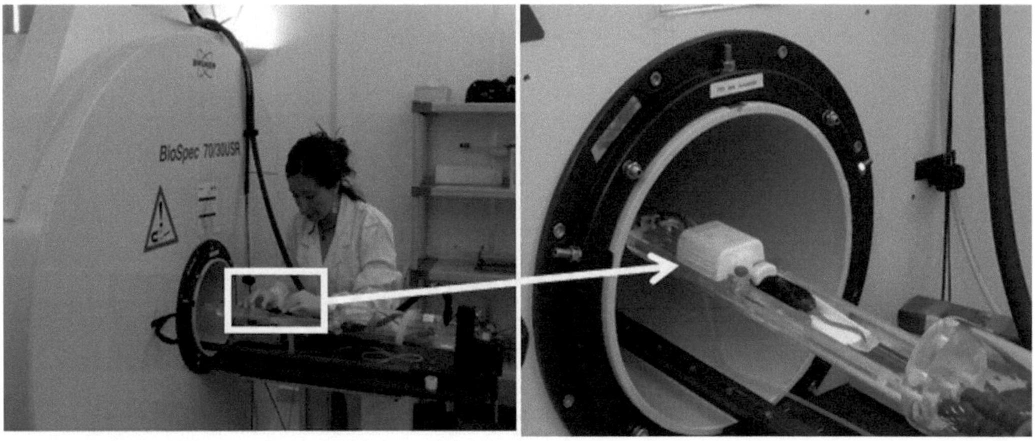

Fig. 4. (*Left*) Overview of the animal holder used for preclinical studies (mice) in the Biospec 70/30 scanner. (*Right*) Detail of the animal holder with the surface coil positioned.

1. The animal is moved from its cage to the anaesthesia induction chamber, where isoflurane gas mixture (4% in O_2, 1 L/min, for about 1 min) is used to put it to sleep.
2. The animal is transferred, while asleep, to the MR holder (should be performed fast), where isoflurane is already circulating at 1.5–2% and 0.8 L/min, and water, heated at about 50°C (depends on the specific configuration used and should be adjusted to keep the body temperature at 37°C unless otherwise indicated), is also circulating.
3. The animal is placed in the MR holder, as detailed in Fig. 4 and Sect. 3.4, and pushed inside the magnet in a way that the brain (tumour lesion) rests in the iso-centre.
4. The PRESS ^{1}H-MRSI protocol is loaded after syntonising the probe and acquiring standard multislice and multidirection localisation MRI scans.
5. The acquisition-weighted mode is selected (Hanning window) and the acquisition and reconstructed MRSI matrix sizes are defined (e.g. 8 × 8 and 32 × 32, respectively).
6. Select VOI visualisation mode and position the VOI box in transversal plane (e.g. 5.5 × 5.5 mm in-plane and 10 mm thickness) in the brain region of interest, using the localisation MRI scans.
7. Change to FOV mode and position the FOV window (e.g. 1.76 × 1.76 cm in transversal plane and 1 cm slice thickness) in a way that it covers most of the animal head and includes the VOI inside the brain region of interest and without reaching the skull.

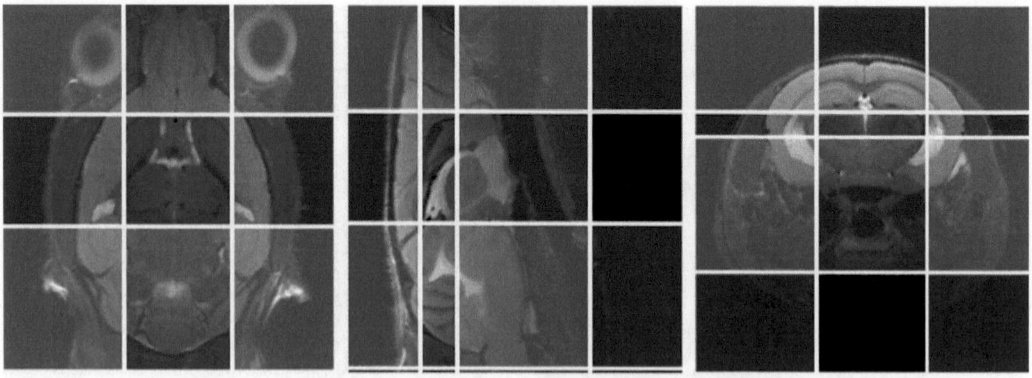

Fig. 5. The six OVS slices (two for each plane, transversal, sagital and coronal) overlaid on a scout MR image.

8. Load a RARE T2-w sequence, use the same FOV geometry and position as in the MRSI experiment, and acquire it—this will be the MRSI reference image.
9. Go back to the MRSI experiment and select six OVS slices, two for each plane with 10 mm thickness each (sech-shaped pulses: 1.0 ms/ 20,250.0 Hz), and position them all around the VOI (Fig. 5).
10. Load a FASTMAP experiment, position its voxel (e.g. 5.8 × 5.8 × 5.8 cm for mice) inside the brain in a way that includes the VOI and carry out the automatic linear and second-order shim adjustments.
11. Load a PRESS-MRS experiment, with the same geometry and position as the MRSI VOI, run an additional adjustment of first-order shims (this time specifically inside the VOI region), optimise PRESS pulse powers (hermite-shaped pulses: excitation, 0.6 ms/ 9,000 Hz; refocusing, 0.6 ms/ 5,700 Hz) previously used for MRSI, and acquire a single scan of non-suppressed water signal using the TR, TE, spectral width and number of points in the time domain chosen for MRSI (e.g. 2.5 s, 12 ms, 4,006.41 Hz (13.34 ppm) and 2 k, respectively)—expected waterline widths at half height should be as low as possible, and normally around 15–22 Hz (if higher try to reposition the VOI and re-shim).
12. Go back to the MRSI experiment, turn on water suppression and adjust VAPOR pulses (hermite-shaped pulses: excitation, 18.0 ms/300 Hz; 11.4 ms, 300 Hz)—see Sect. 3.3.
13. Define number of scans (e.g. 512) and dummy scans (e.g. 4), and acquire.
14. The reconstruction pipeline (see also Sect. 3.5) should automatically Fourier interpolate the FID signal acquired to the reconstructed matrix size, defined in step 5.

This protocol, and orientative values provided thus far, will generate ^{1}H-MRSI data with nominal spatial resolution of 2.2 mm, i.e. 4.84 μL voxel volume, and, after post-acquisition automated Fourier interpolation, digital in-plane voxel sizes of 0.55 mm, i.e. 0.30 μL—see Sect. 3.3. In case artefacts are detected in the individual spectral patterns, these may be due to the spoiler gradients, which crush the residual magnetisation at the end of each scan. If so, they should be adjusted (intensity and/or duration) to minimise those artefacts.

3.4. ^{1}H-MRSI, Useful Considerations

Other parameters, such as heart rate (with, e.g. finger paws) or expired gases (e.g. CO_2), may also be of interest to monitor, depending on the experimental procedure being carried out.

The mouse head is fixed without forcing any of the bars; a lubricant jelly is applied directly into the eyes to prevent them from drying throughout the experiment; a rectal probe is lubricated (Vaseline) and inserted into the anus for body temperature monitoring; the animal can be covered with standard laboratory bench paper to help prevent heat losses, along with the heating water system.

VAPOR pulses (gains) should be adjusted until achieving maximum suppression of the water peak; however, for post-processing generation of certain maps, e.g. temperature (detailed in Sect. 3.6), spectra should only be partially water-suppressed, so that the residual water signal is clearly visible throughout the VOI region.

Due to the non-Cartesian sampling of k-space (acquisition-weighting), the original MRSI signal is acquired beyond the FOV limits, from a grid determined by the Hanning window (13 × 13 in the case described in this protocol: 12 accumulations in the centre of k-space; total of 113 phase encoding steps). Hence, at least one interpolation step (Markus von Kienlin, personal communication) is required to properly visualise the data acquired inside the FOV.

3.5. Post-processing for MRSI

3.5.1. Producing MRSI Maps: Options Available

After successful application of the acquisition protocols described in Sect. 3.3 to living mice, similar MRSI grids to those shown in Figs. 3 and 6 can be obtained.

The interpretation of one of these grids (about 100 spectra) superimposed to a brain image is however, not straightforward, let alone the assignment of metabolites to peaks and their quantification. Therefore, performing this step by visual inspection alone on a regular basis is not feasible. It is strongly recommended that each MRSI grid is visually inspected by the operator just after having performed the experiment, in order to identify whether there is any evident artefact and to decide whether the acquisition should be repeated or not. A good guide to MRS artefacts can be found in (51). In general, the aim is to obtain an MRSI grid in which all spectra are well resolved (good shimming), have good

SNR (enough number of accumulations), water suppression is well achieved (pulses well calibrated) and no outer volume contaminations are detected in the individual MRS patterns. The latter are frequently found in PRESS-MRSI of the brain, such as "ghost" artefacts and lipid contamination from the skull (51), and can be compensated by adjusting the crusher gradients and OVS pulses, as well as by keeping the VOI away from the skull.

Once the experimenter has ensured that the MRSI is of sufficient quality, interpretation should follow. To do that, the most common methods consist in obtaining metabolite or metabolite ratio maps. Metabolite maps are images in which colour intensities encode either the spatial distribution of a certain metabolite peak (2, 3, 5) or alternatively, intra-voxel ratios of selected peak intensities. The latter option may be used within the same MRSI grid, to compare different metabolites, e.g. choline/creatine or NAA/choline (21, 52–54), or to monitor time-course changes of a specific metabolite (55).

Alternatively, maps of normalised intensities can also be generated to have a quick overview of the metabolite distributions over the MRSI grid (Fig. 6). Whatever method is chosen, the best way to interpret a metabolic map is to overlay it on a reference MR anatomical image, with the same geometry and position as the MRSI scan. Some examples are shown in Fig. 6 for brain tumour-bearing mice.

When there is an underlying pathological state, it is expected that the spatial distribution of the different metabolites will not be homogeneous (7, 56–58) and the metabolite map of specific metabolites will have different intensities. These intensities will colocalise with certain areas, such as diseased or injured ones (1, 3, 59) (Fig. 6).

Maps of dynamic metabolite changes (21) can provide additional information about the tumour progression stage (type and grade) and its heterogeneity.

As it can be seen in Fig. 6, depending on the metabolite map represented, the image changes and it is self-evident for an experienced biochemist that all these maps give complementary information to delimitate the tumoural or the unaffected brain tissue. But none of those in fact, unequivocally segment the abnormal area. Why then not try using the information provided not by one metabolite but the entire spectral pattern at the same time, in all voxels, to make a map? To meet this end, MRSI data analysis may also be performed by pattern recognition (60, 61). Pattern recognition is a statistical technique for multivariate analysis which allows analysing several peak areas or heights at the same time (i.e. several variables) and weighs them according to a statistical formula. Pattern recognition, therefore, aims to assign, in an objective way, output values (i.e. tissue type, pathologic state) to certain input features, e.g. the peaks in a 10 × 10 MRSI grid. Generally

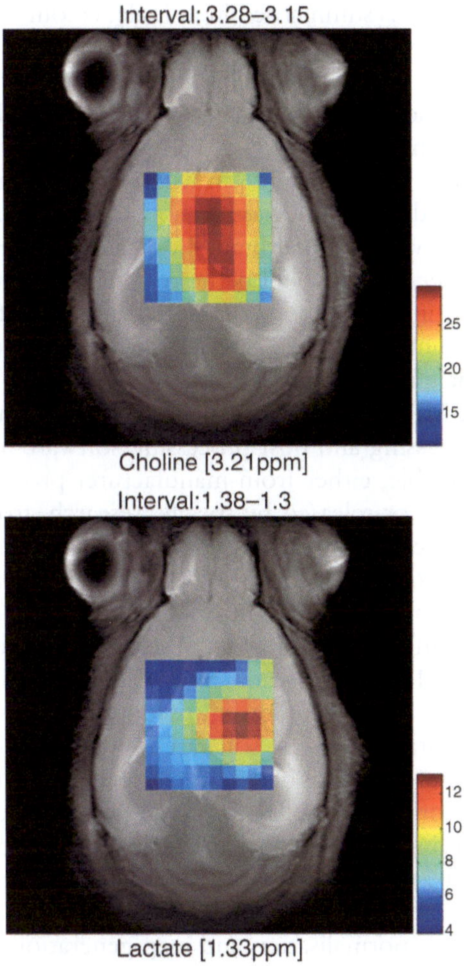

Fig. 6. Metabolite maps superimposed over a T2 reference image in a C57 mouse (code number, C69) bearing a GL261 tumour, from MRSI acquired at long TE (136 ms) obtained using the SC3.0 software and the other modules available from http:/gabrmn.uab.es/ (see text). The colour scale to the right of each plot shows the relative intensity, in absolute value, of each peak height analysed from unit length normalised spectra (UL2, (74, 105)) (choline, creatine, lactate and NAA; the interval above individual figures indicates the ppm range where the absolute peak maxima were located by the software, see Sect. 3.5.1). It can be seen that different metabolites have a distinct distribution in the grid, hence in the tumoural (high choline, high lactate), peritumoural or non-affected areas (high NAA, high creatine).

speaking, pattern recognition methods can be divided into supervised and unsupervised. Unsupervised methods, such as principal component analysis (PCA) have been, and still are widely used, but we prefer supervised approaches, such as linear discriminant analysis (LDA). Unsupervised methods are a good choice for performing an initial analysis of the data, but are not well suited for predicting these output values on a new dataset (test set) on the basis of a previous set of data of known class (training set).

Therefore, imaging tumours based on objective classification of MRSI data brings us closer to the long-pursued concept of nosologic imaging (62). A nosologic image is one in which the presence of different tissues and lesions is summarised in a single, colour-coded image, where each pixel or voxel is coded according to its histopathological class (the output value) (63). This methodology is well reported for human tumour MRSI data (64, 65) and has been shown to be feasible in animal models (12, 13, 66). In this way, a classification image would be a general term with which we simply designate the image produced after classification, i.e. two-colour images of "normal brain vs. abnormal brain," or "brain vs. ventricles," whereas a nosologic image would be more appropriate when talking about tumours, in which the pathological diversity of the tissue is taken into account, i.e. a three-class classification image for "necrosis vs. high-grade glioma vs. low-grade glioma."

But generating the MRSI maps described in Sect. 3.5.1 requires access to advanced processing and post-processing software tools. Several options are available, either from manufacturer providers, e.g. *ParaVision* (Bruker), or released by specific research groups. Some examples of advanced multipurpose NMR processing software include: *LC Model* (67), *jMRUI*(68), *CSIAPO* (63), *3DiCSI* (http://mrs.cpmc.columbia.edu/3dicsi.html) that allow reading data from different manufactures, and allow overlying MRSI data, either 2D or 3D, on MRI reference scans. *jMRUI* (AMARES and QUEST) and *LC Model* are two of the most popular tools for quantification of MRS data (68). These software tools are based on time- and frequency-domain analysis of the data, respectively, allowing line-fitting and deconvolution of the spectral peaks detected. Another option for carrying out individual line-fitting integration of MRSI data is *XSOS* (69). Other post-processing tools, such as *dynamic MRSI processing module (DMPM)* (http://gabrmn.uab.es/dmpm) for alignment, normalisation and map generation, and *SpectraClassifier* (70) (http://gabrmn.uab.es/sc) for performing pattern recognition analysis were developed by our research group and are free. Pattern recognition can also be performed with standard programs such as *R* (http://www.r-project.org/) or *Matlab* (http://www.mathworks.com/products/matlab/) and its toolboxes, but using these requires of a previous learning curve of a set of basic commands and is not suitable for those not familiar with these types of interfaces or with scarce time. The *SpectraClassifier* interface was designed to meet the needs of a typical biochemist with no special expertise in these programs.

3.5.2. MRSI Post-processing Protocol

The steps we follow for producing maps like those shown in Fig. 6 are detailed below.

1. Fourier interpolate the original MRSI matrix to 32 × 32 voxels with *ParaVision 4.0* or *5.0* (see Fig. 7) or directly with *CSIAPO* by voxel shifting, both enabling line broadening adjustments, with a 4 Hz Lorentzian filter, and zero-order phase correction. The result will be an ASCII file containing the processed MRSI.

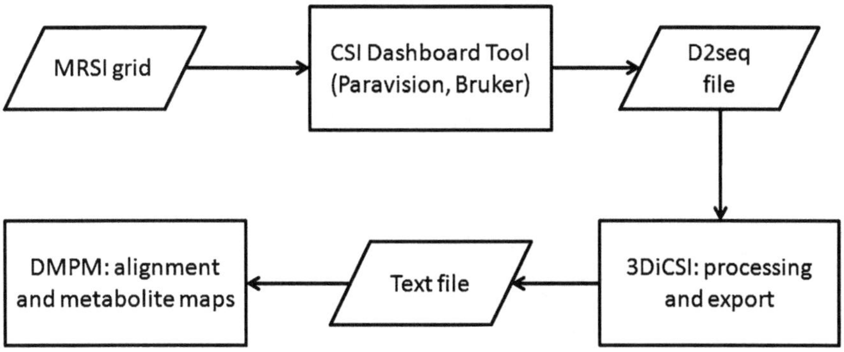

Fig. 7. Diagram of the post-processing strategy used to generate colour-coded maps from MRSI scans.

2. Feed the ASCII file into an additional post-processing module, *DMPM* (http://gabrmn.uab.es/dmpm), to ensure proper alignment and to produce metabolite and metabolite ratio maps.
3. It is very important to normalise the spectra prior to classification. We take the 4.5–0 ppm region of each spectrum and normalise it individually to unit length (UL2), as previously described (55), with *DMPM*, prior to exporting for classification.
4. Load the aligned, normalised spectra into the *SpectraClassifier* software, to classify the individual voxels in one or several MRSI grids.
   (a) Build the training set, i.e. the dataset that will be used to train the classifier. For this, it is necessary to tag each spectrum in each MRSI grid that is to be loaded (Fig. 8).
   (b) Build the test set, i.e. the dataset that the user wants to predict, e.g. the MRSI of one or more different new mice. It is very important that both training and test sets have been obtained under the same experimental conditions and that post-processing has been done in exactly the same way. Otherwise, different number of points or different point/ppm ratios, or different normalisations may produce unreliable results.
   (c) Perform feature selection. Do not overtrain the classifier. Overtraining is easily recognised when a near-perfect classification performance in the training set drops to less than chance in the test set. This is normally caused by using too many features for a small sample, which is called "the curse of dimensionality" (71). A rule of thumb for an ideal feature number is to use no more than one-third of the number of cases available for training (72).
   - Perform classification. Evaluate results both in the training and in the test set. Receiver-operating curves (ROC)(73) and bootstrapping, as well as confusion

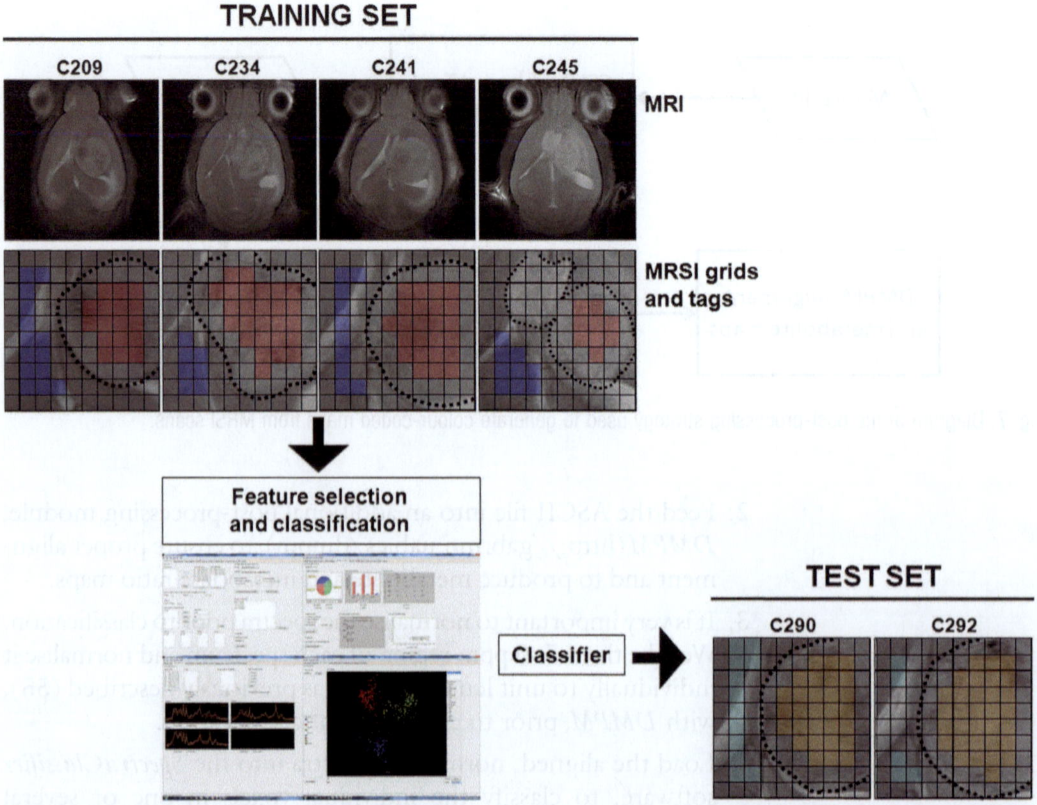

Fig. 8. An easy example in which classification between normal brain tissue and tumour is performed. The training set is composed of four mice (C209, C234, C241 and C245), injected with GL261 glioblastoma cells, and the test set of two additional mice (C290 and C292). The four MRSI grids are tagged according to the reference MRI: *blue*, normal/peritumoural brain parenchyma; *red*, tumour. A classifier is obtained based on these data and then tested with the two additional mice: *light blue*, normal/peritumoural brain; *yellow*, tumour.

matrices, are good tools for this. Repeat the process as many times as needed, changing the number of features in order to obtain the best classification results both in the training and in the test, with the minimum number of features. If the aim is to obtain the best classifier and the test results are not needed for any immediate reason, the following trick can be used (74): Divide all MRSI in two sets of approximately equal size.

- Use set 1 for training and set 2 for testing. Evaluate results.
- Use set 2 for training and set 1 for testing. Evaluate results.

(d) If results are comparable, both with respect to the classification performance and the features chosen, the classifier is representative of the whole population.

3.6. MRSI of Other Nuclei (Hyperpolarised ^{13}C), Basal Pattern Versus Dynamic MRSI (PE-MRSI) and Thermometry by MRSI

Besides proton, other nuclei have been used for MRSI-monitoring of tumours in mice, but studies have been scarce and mostly in subcutaneous or mammary tumours (75, 76) One of the major problems for this has been the low sensitivity of the heteronuclei, compared to proton spectroscopy—natural abundance of, e.g. ^{13}C is only 1.1%. This is changing for ^{13}C due to the large increases in sensitivity produced by the hyperpolarisation methodology (dynamic nuclear polarisation, DNP) (77). For this, in vivo studies in mouse tumour models are carried out by administering hyperpolarised ^{13}C-labelled substrates to the animals of interest, increasing their sensitivity for detection by up to 10,000-fold. DNP-^{13}C- MRSI can therefore be acquired in a single breath, and this is possible by greatly reducing the repetition time (TR) and using pulses with very low flip angles. The main limitation of this technique is the T1 of the hyperpolarised substrate—the shorter the T1 the faster the polarisation is lost. Because the hyperpolarised-enhanced sensitivity is normally lost in just a few seconds, this is the actual time-window for the dynamic MRSI studies to be carried out, thus enabling to snapshot only fast metabolic process, mostly by simple exchange labelling. A full description of this molecular imaging strategy is beyond the purpose of this chapter, although essential details can be found in (14).

The tumour microenvironment can also be dynamically monitored by ^{1}H-MRSI, as reported with the methodology described in Sect. 3. In this case, the main limitation with respect to the hyperpolarisation-based images is the longer scan time: due to the low sensitivity of the technique to the regional detection of ^{1}H visible-metabolites in tissues, either endogenous or exogenous. Thus, several scans need to be acquired in order to generate MRSI data with sufficient SNR, and this will normally take between 20 min and 1 h, mostly depending on the TR, number of scans and matrix size used (48, 78). The advantage of the ^{1}H-MRSI approach over hyperpolarised ^{13}C-MRSI is the possibility to monitor slower metabolome changes, such as those induced by continuous infusion of substrates, e.g. glucose (21, 79). In additional, the possibility of monitoring regional time-course changes in tumour metabolism, as detected by MRSI, induced by externally administered substances (PE-MRSI) opens a new window for investigating inter- and intra-tumour heterogeneity, with promising results as compared to basal MRSI patterns in preclinical brain tumour mouse models(22).

Finally, another useful application of ^{1}H-MRSI is thermometry, i.e. generation of regional temperature maps, which can be employed for, e.g. monitoring tumour thermal therapies (80). Although other MR approaches are available to study local temperatures in tissues (81), the linear dependence of the proton (water) resonance frequency shift with temperature is conceptually simple (18) and highly accurate at high fields: below 0.1°C error at 12 T in the rat brain (19). Early reports were based on ^{1}H-MRS and used the NAA peak as internal reference (16, 82, 83) but for brain

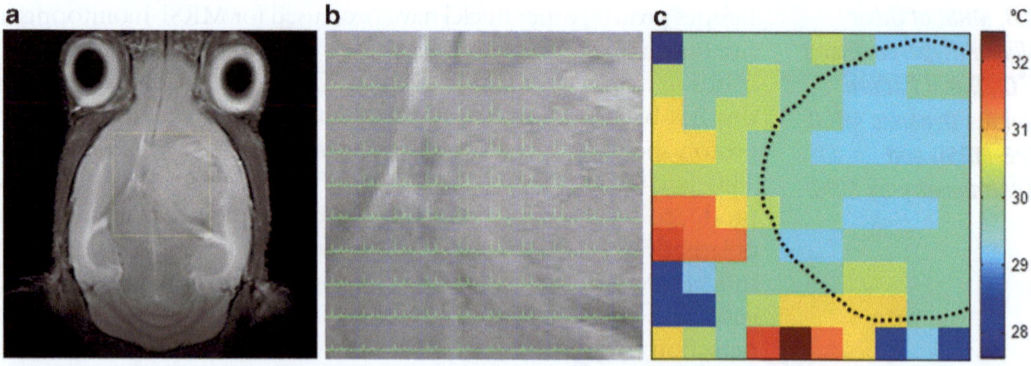

Fig. 9. Local temperature map in a C57BL6 mouse brain, bearing a GL261 tumour. The animal was kept at moderate hypothermia (~30°C rectal temperature) and studied by ^{1}H-MRSI, with PRESS localisation and 136 ms echo time. (**a**) T2 reference image of the mouse brain displaying the VOI position (*yellow square*). (**b**) Enlarged view of the VOI region with MRSI data overlaid in the T2 reference image—three major peaks are visible in all voxels, from left to right: partially suppressed water (~4.75 ppm), choline (3.21 ppm) and lactate (1.32 ppm, inverted due to J-coupling modulation). (**c**) Temperature colour-coded map prepared using SC 2.0 (http:/gabrmn.uab.es/), see Sect. 3.5, that corresponds to the VOI region, with tumour boundaries manually drawn (*dashed line*). The temperature range inside the tumour is 29–30.5°C, slightly below the ipsilateral brain regions at the tumour periphery, range 28–32.5°C.

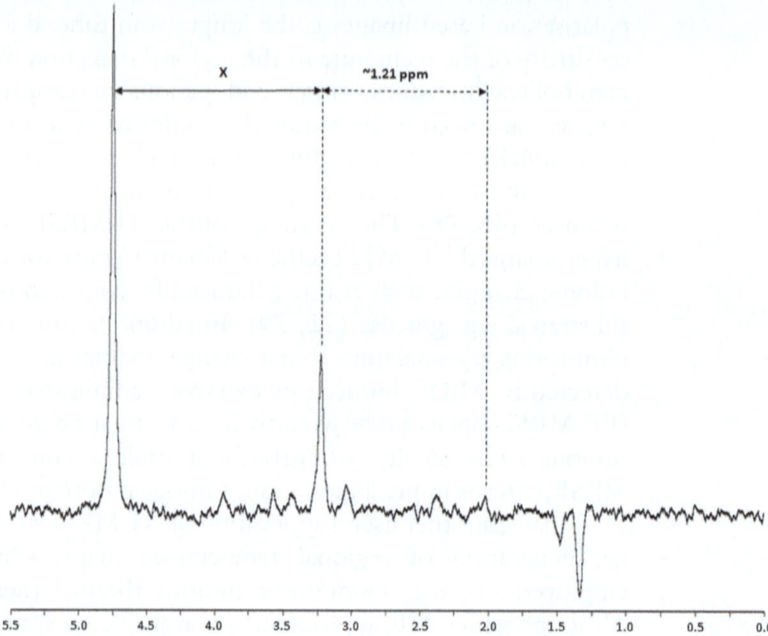

Fig. 10. Enlarged view of one of the MRSI voxels shown in Fig. 9. The calibration curve used was described by others for another animal model (dog) (17). Due to the absence of the NAA peak in tumours, total choline was used here as an intermediate reference, as explained in (21): T_{brain} (°C) $= -82.33$ (X + 1.21) + 255.94, being X and 1.21 the ppm offsets from residual water peak to choline peak and from choline peak to NAA peak (measured in the healthy mouse brain), respectively.

tumours, where NAA may be undetectable, other brain metabolites, e.g. total choline, should be used instead(21). The method has been reproduced by ^{1}H-MRSI in the human brain and in animals (20, 82, 84–87), and can also be used to monitor regional temperatures in brain tumour-bearing mice (Fig. 9). These maps are calculated by measuring, in each voxel, both the frequency of the partially suppressed water resonance and the frequency of a reference peak, e.g. choline (3.21 ppm), Fig. 10. Since the proton frequency shifts are linearly dependent on temperature, the temperature in each MRSI voxel can be calculated from a calibration curve.

4. Technical Requirements for Successful HRMAS Acquisition from Tumour Biopsies. Sample Obtention, Storage and Preparation

4.1. Methods for Animal Sacrifice

The election of the sacrifice method and sample preservation may have a non-negligible influence in the recorded HRMAS pattern. For example, it is well known that even short delay times could lead to significant changes in the spectral pattern due to post-mortem changes (31). Then, if the sample cannot be immediately analysed after sacrifice, tissue is usually snap-frozen and kept at sub-zero storage temperatures prior to further NMR analysis. This may increase the speed of observed biochemical changes after freezing and thawing cycles (88–90). In addition, long HRMAS experimental times may be needed if two dimensional (2D) NMR techniques are used. In (30, 91), e.g. total acquisition time range for 2D experiments was 16–21 h. Those changes are usually minimised by recording HRMAS spectra at temperatures between 0 and 4°C. Nevertheless, recent results indicate that the use of physiological temperatures could be relevant in the analysis of some metabolites, such as mobile lipids and choline-containing compounds (92), but these higher temperatures will increase the speed of changes in the biopsy HRMAS pattern.

In order to take into account the need to minimise post-mortem changes due to ischaemia while allowing for long acquisition periods at physiological temperatures, and additionally having a stable spectral HRMAS pattern, we profit from the FMW irradiation sacrifice method. This method has been previously described by others (55, 93, 94). The FMW rapidly heats the mouse brain to 82–85°C in milliseconds, causing enzymatic inactivation and preventing further metabolism.

A high power microwave fixation system is required (i.e. Muromachi, 5 kW) with a power setting of 5 kW applied for 1.0–1.5 s, although the power settings and irradiation duration could slightly change in different studies. Animal death is achieved in less than 1 s. Use of FMW as euthanising method allows the detection of metabolites such as phosphocreatine that would be undetectable with the anaesthesia overdose sacrifice protocol (Fig. 1).

The steps for euthanizing an animal with FMW are:

- Set up the recirculating bath for refrigeration.
- Prepare the mouse accessory (Fig. 11) filling the "heat sink" with water, and immobilise the anesthetised animal inside the accessory.
- Optimise the irradiation time (for a 20–30 g weight mouse, the adequate irradiation time would be about 1.10 s) and start the irradiation.
- Remove the animal from the accessory and dissect the brain/tumour onto a cold surface (body would be still hot due to the FMW irradiation protocol).
- Change all the "heat sink" water inside the accessory, and wait at least 3 min before starting a new irradiation.

4.2. Frozen Samples. Use of FMW in Frozen Samples

As stated previously, the FMW sacrifice method is effective in order to avoid post-mortem changes that could alter the spectral pattern during the HRMAS recording period (see Sect. 4.4) due to tissue degradation. Nevertheless, although it is possible to use this method for euthanasia in animal models, in humans this is impossible for obvious reasons. Nevertheless, a compromise solution is to apply the FMW irradiation to a previously frozen biopsy sample. This does not eliminate ischaemia effects prior to freezing times, but avoids further changes during the HRMAS recording time (95), its consequence being the stabilisation of the HRMAS NMR pattern.

The experimental procedure is as follows:

- Biopsy tissue obtained is quickly frozen in liquid N_2 and stored until further work at liquid N_2 temperature.
- When preparing the sample (range 11.3 ± 5.3 mg) for HRMAS experiments, it is allowed to thaw in a Petri dish until it reaches 0°C. The biopsy temperature is monitored by a digital probe such as the one used in Sect. 3.4 for mouse rectal temperature monitoring.
- After that, the sample is FMW irradiated (5 kW during 2.6 s, due to the low temperature of the sample). The cryogenic tube should not be fully capped to allow for possible water vapour to escape without building up pressure inside the tube. The power and time settings may need adjustment for every combination sample weight/equipment.
- It is very important to ensure that the biopsy sample is in the optimal position, because the FMWs are directed and optimised for the animal (mouse or rat) accessory in order to concentrate themselves in the animal brain. In our case, the biopsy is placed in a cryogenic tube at the farther end of the mouse accessory.

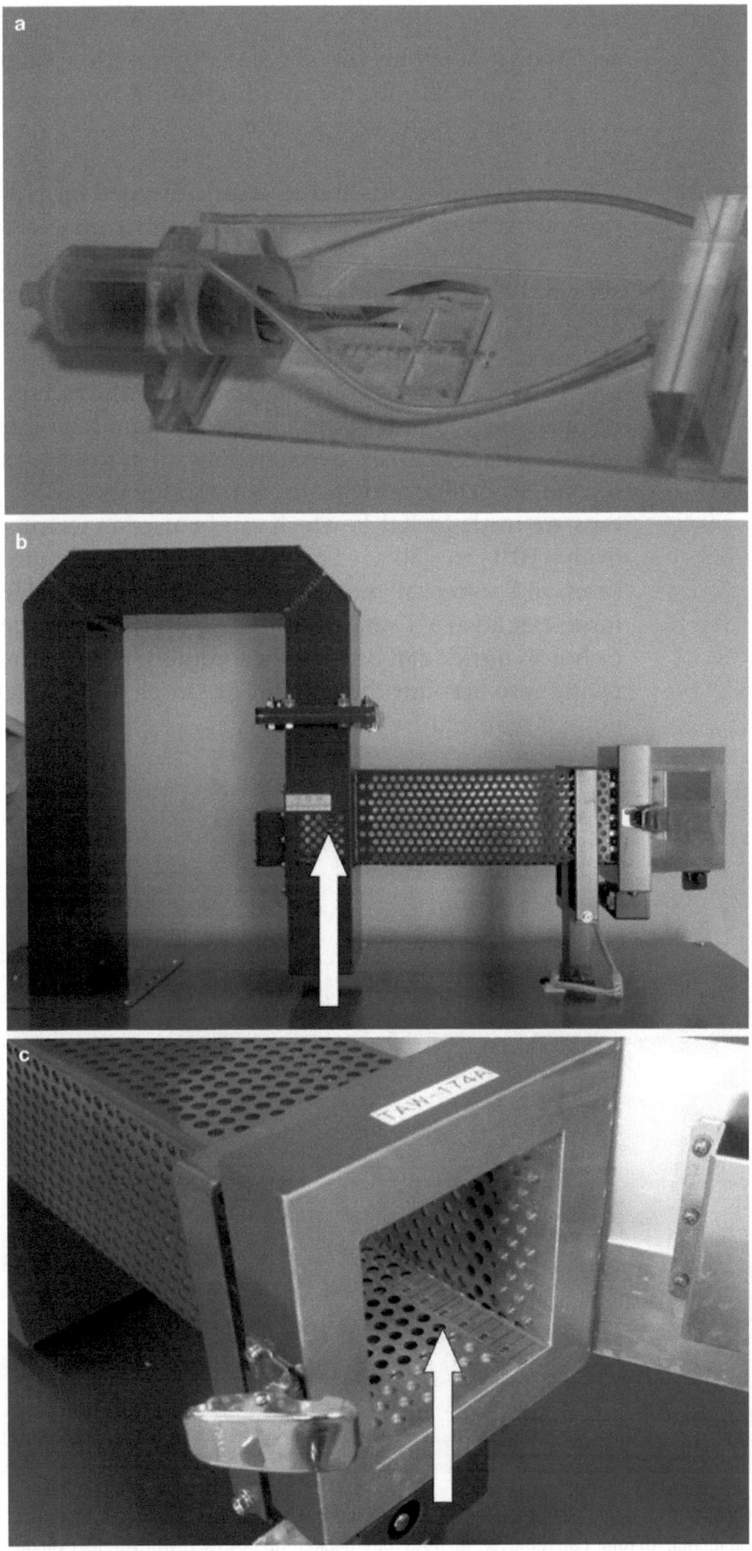

Fig. 11. (**a**) FMW mouse container accessory for animal restrain and its "heat sink" water filling system, (**b**) irradiation chamber of the FMW system (the *white arrow* points the exact position of the irradiation point), and (**c**) scale (*white arrow*) to help positioning the mouse container shown in (**a**).

- After irradiation, the sample is split with a scalpel into small pieces that fit into the 50 μL HRMAS rotor and assembled as described in the next section.

4.3. Rotor Preparation

The rotor components and accessories needed for HRMAS acquisitions are very small and require specific instrumental to manipulate it. In Fig. 12, a typical rotor and associated tools required are shown. Rotors come in different sizes/sample capacities, between 12 and 92 μL. Furthermore, different manufacturing materials are also used. Then, rotors are usually made of zirconium oxide (zirconium). The use of rotor internal spacers produces a spherical sample compartment in order to facilitate the shimming of the probe. The rotor cap has essentially two functions: first, to close the rotors, and second, to facilitate driving the rotor inside the probe. The standard caps are made of Kel-F, which can be used in a temperature range from +10°C to +50°C. This material will shrink at lower temperatures and soften at more elevated temperatures. However, for a more extended VT range (−30 to +70°C), caps made from macor or boron nitride can be used. Accordingly, rotors should be chosen taking into account the amount of sample to be analysed and the desired temperature of analysis.

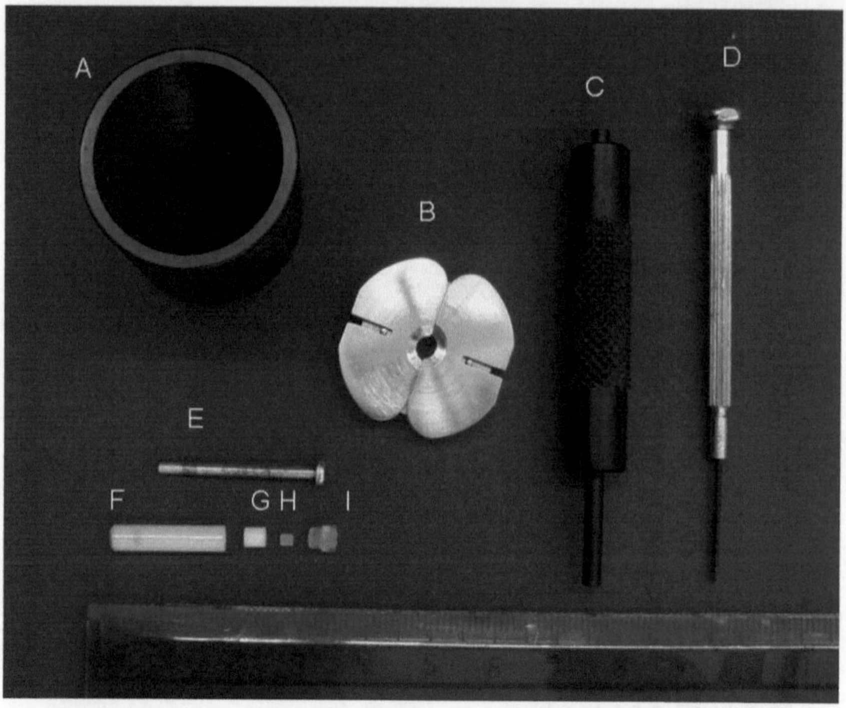

Fig. 12. Rotor and associated components required for HRMAS work. The ruler gives a good indication of the size of these parts. (**a**) MAS filling funnel; (**b**) rotor cap remover; (**c**) MAS rotor packer; (**d**) MAS screwdriver; (**e**) cylinder head screw; (**f–I**) Zirconium rotor, upper spacer, sealing grub screw and Kel-F cap.

The protocol used for sample preparation is detailed next:

- Frozen samples should be split into small pieces inside liquid N_2 pre-cooled porcelain. The size of the pieces has to be suitable for their placement into HRMAS zirconium rotors with a 50 μL spacer.
- Weigh the empty rotor.
- Fill the rotor with sample.
- Weight the rotor again with the sample inside and subtract the initial weight of the empty rotor, in order to have an accurate estimation of the real weight of the sample.
- Pre-cooled D_2O-saline (0.15 M NaCl) is added (about 15 μL) into the rotor to allow for lock signal detection.
- In case of FMW processing (Sect. 4.2), the sample is also split into small pieces to fit in the HRMAS rotor and D_2O-saline is added.

4.4. Standard HRMAS Acquisition Conditions

This section briefly details some of the requirements for good HRMAS spectra acquisition from tissue biopsies, but further details can also be obtained from (25). The basic principles of HRMAS are essentially the same as for high-resolution spectra from liquids, being the main difference the *need of a special probe which allows positioning the sample at the so-called magic angle.* One of the great advantages of performing HRMAS experiments is the elimination of dipolar broadening interactions from the spectrum, leaving mostly narrow lines of the type found in high-resolution spectroscopy. These interactions in solids are time-dependent and can be averaged by spinning the sample (usually at a frequency between 1 and 15 kHz—in case of HRMAS) at the magic angle θ (ca. 54.74°, where $(3\cos^2\theta - 1)/2 = 0$) with respect to the direction of the magnetic field. This spinning replaces molecular tumbling motion as the source for the line-narrowing effect in liquids, and this is the reason for the need of an optimal adjustment of the magic angle at the HRMAS probe.

4.4.1. HRMAS, Equipment

The following equipment configuration is advised in order to acquire routinely good quality HRMAS data from solid biopsy samples:

- High-field magnet, 9.4 T or higher. Care should be taken when choosing the magnetic field, because higher magnetic fields require higher spinning rates to avoid residual water spinning side bands getting into the spectral range of interest.
- HRMAS probe that should ideally allow combination of ^1H NMR experiments with ^{13}C decoupling or ^{13}C NMR experiments with ^1H decoupling with a separate ^2H lock channel.

- A variable temperature unit (VT) for temperature control, operating in a suitable temperature range (e.g. −20°C to +70°C).
- Dry nitrogen gas for low temperature operation, preferably from a pressurised liquid nitrogen tank.
- Air supply with a flow of 4 m^3/h at more than 6 bar for stable fast spinning.
- MAS pneumatic control unit for sample spinning, display of the spinning rate, as well as sample insert and eject.
- Rotors used for HRMAS analysis which are described in Sect. 4.3.
- HRMAS pulse sequences, including at least:
  - Pulse-acquisition sequence
  - Hahn spin-echo sequence

4.4.2. ^{1}H-HRMAS, Protocols

The following protocol summarises the different steps required to generate highly resolved ^{1}H-HRMAS data from solid biopsies:

1. Calibration of the magic angle (for Bruker probes).

   This needs to be performed only when the HRMAS probe is used for the first time, whenever the probe is changed in a multiuser/multiprobe NMR facility environment or when the spectra quality degrades unexpectedly. In order to adjust the magic angle, a sample is needed with a single NMR line which is very sensitive to angle misadjustment. A good example of a suitable sample is powdered KBr.

   To adjust the angle the sample has to be spun at about 5–6 kHz. The following procedures are suitable for a Bruker spectrometer:

   - With a rotor completely filled with KBr, the angle setting procedure can be performed in "gs" mode on the FID.
   - In "gs" mode, go to the "acqu" window, separate the real and imaginary part of the FID and adjust for an on resonance decay. This can be done by changing the offset or adjusting the field.
   - Make sure that the field sweep is off and remains off.
   - Adjust the angle with the probe micrometre screw. The deviation of the angle from the magic angle leads to a splitting of the centre band as well as to a splitting of each sideband. The angle adjusted with the micrometre screw at the probe bottom is correctly set when the splitting disappears and the centre band, as well as at the sidebands, and both have minimum width (Fig. 13).

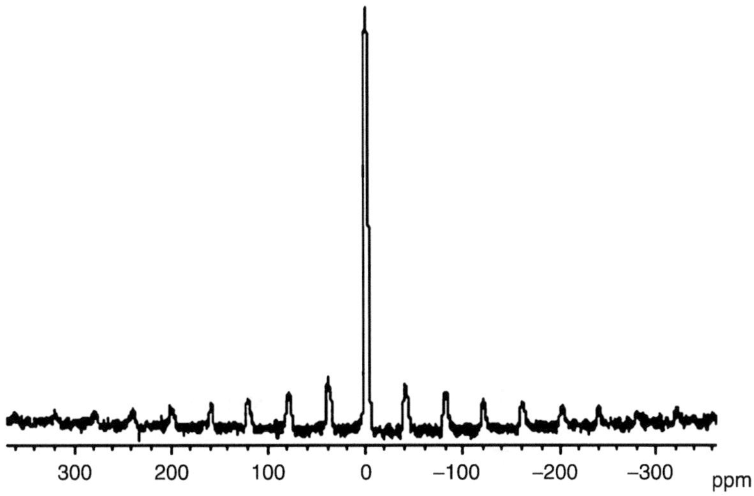

Fig. 13. Example of KBr spectra in the adjustment of the magic angle in a Bruker spectrometer. Experimental parameters for acquisition were sweep width: 75 kHz (748 ppm), spinning rate: 4,000 Hz, number of transients: 1. The nucleus observed in these experiments was ^{79}Br at 100.2 MHz frequency (9.4T).

2. Calibration of the biopsy temperature.

   This is the second step to perform when starting an HRMAS experiment series, although it is also important to perform it if the work temperature has to be changed (e.g. physiological temperature spectra instead of low temperature spectra). This is relevant not only to ensure the correct temperature setting in the spectrometer, but also to correct for unwanted sample heating due to high spinning rates.

   - The substances currently used are: methanol for the low temperature range (180–300 K) or ethylene glycol for the higher temperature range (300–420 K). Accordingly, prepare the HRMAS rotor with methanol or ethylene glycol depending on the temperature range to be controlled.
   - For a given spinning rate, acquire several spectra changing temperature values each time (e.g. steps of 2°C). The values set and controlled by the VT unit will be compared with the methanol/ethylene glycol signal distance between OH signals and methylene signals and, in consequence, the actual temperature registered inside the sample (96). This should be repeated for each desired sample temperature for a predefined spinning rate.
   - Plot the values obtained (sample real temperature vs. probe set temperature), as in Fig. 14, and use always this correction when working with this spinning rate and temperature range. In our hands, the calculation formula to

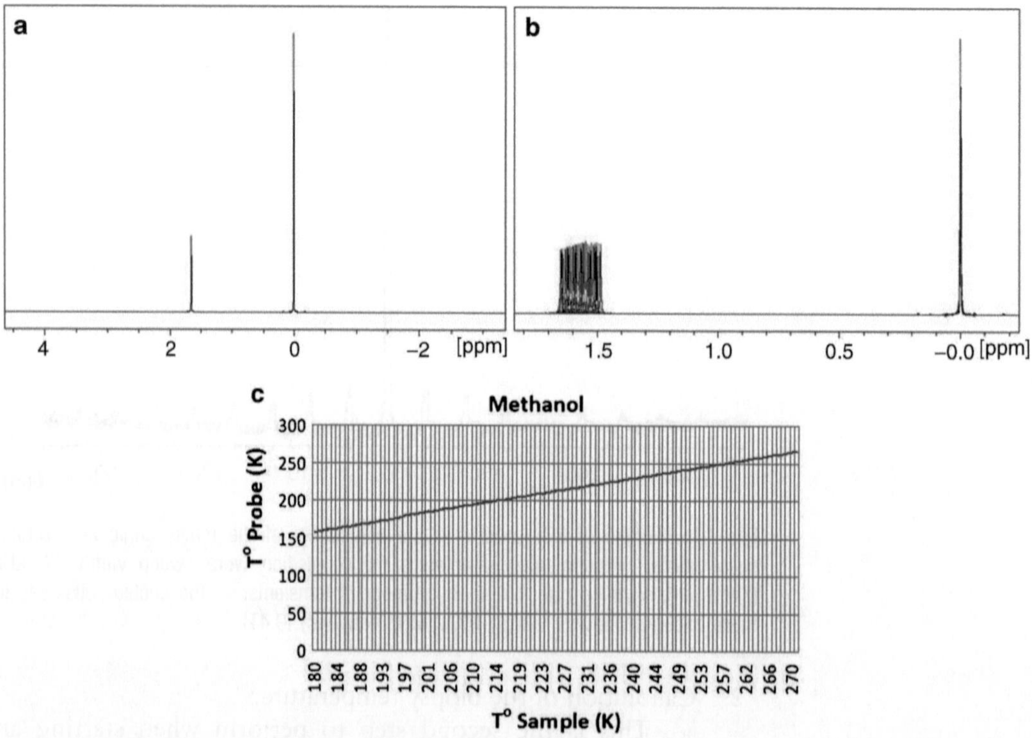

Fig. 14. (**a**) Methanol spectrum acquired with the pulse-and-acquire sequence at −13°C and (**b**) the shift in the ppm difference between CH_3 and OH signals observed with increasing temperature from −13°C to 27°C (260–300 K). (**c**) Example of plot (sample temperature versus probe temperature) obtained for a given spinning rate (3,000 Hz).

be used for acquiring spectra of tumour biopsies at low range temperature is the following:

$$T(°C) = (4{,}637 - \Delta\delta)/0.009967,$$

where $\Delta\delta$ = ppm distance between CH_3 singlet and OH in methanol.

3. Sample analysis.
   - The sample should be carefully prepared and assembled into the rotor as described in Sects. 4.1 and 4.2.
   - For sample preparation, frozen tumour tissue samples should be weighted (in our experiments with preclinical models, the mean weight has been 11.3 ± 5.3 mg) and either FMW irradiated or non-irradiated (please refer to Sect. 4.1 for differences) and packed into the rotor as described in Sect. 4.3.
   - Insert sample:
     - In order to make sure that the rotor will properly fit into the stator, press the EJECT button on the pneumatic unit first. This will set the stator into a vertical position and ensure that a possible rotor which might be inside

the probe is ejected. Then, the packed rotor can be lowered into the probe via the transfer tube with the cap up. Close the insert tube and press the button INSERT on the pneumatic unit. This will automatically set the stator into the magic angle position after 10 s.

- Set temperature, observing any corrections required as described in step "2".
- Start spinning the sample at the chosen rate (3,000 Hz is the standard for 9.4 T and 4,000 Hz for 11.7 T).
- Shimming the sample:
- A one-scan spectrum should be acquired with the pulse-and-acquire sequence, then transformed and phase corrected. The lineshape of the chosen metabolite signal (use of the residual water signal should be avoided because of its large intensity and possible distortion) will be observed and shim coils used for correction in consequence. The expansion in the screen should be adjusted such that the lineshape of the chosen signal can be judged correctly. Measuring the data heights of the lactate doublet peaks and at their point of overlap allows a ratio to be calculated of peak height to mid-point data height. This ratio is a simple quantitation of the spectrual resolution that complements the line width measurements. The 7 Hz separation of the doublet should be clear and the line widths not so broad that they cause too much overlap. This measurement is dependent on the tuning, shimming (field homogeneity), RF pulses, receiver coils and software of the spectrometer and it could provide a basic assessment of spectral quality control.
- Load the desired pulse programs in new experiments.
- Adjust parameters, e.g. 90° pulse and water presaturation, for each sample. Refer to the user's manual in each case. In our experiments, the parameters set were:
  - Water presaturation: 2 s at 0.042 mW (55 dB attenuation).
  - Sweep width: 10 ppm (4,000 Hz).
  - Time domain: 16 k points (8 k real).
  - Number of transients: 512 (although this can be modulated related to the sample amount).
  - Acquisition time: 2.04 s. An additional delay of 1 s is added resulting in total recycling time of 3.04 s.
  - Spinning rate: 3,000 Hz.
  - Echo time for Hahn spin-echo sequence: 136 ms.
- Acquire spectra. We show an example of a preclinical brain tumour sample spectrum (GL261 cells stereotaxically

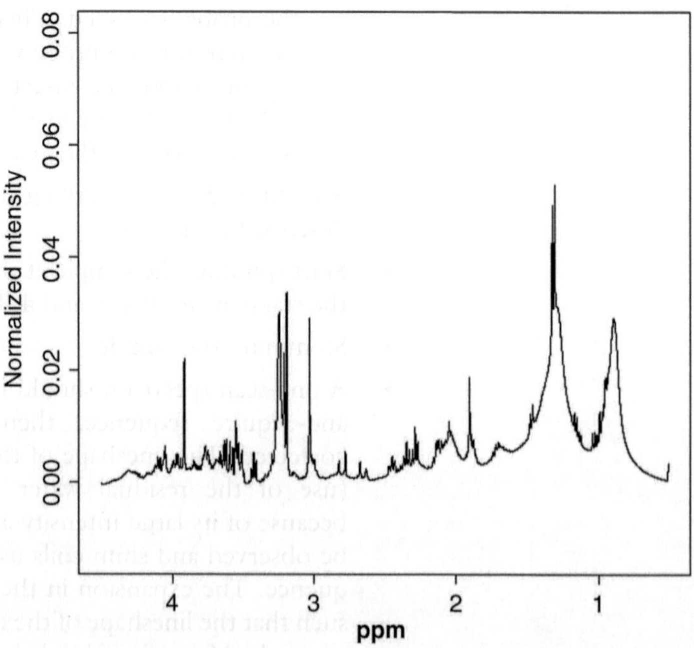

Fig. 15. HRMAS spectrum recorded with a pulse-and-acquire sequence, 9.4 T and 37°C from a GL261 grade IV glioma tumour model biopsy, processed as described in Sect. 4.2.

injected into C57 mice brain and producing a grade IV glioma tumour excised at day 12 post-injection) in Fig. 15.

4.5. Sample Conservation for Post-HRMAS Analysis

The tissue analysed by histopathology to characterise a brain tumour cannot be used for HRMAS analysis. For homogeneous tumours, this should not be a problem, but in case of heterogeneous tumours it could lead to discrepancies between the histopathology derived information and the metabolomic pattern derived from HRMAS, even when the sample used for both methodologies is from adjacent positions. On the other hand, taking into account that the HRMAS analysis does not essentially modify or contaminate the sample studied with chemicals, it can be frozen again and stored until further analysis with histopathology. Early work with different sample types (97, 98) demonstrated that it is possible to perform a post-HRMAS histopathology analysis of the studied sample. We have extended this to investigate possible effects after short-term 37°C HRMAS spectra recording. Results obtained show that tissue architecture is preserved well enough for tumour biopsy typing and grading (92).

The steps for sample conservation for post-HRMAS histopathological analysis are:

- Recovery of the sample from the rotor. A freezing spray can be used to facilitate this recovery. Sample is kept in liquid N_2 until further processing.

- Fixation is carried out using 20 volumes of 4% buffered formaldehyde during 6–24 h.
- Embedding in paraffin and cutting. About 5–6 slices can be obtained from each sample.
- Staining of tissue slices can then be performed with classical Hematoxilin–Eosin protocols or other immunohistochemistry for further histopathological analysis.

4.6. Correlating HRMAS Pattern with Histopathology and Molecular Sub-types, and Evaluating the Robustness of the Results Obtained

4.6.1. Advantages of Pattern Analysis of Spectral Vectors over Conventional Quantification for Tissue Typing

The objective of quantitative histopathology by acquiring HRMAS spectra from preclinical brain tumour biopsies is to recognise tumour type and grade and even molecular subtype from objective analysis of the pattern of the metabolites detectable in the recorded spectral pattern. For this, visual inspection alone is not usually acceptable and some type of multiparametric or pattern recognition analysis is required. Work in this respect is still scarce in preclinical brain tumour models, while much more is available from human brain tumour biopsies. In this respect, it has been shown that robust classifiers can be made to semiautomate the recognition of extreme tumour types (benign, meningothelial meningioma and malignant, glioblastoma multiforme) (92) or major childhood brain tumours (99) using spectral features extracted from HRMAS spectra of those biopsies. Other approaches have demonstrated statistically significant differences in the concentration of several metabolites (then, potential tumour type biomarkers) among astrocytomas (grades II–IV), metastases, meningiomas and lymphomas (28), ability to discriminate low- and high-grade astrocytomas (100) or to correlate survival and HRMAS detected pattern in metastases (30). Besides, recognition of molecular biopsy sub-types has been hinted for gliomas grades II–IV (27). Nevertheless, it should be taken into account that demonstrating significant quantitative differences for certain metabolites among tumour types is not equivalent to being able to successfully predict new cases. For this, a "classifier" based in those quantitative values or in more easily accessible spectral features, such as peak heights or peak ratios must be trained and its performance evaluated, ideally with new samples (independent test set) (101).

4.6.2. Steps Used for Classifier Development with SC 2.0

Prepare the sample as described in Sect. 4.1.

- Assemble the sample into the rotor as described in Sect. 4.2.
- Acquire spectra, taking into account equipment and parameters described in Sect. 4.3.
- Spectra acquired should be processed with software that could produce a compatible output (e.g. Topspin from Bruker), Fourier transformed, phase corrected and calibrated).
- Select the spectral zone of interest (e.g. from 4.5 to 0.5 ppm) and save the desired zone in an ASCII format. Be sure that all spectra have the same number of points.

- Normalise each spectrum (e.g. to unit length). There are several ways to do it, in our case, we use a home-made script for R-program.
- Assign each case to a group in order to organise your training set.
- Enter cases in the SC2.0 assigning the corresponding tag and choose the desired classification system (PCA, sequential forward, etc.).
- Run classifier, optimise the number of variables chosen and be sure that you are working in a PC with enough RAM memory to perform the classification with files with a large number of points.

4.6.3. Software Available

The software actually available to process and quantify 1D HRMAS files, could come either from manufacturer providers, e.g. TOPSPIN (Bruker), or be released by specific research groups. Some examples of advanced processing software include: LC Model (67), jMRUI (68), Mestrenova (Mestrelab Research, http://mestrelab.com/), Tarquin (102), AQSES (103), 3DiCSI (http://mrs.cpmc.columbia.edu/3dicsi.html), or some extensions and complements to known existing programs as "R" (http://rnmr.nmrfam.wisc.edu/).

Post-processing tools, such as SpectraClassifier for performing pattern recognition analysis, were developed by our research group and are also available, but in order to use HRMAS as input files, we should convert first our raw data into ASCII data (presently, only the ASCII files generated by TOPSPIN program are supported). The high number of spectral vectors, if compared with MRSI in vivo spectra, requires powerful processors with large RAM capacity.

Acknowledgements

Authors are supported in their work by: Ministerio de Ciencia e Innovación, MICINN (Spain), SAF 2008-03323; Centro de Investigación Biomédica en Red—Bioingeniería Biomateriales y Nanomedicina (CIBER-BBN) and the intramural project PROGLIO, which are an initiative of the *Instituto de Salud Carlos III* (Spain), co-funded with EU FEDER funds. Authors thank Juana Martín-Sitjar for acquiring data used in Fig. 1, Sandra Ortega-Martorell for obtaining maps in Fig. 6 and Myriam Dávila for acquiring data used in Sect. 4.4.2, as well as in Figs. 13–15.

References

1. Ziegler A, von Kienlin M, Decorps M, Remy C (2001) High glycolytic activity in rat glioma demonstrated in vivo by correlation peak 1 H magnetic resonance imaging. Cancer Res 61:5595–5600
2. von Kienlin M, Ziegler A, Le Fur Y, Rubin C, Décorps M, Rémy C (2000) 2D-spatial/2D-spectral spectroscopic imaging of intracerebral gliomas in rat brain. Magn Reson Med 43:211–219
3. García-Martín ML, Herigault G, Rémy C, Farion R, Ballesteros P, Coles JA, Cerdán S, Ziegler A (2001) Mapping extracellular pH in rat brain gliomas in vivo by 1 H magnetic resonance spectroscopic imaging: comparison with maps of metabolites. Cancer Res 61:6524–6531
4. Provent P, Benito M, Hiba B, Farion R, Lopez-Larrubia P, Ballesteros P, Remy C, Segebarth C, Cerdan S, Coles JA, Garcia-Martin ML (2007) Serial in vivo spectroscopic nuclear magnetic resonance imaging of lactate and extracellular pH in rat gliomas shows redistribution of protons away from sites of glycolysis. Cancer Res 67:7638–7645
5. Liimatainen TJ, Erkkila AT, Valonen P, Vidgren H, Lakso M, Wong G, Grohn OH, Yla-Herttuala S, Hakumaki JM (2008) 1 H MR spectroscopic imaging of phospholipase-mediated membrane lipid release in apoptotic rat glioma in vivo. Magn Reson Med 59:1232–1238
6. Heerschap A, Sommers MG, In 't Zandt HJ, Renema WK, Veltien AA, Klomp DW (2004) Nuclear magnetic resonance in laboratory animals. Methods Enzymol 385:41–63
7. Miyasaka N, Takahashi K, Hetherington HP (2006) 1 H NMR spectroscopic imaging of the mouse brain at 9.4 T. J Magn Reson Imaging 24:908–913
8. Boska MD, Lewis TB, Destache CJ, Benner EJ, Nelson JA, Uberti M, Mosley RL, Gendelman HE (2005) Quantitative 1 H magnetic resonance spectroscopic imaging determines therapeutic immunization efficacy in an animal model of Parkinson's disease. J Neurosci 25:1691–1700
9. Nelson JA, Dou H, Ellison B, Uberti M, Xiong H, Anderson E, Mellon M, Gelbard HA, Boska M, Gendelman HE (2005) Coregistration of quantitative proton magnetic resonance spectroscopic imaging with neuropathological and neurophysiological analyses defines the extent of neuronal impairments in murine human immunodeficiency virus type-1 encephalitis. J Neurosci Res 80:562–575
10. Weiss K, Melkus G, Jakob PM, Faber C (2009) Quantitative in vivo 1 H spectroscopic imaging of metabolites in the early postnatal mouse brain at 17.6 T. Magn Reson Mater Phy 22:53–62
11. Gao H-X, Campbell SR, Cui M-H, Zong P, Hwang J-H, Gulinello M, Putterman C (2009) Depression is an early disease manifestation in lupus-prone MRL/lpr mice. J Neuroimmunol 207:45–56
12. Diekmann C, Simões RV, Pohman R, Cerdán S, Arús C (2006) Proton chemical shift imaging of mouse brain tumors at 7 T. In: Hull WE (ed.) Bruker spin report, Gerd Wolff, Bruker BioSpin GMbH, Germany, pp 18–21
13. Simões RV, Delgado-Goñi T, Lope-Piedrafita S, Arús C (2010) 1 H-MRSI pattern perturbation in a mouse glioma: the effects of acute hyperglycemia and moderate hypothermia. NMR Biomed 23:23–33
14. Day SE, Kettunen MI, Gallagher FA, Hu DE, Lerche M, Wolber J, Golman K, Ardenkjaer-Larsen JH, Brindle KM (2007) Detecting tumor response to treatment using hyperpolarized 13 C magnetic resonance imaging and spectroscopy. Nat Med 13:1382–1387
15. Arús C, Chang Y, Barany M (1985) N-acetylaspartate as an intrinsic thermometer for H-1-NMR of brain slices. J Magn Reson 63:376–379
16. Cady EB, D'Souza PC, Penrice J, Lorek A (1995) The estimation of local brain temperature by in vivo 1 H magnetic resonance spectroscopy. Magn Reson Med 33:862–867
17. Corbett RJ, Purdy PD, Laptook AR, Chaney C, Garcia D (1999) Noninvasive measurement of brain temperature after stroke. AJNR Am J Neuroradiol 20:1851–1857
18. Hindman JC (1966) Proton resonance shift of water in the gas and liquid states. J Chem Phys 44:4582–4592
19. Zhu M, Bashir A, Ackerman JJ, Yablonskiy DA (2008) Improved calibration technique for in vivo proton MRS thermometry for brain temperature measurement. Magn Reson Med 60:536–541
20. Parry-Jones AR, Liimatainen T, Kauppinen RA, Grohn OH, Rothwell NJ (2008) Interleukin-1 exacerbates focal cerebral ischemia and reduces ischemic brain temperature in the rat. Magn Reson Med 59:1239–1249

21. Simoes RV, Delgado-Goni T, Lope-Piedrafita S, Arus C (2010) 1 H-MRSI pattern perturbation in a mouse glioma: the effects of acute hyperglycemia and moderate hypothermia. NMR Biomed 23:23–33
22. Simões R, Ortega-Martorell S, Delgado-Goñi T, Fur YL, Pumarola M, Candiota AP, Martín J, Stoyanova R, Cozzone P, Julià-Sapé M, Arús C (2012) Improving the classification of brain tumors in mice with perturbation enhanced (PE)-MRSI. Integr Biol (Camb). 4 (2):183–191
23. Delgado-Goñi T, Martin-Sitjar J, Simões R, Acosta M, Lope-Piedrafita S, Arús C (2011) DMSO-based contrast as a potential intermediate endpoint biomarker of GBM response to therapy. In: Proceedings of 19th scientific meeting, International society for magnetic resonance in medicine, Montréal, p 19
24. Delgado-Goñi T, Simões R, Acosta M, Martín-Sitjar J, Lope-Piedrafita S, Arús C (2010) DMSO as a potential contrast agent for brain tumours. In: Proceedings of 18th scientific meeting, International society for magnetic resonance in medicine, Stockholm, p 3495
25. Beckonert O, Coen M, Keun H, Wang Y, Ebbels T, Holmes E, Lindon J, Nicholson J (2010) High-resolution magic-angle-spinning NMR spectroscopy for metabolic profiling of intact tissues. Nat Protoc 5:1019–1032
26. Hekmatyar SK, Wilson M, Jerome N, Salek RM, Griffin JL, Peet A, Kauppinen RA (2010) (1)H nuclear magnetic resonance spectroscopy characterisation of metabolic phenotypes in the medulloblastoma of the SMO transgenic mice. Br J Cancer 103:1297–1304
27. Croitor Sava A, Martinez-Bisbal MC, Van Huffel S, Cerda JM, Sima DM, Celda B (2011) Ex vivo high resolution magic angle spinning metabolic profiles describe intratumoral histopathological tissue properties in adult human gliomas. Magn Reson Med 65:320–328
28. Wright AJ, Fellows GA, Griffiths JR, Wilson M, Bell BA, Howe FA (2010) Ex-vivo HRMAS of adult brain tumours: metabolite quantification and assignment of tumour biomarkers. Mol Cancer 9:66
29. Righi V, Lopez-Larrubia P, Schenetti L, Tugnoli V, Garcia-Martin M, Cerdan S (2009) High resolution 13 C HR-MAS spectroscopy analysis of different brain regions from rats bearing C6 implanted gliomas. In: Proceedings of 17th scientific meeting, International society for magnetic resonance in medicine, Honolulu, p 1018
30. Sjobakk TE, Johansen R, Bathen TF, Sonnewald U, Juul R, Torp SH, Lundgren S, Gribbestad IS (2008) Characterization of brain metastases using high-resolution magic angle spinning MRS. NMR Biomed 21:175–185
31. Opstad KS, Bell BA, Griffiths JR, Howe FA (2008) An assessment of the effects of sample ischaemia and spinning time on the metabolic profile of brain tumour biopsy specimens as determined by high-resolution magic angle spinning (1)H NMR. NMR Biomed 21:1138–1147
32. Martinez-Bisbal MC, Esteve V, Martinez-Granados B, Celda B. (2011) Magnetic resonance microscopy contribution to interpret high-resolution magic angle spinning metabolomic data of human tumor tissue. J Biomed Biotechnol, Article ID 763684, p 8, doi:10.1155/2011/763684
33. Valverde-Saubi D, Candiota AP, Molins MA, Feliz M, Godino O, Davila M, Acebes JJ, Arus C (2010) Short-term temperature effect on the HRMAS spectra of human brain tumor biopsies and their pattern recognition analysis. Magn Reson Mater Phy 23:203–215
34. Green CJ (1982) Animal anaesthesia, vol 8. Laboratory Animals LTD, London
35. Sonner JM, Gong D, Li J, Eger EI 2nd, Laster MJ (1999) Mouse strain modestly influences minimum alveolar anesthetic concentration and convulsivity of inhaled compounds. Anesth Analg 89:1030–1034
36. Colby LA, Morenko BJ (2004) Clinical considerations in rodent bioimaging. Comp Med 54:623–630
37. McIntyre JWR (1971) An introduction to general anaesthesia of experimental animals. Lab Anim 5:99–114
38. Bottomley PA, Edelstein WA, Hart HR, Schenck JF, Smith LS (1984) Spatial localization in 31P and 13 C NMR spectroscopy in vivo using surface coils. Magn Reson Med 1:410–413
39. Frahm J, Merboldt K, Hanicke W (1987) Localized proton spectroscoply using stimulated echos. J Magn Reson 72:502–508
40. De Graaf RA, Nicolay K (1997) Adiabatic rf pulses: applications to in vivo NMR. Concepts Magn Reson 9:247–268
41. Garwood M, DelaBarre L (2001) The return of the frequency sweep: designing adiabatic pulses for contemporary NMR. J Magn Reson 153:155–177
42. Mlynarik V, Gambarota G, Frenkel H, Gruetter R (2006) Localized short-echo-time

proton MR spectroscopy with full signal-intensity acquisition. Magn Reson Med 56:965–970
43. Scheenen TW, Klomp DW, Wijnen JP, Heerschap A (2008) Short echo time 1 H-MRSI of the human brain at 3 T with minimal chemical shift displacement errors using adiabatic refocusing pulses. Magn Reson Med 59:1–6
44. Haase A et al (1985) 1 H NMR chemical shift selective (CHESS) imaging. Phys Med Biol 30:341
45. Haase A, Frahm J, Hanicke W, Matthaei D (1985) 1 H NMR chemical shift selective (CHESS) imaging. Phys Med Biol 30:341–344
46. Tkac I, Starcuk Z, Choi IY, Gruetter R (1999) In vivo 1 H NMR spectroscopy of rat brain at 1 ms echo time. Magn Reson Med 41:649–656
47. Gruetter R (1993) Automatic, localized in vivo adjustment of all first- and second-order shim coils. Magn Reson Med 29:804–811
48. Miyasaka N, Takahashi K, Hetherington HP (2006) Fully automated shim mapping method for spectroscopic imaging of the mouse brain at 9.4 T. Magn Reson Med 55:198–202
49. Pohmann R, Rommel E, von Kienlin M (1999) Beyond k-space: spectral localization using higher order gradients. J Magn Reson 141:197–206
50. Pohmann R, von Kienlin M (2001) Accurate phosphorus metabolite images of the human heart by 3D acquisition-weighted CSI. Magn Reson Med 45:817–826
51. Kreis R (2004) Issues of spectral quality in clinical 1 H-magnetic resonance spectroscopy and a gallery of artifacts. NMR Biomed 17:361–381
52. Vigneron D, Bollen A, McDermott M, Wald L, Day M, Moyher-Noworolski S, Henry R, Chang S, Berger M, Dillon W, Nelson S (2001) Three-dimensional magnetic resonance spectroscopic imaging of histologically confirmed brain tumors. Magn Reson Imaging 19:89–101
53. McKnight TR, Noworolski SM, Vigneron DB, Nelson SJ (2001) An automated technique for the quantitative assessment of 3D-MRSI data from patients with glioma. J Magn Reson Imaging 13:167–177
54. Rodrigues TB, Lopez-Larrubia P, Cerdan S (2009) Redox dependence and compartmentation of [13 C]pyruvate in the brain of deuterated rats bearing implanted C6 gliomas. J Neurochem 109(Suppl 1):237–245
55. Simões RV, García-Martín ML, Cerdán S, Arús C (2008) Perturbation of mouse glioma MRS pattern by induced acute hyperglycemia. NMR Biomed 21:251–264
56. Macrì MA, D'Alessandro N, Di Giulio C, Di Iorio P, Di Luzio S, Giuliani P, Bianchi G, Esposito E (2006) Regional changes in the metabolite profile after long-term hypoxia-ischemia in brains of young and aged rats: a quantitative proton MRS study. Neurobiol Aging 27:98–104
57. Provent P, Farion R, Benito M, Hiba B, Ballesteros P, Rémy C, Segebarth C, Coles J, Cerdán S, García-Martín M, López Larrubia P (2006) Improved mapping of extracellular pH in C6 gliomas by 1 H MRSI shows low correlation between Lactate concentration and pHe changes induced by infusion of glucose. In: Proceedings 14th scientific meeting, International society for magnetic resonance in medicine, Seattle, p 1264
58. Zoula S, Hérigault G, Ziegler A, Farion R, Décorps M, Rémy C (2003) Correlation between the occurrence of 1 H-MRS lipid signal, necrosis and lipid droplets during C6 rat glioma development. NMR Biomed 16:199–212
59. Weidensteiner C, Lanz T, Horn M, Neubauer S, Haase A, von Kienlin M (2000) Three-dimensional 13 C-spectroscopic imaging in the isolated infarcted rat heart. J Magn Reson 143:17–23
60. Duda RO, Hart PE, Stork DG, Duda ROPC, Scene A (2001) Pattern classification, 2nd edn. In: Duda RO, Hart PE, Stork DG (eds) Wiley, New York
61. Fukunaga K (1990) Introduction to statistical pattern recognition, 2nd edn. Academic, San Diego
62. De Edelenyi FS, Rubin C, Esteve F, Grand S, Decorps M, Lefournier V, Le Bas JF, Remy C (2000) A new approach for analyzing proton magnetic resonance spectroscopic images of brain tumors: nosologic images. Nat Med 6:1287–1289
63. Le Fur Y, Nicoli F, Guye M, Confort-Gouny S, Cozzone PJ, Kober F (2010) Grid-free interactive and automated data processing for MR chemical shift imaging data. Magn Reson Mater Phy 23:23–30
64. Simonetti A, Van Huffel S, Laudadio T, Heerschap A, De Vos M (2006) Fast nosologic imaging of the brain. In: Proceedings of 14th scientific meeting, International society

for magnetic resonance in medicine, Seattle, p 1780

65. Laudadio T, Pels P, De Lathauwer L, Van Hecke P, Van Huffel S (2005) Tissue segmentation and classification of MRSI data using canonical correlation analysis. Magn Reson Med 54:1519–1529
66. Simoes RV, Martinez-Aranda A, Martin B, Cerdan S, Sierra A, Arus C (2008) Preliminary characterization of an experimental breast cancer cells brain metastasis mouse model by MRI/MRS. Magn Reson Mater Phy 21:237–249
67. Provencher SW (1993) Estimation of metabolite concentrations from localized in vivo proton NMR spectra. Magn Reson Med 30:672–679
68. Stefan D, Cesare FD, Andrasescu A, Popa E, Lazariev A, Vescovo E, Strbak O, Williams S, Starcuk Z, Cabanas M, Ormondt DV, Graveron-Demilly D (2009) Quantitation of magnetic resonance spectroscopy signals: the jMRUI software package. Meas Sci Technol 20:104035
69. Shen W, Mao X, Wang Z, Punyanitya M, Heymsfield SB, Shungu DC (2003) Measurement of intramyocellular lipid levels with 2-D magnetic resonance spectroscopic imaging at 1.5 T. Acta Diabetol 40:s51–s54
70. Ortega-Martorell S, Olier I, Julia-Sape M, Arus C (2010) SpectraClassifier 1.0: a user friendly, automated MRS-based classifier-development system. BMC Bioinformatics 11:106
71. Somorjai RL, Dolenko B, Baumgartner R (2003) Class prediction and discovery using gene microarray and proteomics mass spectroscopy data: curses, caveats, cautions. Bioinformatics 19:1484–1491
72. Tate AR, Underwood J, Acosta DM, Julia-Sape M, Majos C, Moreno-Torres A, Howe FA, van der Graaf M, Lefournier V, Murphy MM, Loosemore A, Ladroue C, Wesseling P, Luc Bosson J, Cabanas ME, Simonetti AW, Gajewicz W, Calvar J, Capdevila A, Wilkins PR, Bell BA, Remy C, Heerschap A, Watson D, Griffiths JR, Arus C (2006) Development of a decision support system for diagnosis and grading of brain tumours using in vivo magnetic resonance single voxel spectra. NMR Biomed 19:411–434
73. Zhou X-H, Obuchowski NA, McClish DK (2002) Statistical methods in diagnostic medicine. Wiley, New York
74. Tate AR, Majos C, Moreno A, Howe FA, Griffiths JR, Arus C (2003) Automated classification of short echo time in in vivo 1 H brain tumor spectra: a multicenter study. Magn Reson Med 49:29–36
75. Artemov D (2008) Novel imaging strategies preclinical: molecular readouts. In: Proceedings of 16th scientific meeting, International society for magnetic resonance in medicine, Toronto, p 413
76. Koutcher J, Rosen N, Lupu M, Matei C, Solit D, Le H (2006) Metabolic response of human prostate cancer post 17 AAG treatment in a mouse model. In: Proceedings of 14th scientific meeting, International society for magnetic resonance in medicine, Seattle, p 1268
77. Ardenkjaer-Larsen JH, Fridlund B, Gram A, Hansson G, Hansson L, Lerche MH, Servin R, Thaning M, Golman K (2003) Increase in signal-to-noise ratio of > 10,000 times in liquid-state NMR. Proc Natl Acad Sci U S A 100:10158–10163
78. Hetherington H, Kuznetsov A, Avdievich N, Pan J (2009) Short TE (15 ms) spectroscopic imaging of the human brain at 7 T using transceiver arrays and B1 shimming based localization. In: Proceedings of 17th scientific meeting, International society for magnetic resonance in medicine, Honolulu, p 327
79. Simoes RV, Garcia-Martin ML, Cerdan S, Arus C (2008) Perturbation of mouse glioma MRS pattern by induced acute hyperglycemia. NMR Biomed 21:251–264
80. Germain D, Chevallier P, Laurent A, Saint-Jalmes H (2001) MR monitoring of tumour thermal therapy. Magn Reson Mater Phy 13:47–59
81. Rieke V, Butts Pauly K (2008) MR thermometry. J Magn Reson Imaging 27:376–390
82. Ishihara Y, Calderon A, Watanabe H, Okamoto K, Suzuki Y, Kuroda K (1995) A precise and fast temperature mapping using water proton chemical shift. Magn Reson Med 34:814–823
83. McCoy CL, Parkins CS, Chaplin DJ, Griffiths JR, Rodrigues LM, Stubbs M (1995) The effect of blood flow modification on intra- and extracellular pH measured by 31P magnetic resonance spectroscopy in murine tumours. Br J Cancer 72:905–911
84. Parry-Jones A, Liimatainen T, Grohn O, Kauppinen R, Rothwell N (2006) The effect of interleukin-1 on local brain temperature during focal cerebral ischemia in the rat: a 1 H magnetic resonance spectroscopic imaging study. In: Proceedings of 14th scientific meeting, International society for magnetic resonance in medicine, Seattle, p 1452

85. Kuroda K, Suzuki Y, Ishihara Y, Okamoto K (1996) Temperature mapping using water proton chemical shift obtained with 3D-MRSI: feasibility in vivo. Magn Reson Med 35:20–29
86. Weis J, Covaciu L, Rubertsson S, Allers M, Lunderquist A, Ortiz-Nieto F, Ahlström H (2009) MRS thermometry of the brain using calibration results of aqueous metabolite solutions. In: Proceedings of 17th scientific meeting, International society for magnetic resonance in medicine, Honolulu, p 2525
87. Kuroda K (2005) Non-invasive MR thermography using the water proton chemical shift. Int J Hyperthermia 21:547–560
88. Wu C, Taylor J, He W, Zepeda A, Halpern E, Bielecki A, Rg G, LL Cheng (2003) Proton high-resolution magic angle spinning NMR analysis of fresh and previously frozen tissue of human prostate. Magn Reson Med 50:1307–1311
89. Bourne R, Dzendrowskyj T, Mountford C (2003) Leakage of metabolites from tissue biopsies can result in large errors in quantitation by MRS. NMR Biomed 16:96–101
90. Waters N, Garrod S, Farrant R, Haselden J, Connor S, Connelly J, Lindon J, Holmes E, Nicholson J (2000) High-resolution magic angle spinning (1)H NMR spectroscopy of intact liver and kidney: optimization of sample preparation procedures and biochemical stability of tissue during spectral acquisition. Anal Biochem 282:16–23
91. Erb G, Elbayed K, Piotto M, Raya J, Neuville A, Mohr M, Maitrot D, Kehrli P, Namer I (2008) Toward improved grading of malignancy in oligodendrogliomas using metabolomics. Magn Reson Med 59:959–965
92. Valverde-Saubí D, Candiota AP, Molins MA, Feliz M, Godino O, Davila M, Acebes JJ, Arus C (2010) Short-term temperature effect on the HRMAS spectra of human brain tumor biopsies and their pattern recognition analysis. Magn Reson Mater Phy 23:203–215
93. O'Callaghan JP, Sriram K (2004) Focused microwave irradiation of the brain preserves in vivo protein phosphorylation: comparison with other methods of sacrifice and analysis of multiple phosphoproteins. J Neurosci Methods 135:159–168
94. Risa O, Melø T, Sonnewald U (2009) Quantification of amounts and (13)C content of metabolites in brain tissue using high-resolution magic angle spinning (13)C NMR spectroscopy. NMR Biomed 22:266–271
95. Davila M, Candiota AP, Pumarola M, Arus C (2012) Minimization of spectral pattern changes during HRMAS experiments at 37 degrees celsius by prior focused microwave irradiation. Magn Reson Mater Phy. [Epub ahead of print]. PMID: 22286777
96. Berger S, Braun S, more basic, NMR e. (2004) 200 and more NMR experiments : a practical course, New edn. Wiley, Weinheim
97. Cheng LL, Anthony DC, Comite AR, Black PM, Tzika AA, Gonzalez RG (2000) Quantification of microheterogeneity in glioblastoma multiforme with ex vivo high-resolution magic angle spinning (HRMAS) proton magnetic resonance spectroscopy. Neuro Oncol 2:87–95
98. Mahon M, Williams A, Soutter W, Cox I, McIndoe G, Coutts G, Dina R, deSouza N (2004) 1 H magnetic resonance spectroscopy of invasive cervical cancer: an in vivo study with ex vivo corroboration. NMR Biomed 17:1–9
99. Wilson M, Davies NP, Grundy RG, Peet AC (2009) A quantitative comparison of metabolite signals as detected by in vivo MRS with ex vivo 1 H HR-MAS for childhood brain tumours. NMR Biomed 22:213–219
100. Righi V, Roda JM, Paz J, Mucci A, Tugnoli V, Rodriguez-Tarduchy G, Barrios L, Schenetti L, Cerdan S, Garcia-Martin ML (2009) 1 H HR-MAS and genomic analysis of human tumor biopsies discriminate between high and low grade astrocytomas. NMR Biomed 22:629–637
101. Altman DG, Royston P (2000) What do we mean by validating a prognostic model? Stat Med 19:453–473
102. Reynolds G, Wilson M, Peet A, Arvanitis T (2006) An algorithm for the automated quantitation of metabolites in in vitro NMR signals. Magn Reson Med 56:1211–1219
103. Poullet JB, Sima DM, Simonetti AW, De Neuter B, Vanhamme L, Lemmerling P, Van Huffel S (2007) An automated quantitation of short echo time MRS spectra in an open source software environment: AQSES. NMR Biomed 20:493–504
104. Tkáč I, Rao R, Georgieff MK, Gruetter R (2003) Developmental and regional changes in the neurochemical profile of the rat brain determined by in vivo 1 H NMR spectroscopy. Magn Reson Med 50:24–32
105. Tate AR, Griffiths JR, Martinez-Perez I, Moreno A, Barba I, Cabanas ME, Watson D, Alonso J, Bartumeus F, Isamat F, Ferrer I, Vila F, Ferrer E, Capdevila A, Arus C (1998) Towards a method for automated classification of 1 H MRS spectra from brain tumours. NMR Biomed 11:177–191

Neuromethods (2013) 77: 367–382
DOI 10.1007/7657_2011_32

Published online: 13 March 2012

Single Photon Emission Computed Tomography in Small Animal CNS Research

André Constantinesco, Christian Goetz, and Philippe Choquet

Abstract

Micro-single photon emission computed tomography (SPECT) imaging is a powerful imaging technique able to achieve the temporal and spatial resolution needed for obtaining detailed functional data of different brain regions in small rodents. Multiple radiochemical strategies to develop SPECT probes able to target brain neurophysiology and neuropathology are developed. When combined to morphological imaging modalities such as micro-CT or micro-MRI, structural and functional information are merged improving, for translational CNS research, the understanding of neurological disorders in small animal models.

Key words: Micro-SPECT, Micro-CT, MRI, Small animal, CNS

1. Introduction

Single photon emission computed tomography (SPECT) is a highly sensitive clinical routine technique using radioactively labeled substances able to assess in vivo physiological and biochemical processes in 4D (3D isotropic space and time). Micro-SPECT imaging in small laboratory animals uses dedicated scanners with half millimeter to millimeter spatial resolution and time resolutions of minutes with sensitivities of about 1,000–2,000 Cps/MBq; thanks to different configurations of multi-pinhole or multi-slit collimators and scintillation crystal or semiconductor pixellated detectors (1–6) (Note 1). High-resolution micro-SPECT imaging in small animal preclinical brain research offers the advantages of detailed functional information of CNS diseases at the tissue, cellular, molecular, and genetic levels. Micro-SPECT imaging has been used in mouse and rat to perform in vivo imaging and quantification of brain–blood flow, dopaminergic, serotoninergic, cholinergic neuroreceptors and neurotransmitters in normal, degeneration or pharmacological stress, abuse or anxiety situations as well as in

Table 1
Principal SPECT isotopes that are used with corresponding main gamma energies and periods with a list of labeled agents and targets that can be employed

Isotope source of signal	Main γ energy (keV)	Period (h)	Labeled agent category	Target category
^{99m}Tc	140	6	• Antibodies	• Antigens
^{123}I	159	13.2	• Cells	• Adhesion molecules
^{111}In	171 and 245	67.2	• Ligands	• Cells
^{201}Tl	71	72.9	• Nucleic acids	• Enzymes
^{125}I	27 and 30	1,437	• Peptides	• Lipids
			• Proteins	• Receptors
			• Nanoparticles compounds	• Transporters
				• Nucleic acids

gene-engineered animal models of CNS disorders (7–20). SPECT agent candidates for β-amyloid plaques imaging, cell apoptosis, or viability of systemic or transplanted stem cells in CNS have also been reported in animal models (21–25). Tumor-imaging agents based on cell proliferation properties of malignant tissue, or aberrant expression of certain receptors, were also developed for SPECT molecular imaging in nude mouse CNS xenograft tumor models (26–35) (Note 2).

The most widely used small animal nuclear neuroimaging technique is nevertheless positron emission tomography (micro-PET) (36). However, SPECT overcomes the physical spatial resolution limitation of PET due to positron range in tissues and presents another interesting advantage, which is its ability to perform multi-spectral imaging. Different radiopharmaceuticals labeled with isotopes having different main gamma emission energies can be registered at the same time with SPECT. Moreover, with half-lives that are in the range of hours to several days, SPECT imaging is able to follow slow biological and biochemical tissue kinetics. Table 1 illustrates the principal single photon emission isotopes that can be used as well as their main gamma energies and corresponding periods with a list of labeled agents and targets that can be employed. Table 2 shows examples of radiopharmaceuticals for CNS SPECT and corresponding targeted small animal and/or human metabolisms and functions in a translational perspective (37). There are also great efforts in order to simplify the chemical strategies and technical chemistry steps for radio-labeling biological agents for an in vivo easy use for human and animal SPECT targeted applications (Note 2) (38). SPECT imaging is economical compared to PET, providing also less radiation burden

Table 2
Examples of SPECT neurotracers with their radiopharmaceuticals acronyms, corresponding isotopes, targeted metabolisms and functions, commercial or laboratory availability and human or animal applications

Tracer name or acronym	Isotope	Metabolism and function [target]	Commercial, Laboratory (human, animal)	References
SESTAMIBI	^{99m}Tc	Tumor (glioma) [cellular viability: P-glycoprotein, MDR-1]	c (h)	(27)
TETROFOSMIN	^{99m}Tc	Tumor (glioma) [cellular viability: cytosol and mitochondria]	c (h)	(30)
THALLIUM	^{201}Tl	Tumor (glioma) [cellular viability: Na/K-ATPase pump and membrane potential]	c (h)	(26, 27, 30)
DTPA-Bis	^{99m}Tc	Tumor (glioma) [methionine]	Laboratory (a)	(35)
CXCR4	^{125}I	Tumor (glioma) [chemokine receptors]	Laboratory (a)	(33)
RGD	^{99m}Tc	Tumor angiogenesis (glioma) [αvβ3 integrin expression]	Laboratory (a)	(32)
IMT	^{123}I	Tumor (glioma) [tyrosine]	Laboratory (h)	(29)
HMPAO	^{99m}Tc	Brain perfusion [astrocytes binding]	c (h, a)	(11)
ADAM	^{123}I	Neurological and psychiatric disorders [serotonin transporters]	c (h, a)	(14, 16)
I-CO	^{123}I	Neurological and psychiatric disorders [5-HT2a serotonin receptor]	Laboratory (h, a)	(19)
BOMBESIN	^{99m}Tc	Neurotransmitter Tumor Transplanted cells [bN receptors]	Laboratory (a)	(23, 34)
ANNEXIN-V	^{99m}Tc	Neurodegenerative disorders Apoptosis-cell stress Ischemia-stroke [phosphatidylserine]	c (h, a)	(21)
IMPY	^{123}I	Alzheimer disease [β amyloid plaques: thioflavine-T-derivative]	Laboratory (a)	(22, 24)
IBZM	^{123}I	Neuronal dopaminergic system [D2 dopamine receptors]	c (h, a)	(8, 17, 18)
FP-CIT	^{123}I	Neuronal dopaminergic system [presynaptic dopamine transporters]	c (h, a)	(11, 12, 15)
IOMAZENIL	^{123}I	Neuronal activity [benzodiazepine receptors]	c (h, a)	(7)

and has potential for the quantitative studies of biodistribution kinetics of new diagnostic molecules and radio-ligands or control of efficacy of treatment with clinical impact in translational research of CNS disorders (20). However, general drawbacks of SPECT imaging are the need of hot lab installations and radioactive waste control. In the following, we will succinctly describe the micro-SPECT instrumentation as well as the CT and MRI instruments used in our laboratory for co-registration of anatomical and functional images. As examples, essential preparation steps of radiopharmaceuticals targeted for brain perfusion and dopamine transporter with corresponding acquisition protocols and images will be presented together with their resulting micro-SPECT and -CT and MRI fusion images in rat and mouse.

2. Materials

2.1. SPECT Imaging Instrumentation

SPECT images were acquired using the eXplore speCZT Vision 120 preclinical imaging system (GE Healthcare, Waukesha, USA) illustrated in Fig. 1. The SPECT part consists in a full ring of semiconductors, CZT detectors (10 heads of 8 cm × 8 cm, pixel size 2.46 mm, energy range 20–200 keV, and energy resolution <7% at 140 keV), and rotating multi-pinhole dedicated tungsten collimators for mouse or rat. For mouse imaging, radial field of view (FOV) is 34 mm with a cylindrical multi-pinhole collimator of seven pinholes of 1 mm diameter each (Note 3). For rat imaging, radial FOV is 78 mm using a cylindrical multi-pinhole collimator of five pinholes of 1.5 mm diameter each. Using the helical acquisition mode, the axial FOV can be adjusted to the size of the head and neck of the small rodent but at the expend of longer acquisition times.

2.2. Anesthesia, Imaging Cell and Animal Monitoring Instrumentation

Small animals (mice and rats) are anesthetized with a gas mixture (air and 1–2% isoflurane) in an air warmed-up multimodality imaging chamber (Minerve, Esternay, France) to keep constant at 37°C their temperature. Heart rate is obtained through three paws carbon tube electrodes (Note 4) fitted with conductive gel and insensitive to both X-rays or magnetic field as previously described (39) and connected with an electrocardiograph (ODAM Physiogard RSM 784, Wissembourg, France) delivering a physiological trigger corresponding to the QRS of the ECG signal. Monitoring of breathing rate is obtained via a small differential pressure transducer (TSD 110-MRI; Biopac Systems Inc., CA, USA) placed under the thorax of the animal lying in prone position (Note 5).

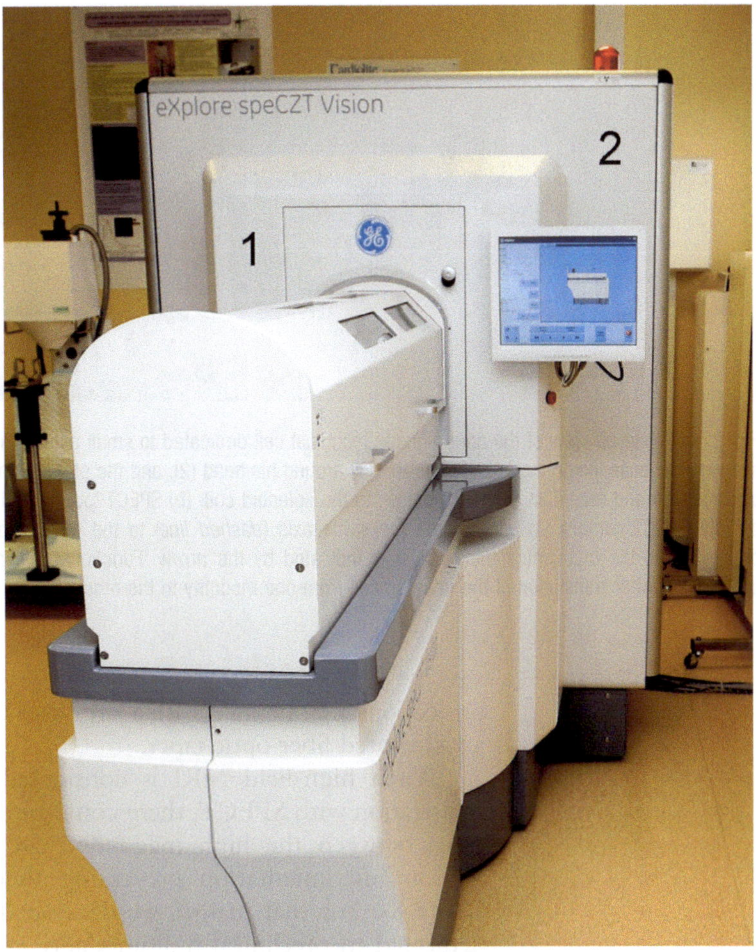

Fig. 1. View of the combined SPECT/CT preclinical imaging system used (eXplore SPECZT vision 120) with the SPECT part (1) in front and CT part (2) in the back. Functional and anatomical images are acquired sequentially after translation on the same axis of the imaging cell from one modality to the other.

2.3. Multimodality Instrumentation for Image Co-registration (SPECT/CT and SPECT/MRI)

Micro-SPECT functional images can be fused with anatomic images issued from other imaging techniques as micro-CT (Note 6) or micro-MRI (Note 7). The image fusion allows a correct anatomic location of the radioactivity uptake, which is of particular importance in brain imaging. Translation of the animal from one modality to the other is achieved using the imaging cell and without moving the animal, avoiding then the use of fiducial markers. A close proximity of both imaging systems or better a co-axially dual modality device configuration is therefore recommended.

The CT part of the eXplore speCZT Vision 120 illustrated in Fig. 1 and which we used for multimodal imaging is placed on the same axis and back-to-back with the SPECT and has a high-power rotating anode X-ray tube (focal spot of 50 μ, 120 kV maximum potential, and 50 mA maximum current). X-rays attenuation

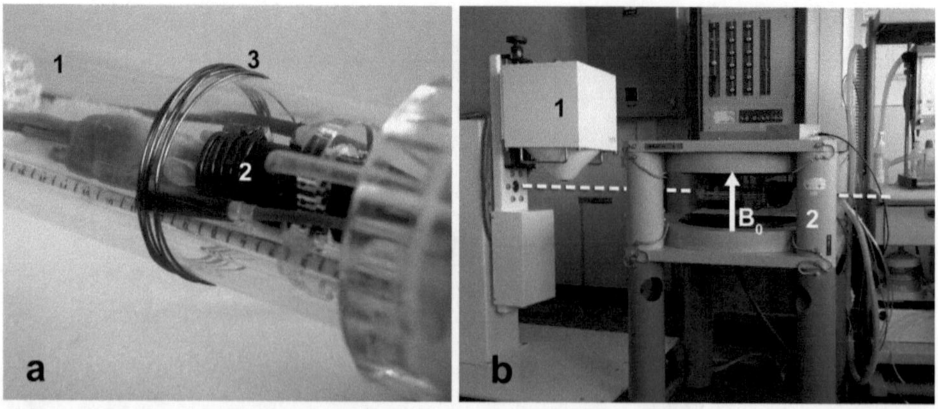

Fig. 2. (**a**) Close-up view of the nonmagnetic technical cell dedicated to small animal imaging adapted for multimodality showing a mouse inside the cell (1), the RF coil around his head (2), and the pick-up coil (3) around the imaging cell for transmission and reception of the RF pulses to the solenoid coil. (**b**) SPECT/low-field MRI dual imaging system with the pinhole SPECT camera (1) adjacent on the same axis (*dashed line*) to the magnet of the low-field MRI system (2), orientation of the main magnetic field B_0 is indicated by the *arrow*. Functional and anatomical images are acquired sequentially after translation of the imaging cell from one modality to the other.

projections are obtained with a large area CCD detector bonded to a CsI-coated fiber optic taper.

When high-field MRI is considered for morphological co-registration with SPECT, there could be complex physical interactions between the high magnetic field generally used and the nuclear instrumentation preventing then their close proximity (Note 8). For that reason, we developed a low-field MR imager dedicated to preclinical multimodality imaging, Fig. 2 (40). The main magnetic field B_0 of 0.1 T is oriented vertically and delivered by a small water-cooled open electromagnet (Bouhnik SAS, Velizy, France). The field homogeneity $\Delta B/B_0 = 5 \times 10^{-6}$ is obtained in an imaging volume of $10 \times 10 \times 6$ cm^3 with maximum encoding gradients' strength of 20 mT/m that are reached in 0.5 ms. Radio frequency (RF) transmit-and-receive solenoid coils are made by adjacent loops of copper wire tuned at 4.26 MHz corresponding to the Larmor frequency of the ^{1}H at that field.

RF pulses are transmitted to the solenoid coil by a pick-up copper coil rounded outside the imaging cell (as illustrated in Fig. 2), which also receive the encoded MR signal. For anatomical images, co-registration a T1-weighted 3-dimensional (3D) sequence was developed on a SMIS (Guilford, UK) console to achieve typical isotropic (Note 9) spatial resolutions of, respectively, $1 \times 1 \times 1$ mm^3 for rat and $0.5 \times 0.5 \times 0.5$ mm^3 for mouse. The low-field MR imaging system was used back to back with a small animal SPECT gamma camera (Gaede Medizinsyteme, Freiburg, Germany) equipped with one pinhole of 1.5 mm in diameter and a focal distance of 12 cm (41).

2.4. SPECT Images Reconstruction Software

The Microview software driving the eXplore speCZT Vision 120 permitted the SPECT/CT images reconstruction. A 3D OSEM (11 averages and 5 subsets) algorithm, without Compton and attenuation correction, was applied for micro-SPECT images reconstruction leading to reconstructed voxels of $300 \times 300 \times 300\ \mu^3$. A cone beam filtered back-projection Feldkamp algorithm is used for micro-CT images reconstruction leading to isotropic reconstructed voxels of $100 \times 100 \times 100\ \mu^3$. For the SPECT/low-field MRI experiment, the SPECT images were reconstructed using an algebraic reconstruction technique (ART) enabling an isotropic $1 \times 1 \times 1\ mm^3$ spatial resolution and a voxel size of $470 \times 470 \times 470\ \mu^3$ (42).

2.5. Hot Lab Equipment for Radiopharmaceutical Preparation and Quality Control

Hot lab equipment needs to meet each national nuclear regulatory standards for single photon emitters biological applications including:

- Shielded cabinet(s)
- Dose calibrator for accurate radionuclide activity measurements
- Shielded syringes
- Radiochromatograph and chromatography strips for radiochemical QC
- Shielded waste container
- Low energy shielded decay drums
- Small animal QC SPECT phantoms
- Contamination monitor and personal dosemeter monitor(s)
- Radioactive waste management systems (for collection of radioactive liquid and organic waste)

3. Methods

3.1. Radiopharmaceutical Preparation Steps

Detailed data on numerous SPECT radiopharmaceuticals chemistry and labeling preparation steps can be found in the MICAD database (Note 10). In the following, as typical examples, we will only focus on the principal preparation and quality control steps of two commercial radiopharmaceuticals (respectively, ^{99m}Tc-HMPAO and ^{123}I-FP-CIT) that will be illustrated in small animal SPECT for imaging brain perfusion and dopamine transporters.

3.1.1. ^{99m}Tc-HMPAO: A Brain Perfusion Agent

The lipophilic complex ^{99m}Tc-Exametazime or formerly ^{99m}Tc-Hexamethylpropylene amine oxime (HMPAO) (Ceretec®, GE Healthcare, Little Chalfont, UK) is a pre-dispensed sterile, nonpyrogenic, lyophilized mixture of 0.5 mg [(RR-SS)-4.8-diaza-3, 6,6,9-tetramethilundecane-2,10-dione bisoxime] which is reconstituted with a solution of ^{99m}Tc sodium pertechnetate. Sterile technique must be used throughout.

1. Place the vial in a suitable shielding container and swab the rubber septum with the sterile swab provided or with an appropriate disinfectant (e.g., chlorexidine, benzalkonium chloride).
2. Add 0.37 GBq up to 2 GBq of ^{99m}Tc to the vial using an eluate from a ^{99m}Tc generator which was eluted within 30 min. Use only elute from a technetium generator which has been eluted within 24 h. Before reconstitution, the technetium generator elute may be adjusted to the adequate desired specific activity in a volume of 5 mL by dilution with sodium chloride solution USP for injection.
3. Inject 5 mL of sterile elute into the shielded vial using a 5 mL syringe. Before withdrawing the syringe from the vial, withdraw 5 mL of gas from the space above the solution to normalize the pressure in the vial. Gently invert the shielded vial for 10 s to ensure complete dissolution of the powder.

The useful life of the reconstituted agent is limited to 30 min. Quality controls must be performed according to the following points:

- Preparation must be an aqueous, clear, and colorless solution.
- pH of the prepared injection should be in the range 9–9.8.
- Radiochemical purity greater than 80% is mandatory for product acceptance.

Control of radiochemical purity is performed using chromatography strips (model 150-130, Biodex Medical Systems Inc.) with ethyl acetate as solvent. First, prepare a chromatography tube containing ethyl acetate to saturate it. Then apply at least 5 μL samples of Ceretec® to the origin of the strip within 15 min of reconstitution. Immediately place one strip into the ethyl acetate tube and make sure that the strip is not adhering to the side of the test tube. When elute has reached the solvent front, remove the strip from the tube and count the strip with a suitable radiochromatograph equipment (Minigita, Raytest Isotopenmessgeräte GmBH, Straubenhardt, Germany).

3.1.2. ^{123}I-FP-CIT: A Dopamine Transporter Imaging Probe

Ioflupane or [^{123}I]-*N*-ω-fluoropropyl-2β-carbome-thoxy-3β-(4-iodophenyl)nortropane (Datscan®, GE Healthcare, Little Chalfont, UK) is an injectable "ready to use, 74 MBq/mL" specialty. Vial of 2.5 mL have 4.5×10^{14} specific activity. Before being placed in the shielded enclosure, vial is disinfected with 2-propanol. The septum is disinfected with an appropriate disinfectant (e.g., chlorexidine, benzalkonium chloride) immediately before to take a sample for IV injection. Check for radiochemical purity as detailed above.

3.2. Radiopharmaceutical Administration Routes, Volumes, and Specific Tracer Radioactivity

Intravenous route is the most common way for radiopharmaceutical administration in small rodents; however, care must be taken to keep the injected volume less than 0.3 mL for mouse. For multiple radiopharmaceutical administrations, in case of longitudinal examinations, for instance, intraperitoneal or subcutaneous routes are promising alternative solutions enabling larger injected volumes even if they result in slower tissue or tumor uptake kinetics (43). Amount of administered radioactivity is linked to both the spatial and time resolutions needed for each particular application but also to the sensitivity of the SPECT instrument involving the type and number of detectors and collimator geometries and configurations. Typical values of specific activities of radiopharmaceuticals for small rodents CNS SPECT imaging are in the range 1–15 MBq/g according to our experience and the literature.

3.3. Transfer of Animal in Imaging Cell, Cardiac and Respiratory Monitoring, Control of Temperature

Animal experiments were conducted in compliance with the French guidelines for the care and use of research animals (authorization 67-48220). Animal is transferred to the imaging cell after SPECT tracer administration. Heart and breathing rates are continuously monitored and registered during imaging sessions (Note 8). Control of the animal's temperature is obtained through heating the air inside the imaging chamber. If multimodality is considered, then the same imaging cell holding the animal is simply translated from the SPECT tomograph to the anatomical scanner modality (CT or low-field MRI).

3.4. SPECT Images Acquisition Protocols Using the Explore speCZT Vision 120

For acquisition of brain perfusion micro-SPECT projection images, we used a step and shoot mode (52 steps of 1° and 30 s per step) with an energy window adjusted to the isotope main gamma energy of the radiopharmaceutical employed (between 130 and 150 keV to encompass the ^{99m}Tc photopeak, for instance). Typical total acquisition time for mouse or rat brain perfusion imaging is around 40 min. Dynamic SPECT images were also acquired for 3.2 h at a frame rate of 4 min for dopamine transporter imaging and binding quantification in the mouse striatum.

3.5. CT Images Acquisition Protocol for SPECT/CT Functional/Anatomical Co-registration Using the Explore speCZT Vision 120

The CT fast protocol used 220 views acquired on 220° (Parker mode), at 80 kV and 32 mA, binning 4 × 4, exposure time of 16 ms and one average. The duration of one rat or mouse head imaging acquisition was less than 2 min including bright and dark images for CT Hounsfield unit quantification.

3.6. SPECT/MRI Imaging Protocol for Functional/Anatomical Co-registration

The SPECT images acquisition protocol with the dedicated SPECT/low-field MRI system consisted in acquisition of 32 projections of 1 min over a circular 180° arch with a 128 × 128 image matrix. SPECT is immediately followed by low-field MRI using a 3D gradient echo T1-weighted isotropic image sequence obtained in 42 min, with echo time (TE) of 7 ms, repetition time (TR) of 23 ms, flip angle (FA) of 60°, 128 × 48 × 64 matrix, number of excitations (NEX) of 36, and FOV of 64 × 32 × 32 mm^3 corresponding to a voxel size of 500 × 500 × 500 μ^3. Then, rigid co-registration of SPECT and MRI images is obtained using the AMIDE software (41).

3.7. Examples of SPECT Images Interpretation and Analysis, Images Fusion, and Data Quantification

Brain SPECT data quantification can be performed with standardized or anatomically adjusted ROIs (using brain atlas templates, ex vivo CT contrasted images, or MRI overlay techniques) enabling determination of the relative uptake ratio of radiopharmaceutical in the selected volume. Figure 3 illustrates the in vivo SPECT/CT fusion of brain perfusion in normal mouse. Figure 4 shows, in normal mouse, detailed anatomical images of brain structures that were obtained ex vivo after tissue saturation with iodinated X-rays contrast agent enabling anatomical correlations between functional (brain perfusion) and anatomical images. Quantitative approach of cerebral blood flow (CBF) is also possible following the methodology reported by Matsuda et al. (44) or Zeniya et al. (45, 46).

In vivo uptake of the tracer of dopamine transporter in mouse brain striatum is illustrated in Fig. 5a where left and right striatal structures are clearly resolved by SPECT. For objective assessment of this transporter ligand binding, in addition to visual interpretation, basics of quantitative analysis is also presented in Fig. 5b using a set of dynamic SPECT images acquisition and kinetic modeling based on time–activity curves derived from volume of interest regions (VOIs) encompassing corresponding brain structures. As an illustration of quantitative SPECT data analysis and if STR(t)

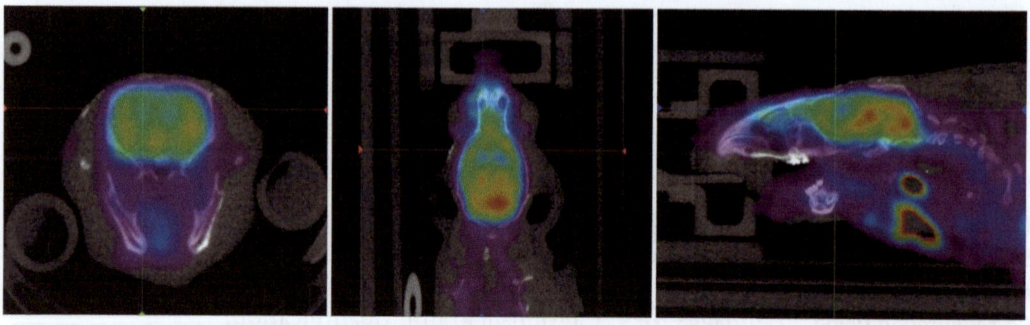

Fig. 3. From left to right: transverse, frontal, and sagittal orthogonal mid-head images corresponding to the fusion of CT images and brain SPECT perfusion in a normal male mouse (Swiss, weighting 45 g). Images were acquired in vivo and sequentially with the eXplore speCZT vision 120. 180 MBq of ^{99m}Tc-HMPAO were IV administered in a volume of 0.2 mL, followed by images acquisition 10 min later. Total images acquisition time was 42 min.

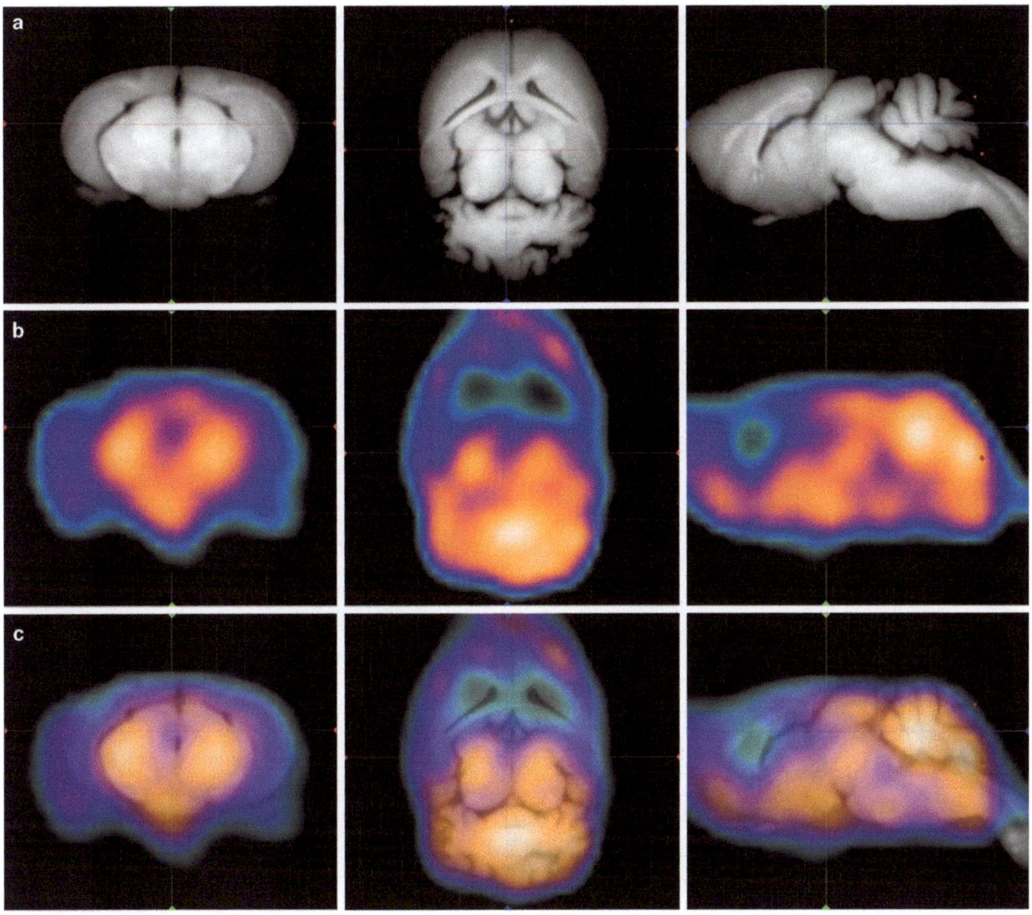

Fig. 4. Row (**a**) showing, from left to right, transverse, frontal, and sagittal orthogonal slice images of ex vivo brain iodine contrasted micro-CT obtained in normal mouse (Swiss, weighting 45 g) after soaking the mouse brain in a KI solution. Anatomical details of the brain structures are clearly visible. Row (**b**) show corresponding slices of in vivo brain SPECT perfusion obtained previously in the same mouse after IV administration of 185 MBq of ^{99m}Tc-HMPAO in 0.2 mL. In row (**c**), anatomical/functional correlations are illustrated by ex vivo/in vivo co-registration images of corresponding slices.

represents the striatum counts as a function of time, CER(t) the background cerebellum counts over time, and DVR the distribution volume ratio, which is a linear function of transporter ligand availability, then the tracer biodistribution and binding compartmental analysis can be simplified as follows (11, 12, 15):

$$\left[\int_0^T \mathrm{STR}(t)\mathrm{d}t\right]/\mathrm{STR}(T) = \mathrm{DVR}\left[\int_0^T \mathrm{CER}(t)\mathrm{d}t\right]/\mathrm{STR}(T)$$

for a sufficiently long period of time T

Hence DVR, which corresponds to the slope of this linear approximation, is obtained by a graphical method that does not require blood sampling. In this example and in the perspective of

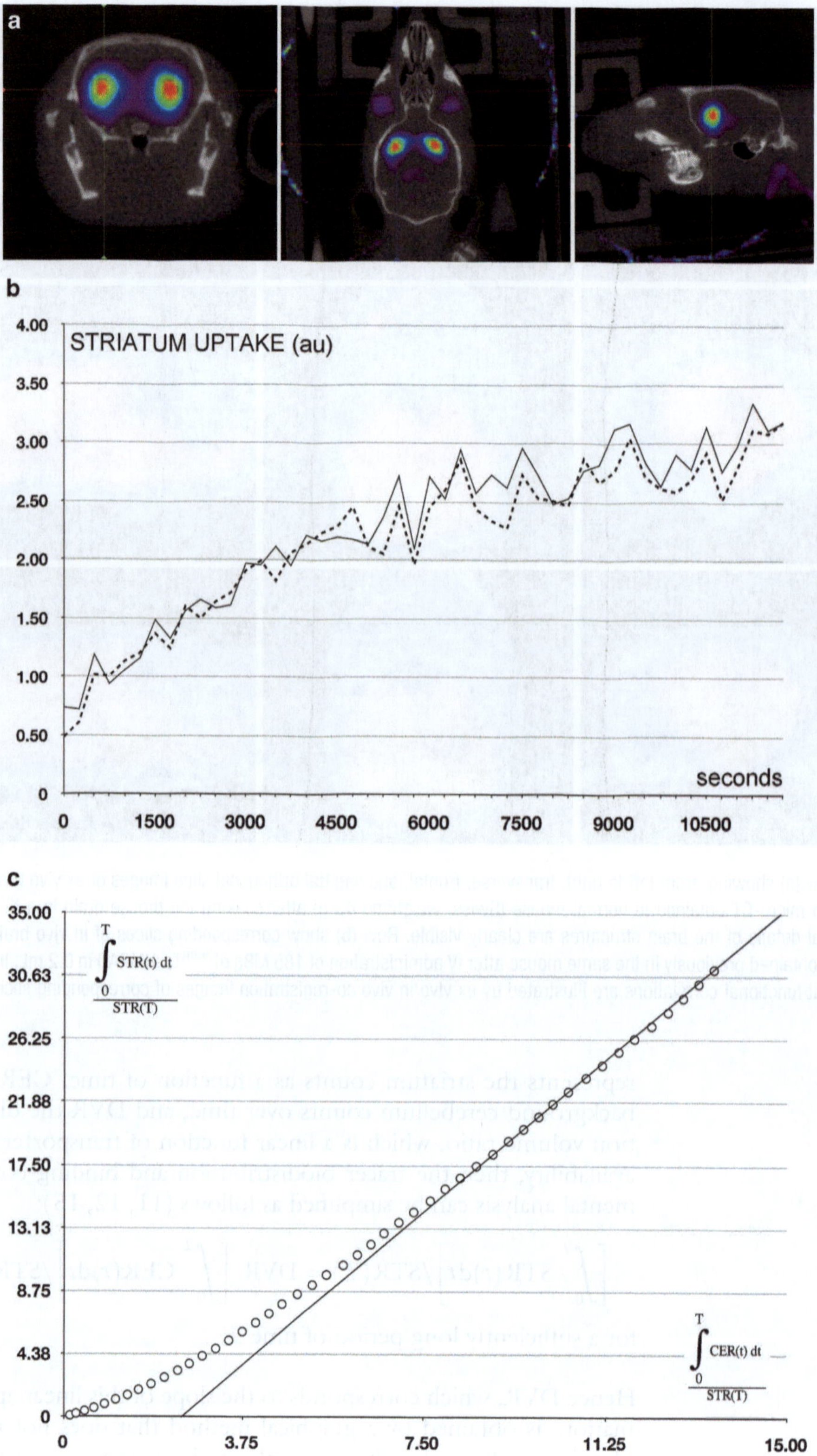
a
b
STRIATUM UPTAKE (au)
4.00
3.50
3.00
2.50
2.00
1.50
1.00
0.50
0
seconds
0
1500
3000
4500
6000
7500
9000
10500
c
35.00
30.63
26.25
21.88
17.50
13.13
8.75
4.38
0
$\int_0^T STR(t)\,dt \,/\, STR(T)$
$\int_0^T CER(t)\,dt \,/\, STR(T)$
0
3.75
7.50
11.25
15.00

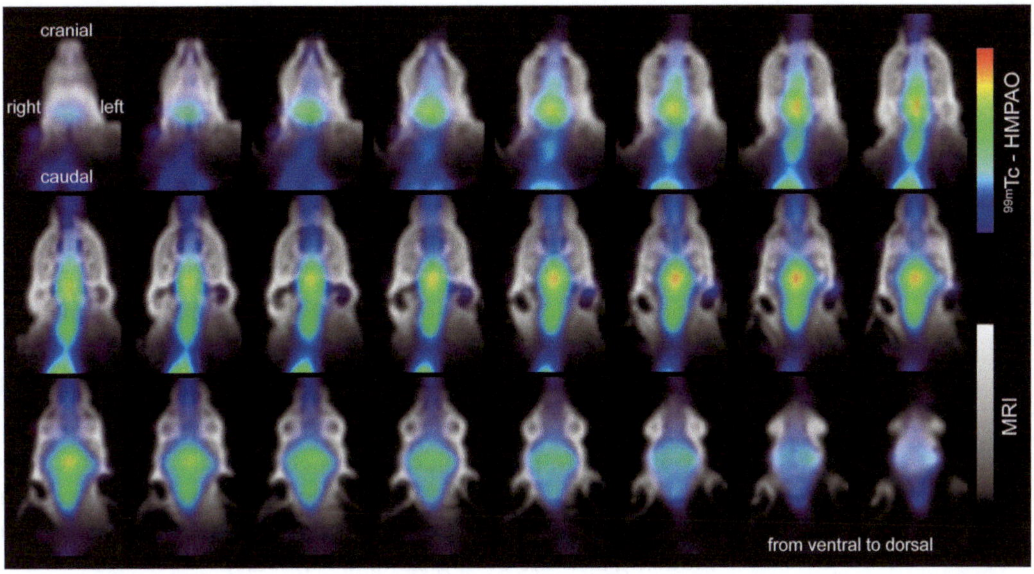

Fig. 6. Set of frontal co-registration images in a nude mouse of 25 g of single pinhole SPECT brain perfusion (in *color*) after IV administration of 0.2 mL of 120 MBq of ^{99m}Tc-HMPAO and low-field MRI for anatomy (in *gray*). Total acquisition time was 1 h 14 min.

translational research where repeated exams in the same animal are expected, the subcutaneous route of tracer administration was preferred resulting, as expected in comparison with the IP route (43), in slower uptake of the tracer by the striatum, Fig. 5b, compared to IV tracer administration (12). Nevertheless, in Fig. 5c, we show that the mouse striatum slower tracer uptake does not affect the linearization of compartmental analysis yielding for the DVR parameter a similar quantitative result compared to the literature (6, 12).

Finally, SPECT/low-field MRI images, acquired sequentially with the dual system described above and in Fig. 2, and presented in Fig. 6, illustrate in nude mouse the fusion of brain perfusion obtained with SPECT and anatomy obtained with low-field MRI.

Fig. 5. Row (**a**) shows, from left to right, transverse, frontal, and sagittal in vivo mid-brain slices of SPECT/CT images of presynaptic dopamine transporters in a normal male mouse (B6-129, weighting 35 g), using the eXplore speCZT vision 120, 3 h post-subcutaneous administration of 194 MBq of $_{123}$I-FP-CIT in a total volume of 2.5 mL. Row (**b**) illustrates right (*solid line*) and left (*dashed line*) time–activity curves obtained in the same mouse after a total of 3.2 h dynamic SPECT images acquisition with a time resolution of 4 min. $_{123}$I-FP-CIT uptake curves are based on measured counts inside VOIs encompassing the right and left striatum, corrected by the nonspecific tracer binding inside a cerebellum VOI. Row (**c**) illustrates graphical determination of the distribution volume ratio of the tracer (DVR) represented by the slope of the linear portion of the graph yielding a value of 2.6. The graph (*circles*) represents the normalized integral of the striatum radioactivity as a function of the normalized integral of the nonspecific radioactivity uptake in cerebellum (see text for details).

4. Notes

1. Technical specifications of different commercial micro-SPECT scanners are available at: http://www.bioscan.com; http://www.gehealthcare.com; http://www.siemens.com; http://www.gammamedica.com; http://www.milabs.com.
2. Human and/or animal radiopharmaceutical specificity.
3. Periodic quality control of imaging instrumentation is mandatory.
4. Pediatric ECG electrodes are also frequently used in preclinical imaging for heart rate monitoring (http://www.biovet.com).
5. Monitoring of physiological parameters (heart and respiratory rates, temperature) is mandatory during images acquisition in order to ensure scientific pertinence of registered imaging data reducing also the biological data variability.
6. In vivo CT for CNS image fusion with SPECT is limited to visualization of head bone structures. Iodinated contrast agents do not cross in vivo the blood–brain barrier except in case of brain tumors. However, the ex vivo use of highly concentrated iodinated compounds could be helpful to contrast the brain tissue itself for morphological correlations with in vivo SPECT images as illustrated in Fig. 4.
7. MRI for CNS image fusion with SPECT for brain tissue anatomic details is more appropriate than CT due to T1 and/or T2 intrinsic contrasts of biological tissues.
8. However specific SPECT inserts that can be placed inside the magnetic field for simultaneous dual images registration are under development (47).
9. Isotropic imaging sequences for MRI are recommended because they are appropriate for images fusion with micro-SPECT, which is as micro-CT, an intrinsically isotropic imaging technique.
10. MICAD (molecular imaging and contrast agent database) http://www.ncbi.nlm.nih.gov/bookshelf/corehtml/homepages/bookshelf/micad.html.

References

1. Strand S, Ivanovic M, Erlandsson FD, Button T, Sjögren K, Weber D (1994) Small animal imaging with pinhole single-photon emission computed tomography. Cancer 73:981–984
2. Schramm N, Ebel G, Engeland U, Schurrat T, Behe M, Behr T (2003) High resolution SPECT using multipinhole collimation. IEEE Trans Nucl Sci 50:315–320
3. Meikle S, Kench P, Kassiou M, Banati R (2005) Small animal SPECT in the matrix of molecular imaging technologies. Phys Med Biol 50: R45–R61

4. Beekman F, van der Have F, Vastenhouw B, Van der Linden A, Van Rijk P, Burbach J, Smidt M (2005) U-SPECT-I: a novel system for submillimeter-resolution tomography with radiolabeled molecules in mice. J Nucl Med 46:1194–1200
5. Walrand S, Jamar F, De Jong M, Pauwels S (2005) Evaluation of novel whole body high resolution rodent SPECT (linoview) based on direct acquisition of linogram projections. J Nucl Med 46:1872–1880
6. Pissarek M, Oros-Peusquens AM, Schramm N (2008) Challenge by the murine brain: multipinhole SPECT of ^{123}I labelled radiopharmaceuticals. J Neurosci Methods 168:282–292
7. Kung H, Pan S, Kung M, Billings J, Kaliswall R, Reilley J, Alavi A (1989) In vitro and in vivo evaluation of ^{123}I-IBZM: a potential CNS D2 dopamine receptor imaging agent. J Nucl Med 30:88–92
8. Kurokawa K, Jibiki I, Matsuda H, Fukushima T, Tsuji S, Yamaguchi N, Hisada K (1994) Comparison of benzodiazepine receptor and regional cerebral blood flow imaging of epileptiform foci in hippocampal kindled rabbits: a study with in vivo double tracer autoradiography using Iomazenil and HMPAO. Brain Res 642:303–310
9. Lan R, Lu Q, Fan P, Gatley J, Volkow N, Fernado S, Pertwee R, Makryannis A (1999) Design and synthesis of the CB1 selective cannabinoid antagonist AM281: a potential human SPECT ligand. AAPS Pharmasci 1 (2):39–45
10. Zerarka S, Pellerin L, Slosman D, Magistretti P (2001) Astrocytes as a predominant cellular site of ^{99m}Tc-HMPAO retention. J Cereb Flow Metab 21:456–468
11. Scherfler C, Donnemiller E, Schocke M, Dierkes K, Decristoforo C, Oberladstatter M, Kolbitsch C, Zschiegner F, Riccabona G, Poewe W, Wenning G (2002) Evaluation of striatal dopamine transporter function in rats by in vivo beta-[^{123}I]CIT pinhole SPECT. Neuroimage 17:128–141
12. Acton P, Choi S, Plössl K, Kung H (2002) Quantification of dopamine transporters in the mouse brain using ultra-high resolution single-photon emission tomography. Eur J Nucl Med 29:691–698
13. De Win M, De Jeu R, De Bruin K, Habraken J, Reneman L, Booij J, Den Heeten G (2004) Validity of in vivo ^{123}I-β-CIT SPECT in detecting MMDA-induced neurotoxicity in rats. Eur Neuropsychopharmacol 14:185–189
14. Chalon S, Bronquard C, Vercouillie J, Kodas E, Garreau L, Bodard S, Emond P, Besnard JC, Guilloteau D (2004) ADAM is an effective tool for in vivo study of serotoninergic function: validation in rat models. Synapse 52:136–142
15. Alvarez-Fischer D, Blessmann G, Trosowski C, Béhé M, Schurrat T, Hartmann A, Behr T, Oertel W, Höglinger G, Höffken H (2007) Quantitative ^{123}I-FP-CIT pinhole SPECT imaging predicts striatal dopamine levels but not number of nigral neurons in different mouse models of Parkinson disease. Neuroimage 38:5–12
16. Hwang L, Chang C, Liu H, Lee S, Jan M, Chen C (2007) Imaging the availability of serotonin transporter in rat brain with ^{123}I-ADAM and small animal SPECT. Nucl Med Commun 28:615–621
17. Meyer P, Salber D, Schiefer J, Cremer M, Schaefer W, Kosinski C, Langen K (2008) Cerebral kinetics of the dopamine D2 receptor ligand ^{123}I-IBZM in mice. Nucl Med Biol 35:467–473
18. Jongen C, de Bruin K, Beekmann F, Booij J (2008) SPECT imaging of D2 dopamine receptors and endogenous dopamine release in mice. Eur J Nucl Med Mol Imaging 35:1692–1698
19. Blanckaert P, Burvenich I, Wyffels L, De Bruyne S, Moerman L, De Vos F (2008) In vivo evaluation in rodents of ^{123}I-3-I-CO as a potential SPECT tracer for the serotonin 5-HT2a receptor. Nucl Med Biol 35:861–867
20. Sharma S, Ebadi M (2008) SPECT neuroimaging in translational research of CNS disorders. Neurochem Int 52:352–362
21. Blankenberg F, Katsikis P, Tait J, Davis R, Naumovski L, Ohtsuki K, Kopiwoda S, Abrams M, Darkes M, Robbins R, Maecker H, Strauss H (1998) In vivo detection and imaging of phosphatidylserine expression during programmed cell death. Proc Natl Acad Sci 95:6349–6354
22. Kung MP, Hou C, Zhuang ZP, Zhang B, Skovronsky D, Trojanowski J, Lee V, Kung H (2002) IMPY: an improved thioflavin-T derivative for in vivo labeling of b-amyloid plaques. Brain Res 956:202–210
23. Scopinaro F, Paschali E, Di Santo G, Antonellis T, Massari R, Trotta C, Gourni H, Bouziotis P, David V, Soluri A, Varvarigou A (2006) Bombesin receptors and transplanted stem cells in rat brain: high resolution scan with ^{99m}Tc-BN1.1. Nucl Inst Methods Phys Res A 569:525–528
24. Ono M (2009) Development of positron-emission tomography/single-photon emission computed tomography imaging probes for in vivo detection of β-amyloid plaques in Alzheimer's brains. Chem Pharm Bull 57:1029–1039

25. Lappalainen R, Narkilahti S, Huhtala T, Liimatainen T, Suuronen T, Närvänen A, Suuronen R, Hovatta O, Jolkkonen J (2008) The SPECT imaging shows the accumulation of neural progenitor cells into internal organs after systemic administration in middle cerebral artery occlusion rats. Neurosci Lett 440:246–250
26. Brismar T, Collins P, Kesselberg M (1989) Thallium 201 uptake relates to membrane potential and potassium permeability in human glioma cells. Brain Res 500:30–36
27. Kim K, Black K, Marciano D, Mazziotta J, Guze B, Grafton S, Hawkins R, Becker D (1990) Thallium-201 SPECT imaging of brain tumors: methods and results. J Nucl Med 31:965–969
28. Soler C, Beauchesne P, Maatougui K, Schmitt T, Barral FG, Michel D, Dubois F, Brunon J (1998) Technetium-99 m sestamibi brain single-photon emission tomography for detection of recurrent gliomas after radiation therapy. Eur J Nucl Med 25:1649–1657
29. Barai S, Bandopadhayaya G, Julka P, Kale S, Malhotra A, Haloi A, Seith A, Sing N (2004) Evaluation of ^{99m}Tc-L-methionine brain SPECT for detection of recurrent brain tumor: a pilot study with radiological and pathological correlation. Acta Radiol 45:649–657
30. Fukumoto M (2004) Single photon agents for tumor imaging: ^{201}Tl, ^{99m}Tc-MIBI, and ^{99m}Tc-tetrofosmin. Ann Nucl Med 18:79–95
31. Alexiou G, Fotopoulos A, Papadopoulos A, Kyritsis A, Polyzoidis K, Tsiouris S (2007) Evaluation of brain tumor recurrence by ^{99m}Tc-Tetrofosmin SPECT: a prospective pilot study. Ann Nucl Med 21:293–298
32. Qihua H, Zhaofei L, Bing J, Xiaoxia L, Jiyun S, Jun Z, Feng L, Zhi Y, Yinan L, Li S, Fan W (2008) In vivo gamma imaging of the secondary tumors of transplanted human fetal striatum neural stem cells-derived primary tumor cell. Neuroreport 19:1009–1014
33. Nimmagadda S, Pullambhatla M, Pomper M (2009) Immunoimaging of CXCR4 expression in brain tumor xenografts using SPECT/CT. J Nucl Med 50:1124–1130
34. De Branco A, Mota L, Ferreira C, Oliveira M, Goés A, Cardoso V (2010) Bombesin derivative radiolabeled with technetium 99 m as agent for tumor identification. Bioorg Med Chem Lett. doi:10.1016/j.bmcl.2010.08.124
35. Hazari P, Shukla G, Goel V, Chuttani K, Kumar N, Sharma R, Mishra A (2010) Synthesis of specific SPECT-radiopharmaceutical for tumor imaging based on methionine: ^{99m}Tc-DTPA-bis (methionine). Bioconjug Chem 21:229–239
36. Waerzeggers Y, Monfared P, Viel T, Winkeler A, Jacobs A (2010) Mouse models in neurological disorders: applications of non-invasive imaging. Biochim Biophys Acta 1802:819–839
37. Hagooly A, Rossin R, Welch J (2008) Small molecule receptors as imaging targets. In: Semmler W, Schwaiger M (eds) Molecular imaging II, handbook of experimental pharmacology 185/II. Springer, Berlin Heidelberg
38. Mindt T, Struthers H, Spingler B, Brans L, Tourwé D, Garcia-Garayoa E, Schibli R (2010) Molecular assembly of multifunctional 99mTc radiopharmaceuticals using "clickable" amino acid derivatives. Chem Med Chem. doi:10.1002/cmdc.201000342
39. Choquet P, Goetz C, Aubertin G, Hubele F, Sannié S, Constantinesco A (2011) Carbon tube electrodes for ECG gated multimodality imaging in mice. JAALAS 50:61–64
40. Choquet P, Breton E, Goetz C, Marin C, Constantinesco A (2009) Dedicated low-field MRI in mice. Phys Med Biol 54:5287–5299
41. Goetz C, Breton E, Choquet P, Israel-Jost V, Constantinesco A (2008) SPECT Low field MRI system for small animal imaging. J Nucl Med 49:88–93
42. Israel-Jost V, Choquet P, Salmon S, Blondet C, Sonnendrucker E, Constantinesco A (2006) Pinhole SPECT imaging: compact projection/backprojection operator for efficient algebraic reconstruction. IEEE Trans Med Imaging 25:158–167
43. Meyer P, Salber D, Schiefer J, Cremer M, Schaeffer W, Kosinski C, Langen K (2008) Comparison of intravenous and intraperitoneal ^{123}I-IBZM injection for dopamine D2 receptor imaging in mice. Nucl Med Biol 35:543–548
44. Matsuda H, Tsuji S, Shuke N, Sumiya H, Tonami N, Hisada K (1992) A quantitative approach to technetium-99 m hexamethylpropylene amine oxime. Eur J Nucl Med 19:195–200
45. Zeniya T, Watabe H, Hayashi T, Teramoto N, Myogin K, Taguchi A, Sato H, Yamamoto A, Sohlberg A, Inomata T, Iida H (2007) Absolute quantitation of regional cerebral blood flow in mouse using ^{123}I-iodoamphetamine and pinhole SPECT. J Cereb Flow Metab 27 (Suppl 1):BP20-04 H
46. Choquet P, Israel-Jost V, Namer I, Bilbaut P, Schneider F, Constantinesco A (2005) CBF imaging and global CBF quantification in normal rats using dedicated pinhole SPECT system. J Cereb Flow Metab 25:S527
47. Hamamura M, Ha S, Roeck W, Muftuler L, Wagenaar D, Meier D, Patt B, Nalcioglu O (2010) Development of an MR-compatible SPECT system (MRSPECT) for simultaneous data acquisition. Phys Med Biol 55:1563–1575

Neuromethods (2013) 77: 383–391
DOI 10.1007/7657_2012

INDEX

A

B

C

D

Q

R

S

If you have any concerns about our products, you can contact us on
ProductSafety@springernature.com

In case Publisher is established outside the EU, the EU authorized representative is:
Springer Nature Customer Service Center GmbH
Europaplatz 3, 69115 Heidelberg, Germany

Printed by Libri Plureos GmbH
in Hamburg, Germany